Communications in Computer and Information Science 1275

Commenced Publication in 2007
Founding and Former Series Editors:
Simone Diniz Junqueira Barbosa, Phoebe Chen, Alfredo Cuzzocrea,
Xiaoyong Du, Orhun Kara, Ting Liu, Krishna M. Sivalingam,
Dominik Ślęzak, Takashi Washio, Xiaokang Yang, and Junsong Yuan

More information about this series at http://www.springer.com/series/7899

Yury Kochetov · Igor Bykadorov ·
Tatiana Gruzdeva (Eds.)

Mathematical Optimization Theory and Operations Research

19th International Conference, MOTOR 2020
Novosibirsk, Russia, July 6–10, 2020
Revised Selected Papers

Springer

Editors
Yury Kochetov (iD)
Sobolev Institute of Mathematics
Novosibirsk, Russia

Igor Bykadorov (iD)
Sobolev Institute of Mathematics
Novosibirsk, Russia

Tatiana Gruzdeva (iD)
Matrosov Institute for System Dynamics
and Control Theory
Irkutsk, Russia

ISSN 1865-0929 ISSN 1865-0937 (electronic)
Communications in Computer and Information Science
ISBN 978-3-030-58656-0 ISBN 978-3-030-58657-7 (eBook)
https://doi.org/10.1007/978-3-030-58657-7

This Springer imprint is published by the registered company Springer Nature Switzerland AG
The registered company address is: Gewerbestrasse 11, 6330 Cham, Switzerland

Preface

This volume contains the refereed and selected papers presented at 19th International Conference on Mathematical Optimization Theory and Operations Research (MOTOR 2020)[1] held during July 6–10, 2020.

It was originally planned that the conference would be held near Novosibirsk Scientific Center, Russia. But due to the difficult situation all around the world related to the COVID-19 pandemic, MOTOR 2020 conference only took place online, via Zoom.

MOTOR 2020 is the second joint scientific event[2] unifying a number of well-known international and Russian conferences held in Ural, Siberia, and the Far East for a long time:

- Baikal International Triennial School Seminar on Methods of Optimization and Their Applications (BITSS MOPT), established in 1969 by academician N.N. Moiseev; the 17th event[3] in this series was held on 2017, in Buryatia.
- All-Russian Conference on Mathematical Programming and Applications (MPA), established in 1972 by academician I.I. Eremin; the 15th conference[4] in this series was held in 2015, near Ekaterinburg.
- The International Conference on Discrete Optimization and Operations Research (DOOR) was organized 9 times since 1996; the last event[5] was held in 2016 in Vladivostok.
- The International Conference on Optimization Problems and Their Applications (OPTA) was organized regularly in Omsk since 1997; the 7th event[6] in this series was held in 2018.

As per tradition, the main conference scope included, but was not limited to, mathematical programming, bi-level and global optimization, integer programming and combinatorial optimization, approximation algorithms with theoretical guarantees and approximation schemes, heuristics and meta-heuristics, game theory, optimization in machine learning and data analysis, and valuable practical applications in operations research and economics.

In response to the call for papers, MOTOR 2020 received 175 submissions. Out of 102 full papers considered for reviewing (73 abstracts and short communications were excluded because of formal reasons), only 31 papers were selected by the Program

---

[1] http://math.nsc.ru/conference/motor/2020/.

[2] The first conference of this series, MOTOR 2019, http://motor2019.uran.ru, was held on July, 2019, in Ekaterinburg.

[3] http://isem.irk.ru/conferences/mopt2017/en/index.html.

[4] http://mpa.imm.uran.ru/96/en.

[5] http://www.math.nsc.ru/conference/door/2016/.

[6] http://opta18.oscsbras.ru/en/.

Committee (PC) for publication in the first volume of proceedings (published in Springer LNCS, Vol. 12095). Out of the remaining, the PC selected 33 revised papers for publication in this volume. Each submission was reviewed by at least three PC members or invited reviewers, experts in their fields, in order to supply detailed and helpful comments.

The conference featured nine invited lectures:

- Prof. Aida Abiad (Eindhoven University of Technology, The Netherlands, and Ghent University, Belgium), "On graph invariants and their application to the graph isomorphism problem"
- Prof. Evripidis Bampis (Sorbonne Université, France), "Multistage optimization problems"
- Prof. Bo Chen (University of Warwick, UK), "Capacity auctions: VCG mechanism vs. submodularity"
- Prof. Sergei Chubanov (Bosch Research, Germany), "Convex geometry in the context of artificial intelligence"
- Prof. Igor Konnov (Kazan Federal University, Russia) "Equilibrium formulations of relative optimization problems"
- Prof. Alexander Kostochka (University of Illinois at Chicago, USA), "Long cycles in graph and hypergraphs"
- Prof. Panos Pardalos (University of Florida, USA), "Inverse combinatorial optimization problems"
- Prof. Soumyendu Raha (Indian Institute of Science, Bangalore, India) "Partitioning a reaction-diffusion ecological network for dynamic stabilitys"
- Prof. Yakov Zinder (University of Technology Sydney, Australia), "Two-stage scheduling models with limited storage"

The following four tutorials were given by outstanding scientists:

- Prof. Alexander Grigoriev (Maastricht University, The Netherlands), "Evolution of sailor and surgical knots"
- Prof. Michael Khachay (Krasovsky Institute of Mathematics and Mechanics, Ekaterinburg, Russia), "Metrics of a fixed doubling dimension: an efficient approximation of combinatorial problems"
- Prof. Vladimir Mazalov (Institute of Applied Mathematical Research, Petrozavodsk, Russia), "Game theory and social networks"
- Dr. Andrey Melnikov (Sobolev Institute of Mathematics, Russia), "Practice of using the Gurobi optimizer"

We thank the authors for their submissions, members of the PC and external reviewers for their efforts in providing exhaustive reviews. We thank our sponsors and partners: Mathematical Center in Akademgorodok, Sobolev Institute of Mathematics, Novosibirsk State University, Krasovsky Institute of Mathematics and Mechanics, Higher School of Economics, and Melentiev Energy Systems Institute. We are grateful to Alfred Hofmann, Aliaksandr Birukou, Anna Kramer, and colleagues from Springer LNCS and CCIS editorial boards for their kind and helpful support.

August 2020

Yury Kochetov
Igor Bykadorov
Tatiana Gruzdeva

Organization

Program Committee Chairs

Vladimir Beresnev	Sobolev Institute of Mathematics, Russia
Alexander Kononov	Sobolev Institute of Mathematics, Russia
Michael Khachay	Krasovsky institute of Mathematics and Mechanics, Russia
Valery Kalyagin	Higher School of Economics, Russia
Panos Pardalos	University of Florida, USA

Program Committee

Edilkhan Amirgaliev	Suleyman Demirel University, Kazakhstan
Anatoly Antipin	Dorodnicyn Computing Centre FRC CSC RAS, Russia
Evripidis Bampis	Sorbonne Université, France
Olga Battaïa	ISAE-Supaero, France
René van Bevern	Novosibirsk State University, Russia
Oleg Burdakov	Linköping University, Sweden
Maxim Buzdalov	ITMO University, Russia
Igor Bykadorov	Sobolev Institute of Mathematics, Russia
Tatjana Davidović	Mathematical Institute SANU, Serbia
Stephan Dempe	Freiberg University, Germany
Anton Eremeev	Sobolev Institute of Mathematics, Russia
Adil Erzin	Novosibirsk State University, Russia
Stefka Fidanova	Institute of Information and Communication Technologies, Bulgaria
Alexander Filatov	Far Eastern Federal University, Russia
Fedor Fomin	University of Bergen, Norway
Edward Gimadi	Sobolev Institute of Mathematics, Russia
Alexander Grigoriev	Maastricht University, The Netherlands
Evgeny Gurevsky	University of Nantes, France
Klaus Jansen	Kiel University, Germany
Vadim Kartak	Ufa State Aviation Technical University, Russia
Alexander Kazakov	Matrosov Institute of System Dynamics and Control Theory, Russia
Lev Kazakovtsev	Siberian State Aerospace University, Russia
Oleg Khamisov	Melentiev Energy Systems Institute, Russia
Andrey Kibzun	Moscow Aviation Institute, Russia
Donghyun (David) Kim	Kennesaw State University, USA
Yury Kochetov	Sobolev Institute of Mathematics, Russia
Igor Konnov	Kazan Federal University, Russia
Vadim Levit	Ariel University, Israel

Bertrand M. T. Lin	National Chiao Tung University, Taiwan
Vittorio Maniezzo	University of Bologna, Italy
Vladimir Mazalov	Institute of Applied Mathematical Research, Russia
Nenad Mladenović	Khalifa University, UAE
Rolf Niedermeier	Technical University of Berlin, Germany
Yury Nikulin	University of Turku, Finland
Evgeni Nurminski	Far Eastern Federal University, Russia
Leon Petrosyan	Saint Petersburg State University, Russia
Petros Petrosyan	Yerevan State University, Armenia
Alexander Petunin	Ural Federal University, Russia
Sergey Polyakovskiy	Deakin University, Australia
Leonid Popov	Krasovsky Institute of Mathematics and Mechanics, Russia
Mikhail Posypkin	Dorodnicyn Computing Centre, Russia
Artem Pyatkin	Sobolev Institute of Mathematics, Russia
Soumyendu Raha	Indian Institute of Science, Bangalore, India
Yaroslav Sergeyev	University of Calabria, Italy
Sergey Sevastyanov	Sobolev Institute of Mathematics, Russia
Natalia Shakhlevich	University of Leeds, UK
Alexander Shananin	Moscow Institute of Physics and Technology, Russia
Angelo Sifaleras	University of Macedonia, Greece
Vladimir Skarin	Krasovsky Institute of Mathematics and Mechanics, Russia
Petro Stetsyuk	Glushkov Institute of Cybernetics, Ukraine
Alexander Strekalovsky	Matrosov Institute for System Dynamics and Control Theory, Russia
Vitaly Strusevich	University of Greenwich, UK
Maxim Sviridenko	Yahoo! Labs, USA
Tatiana Tchemisova	University of Aveiro, Portugal

Additional Reviewers

Agafonov, Evgeny	Bulanova, Nina
Alazmi, Hakem	Buzhinsky, Igor
Alekseeva, Ekaterina	Bykadorov, Igor
Almeida, Ricardo	Camacho, David
Antipin, Anatoly	Chandramouli, Shyam
Antipov, Denis	Chentsov, Alexander
Ayzenberg, Natalya	Chicano, Francisco
Benmansour, Rachid	Chivilikhin, Daniil
Bentert, Matthias	Dang, Duc-Cuong
Berikov, Vladimir	Davydov, Ivan
Berndt, Sebastian	Dobrynin, Andrey
Boehmer, Niclas	Filchenkov, Andrey
Borisovsky, Pavel	Filippova, Tatiana

Forghani, Majid
Gluschenko, Konstantin
Gomoyunov, Mikhail
Grage, Kilian
Gromova, Ekaterina
Gruzdeva, Tatiana
Gubar, Elena
Gudyma, Mikhail
Gusev, Mikhail
Himmel, Anne-Sophie
Jaksic Kruger, Tatjana
Karavaev, Vitalii
Khandeev, Vladimir
Khoroshilova, Elena
Khutoretskii, Alexander
Kobylkin, Konstantin
Kononova, Polina
Kovalenko, Yulia
Krylatov, Alexander
Kumacheva, Suriya
Latyshev, Aleksei
Lempert, Anna
Lengler, Johannes
Levanova, Tatyana
Lezhnina, Elena
Liberti, Leo
Maack, Marten
Martins, Natália
Masich, Igor
Melnikov, Andrey
Mironovich, Vladimir
Molter, Hendrik
Muravyov, Sergey
Naumova, Nataliya
Neznakhina, Ekaterina
Ogorodnikov, Yuri

Olenev, Nicholas
Orlov, Andrei
Parilina, Elena
Plotnikov, Roman
Plyasunov, Alexander
Polyakova, Anastasiya
Prolubnikov, Alexander
Renken, Malte
Rettieva, Anna
Sedakov, Artem
Servakh, Vladimir
Shary, Sergey
Shenmaier, Vladimir
Shkaberina, Guzel
Sidorov, Alexander
Simanchev, Ruslan
Sleptchenko, Andrei
Smetannikov, Ivan
Sopov, Evgenii
Stanimirovic, Zorica
Stashkov, Dmitry
Stupina, Alena
Suslov, Nikita
Suvorov, Dmitrii
Tsoy, Yury
Urazova, Inna
Vasilyev, Igor
Vasin, Alexander
Yanovskaya, Elena
Zabashta, Alexey
Zabudsky, Gennady
Zalyubovskiy Vyacheslav
Zaozerskaya, Lidia
Zenkevich, Anatolii
Zeufack, Vannel
Zschoche, Philipp

Industry Section Chairs

Igor Vasilyev	Matrosov Institute for System Dynamics and Control Theory, Russia
Alexander Kurochkin	Sobolev Institute of Mathematics, Russia

Organizing Committee

Yury Kochetov	Sobolev Institute of Mathematics, Russia
Nina Kochetova	Sobolev Institute of Mathematics, Russia
Polina Kononova	Sobolev Institute of Mathematics, Russia
Timur Medvedev	Higher School of Economics, Russia
Tatyana Gruzdeva	Matrosov Institute for System Dynamics and Control Theory, Russia
Ivan Davydov	Sobolev Institute of Mathematics, Russia
Adil Erzin	Sobolev Institute of Mathematics, Russia
Sergey Lavlinsky	Sobolev Institute of Mathematics, Russia
Vladimir Khandeev	Sobolev Institute of Mathematics, Russia
Igor Kulachenko	Sobolev Institute of Mathematics, Russia
Andrey Melnikov	Sobolev Institute of Mathematics, Russia
Anna Panasenko	Sobolev Institute of Mathematics, Russia
Artem Panin	Sobolev Institute of Mathematics, Russia
Roman Plotnikov	Sobolev Institute of Mathematics, Russia

Organizers

Sobolev Institute of Mathematics, Russia
Novosibirsk State University, Russia
Krasovsky Institute of Mathematics and Mechanics, Russia
Higher School of Economics (Campus Nizhny Novgorod), Russia
Melentiev Energy Systems Institute, Russia

Sponsor

Mathematical Center in Akademgorodok, Russia

Contents

Mathematical Programming

Global Optimization

Game Theory and Mathematical Economics

Heuristics and Metaheuristics

Machine Learning and Data Analysis

Combinatorial Optimization

A 0.3622-Approximation Algorithm for the Maximum k-Edge-Colored Clustering Problem

Alexander Ageev$^{(\boxtimes)}$ and Alexander Kononov

Sobolev Institute of Mathematics, Novosibirsk, Russia
{ageev,alvenko}@math.nsc.ru

Abstract. In the Max k-Edge-Colored Clustering problem (abbreviated as MAX-k-EC) we are given an undirected graph and k colors. Each edge of the graph has a color and a nonnegative weight. The goal is to color the vertices so as to maximize the total weight of the edges whose colors coincide with the colors of their endpoints. The problem was introduced by Angel et al. [3]. In this paper we give a polynomial-time algorithm for MAX-k-EC with an approximation factor $\frac{4225}{11664} \approx 0.3622$ which significantly improves the best previously known approximation bound $\frac{49}{144} \approx 0.3402$ established by Alhamdan and Kononov [2].

Keywords: Clustering problem · Edge-colored graph · Linear relaxation · Approximation algorithm · Worst-case analysis

1 Introduction

In the Max k-Edge-Colored Clustering problem we are given an undirected graph $G = (V, E)$ whose edges have colors $c : E \to \{1, \ldots, k\}$ and weights $w : E \to Q^+$. The goal is to color vertices of G so as to maximize the total weight of edges whose colors coincide with the colors of their endpoints. Cai and Leung [6] call the edges whose colors coincide with the colors of their endpoints stable. In these terms the problem is to color the vertices of G so as to maximize the total weight of stable edges.

The problem was introduced by Angel et al. in [3]. It is easy to see that the case when each edge of G has it own color is nothing but the Maximum Weight Matching Problem. Cai and Leung [6] observed that the MAX-k-EC problem can be considered as the optimization counterpart of the Vertex-Monochromatic Subgraph problem or the Alternating Path Removal problem studied in a series of research papers [5,8].

The MAX-k-EC problem can also be interpreted as an extension of the centralized version of the information-sharing model introduced by Kleinberg and Ligett [9] and as a special case of the combinatorial allocation problem [7] (for a more detailed discussion see [2,3]).

© Springer Nature Switzerland AG 2020
Y. Kochetov et al. (Eds.): MOTOR 2020, CCIS 1275, pp. 3–15, 2020.
https://doi.org/10.1007/978-3-030-58657-7_1

1.1 Related Work

Angel et al. [3,4] showed that the MAX-k-EC problem is strongly NP-hard in the case of edge-tricoloured bipartite graphs. Cai and Leung [6] strengthened this result by establishing that MAX-k-EC is NP-hard in the strong sense even on edge-tricoloured planar bipartite graphs of maximum degree four.

On the other hand, Angel et al. [3,4] showed that the MAX-k-EC problem is polynomially solvable in the case of edge-bicoloured graphs by a reduction to the maximum independent set problem on bipartite graphs.

Cai and Leung [6] presented two FPT algorithms for the MAX-k-EC problem under the assumption that the number of stable edges is a fixed parameter.

Angel et al. [3,4] also derived the first constant-factor approximation algorithm for the MAX-k-EC problem. It is base on randomized rounding a linear relaxation of the problem and finds a set of stable edges with weight $\frac{1}{e^2} \approx 0.1353$ of the optimum. Later Ageev and Kononov [1] showed that a refined worst-case analysis of this algorithm gives an approximation factor of 0.25. They also presented an approximation algorithm with a factor of $\frac{7}{23} \approx 0.3043$ based on rounding the same LP relaxation. Very recently, Alhamdan and Kononov [2] further improved this bound to $\frac{49}{144} \approx 0.3402$ by applying a modified rounding technique.

1.2 Our Results

We present a modified version of the algorithm by Alhamdan and Kononov [2] for the MAX-k-EC problem. The algorithm retrieves a set of stable edges whose weight is a factor of $\frac{4225}{11664} \approx 0.3622$ of the optimum. This is achieved through the use a slightly more sophisticated rounding scheme. Though the main ideas behind our approach are the same as in [3,4]. We use a similar two-phase scheme. On the first phase the algorithm chooses a set of desired stable edges randomly and independently for each color. On the second phase the algorithm colors vertices taking into account the selection of the edges made on the first phase.

2 Algorithm

Angel et al. [3,4] suggest the following integer linear program (ILP) for MAX-k-EC:

$$\text{maximize} \sum_{e \in E} w_e z_e \tag{1}$$

$$\text{subject to} \sum_{i \in \mathcal{C}} x_{vi} = 1, \qquad\qquad \forall v \in V \tag{2}$$

$$z_e \leq \min\{x_{vc(e)}, x_{uc(e)}\} \qquad \forall e = [v,u] \in E \tag{3}$$

$$x_{vi}, z_e \in \{0,1\}, \qquad\qquad \forall v \in V, i \in \mathcal{C}, e \in E \tag{4}$$

In this program, the variables x_{vi}, $v \in V$, $i \in \mathcal{C}$ specify the colors assigned to the vertices: $x_{vi} = 1$ if v is colored with color i and $x_{vi} = 0$ otherwise. The variables

z_e, $e \in E$ indicate the stable edges: $z_e = 1$ if both endpoints of e are colored with the same color as e and $z_e = 0$ otherwise.

The first set of constraints ensures that each vertex is colored exactly by one color, and the second ensures that an edge e is stable if its color coincides with the colors of its endpoints.

The LP-relaxation (LP) of (1)–(4) arises after replacing the constraints $x_{vi} \in \{0, 1\}$ and $z_e \in \{0, 1\}$ by $x_{vi} \geq 0$ and $z_e \geq 0$, respectively.

The following two-phase randomized algorithm was first presented in [3] and analyzed in [1–3]. In the first phase, it starts with solving LP and then works in k iterations, by considering each color i, $1 \leq i \leq k$, independently from the others. For each color, the algorithm picks a threshold r at random in $(0, 1)$ and selects all edges of this color with $z_e^* \geq r$. When an edge is selected, this means that both its endpoints receive the color of this edge. Since a vertex can be adjacent to differently colored edges, it may receive more than one colors. In the second phase, the algorithm chooses randomly one of these colors. Denote by $\lambda_v(l, c)$ the probability with which the algorithm chooses the color c if l colors were assigned to v at the first phase of the algorithm. We present the algorithm below.

Algorithm 1. Algorithm 2-PHASE

1: **Phase I**:
2: Solve LP and let z_e^* be the values of variables z_e.
3: **for** each color $c \in C$ **do**
4:    Let r be a random value in **[0,1]**.
5:    Choose the c-colored edges e with $z_e^* \geq r$ and give color c to both of e's endpoints.
6: **end for**
7: **Phase II**:
8: **for** each vertex $v \in V$ **do**
9:    Let vertex v got l colors.
10:    assign randomly one of l colors to v, each with the probability $\lambda_v(l, c)$.
11: **end for**

Let (x^*, z^*) be an optimal solution of the LP. Following [2] we say that an edge e is *big* if $z_e^* > \frac{1}{3}$; otherwise an edge e is *small*. We say that a vertex v is *heavy* if it is incident to at least one big edge; otherwise vertex v is *light*. Given a vertex $v \in V$, we say that color i is *heavy* for v if v is incident to an i-colored big edge, otherwise color i is *light* for v. We note that each vertex has at most two heavy colors. If the vertex v got two colors: a heavy color i and a light color j then we set $\lambda_v(2, i) \doteq \lambda_v(\{j\}, i) = \frac{1}{3}$ and $\lambda_v(2, j) \doteq \lambda_v(\{i\}, j) = \frac{2}{3}$ else we set $\lambda_v(l, c) = \frac{1}{l}$ for all l colors assigned to v at the first phase of the algorithm. Here $\lambda_v(\{i\}, q)$ means the probability with which the algorithm assigns color q to vertex v on Phase 2 if on Phase 1 v receives two colors q and i. For convenience, we will use two notation for the probability with which the algorithm chooses the color c if two colors were assigned to v.

Our rounding scheme differs from that in [2] by the definition of big edge. In [2], an edge e is big if $z_e^* > \frac{1}{2}$. This implies that each vertex can have at most one heavy color, which significantly simplifies the analysis but results in a weaker approximation factor. Moreover, in our analysis we need a bit stronger probability lemma than in [2] (Lemma 9).

3   Analysis

In this subsection, we give a worst-case analysis of Algorithm 2-PHASE. Let X_{vi} denote the event where vertex v gets color i after **Phase I** of the algorithm. Since the first phase of the algorithm coincides with the first phases of the algorithms RR and RR2, presented in [3] and [1], respectively, the following simple statements are valid.

Lemma 1. *[3] For any edge $e \in E$, the probability that e is chosen in **Phase I** is z_e^*.*

Lemma 2. *[3] For every vertex $v \in V$ and for all $i \in C$ we have:*

$$Pr[X_{vi}] = \max\{z_e^* : e = [v, u] \in E \ \& \ c(e) = i\}.$$

Lemma 3. *[3] For every vertex $v \in V$, $\sum_{i \in C} Pr[X_{vi}] \leq 1$.*

Recall that the vertex v can get several colors after **Phase I**. However, in general this number will be small. Let Y_{vi} denote the event where vertex v is colored with i after **Phase II** of the algorithm.

Assume that a vertex v gets a color q in **Phase I** of Algorithm 2-Phase. The probability that a vertex v is colored with a color q in **Phase II** depends on how many colors a vertex v received in **Phase I**. Without loss of generality, assume that the edges with colors $1, \ldots, t$ and q are incident to the vertex v. By the law of total probability we have

$$Pr[Y_{vq}|X_{vq}] \geq \prod_{i=1}^{t}(1 - Pr[X_{vi}]) + \sum_{i=1}^{t}\lambda_v(\{i\}, q)Pr[X_{vi}]\prod_{j \neq i}(1 - Pr[X_{vj}])$$

$$+ \frac{1}{3}\sum_{i,j}Pr[X_{vi}]Pr[X_{vj}]\prod_{l \neq i, l \neq j}(1 - Pr[X_{vl}])$$

$$+ \frac{1}{4}\sum_{i,j,k}Pr[X_{vi}]Pr[X_{vj}]Pr[X_{vk}]\prod_{l \neq i, l \neq j, l \neq k}(1 - Pr[X_{vl}]) \quad (5)$$

The following lemmas give a lower bound for the probability that color q was assigned to vertex v in **Phase II**.

Lemma 4. *Assume that a heavy vertex v gets the only heavy color q in **Phase I** of Algorithm 2-PHLV, then $Pr[Y_{vq}|X_{vq}] \geq \frac{17}{27} > 0.62$.*

We assume that the vertex v has no heavy colors except q. It follows that $\lambda_v(\{i\}, q) = \frac{1}{3}$ for all $i = 1, \ldots, t$ and we can rewrite (5) as

$$Pr[Y_{vq}|X_{vq}] \geq \prod_{i=1}^{t}(1 - Pr[X_{vi}]) + \frac{1}{3}\sum_{i=1}^{t} Pr[X_{v1}]\prod_{i=2}^{t}(1 - Pr[X_{vi}])$$

$$+ \frac{1}{3}\sum_{i=1}^{t-1}\sum_{j=i+1}^{t} Pr[X_{vi}]Pr[X_{vj}]\prod_{l \neq i, l \neq j}(1 - Pr[X_{vl}]) \qquad (6)$$

We drop all the remaining terms of the formula because they are equal to zero in the worst case.

To simplify computations we set $X_i = Pr[X_{vi}]$ and consider the right-hand-size of (6) as a function f_{vq} of variables $X_1, X_2, ..., X_t$. We have

$$f_{vq} = \prod_{i=1}^{t}(1 - X_i) + \frac{1}{3}\sum_{i=1}^{t} X_i\prod_{j \neq i}(1 - X_j) + \frac{1}{3}\sum_{i=1}^{t-1}\sum_{j=i+1}^{t} X_iX_j\prod_{l \neq i, l \neq j}(1 - X_l).$$

Taking into account that color q is heavy and other colors are light, we have $\sum_{i=1}^{t} X_i \leq \frac{2}{3}$ and $X_i \leq \frac{1}{3}$, $i = 1, \ldots, t$.

Putting the first two variables out of the summation and the product, we get

$$f_{vq} = (1 - X_1)(1 - X_2)\prod_{i=3}^{t}(1 - X_i) + \frac{1}{3}(X_1(1 - X_2) + X_2(1 - X_1))\prod_{i=3}^{t}(1 - X_i)$$

$$+ \frac{1}{3}(1 - X_1)(1 - X_2)\sum_{i=3}^{t} X_i\prod_{j \geq 3, j \neq i}(1 - X_j) + \frac{1}{3}X_1X_2\prod_{l=3}^{t}(1 - X_l)$$

$$+ \frac{1}{3}(X_1(1 - X_2) + (1 - X_1)X_2)\sum_{i=3}^{t} X_i\prod_{j \geq 3, j \neq i}(1 - X_j)$$

$$+ \frac{1}{3}(1 - X_1)(1 - X_2)\sum_{i=3}^{t}\sum_{j=i+1}^{t} X_iX_j\prod_{l \geq 3, l \neq i, l \neq j}(1 - X_l)$$

$$\geq (1 - X_1)(1 - X_2)\prod_{i=3}^{t}(1 - X_i) + \frac{1}{3}(X_1(1 - X_2) + X_2(1 - X_1))\prod_{i=3}^{t}(1 - X_i)$$

$$+ \frac{1}{3}X_1X_2\prod_{i=3}^{t}(1 - X_i) = (1 - \frac{2}{3}X_1 - \frac{2}{3}X_2)\prod_{i=3}^{t}(1 - X_i) + \frac{2}{3}X_1X_2\prod_{i=3}^{t}(1 - X_i).$$

$$(7)$$

Consider f_{vq} as a function of two variables X_1 and X_2. Assume that $X_1 + X_2 = \gamma$, where $\gamma \leq \frac{2}{3}$ is a constant. Let $X_1 \geq X_2 > 0$. If we increase X_1 and decrease X_2 by δ, $0 < \delta \leq X_2$, then the first term of (7) does not change and the last term decreases and therefore the function f_{vq} decreases as well. It follows that the minimum of f_{vq} is attained at $X_1 = \min\{1/3, \gamma\}$ and $X_2 = \max\{0, \gamma - 1/3\}$. By repeating this argument we get that the minimum of f_{vq} is attained when $X_1 = \frac{1}{3}$, $X_2 = \frac{1}{3}$, and $X_i = 0$, $i = 3, \ldots, t$. Finally, we get $Pr[Y_{vq}|X_{vq}] \geq f_{vq} \geq \frac{17}{27} > 0.62$.

Lemma 5. *Assume that a heavy vertex v has only one heavy color and gets a light color q in* **Phase I** *of Algorithm 2-PHLV, then $Pr[Y_{vq}|X_{vq}] \geq \frac{50}{81} > 0.61$.*

Proof. Let color 1 be the heavy color. It follows that $\lambda_v(\{1\}, q) = \frac{2}{3}$ and $\lambda_v(\{i\}, q) = \frac{1}{2}$ for all $i = 1, \ldots, t$ and we can rewrite (5) as

$$Pr[Y_{vq}|X_{vq}] \geq \prod_{i=1}^{t}(1 - Pr[X_{vi}]) + \frac{2}{3}Pr[X_{v1}]\prod_{i=2}^{t}(1 - Pr[X_{vi}])$$

$$+ \frac{1}{2}(1 - Pr[X_{v1}])\sum_{i=2}^{t}Pr[X_{vi}]\prod_{j\geq 2, j\neq i}(1 - Pr[X_{vj}])$$

$$+ \frac{1}{3}\sum_{i=1}^{t}\sum_{j=i+1}^{t}Pr[X_{vi}]Pr[X_{vj}]\prod_{l\neq i, l\neq j}(1 - Pr[X_{vl}])$$

By setting $A = \prod_{i=3}^{t}(1 - Pr[X_{vi}])$, $B = \sum_{i=3}^{t}Pr[X_{vi}]\prod_{j\geq 3, j\neq i}(1 - Pr[X_{vj}])$ and $C = \sum_{i=3}^{t}\sum_{j=i+1}^{t}Pr[X_{vi}]Pr[X_{vj}]\prod_{l\geq 3, l\neq i, l\neq j}(1 - Pr[X_{vl}])$ we can rewrite this expression as

$$Pr[Y_{vq}|X_{vq}] \geq (1 - Pr[X_{v1}])(1 - Pr[X_{v2}])A + \frac{2}{3}Pr[X_{v1}](1 - Pr[X_{v2}])A +$$

$$\frac{1}{2}(1 - Pr[X_{v1}])Pr[X_{v2}]A + \frac{1}{2}(1 - Pr[X_{v1}])(1 - Pr[X_{v2}])B + \frac{1}{3}Pr[X_{v1}]Pr[X_{v2}]A$$

$$+ \frac{1}{3}(Pr[X_{v1}](1 - Pr[X_{v2}]) + (1 - Pr[X_{v1}])Pr[X_{v2}])B + \frac{1}{3}(1 - Pr[X_{v1}])(1 - Pr[X_{v2}])C.$$

Discarding the last term and setting $X_i = Pr[X_{vi}]$ we get

$$Pr[Y_{vq}|X_{vq}] \geq f_{vq} \doteq (1 - X_1)(1 - X_2)A + \frac{2}{3}X_1(1 - X_2)A + \frac{1}{2}(1 - X_1)X_2A$$

$$+ \frac{1}{2}(1 - X_1)(1 - X_2)B + \frac{1}{3}X_1X_2A + \frac{1}{3}(X_1(1 - X_2) + (1 - X_1)X_2)B.$$

In order to obtain a lower bound for $Pr[Y_{vq}|X_{vq}]$, we first show that the minimum of f_{vq} is attained when $X_1 = \frac{1}{3}$. After multiplying the terms with each other, we get

$$f_{vq} = (1 - \frac{1}{3}X_1 - \frac{1}{2}X_2 + \frac{1}{6}X_1X_2)A + (\frac{1}{2} - \frac{1}{6}X_1 - \frac{1}{6}X_2 - \frac{1}{6}X_1X_2)B$$

$$= (1 - \frac{1}{3}X_1 - \frac{1}{3}X_2)A + (\frac{1}{2} - \frac{1}{6}X_1 - \frac{1}{6}X_2)B - \frac{1}{6}(X_1X_2B + X_2(1 - X_1)A).$$

Let us consider f_{vq} as a function of two variables X_1 and X_2. Assume that $X_1 + X_2 = \gamma$. Since color 1 is heavy then $\frac{1}{3} \leq X_1 \leq \gamma$ and $X_2 \leq \frac{1}{3}$. If we decrease X_1 and increase X_2 by δ, $0 < \delta \leq X_1 - \frac{1}{3}$, then the expression $X_1X_2B + X_2(1 - X_1)A$ increases. It follows that f_{vq} reaches a minimum when $X_1 = \frac{1}{3}$.

Now, substitute X_1 by $\frac{1}{3}$. Thus, we obtain

$$f_{vq} \geq (\frac{8}{9} - \frac{4}{9}X_2)\prod_{i=3}^{t}(1 - X_i) + (\frac{4}{9} - \frac{2}{9}X_2)\sum_{i=3}^{t}X_i\prod_{j\geq 3, j\neq i}(1 - X_j). \quad (8)$$

Rewrite the right-hand side of (8) as

$$f_{vq} = (\frac{8}{9} - \frac{4}{9}X_2)(1 - X_3) \prod_{i=4}^{t}(1 - X_i) + (\frac{4}{9} - \frac{2}{9}X_2)X_3 \prod_{i=4}^{t}(1 - X_i)$$

$$+ (\frac{4}{9} - \frac{2}{9}X_2)(1 - X_3) \sum_{i=4}^{t} X_i \prod_{j=4, j \neq i}^{t} (1 - X_j)$$

$$\geq (\frac{8}{9} - \frac{4}{9}X_2)(1 - X_3) \prod_{i=4}^{t}(1 - X_i) + (\frac{4}{9} - \frac{2}{9}X_2)X_3 \prod_{i=4}^{t}(1 - X_i)$$

$$= \frac{8}{9} \prod_{i=4}^{t}(1 - X_i)(1 - \frac{X_2}{2} - \frac{X_3}{2} + \frac{X_2 X_3}{4}) \quad (9)$$

Consider f_{vq} as a function of two variables X_2 and X_3. Assume that $X_2 + X_3 = \gamma$, where $\gamma \leq \frac{2}{3}$ is a constant. Let $X_2 \geq X_3 > 0$. If we increase X_2 and decrease X_3 by δ, $0 < \delta \leq X_3$, then the last term decreases and therefore the function f_{vq} decreases as well. It follows that the minimum of f_{vq} is attained at $X_2 = \min\{\gamma, \frac{1}{3}\}$. By repeating this argument we get that the minimum of f_{vq} is attained when $X_2 = \frac{1}{3}$, $X_3 = \frac{1}{3}$, and $X_i = 0$, $i = 4, \ldots, t$. Finally, we get $Pr[Y_{vq}|X_{vq}] \geq f_{vq} \geq \frac{50}{81} \approx 0.6172839507$.

Lemma 6. *Assume that a vertex v has two heavy color and a color q is heavy. Then $Pr[Y_{vq}|X_{vq}] \geq \frac{2}{3}$.*

Proof. Let color 1 be the second heavy color. It follows that $\lambda_v(\{1\}, q) = \frac{1}{2}$ and $\lambda_v(\{i\}, q) = \frac{1}{3}$ for all $i = 2, \ldots, t$ and we can rewrite (5) as

$$Pr[Y_{vq}|X_{vq}] \geq \prod_{i=1}^{t}(1 - Pr[X_{vi}]) + \frac{1}{2}Pr[X_{v1}] \prod_{i=2}^{t}(1 - Pr[X_{vi}])$$

$$+ \frac{1}{3}(1 - Pr[X_{v1}]) \sum_{i=2}^{t} Pr[X_{vi}] \prod_{j \geq 2, j \neq i} (1 - Pr[X_{vj}])$$

$$+ \frac{1}{3} \sum_{i=1}^{t} \sum_{j=i+1}^{t} Pr[X_{vi}] Pr[X_{vj}] \prod_{l \neq i, l \neq j} (1 - Pr[X_{vl}])$$

By setting $A = \prod_{i=4}^{t}(1 - Pr[X_{vi}])$ and $X_i = Pr[X_{vi}]$ we obtain

$$Pr[Y_{vq}|X_{vq}] \geq f_{vq} \doteq (1 - X_1)(1 - X_2)(1 - X_3)A + \frac{1}{2}X_1(1 - X_2)(1 - X_3)A$$

$$+ \frac{1}{3}(1 - X_1)X_2(1 - X_3)A + \frac{1}{3}(1 - X_1)(1 - X_2)X_3 A + \frac{1}{3}X_1 X_2(1 - X_3)A$$

$$+ \frac{1}{3}X_1(1 - X_2)X_3 A + \frac{1}{3}(1 - X_1)X_2 X_3 A$$

$$= A(1 - X_1)(1 - \frac{2}{3}X_2 - \frac{2}{3}X_3 + \frac{2}{3}X_2 X_3) + AX_1(\frac{1}{2} - \frac{1}{6}X_2 - \frac{1}{6}X_3 - \frac{1}{6}X_2 X_3)$$

Finally we obtain

$$f_{vq} = A(1 - X_1)(1 - \frac{2}{3}X_2 - \frac{2}{3}X_3) + AX_1(\frac{1}{2} - \frac{1}{6}X_2 - \frac{1}{6}X_3)$$

$$+ AX_2X_3(\frac{2}{3} - \frac{5}{6}X_1). \quad (10)$$

Consider f_{vq} as a function of two variables X_2 and X_3. Assume that $X_2 + X_3 = \gamma$, where $\gamma \leq \frac{1}{3}$ is a constant. Let $X_2 \geq X_3 > 0$. If we increase X_2 and decrease X_3 by δ, $0 < \delta \leq X_3$, then the first two terms of (10) do not change. Since $\frac{2}{3} - \frac{5}{6}X_1 \geq 0$ for $X_1 \leq \frac{2}{3}$ then the last term decreases and therefore the function f_{vq} decreases as well. It follows that the minimum of f_{vq} is attained at $X_2 = \gamma$ and $X_3 = 0$. By repeating this argument we get that the minimum of f_{vq} is attained when $X_i = 0$, $i = 3, \ldots, t$. It follows that

$$f_{vq} \geq (1 - X_1)(1 - \frac{2}{3}X_2) + X_1(\frac{1}{2} - \frac{1}{6}X_2) = 1 - \frac{1}{2}X_1 - \frac{2}{3}X_2 + \frac{1}{2}X_1X_2 \geq \frac{2}{3},$$

where the last inequality follows from $X_1 + X_2 \leq \frac{2}{3}$ and $X_1 \geq \frac{1}{3}$.

Lemma 7. *Assume that a vertex v has two heavy color and a color q is light. Then $Pr[Y_{vq}|X_{vq}] \geq \frac{211}{324}$.*

Proof. Let colors 1 and 2 be the heavy colors. It follows that $\lambda_v(\{1\}, q) = \lambda_v(\{2\}, q) = \frac{2}{3}$ and $\lambda_v(\{i\}, q) = \frac{1}{2}$ for all $i = 3, \ldots, t$ and we can rewrite (5) as

$$Pr[Y_{vq}|X_{vq}] \geq \prod_{i=1}^{t}(1 - Pr[X_{vi}]) + \frac{2}{3}Pr[X_{v1}](1 - Pr[X_{v2}])\prod_{i=3}^{t}(1 - Pr[X_{vi}])$$

$$+ \frac{2}{3}Pr[X_{v2}](1 - Pr[X_{v1}])\prod_{i=3}^{t}(1 - Pr[X_{vi}])$$

$$+ \frac{1}{2}(1 - Pr[X_{v1}])(1 - Pr[X_{v2}])\sum_{i=3}^{t} Pr[X_{vi}] \prod_{j \geq 2, j \neq i}(1 - Pr[X_{vj}])$$

$$+ \frac{1}{3}\sum_{i=1}^{t}\sum_{j=i+1}^{t} Pr[X_{vi}]Pr[X_{vj}] \prod_{l \neq i, l \neq j}(1 - Pr[X_{vl}])$$

$$+ \frac{1}{4}\sum_{i,j,k} Pr[X_{vi}]Pr[X_{vj}]Pr[X_{vk}] \prod_{l \neq i, l \neq j, l \neq k}(1 - Pr[X_{vl}]).$$

By setting $A = \prod_{i=5}^{t}(1 - Pr[X_{vi}])$ and $X_i = Pr[X_{vi}]$ we obtain

$$f_{vq} = A\prod_{i=1}^{4}(1 - X_i) + \frac{2A}{3}(X_1(1 - X_2) + X_2(1 - X_1))(1 - X_3)(1 - X_4)$$

$$+ \frac{A}{2}(X_3(1 - X_4) + X_4(1 - X_3))(1 - X_1)(1 - X_2)$$

$$+ \frac{A}{3}(X_1X_2(1 - X_3)(1 - X_4) + X_1X_3(1 - X_2)(1 - X_4) + X_1X_4(1 - X_2)(1 - X_3)$$

$$+ X_2X_3(1 - X_1)(1 - X_4) + X_2X_4(1 - X_1)(1 - X_3) + X_3X_4(1 - X_1)(1 - X_2))$$

$$+ \frac{A}{4}(X_1X_2X_3(1 - X_4) + X_1X_2X_4(1 - X_3) + X_1X_3X_4(1 - X_2) + X_2X_3X_4(1 - X_1))$$

$$(11)$$

After transforming the expression, we get

$$f_{vq}(X_3, X_4) = A(1 - X_1)(1 - X_2)[1 - \frac{X_3}{2} - \frac{X_4}{2}] + AX_1(1 - X_2)[\frac{2}{3} - \frac{1}{3}X_3 - \frac{1}{3}X_4]$$

$$+ AX_2(1 - X_1)[\frac{2}{3} - \frac{1}{3}X_3 - \frac{1}{3}X_4] + AX_1X_2[\frac{1}{3} - \frac{X_3}{12} - \frac{X_4}{12}]$$

$$+ AX_3X_4[\frac{1}{3}(1 - X_1)(1 - X_2) + \frac{1}{4}X_1(1 - X_2) + \frac{1}{4}X_2(1 - X_1) - \frac{1}{6}X_1X_2]\quad(12)$$

Consider f_{vq} as a function of two variables X_3 and X_4. Assume that $X_3 + X_4 = \gamma$, where $\gamma \le \frac{1}{3}$ is a constant and $X_3 \ge X_4 > 0$. If we increase X_3 and decrease X_4 by δ, $0 < \delta \le X_3$, then the first four terms of (12) do not change. Transforming the expression in the last term in square brackets, we obtain

$$\frac{1}{3}(1 - X_1)(1 - X_2) + \frac{1}{4}X_1(1 - X_2) + \frac{1}{4}X_2(1 - X_1) - \frac{1}{6}X_1X_2$$

$$= \frac{1}{3} - \frac{1}{3}X_1X_2 - \frac{1}{12}X_1 - \frac{1}{12}X_2 > 0,\quad(13)$$

where the last inequality follows from $X_1 + X_2 \le 1$. It follows that the minimum of f_{vq} is attained at $X_3 = \gamma$ and $X_4 = 0$. By repeating this argument we get that the minimum of

$$f_{vq} \ge \prod_{i=1}^{3}(1 - X_i) + \frac{2}{3}(X_1(1 - X_2) + X_2(1 - X_1))(1 - X_3)$$

$$+ \frac{1}{2}X_3(1 - X_1)(1 - X_2) + \frac{1}{3}(X_1X_2(1 - X_3) + X_1X_3(1 - X_2)$$

$$+ X_2X_3(1 - X_1)) + \frac{1}{4}X_1X_2X_3\quad(14)$$

Taking into account that $X_1 + X_2 + X_3 \le 1$, $X_1 \ge \frac{1}{3}$, and $X_2 \ge \frac{1}{3}$, we obtain that the minimum of f_{vq} is attained at $X_1 = X_2 = X_3 = \frac{1}{3}$ and $f_{vq} \ge \frac{211}{324} > 0.65$.

Lemma 8. *Assume that a light vertex v gets a color q in **Phase I** of Algorithm 2-PHLV, then $Pr[Y_{vq}|X_{vq}] \ge \frac{65}{108} > 0.6$.*

Proof. Since the vertex v has only light colors then $\lambda_v(\{i\}, q) = \frac{1}{2}$ for all $i = 2, \ldots, t$ and we can rewrite (5) as

$$Pr[Y_{vq}|X_{vq}] \geq \prod_{i=1}^{t}(1 - Pr[X_{vi}]) + \frac{1}{2}\sum_{i=1}^{t}Pr[X_{vi}]\prod_{j\neq i}(1 - Pr[X_{vj}])$$

$$+ \frac{1}{3}\sum_{i,j}Pr[X_{vi}]Pr[X_{vj}]\prod_{l\neq i, l\neq j}(1 - Pr[X_{vl}])$$

$$+ \frac{1}{4}\sum_{i,j,k}Pr[X_{vi}]Pr[X_{vj}]Pr[X_{vk}]\prod_{l\neq i, l\neq j, l\neq k}(1 - Pr[X_{vl}]) \quad (15)$$

By setting $A = \prod_{i=3}^{t}(1 - Pr[X_{vi}])$, $B = \sum_{i=3}^{t}Pr[X_{vi}]\prod_{j\geq 3, j\neq i}(1 - Pr[X_{vj}])$, setting $X_i = Pr[X_{vi}]$ and discarding some terms we get

$$Pr[Y_{vq}|X_{vq}] \geq f_{vq} \doteq (1 - X_1)(1 - X_2)A + \frac{1}{2}X_1(1 - X_2)A + \frac{1}{2}(1 - X_1)X_2 A$$

$$+ \frac{1}{2}(1 - X_1)(1 - X_2)B + \frac{1}{3}X_1 X_2 A + \frac{1}{3}(X_1(1 - X_2) + (1 - X_1)X_2)B + \frac{1}{4}X_1 X_2 B$$

$$= A(1 - \frac{1}{2}X_1 - \frac{1}{2}X_2 + \frac{1}{3}X_1 X_2) + B(\frac{1}{2} - \frac{1}{6}X_1 - \frac{1}{6}X_2 + \frac{1}{12}X_1 X_2).$$

Consider f_{vq} as a function of two variables X_1 and X_2. Assume that $X_1 + X_2 = \gamma$. Remind that $X_1 \leq \frac{1}{3}$ and $X_2 \leq \frac{1}{3}$. Let $X_1 \geq X_2 > 0$. If we increase X_1 and decrease X_2 by δ, $0 < \delta \leq X_2$, the function f_{vq} decreases. It follows that the minimum of f_{vq} is attained at $X_1 = \min\{\frac{1}{3}, \gamma\}$ and $X_2 = 0$. Repeating this argument for any pair of variables not equal to 0 or $\frac{1}{3}$ we get that the minimum of f_{vq} is attained when $X_1 = X_2 = X_3 = \frac{1}{3}$ and $X_i = 0$, $i = 4, \ldots, t$. Substituting these values in (15), we get $Pr[Y_{vq}|X_{vq}] \geq f_{vq} \geq \frac{65}{108} \approx 0.6018518519 > 0.6$.

Denote by $\lambda_v(\mathcal{Y}, c)$ the probability with which the algorithm chooses the color c if a set $\mathcal{Y} \subseteq \mathcal{C}$ colors were assigned to v at the first phase of the algorithm.

Lemma 9. *Let $e = (u, v)$ has a color c and it is chosen in the first phase of Algorithm 2-PHASE. If for any two subsets of colors $\mathcal{Y}$ and $\mathcal{Y}'$ such that $\mathcal{Y} \subseteq \mathcal{Y}' \subseteq \mathcal{C}$ we have $\lambda_v(\mathcal{Y}, c) \geq \lambda_v(\mathcal{Y}', c)$ then $Pr[e \text{ is stable}] \geq Pr[Y_{uc}|X_{uc}]Pr[Y_{vc}|X_{vc}]$.*

Proof. In order to prove this lemma, we consider a sequence of algorithms denoted by $\Sigma_0, \Sigma_1, \ldots, \Sigma_k$ where Σ_0 is algorithm 2-PHASE. The difference among these algorithms comes from the way in which the vertices get a color in **Phase I**. Let us fix a color x. We consider two different procedures for assigning colors to the vertices. *Procedure I* assigns the colors in the same way as algorithm 2-PHASE. *Procedure II* colors the vertices with color x independently.

Let us look at how these two procedures work for just two vertices. Suppose there is an edge e' that is incident with u colored by x and another edge e'' incident with v colored by x. Assume that e' and e'' are the edges with the maximal values of $z_{e'}^*$ and $z_{e''}^*$ among all x-colored edges incident with u and v, respectively. Let $z_{e'}^* \leq z_{e''}^*$. To simplify the notation we set $p = z_{e'}^*$ and $q = z_{e''}^*$.

Lemma 2 implies that *Procedure I* assigns the color x to both vertices u and v with probability p and only to the vertex v with probability $q - p$. *Procedure I* doesn't assign the color x to both vertices u and v with probability $1 - q$. Using *Procedure II*, we color the vertex u with probability p and the vertex v with probability q, each vertex independently.

In the algorithm Σ_0, for each color x, $1 \le x \le k$ we use *Procedure I* to assign colors to vertices. In the algorithm Σ_i, $1 \le i \le k$, for colors x such that $1 \le x \le i$ (resp. $i+1 \le x \le k$) we use *Procedure II* (resp. *Procedure I*) for assigning those colors to vertices. Thus, in algorithm Σ_k, all colors are assigned to vertices using *Procedure II*. Denote by $p_e(\Sigma_i)$ the probability that both extremities of e get the color c by algorithm Σ_i.

Let Σ_{i-1} and Σ_i be two consecutive algorithms; i.e., the algorithm Σ_{i-1} assigns colors from $[1, i-1]$ using *Procedure II* and colors from $[i, k]$ using *Procedure I*. While for Σ_i, it assigns colors from $[1, i]$ using *Procedure II*, and colors from $[(i+1), k]$ using *Procedure I*. Thus, they differ only in the way they assign color i to vertices. If there is no i-colored edge incident to either u or v, then $p_e(\Sigma_{i-1}) = p_e(\Sigma_i)$. Recall that we denote by X_{vc} (resp. $\overline{X_{vc}}$) the event that v gets (resp. does not get) color c after **Phase I** of the algorithm.

Let $\mathcal{C}' = \mathcal{C} \setminus \{c, i\}$. Denote by $A_v^{(\mathcal{Y})}$ the event which corresponds to the situation where vertex v gets a set $\mathcal{Y}$ of colors after **Phase I**. Let $\mathcal{Y} \subseteq \mathcal{C}'$ and $\mathcal{Y}' \subseteq \mathcal{C}'$, then the probability of the event $A_v^{(\mathcal{Y})} \wedge A_u^{(\mathcal{Y}')}$ are the same for both algorithms Σ_i and Σ_{i+1}, i.e. $Pr_{\Sigma_i}[A_v^{(\mathcal{Y})} \wedge A_u^{(\mathcal{Y}')}] = Pr_{\Sigma_{i+1}}[A_v^{(\mathcal{Y})} \wedge A_u^{(\mathcal{Y}')}]$.

For $\Sigma \in \{\Sigma_{i-1}, \Sigma_i\}$ we have

$$p_e(\Sigma) = \sum_{\mathcal{Y} \subseteq \mathcal{C}', \mathcal{Y}' \subseteq \mathcal{C}'} Pr_\Sigma[A_v^{(\mathcal{Y})} \wedge A_u^{(\mathcal{Y}')}] \phi_{\mathcal{Y}, \mathcal{Y}'}(\Sigma),$$

where

$$
\begin{aligned}
\phi_{\mathcal{Y}, \mathcal{Y}'}(\Sigma) = {} & \lambda_u(\mathcal{Y}, c)\lambda_v(\mathcal{Y}', c)(Pr[\overline{X_{u,i}} \wedge \overline{X_{v,i}}]) \\
& + \lambda_u(\mathcal{Y} \cup \{i\}, c)\lambda_v(\mathcal{Y}', c)(Pr[X_{u,i} \wedge \overline{X_{v,i}}]) \\
& + \lambda_u(\mathcal{Y}, c)\lambda_v(\mathcal{Y} \cup \{i\}, c)(Pr[\overline{X_{u,i}} \wedge X_{v,i}]) \\
& + \lambda_u(\mathcal{Y} \cup \{i\}, c)\lambda_v(\mathcal{Y}' \cup \{i\}, c)(Pr[X_{u,i} \wedge X_{v,i}])
\end{aligned}
$$

We claim that $\phi(\sum_{i-1}) \ge \phi(\sum_i)$ (from now on we omit subindices of ϕ for shortness). Taking into account the notation introduced we have

$$
\begin{aligned}
\phi(\Sigma_{i-1}) = {} & (1-q)\lambda_u(\mathcal{Y}, c)\lambda_v(\mathcal{Y}', c) + (q-p)\lambda_u(\mathcal{Y}, c)\lambda_v(\mathcal{Y}' \cup \{i\}, c) \\
& + p\lambda_u(\mathcal{Y} \cup \{i\}, c)\lambda_v(\mathcal{Y}' \cup \{i\}, c)
\end{aligned}
$$

and

$$
\begin{aligned}
\phi(\Sigma_i) = {} & (1-p)(1-q)\lambda_u(\mathcal{Y}, c)\lambda_v(\mathcal{Y}', c) + (1-q)p\lambda_u(\mathcal{Y} \cup \{i\}, c)\lambda_v(\mathcal{Y}', c) \\
& + (1-p)q\lambda_u(\mathcal{Y}, c)\lambda_v(\mathcal{Y}' \cup \{i\}, c) + pq\lambda_u(\mathcal{Y} \cup \{i\}, c)\lambda_v(\mathcal{Y}' \cup \{i\}, c)
\end{aligned}
$$

Thus,

$$
\begin{aligned}
\phi(\Sigma_{i-1}) - \phi(\Sigma_i) &= (1-q)p\lambda_u(\mathcal{Y},c)\lambda_v(\mathcal{Y}',c) - (1-q)p\lambda_u(\mathcal{Y}\cup\{i\},c)\lambda_v(\mathcal{Y}',c) \\
&\quad - (1-q)p\lambda_u(\mathcal{Y},c)\lambda_v(\mathcal{Y}'\cup\{i\},c) + (1-q)p\lambda_u(\mathcal{Y}\cup\{i\},c)\lambda_v(\mathcal{Y}'\cup\{i\},c) \\
&= (1-q)p[\lambda_u(\mathcal{Y},c)\lambda_v(\mathcal{Y}',c) - \lambda_u(\mathcal{Y}\cup\{i\},c)\lambda_v(\mathcal{Y}',c) \\
&\quad - \lambda_u(\mathcal{Y},c)\lambda_v(\mathcal{Y}'\cup\{i\},c) + \lambda_u(\mathcal{Y}\cup\{i\},c)\lambda_v(\mathcal{Y}'\cup\{i\},c)] \\
&= (1-q)p(\lambda_u(\mathcal{Y},c) - \lambda_u(\mathcal{Y}\cup\{i\},c))(\lambda_v(\mathcal{Y}',c) - \lambda_v(\mathcal{Y}'\cup\{i\},c)) \geq 0
\end{aligned}
$$

where the last inequality follows from $\lambda_u(\mathcal{Y},c) \geq \lambda_u(\mathcal{Y}\cup\{i\},c)$ and $\lambda_v(\mathcal{Y}',c) \geq \lambda_v(\mathcal{Y}'\cup\{i\},c)$.

Theorem 1. *The expected approximation ratio of Algorithm 2-PHASE is greater than 0.3622.*

Proof. Let OPT denote the sum of the weights of the stable edges in an optimal solution. Since z^* is an optimal solution of the LP, we have $OPT \leq \sum_{e\in E} w_e z_e^*$.

Consider an edge $e \in E$ is chosen in **Phase I** of Algorithm 2-PHLV. This occurs with probability z_e^* by Lemma 1. Suppose an edge $e = [u,v]$ has a color c, then by Lemma 9, the probability that the both endpoints of e are colored with c at least $Pr[Y_{vc}|X_{vc}]Pr[Y_{uc}|X_{uc}]$. Lemmata 4-8 imply that the expected contribution of the edge e is greater than $(65/108)^2 w_e z_e^* > 0.3622 w_e z_e^*$.

The expected weight of the stable edges in a solution obtained by Algorithm 2-PHLV is

$$
W = \sum_{e\in E} w_e Pr[e \text{ is stable}] > 0.3622 \sum_{e\in E} w_e z_e^* \geq 0.3622 \cdot \text{OPT}.
$$

4 Concluding Remarks

The goal of this paper was to exhaust the limits of the method first presented in [3]. We think that further significant improvements in approximation bounds will need some different approaches.

Acknowledgments. The authors thank the anonymous referees for their helpful comments and suggestions. A. Ageev was supported by Program no. I.5.1 of Fundamental Research of the Siberian Branch of the Russian Academy of Sciences (project no. 0314-2019-0019). A. Kononov was supported by RFBR, project number 20-07-00458.

References

1. Ageev, A., Kononov, A.: Improved approximations for the max k-colored clustering problem. In: Bampis, E., Svensson, O. (eds.) WAOA 2014. LNCS, vol. 8952, pp. 1–10. Springer, Cham (2015). https://doi.org/10.1007/978-3-319-18263-6_1
2. Alhamdan, Y.M., Kononov, A.: Approximability and Inapproximability for Maximum k-Edge-Colored Clustering Problem. In: van Bevern, R., Kucherov, G. (eds.) CSR 2019. LNCS, vol. 11532, pp. 1–12. Springer, Cham (2019). https://doi.org/10.1007/978-3-030-19955-5_1

3. Angel, E., Bampis, E., Kononov, A., Paparas, D., Pountourakis, E., Zissimopoulos, V.: Clustering on k-edge-colored graphs. In: Chatterjee, K., Sgall, J. (eds.) MFCS 2013. LNCS, vol. 8087, pp. 50–61. Springer, Heidelberg (2013). https://doi.org/10. 1007/978-3-642-40313-2_7

4. Angel, E., Bampis, E., Kononov, A., Paparas, D., Pountourakis, E., Zissimopoulos, V.: Clustering on k-edge-colored graphs. Discrete Appl. Math. **211**, 15–22 (2016)

5. Bang-Jensen, J., Gutin, G.: Alternating cycles and pathes in edge-coloured multi-graphs: a servey. Discrete Math. **165–166**, 39–60 (1997)

6. Cai, L., Leung, O.-Y.: Alternating path and coloured clustering (2018). arXiv:1807.10531

7. Feige, U., Vondrak, J.: Approximation algorithm for allocation problem improving the factor of $1 - \frac{1}{e}$. In: FOCS 2006, pp. 667–676 (2006)

8. Kano, M., Li, X.: Monochromatic and heterochromatic subgraphs in edge-colored graphs - a survey. Graphs Comb. **24**(4), 237–263 (2008)

9. Kleinberg, J.M., Ligett, K.: Information-sharing and privacy in social networks. Games Econ. Behav. **82**, 702–716 (2013)

On the Proximity of the Optimal Values of the Multi-dimensional Knapsack Problem with and Without the Cardinality Constraint

Aleksandr Yu. Chirkov[ID], Dmitry V. Gribanov[ID],
and Nikolai Yu. Zolotykh[✉][ID]

Lobachevsky State University of Nizhny Novgorod, Gagarin Avenue 23,
Nizhny Novgorod 603600, Russia
{aleksandr.chirkov,dmitry.gribanov,nikolai.zolotykh}@itmm.unn.ru

Abstract. We study the proximity of the optimal value of the m-dimensional knapsack problem to the optimal value of that problem with the additional restriction that only one type of items is allowed to include in the solution. We derive exact and asymptotic formulas for the precision of such approximation, i.e. for the infimum of the ratio of the optimal value for the objective functions of the problem with the cardinality constraint and without it. In particular, we prove that the precision tends to $0.59136\ldots/m$ if $n \to \infty$ and m is fixed. Also, we give the class of the worst multi-dimensional knapsack problems for which the bound is attained. Previously, similar results were known only for the case $m = 1$.

Keywords: Multi-dimensional knapsack problem · Approximate solution · Cardinality constraints

1 Introduction

In [1,2,4,5] the proximity of the optimal value of the (one-dimensional) knapsack problem to the optimal value of the problem with the cardinality constraints was studied. The cardinality constraint is the additional restriction that only k type of items is allowed to include in the solution (i.e. that only k coordinates of the optimal solution vector can be non-zero). Different upper and lower bounds for the guaranteed precision, i.e. for the infimum of the ratio of the optimal value for the objective functions of the problem with the cardinality constraints and without them, were obtained. Also, in some cases the classes of worst problems were constructed.

The importance of such kind of research is due to the fact that some algorithms for solving the knapsack problems require to find an optimal solution to that problem with the cardinality constraints; see, for example [4,5], where this

This work was performed at UNN Scientific and Educational Mathematical Center.

approach is used for constructing greedy heuristics for the integer knapsack problem. Moreover, the results of research can be potentially useful for constructing new fully polynomial approximation schemes.

Here, from this point of view, we consider the m-dimensional knapsack problem. The solution to that problem with the additional constraint that only 1 coordinate can be non-zero is called the approximate solution. We derive exact and asymptotic formulas for the precision of such approximation. In particular, we prove that the precision tends to $0.59136\ldots/m$ if $n \to \infty$ and m is fixed. Also, we give a class of worst multi-dimensional knapsack problems for which the bound is attained.

2   Definitions

Denote by $\mathbb{Z}_+$, $\mathbb{R}_+$ the sets of all non-negative integer and real numbers respectively. Let

$$L(A, b) = \left\{ x \in \mathbb{Z}_+^n : Ax \leq b \right\}, \qquad A = (a_{ij}) \in \mathbb{R}_+^{m \times n}, \qquad b = (b_i) \in \mathbb{R}_+^m.$$

The *integer m-dimensional knapsack problem* is to find x such that

$$cx \to \max \qquad \text{s.t. } x \in L(A, b), \tag{1}$$

where $c = (c_j) \in \mathbb{R}_+^n$ [3,6].

Denote by $v^{(j)}$ $(j = 1, 2, \ldots, n)$ a point in $L(A, b)$, all of whose coordinates $v_i^{(j)}$ are 0, except for of $v_j^{(j)}$, which is

$$v_j^{(j)} = \min_{i:\, a_{ij} > 0} \lfloor b_i / a_{ij} \rfloor .$$

It is not hard to see that $v^{(j)} \in L(A, b)$ and $cv^{(j)} = c_j v_j^{(j)}$. Denote $V(A, b) = \left\{ v^{(1)}, \ldots, v^{(n)} \right\}$. A point $v^{(j)}$, on which the maximum

$$\max_j cv^{(j)}$$

attained is called an *approximate solution* to the problem (1). The *precision of the approximate solution* is

$$\alpha(A, b, c) = \frac{\max\limits_{x \in V(A,b)} cx}{\max\limits_{x \in L(A,b)} cx} .$$

In this paper we study the value

$$\alpha_{mn} = \inf_{\substack{A \in \mathbb{R}_+^{m \times n} \\ b \in \mathbb{R}_+^m,\, c \in \mathbb{R}_+^n}} \alpha(A, b, c).$$

Table 1. Values of δ_n, ε_n and α_{1n} for small n

n	$\delta_n = \delta_{n-1}(\delta_{n-1}+1)$	$\varepsilon_n = 1 + \varepsilon_{n-1}(\delta_{n-1}+1)$	$\alpha_{1n} = \delta_n/\varepsilon_n$
1	1	1	1.000000000000000
2	2	3	0.666666666666667
3	6	10	0.600000000000000
4	42	71	0.591549295774648
5	1806	3054	0.591355599214145
6	3263442	5518579	0.591355492056923
7	10650056950806	18009568007498	0.591355492056890
8	113423713055421844361000442	191802924939285448393150887	0.591355492056890

3 Previous Work

The precision of the approximate solution to the 1-dimensional ($m = 1$) knapsack problem was studied in [1,2,4,5]. In particular, in [2,4] it was proven that

$$\delta_n = \delta_{n-1}(\delta_{n-1}+1), \qquad \varepsilon_n = 1 + \varepsilon_{n-1}(\delta_{n-1}+1), \qquad \delta_1 = \varepsilon_1 = 1.$$

The sequence $\{\delta_n\}$ is the A007018 sequence in On-Line Encyclopedia of Integer Sequences (OEIS) [7]. The sequence $\{\varepsilon_n\}$ is currently absent in OEIS.

The sequence $\alpha_{1n} = \delta_n/\varepsilon_n$ decreases monotonously and tends to the value $\alpha_{1\infty} = 0.591355492056890\ldots$ The values for δ_n, ε_n and α_{1n} for small n are presented in Table 1.

In [4,5] these results are used in constructing the approximate scheme for the integer knapsack problem. Note that α_{1n} is even higher than the guaranteed precision 0.5 of the greedy algorithm [6].

The infinum for α_{1n} is achieved on the problem (the worst case)

$$\sum_{j=1}^{n} \frac{x_j}{\delta_j} \to \max$$

s.t.

$$\sum_{j=1}^{n} \frac{x_j}{\delta_j + \mu_n} \le 1,$$

where $0 \le \mu_n < 1$ and $\sum_{j=1}^{n} \dfrac{1}{\delta_j + \mu_n} = 1$. In particular,

$$\mu_1 = 1, \quad \mu_2 = \frac{\sqrt{5}-1}{2} = 0.61803\ldots, \quad \mu_3 = 0.93923\ldots, \quad \mu_4 = 0.99855\ldots$$

The optimal solution vector to this problem is $(1,1,\ldots,1)$ and the optimal solution value is ε_n/δ_n, whereas the approximate solution vectors are

$$(1,0,0\ldots,0), \quad (0,\delta_2,0,\ldots,0), \quad (0,0,\delta_3,\ldots,0), \quad \ldots, \quad (0,0,0,\ldots,\delta_n)$$

and the corresponding value of the objective function is 1.

Lower and upper bounds for the guaranteed precision for $k \geq 2$ are obtained in [2].

In this paper we obtain formulas for α_{mn} for $m \geq 1$. In particular, we prove that $\alpha_{mn} \to \dfrac{\alpha_{1\infty}}{m}$ if $n \to \infty$ and m is fixed.

4   Preliminaries

Lemma 1. *For any fixed m the sequence $\{\alpha_{mn}\}$ decreases monotonously.*

Proof. Let $A \in \mathbb{R}_+^{m \times n}$, $h, b \in \mathbb{R}_+^m$, $c \in \mathbb{R}_+^n$ and $h > b$. Consider a matrix $A' = (A \mid h) \in \mathbb{R}_+^{m \times (n+1)}$ and a vector $c' = (c, 0) \in \mathbb{R}_+^{n+1}$. It is not hard to see that all points in $L(A', b)$ are obtained from the points in $L(A, b)$ by writing the zero component to the end. Hence $\alpha(A, b, c) = \alpha(A', b, c') \geq \alpha_{m,n+1}$. Due to the arbitrariness of A, b, c, we get $\alpha_{mn} \geq \alpha_{m,n+1}$.

Lemma 2. $\alpha(A, b, c) = \alpha(A', b, c)$ *for some $A' \leq A$, where each column of A' contains at least one non-zero element.*

Proof. Let for some s, t we have $a_{st} > 0$ and for all $i \neq s$

$$\left\lfloor \frac{b_s}{a_{st}} \right\rfloor \leq \left\lfloor \frac{b_i}{a_{it}} \right\rfloor$$

(if there are no such s, t, then put $A' = A$ and A' has the required form). From the matrix A we construct a matrix A' by setting $a'_{it} = 0$ for all $i \neq s$ and $a'_{ij} = a_{ij}$ otherwise.

For all $x \in \mathbb{R}_+^n$ we have $A'x \leq Ax$. Hence $L(A, b) \subseteq L(A', b)$. Hence

$$\max_{x \in L(A,b)} cx \leq \max_{x \in L'(A,b)} cx.$$

But

$$\min_{k:\, a_{kj}>0} \left\lfloor \frac{b_k}{a_{kj}} \right\rfloor = \min_{k:\, a'_{kj}>0} \left\lfloor \frac{b_k}{a'_{kj}} \right\rfloor \qquad (j = 1, 2, \ldots, n),$$

hence $V(A, b) = V(A', b)$. Now we have

$$\alpha(A, b, c) = \frac{\max\limits_{x \in V(A,b)} cx}{\max\limits_{x \in L(A,b)} cx} \geq \frac{\max\limits_{x \in V(A',b)} cx}{\max\limits_{x \in L(A',b)} cx} = \alpha(A', b, c).$$

To complete the proof we note that the procedure described above can be performed until the matrix A' acquires the required form.

From Lemma 2 it follows that to study α_{mn} it is enough to consider only multi-dimensional knapsack problems with constraints

$$\begin{cases} a_{11}x_1 + \ldots + a_{1l_1}x_{l_1} & \leq b_1, \\ \quad a_{2,l_1+1}x_{l_1+1} + \ldots + a_{2,l_2}x_{l_2} & \leq b_2, \\ \dotfill \\ \quad a_{m,l_{m-1}+1}x_1 + \ldots + a_{mn}x_n \leq b_m, \end{cases}$$

that can be called a *direct product of m knapsack problems*. All inequalities $0 \leq b_i$ have to be deleted due to Lemma 1. Denote $n_i = l_i - l_{k-1}$, where $l_0 = 0$, $l_m = n$ ($i = 1, 2, \ldots, m$). Thus, we have proved the following.

Lemma 3. *For each m, n the infimum α_{mn} is attained on the direct product of knapsack problems.*

5   The Main Result

The main result of the paper is formulated in the following theorem.

Theorem 1. *For each m, n*

$$\alpha_{mn} = \frac{\alpha_{1q}}{m + r\left(\dfrac{\alpha_{1q}}{\alpha_{1,q+1}} - 1\right)}, \tag{2}$$

where $n = qm + r$, $q = \lfloor n/m \rfloor$.

The theorem follows from two lemmas below.

Lemma 4. *For each m, n*

$$\alpha_{mn} \geq \frac{\alpha_{1q}}{m + r\left(\dfrac{\alpha_{1q}}{\alpha_{1,q+1}} - 1\right)}.$$

Proof. Thanks to Lemma 3, it is enough to consider only direct products of m knapsack problems. Let $\tau_i = \gamma_i/\beta_i$ be the precision of approximate solution to the i-th knapsack problem ($i = 1, 2, \ldots, m$), where γ_i is the approximate solution value, β_i is the optimal solution value. For their product we have

$$\alpha(A, b, c) = \frac{\max\limits_{i=1,\ldots,m} \gamma_i}{\sum\limits_{i=1}^{m} \beta_i} = \frac{\gamma_s}{\sum\limits_{i=1}^{m} \beta_i} = \frac{1}{\sum\limits_{i=1}^{m} \dfrac{\beta_i}{\gamma_s}} = \frac{1}{\sum\limits_{i=1}^{m} \dfrac{\gamma_i}{\gamma_s \tau_i}} \geq \frac{1}{\sum\limits_{i=1}^{m} \dfrac{1}{\tau_i}}.$$

The inequality turns into equality if and only if $\gamma_1 = \gamma_2 = \cdots = \gamma_m$. Since $\tau_s \geq \alpha_{1n_1}$ then

$$\alpha(A, b, c) \geq \frac{1}{\sum\limits_{i=1}^{m} \dfrac{1}{\alpha_{1n_i}}}.$$

Thus, we obtain the problem to find $n_1, n_2, \ldots, n_m$ such that

$$\frac{1}{\sum\limits_{i=1}^{m} \dfrac{1}{\alpha_{1n_i}}} \to \min \quad \text{s.t.} \quad \sum_{i=1}^{m} n_i = n. \tag{3}$$

The sequence

$$\frac{1}{\alpha_{1,n+1}} - \frac{1}{\alpha_{1n}} = \frac{\varepsilon_{n+1}}{\delta_{n+1}} - \frac{\varepsilon_n}{\delta_n} = \frac{1 + \varepsilon_n(\delta_n + 1)}{\delta_{n+1}} - \frac{\varepsilon_n(\delta_n + 1)}{\delta_{n+1}} = \frac{1}{\delta_{n+1}}$$

decreases monotonously as $n \to \infty$, hence

$$\frac{1}{\alpha_{1,n+2}} + \frac{1}{\alpha_{1n}} \leq \frac{2}{\alpha_{1,n+1}}.$$

We conclude that the minimum for (3) is reached if $n_1 = \cdots = n_r = q + 1$, $n_{r+1} = \cdots = n_m = q$. Thus,

$$a(A, b, c) \geq \frac{1}{\displaystyle\sum_{i=1}^{m} \frac{1}{\alpha_{1n_i}}} = \frac{1}{\dfrac{r}{\alpha_{1,q+1}} + \dfrac{m-r}{\alpha_{1q}}} = \frac{\alpha_{1q}}{m + r\left(\dfrac{\alpha_{1q}}{\alpha_{1,q+1}} - 1\right)}.$$

In the following lemma we construct a class of (worst) multi-dimensional knapsack problems on which the bound (2) is attained.

Lemma 5. *For each m and n*

$$\alpha_{mn} \leq \frac{\alpha_{1q}}{m + r\left(\dfrac{\alpha_{1q}}{\alpha_{1,q+1}} - 1\right)},$$

where $n = qm + r$, $q = \lfloor n/m \rfloor$.

Proof. Consider the direct product of r knapsack problems of the form

$$\max \sum_{j=1}^{q+1} \frac{x_j}{\delta_j} \to \max \qquad \text{s.t.} \quad \sum_{j=1}^{q+1} \frac{x_j}{\delta_j + \mu_{q+1}} \leq 1$$

and $m - r$ knapsack problems of the form

$$\max \sum_{j=1}^{q} \frac{x_j}{\delta_j} \to \max \qquad \text{s.t.} \quad \sum_{j=1}^{q} \frac{x_j}{\delta_j + \mu_q} \leq 1.$$

The precision of the approximate solutions to these problems is α_{1q} and $\alpha_{1,q+1}$ respectively (see Sect. 3). For the product of these problems the optimal solution value is

$$r \frac{\varepsilon_{q+1}}{\delta_{q+1}} + (m - r)\frac{\varepsilon_q}{\delta_q} = \frac{r}{\alpha_{1,q+1}} + \frac{m-r}{\alpha_{1q}}$$

and the approximate solution value is 1, hence the precision of the approximate solution is

$$a(A, b, c) = \frac{1}{\dfrac{r}{\alpha_{1,q+1}} + \dfrac{m-r}{\alpha_{1q}}} = \frac{\alpha_{1q}}{m + r\left(\dfrac{\alpha_{1q}}{\alpha_{1,q+1}} - 1\right)}.$$

Corollary 1.

$$\frac{\alpha_{1,\lceil n/m \rceil}}{m} \le \alpha_{mn} \le \frac{\alpha_{1,\lfloor n/m \rfloor}}{m}.$$

Proof. The first inequality obviously follows from (2). Let us prove the second one. If $r = 0$ then

$$\alpha_{mn} = \frac{\alpha_{1q}}{m} = \frac{\alpha_{1,\lceil n/m \rceil}}{m}.$$

If $0 < r < m$ then

$$\alpha_{mn} = \frac{\alpha_{1q}}{m + r\left(\dfrac{\alpha_{1q}}{\alpha_{1,q+1}} - 1\right)} > \frac{\alpha_{1q}}{m + m\left(\dfrac{\alpha_{1q}}{\alpha_{1,q+1}} - 1\right)} = \frac{\alpha_{1,q+1}}{m} = \frac{\alpha_{1,\lceil n/m \rceil}}{m}.$$

From Corollary 1 we obtain the following.

Corollary 2. *If $n \to \infty$, $m = o(n)$ then $\alpha_{mn} \sim \dfrac{\alpha_{1,\lfloor n/m \rfloor}}{m}$.*

Corollary 3. *If $n \to \infty$ and m is fixed then $\alpha_{mn} \to \dfrac{\alpha_{1\infty}}{m}$.*

6   Conclusion

In this paper we derived exact and asymptotic formulas for the precision of approximate solutions to the m-dimensional knapsack problem. In particular, we proved that the precision tends to $0.59136\ldots/m$ if $n \to \infty$ and m is fixed. The proof of the attainability of the obtained bounds for the precision is constructive.

In the future, our results can be base for new fully polynomial time approximation schemes.

References

1. Caprara, A., Kellerer, H., Pferschy, U., Pisinger, D.: Approximation algorithms for knapsack problems with cardinality constraints. Eur. J. Oper. Res. **123**, 333–345 (2000)
2. Chirkov, A.Yu., Shevchenko, V.N.: On the approximation of an optimal solution of the integer knapsack problem by optimal solutions of the integer knapsack problem with a restriction on the cardinality. Diskretn. Anal. Issled. Oper. Ser. **213**(2), 56–73 (2006)
3. Kellerer, H., Pferschy, U., Pisinger, D.: Knapsack Problems. Springer, Heidelberg (2004). https://doi.org/10.1007/978-3-540-24777-7
4. Kohli, R., Krishnamurti, R.: A total-value greedy heuristic for the integer knapsack problem. Oper. Res. Lett. **12**, 65–71 (1992)
5. Kohli, R., Krishnamurti, R.: Joint performance of greedy heuristics for the integer knapsack problem. Discrete Appl. Math. **56**, 37–48 (1995)
6. Martello, S., Toth, P.: Knapsack Problems: Algorithms and Computer Implementations. Wiley, New York (1990)
7. OEIS Foundation Inc.: The On-Line Encyclopedia of Integer Sequences (2020). http://oeis.org

An Approximation Algorithm for a Semi-supervised Graph Clustering Problem

Victor Il'ev$^{1,2(\boxtimes)}$ ⓘ, Svetlana Il'eva^1 ⓘ, and Alexander Morshinin^2 ⓘ

^{1} Dostoevsky Omsk State University, Omsk, Russia
iljev@mail.ru
^{2} Sobolev Institute of Mathematics SB RAS, Omsk, Russia
morshinin.alexander@gmail.com

Abstract. Clustering problems form an important section of data analysis. In machine learning clustering problems are usually classified as unsupervised learning. Semi-supervised clustering problems are also considered. In these problems relatively few objects are labeled (i.e., are assigned to clusters), whereas a large number of objects are unlabeled.

We consider the most visual formalization of a version of semi-supervised clustering. In this problem one has to partition a given set of n objects into k clusters ($k < n$). A collection of k pairwise disjoint nonempty subsets of objects is fixed. No two objects from different subsets of this collection may belong to the same cluster and all objects from any subset must belong to the same cluster. Similarity of objects is determined by an undirected graph. Vertices of this graph are in one-to-one correspondence with objects, and edges connect similar objects. One has to partition the vertices of the graph into pairwise disjoint groups (clusters) minimizing the number of edges between clusters and the number of missing edges inside clusters.

The problem is NP-hard for any fixed $k \geq 2$. For $k = 2$ we present a polynomial time approximation algorithm and prove a performance guarantee of this algorithm.

Keywords: Graph clustering · Approximation algorithm · Performance guarantee

1 Introduction

The objective of clustering problems is to partition a given set of objects into a family of subsets (called *clusters*) such that objects within a cluster are more similar to each other than objects from different clusters. In pattern recognition and machine learning clustering methods fall under the section of *unsupervised learning*. At the same time, *semi-supervised* clustering problems are studied. In these problems relatively few objects are labeled (i.e., are assigned to clusters), whereas a large number of objects are unlabeled [1,3].

© Springer Nature Switzerland AG 2020
Y. Kochetov et al. (Eds.): MOTOR 2020, CCIS 1275, pp. 23–29, 2020.
https://doi.org/10.1007/978-3-030-58657-7_3

One of the most visual formalizations of clustering is the *graph clustering*, that is, grouping the vertices of a graph into clusters taking into consideration the edge structure of the graph. In this paper, we consider three interconnected versions of graph clustering, two of which are semi-supervised ones.

We consider only *simple* graphs, i.e., undirected graphs without loops and multiple edges. A graph is called a *cluster graph*, if each of its connected components is a complete graph [6].

Let V be a finite set. Denote by $\mathcal{M}(V)$ the set of all cluster graphs on the vertex set V. Let $\mathcal{M}_k(V)$ be the set of all cluster graphs on V consisting of exactly k nonempty connected components, $2 \le k \le |V|$.

If $G_1 = (V, E_1)$ and $G_2 = (V, E_2)$ are graphs on the same labeled vertex set V, then the *distance* $\rho(G_1, G_2)$ between them is defined as follows

$$\rho(G_1, G_2) = |E_1 \Delta E_2| = |E_1 \setminus E_2| + |E_2 \setminus E_1|,$$

i.e., $\rho(G_1, G_2)$ is the number of noncoinciding edges in G_1 and G_2.

Consider three interconnected graph clustering problems.

GC$_k$ (Graph k-Clustering). Given a graph $G = (V, E)$ and an integer k, $2 \le k \le |V|$, find a graph $M^* \in \mathcal{M}_k(V)$ such that

$$\rho(G, M^*) = \min_{M \in \mathcal{M}_k(V)} \rho(G, M).$$

SGC$_k$ (Semi-supervised Graph k-Clustering). Given a graph $G = (V, E)$, an integer k, $2 \le k \le |V|$, and a set $Z = \{z_1, \dots z_k\} \subset V$ of pairwise different vertices, find $M^* \in \mathcal{M}_k(V)$ such that

$$\rho(G, M^*) = \min_{M \in \mathcal{M}_k(V)} \rho(G, M),$$

where minimum is taken over all cluster graphs $M = (V, E_M) \in \mathcal{M}_k(V)$ with $z_i z_j \notin E_M$ for all $i, j \in \{1, \dots k\}$ (in other words, all vertices of Z belong to different connected components of M).

SSGC$_k$ (Set Semi-supervised Graph k-Clustering). Given a graph $G = (V, E)$, an integer k, $2 \le k \le |V|$, and a collection $\mathcal{Z} = \{Z_1, \dots Z_k\}$ of pairwise disjoint nonempty subsets of V, find $M^* \in \mathcal{M}_k(V)$ such that

$$\rho(G, M^*) = \min_{M \in \mathcal{M}_k(V)} \rho(G, M),$$

where minimum is taken over all cluster graphs $M = (V, E_M) \in \mathcal{M}_k(V)$ such that

1. $zz' \notin E_M$ for all $z \in Z_i, z' \in Z_j, i, j = 1, \dots, k, i \ne j$;
2. $zz' \in E_M$ for all $z, z' \in Z_i, i = 1, \dots, k$

(in other words, all sets of the family $\mathcal{Z}$ are subsets of different connected components of M).

Problem $\mathbf{GC}_k$ is NP-hard for every fixed $k \geq 2$ [6]. It is not difficult to construct Turing reduction of problem $\mathbf{GC}_k$ to problem $\mathbf{SGC}_k$ and as a result to show that $\mathbf{SGC}_k$ is NP-hard too. Thus, problem $\mathbf{SSGC}_k$ is also NP-hard as generalization of $\mathbf{SGC}_k$.

In 2004, Bansal, Blum, and Chawla [2] presented a polynomial time 3-approximation algorithm for a version of the graph clustering problem similar to $\mathbf{GC}_2$ in which the number of clusters doesn't exceed 2. In 2008, Coleman, Saunderson, and Wirth [4] presented a 2-approximation algorithm for this version applying local search to every feasible solution obtained by the 3-approximation algorithm from [2]. They used a switching technique that allows to reduce clustering any graph to the equivalent problem whose optimal solution is the complete graph, i.e., the cluster graph consisting of the single cluster. In [5], we presented a modified 2-approximation algorithm for problem $\mathbf{GC}_2$. In contrast to the proof of Coleman, Saunderson, and Wirth, our proof of the performance guarantee of this algorithm didn't use switchings.

In this paper, we use a similar approach to construct a 2-approximation local search algorithm for the set semi-supervised graph clustering problem $\mathbf{SSGC}_2$. Applying this method to problem $\mathbf{SGC}_2$ we get a variant of 2-approximation algorithm for this problem.

2   Problem SSGC$_2$

2.1   Notation and Auxiliary Propositions

Consider the special case of problem $\mathbf{SSGC}_k$ with $k = 2$. We need to introduce the following notation.

Given a graph $G = (V, E)$ and a vertex $v \in V$, we denote by $N_G(v)$ the set of all vertices adjacent to v in G, and let $\overline{N}_G(v) = V \setminus (N_G(v) \cup \{v\})$.

Let $G_1 = (V, E_1)$ and $G_2 = (V, E_2)$ be graphs on the same labeled vertex set V, $n = |V|$. Denote by $D(G_1, G_2)$ the graph on the vertex set V with the edge set $E_1 \triangle E_2$. Note that $\rho(G_1, G_2)$ is equal to the number of edges in the graph $D(G_1, G_2)$.

Lemma 1. *[5] Let $d_{\min}$ be the minimum vertex degree in the graph $D(G_1, G_2)$. Then*

$$\rho(G_1, G_2) \geq \frac{nd_{\min}}{2}.$$

Let $G = (V, E)$ be an arbitrary graph. For any vertex $v \in V$ and a set $A \subseteq V$ we denote by A_v^+ the number of vertices $u \in A$ such that $vu \in E$, and by A_v^- the number of vertices $u \in A \setminus \{v\}$ such that $vu \notin E$.

For nonempty sets $X, Y \subseteq V$ such that $X \cap Y = \emptyset$ and $X \cup Y = V$ we denote by $M(X, Y)$ the cluster graph in $\mathcal{M}_2(V)$ with connected components induced by X, Y. The sets X and Y will be called *clusters*.

The following lemma was proved in [5] for problem $\mathbf{GC_2}$. Its proof for problem $\mathbf{SSGC_2}$ is exactly the same.

Lemma 2. *Let $G = (V, E)$ be an arbitrary graph, $M^* = M(X^*, Y^*)$ be an optimal solution to problem $\mathbf{SSGC_2}$ on the graph G, and $M = M(X, Y)$ be an arbitrary feasible solution to problem $\mathbf{SSGC_2}$ on the graph G. Then*

$$\rho(G, M) - \rho(G, M^*) =$$

$$\sum_{u \in X \cap Y^*} \left((X \cap X^*)_u^- - (X \cap X^*)_u^+ + (Y \cap Y^*)_u^+ - (Y \cap Y^*)_u^- \right) +$$

$$\sum_{u \in Y \cap X^*} \left((Y \cap Y^*)_u^- - (Y \cap Y^*)_u^+ + (X \cap X^*)_u^+ - (X \cap X^*)_u^- \right).$$

2.2   Local Search Procedure

Let us introduce the following local search procedure.

Procedure $\mathbf{LS}(M, X, Y, Z_1, Z_2)$.
Input: cluster graph $M = M(X, Y) \in \mathcal{M}_2(V)$, Z_1, Z_2 are disjoint nonempty sets, $Z_1 \subset X, Z_2 \subset Y$.
Output: cluster graph $L = M(X', Y') \in \mathcal{M}_2(V)$ such that $Z_1 \subseteq X'$, $Z_2 \subseteq Y'$.

Iteration 0. Set $X_0 = X, Y_0 = Y$.
Iteration $k(k \geq 1)$.
Step 1. For each vertex $u \in V \setminus (Z_1 \cup Z_2)$ calculate the following quantity $\delta_k(u)$ (possible variation of the value of the objective function in case of moving the vertex u to another cluster):

$$\delta_k(u) = \begin{cases} (X_{k-1})_u^- - (X_{k-1})_u^+ + (Y_{k-1})_u^+ - (Y_{k-1})_u^- & \text{for } u \in X_{k-1} \setminus Z_1, \\ (Y_{k-1})_u^- - (Y_{k-1})_u^+ + (X_{k-1})_u^+ - (X_{k-1})_u^- & \text{for } u \in Y_{k-1} \setminus Z_2. \end{cases}$$

Step 2. Choose the vertex $u_k \in V \setminus (Z_1 \cup Z_2)$ such that

$$\delta_k(u_k) = \max_{u \in V \setminus (Z_1 \cup Z_2)} \delta_k(u).$$

Step 3. If $\delta_k(u_k) > 0$, then set $X_k = X_{k-1} \setminus \{u_k\}$, $Y_k = Y_{k-1} \cup \{u_k\}$ in case of $u_k \in X_{k-1}$, and set $X_k = X_{k-1} \cup \{u_k\}$, $Y_k = Y_{k-1} \setminus \{u_k\}$ in case of $u_k \in Y_{k-1}$; **go to iteration $k + 1$.** Else **STOP.** Set $X' = X_{k-1}$, $Y' = Y_{k-1}$, and $L = M(X', Y')$.
End.

2.3   2-Approximation Algorithm for Problem $\mathbf{SSGC_2}$

Consider the following approximation algorithm for problem $\mathbf{SSGC_2}$.

Algorithm A_1.
Input: graph $G = (V, E)$, Z_1, Z_2 are disjoint nonempty subsets of V.
Output: graph $M_1 = M(X, Y) \in \mathcal{M}_2(V)$, sets Z_1, Z_2 are subsets of different clusters.

Step 1. For every vertex $u \in V$ do the following:

Step 1.1. (a) If $u \notin Z_1 \cup Z_2$, then define the cluster graphs $\overline{M}_u = M(\overline{X}, \overline{Y})$ and $\overline{\overline{M}}_u = M(\overline{\overline{X}}, \overline{\overline{Y}})$, where

$$\overline{X} = \{u\} \cup ((N_G(u) \cup Z_1) \setminus Z_2), \overline{Y} = V \setminus \overline{X},$$
$$\overline{\overline{X}} = \{u\} \cup ((N_G(u) \cup Z_2) \setminus Z_1), \overline{\overline{Y}} = V \setminus \overline{\overline{X}}.$$

(b) If $u \in Z_1 \cup Z_2$, then define the cluster graph $M_u = M(X, Y)$, where

$$X = \{u\} \cup ((N_G(u) \cup Z) \setminus \overline{Z}), Y = V \setminus X.$$

Here $Z = Z_1$, $\overline{Z} = Z_2$ in case of $u \in Z_1$, and $Z = Z_2$, $\overline{Z} = Z_1$, otherwise.

Step 1.2. (a) If $u \notin Z_1 \cup Z_2$, then run the local search procedure **LS**$(\overline{M}_u, \overline{X}, \overline{Y}, Z_1, Z_2)$ and **LS**$(\overline{\overline{M}}_u, \overline{\overline{X}}, \overline{\overline{Y}}, Z_1, Z_2)$. Denote resulting graphs by $\overline{L}_u$ and $\overline{\overline{L}}_u$.

(b) If $u \in Z_1 \cup Z_2$, then run the local search procedure **LS**(M_u, X, Y, Z_1, Z_2). Denote resulting graph by L_u.

Step 2. Among all locally-optimal solutions L_u, $\overline{L}_u$, $\overline{\overline{L}}_u$ obtained at step 1.2 choose the nearest to G cluster graph $M_1 = M(X, Y)$.

The following lemma can be proved in the same manner as Remark 1 in [5].

Lemma 3. *Let $G = (V, E)$ be an arbitrary graph, Z_1, Z_2 be arbitrary disjoint nonempty subsets of V, $M^* = M(X^*, Y^*) \in \mathcal{M}_2(V)$ be an optimal solution to problem* **SSGC$_2$** *on the graph G, and $d_{\min}$ be the minimum vertex degree in the graph $D = D(G, M^*)$. Among all graphs M_u, $\overline{M}_u$, $\overline{\overline{M}}_u$ constructed by algorithm* A_1 *at step 1.1 there is the cluster graph $M = M(X, Y)$ such that*

1. *M can be obtained from M^* by moving at most $d_{\min}$ vertices to another cluster;*
2. *If $Z_1 \subset X^*, Z_2 \subset Y^*$, then $Z_1 \subset X \cap X^*, Z_2 \subset Y \cap Y^*$. Otherwise, if $Z_2 \subset X^*, Z_1 \subset Y^*$, then $Z_1 \subset Y \cap Y^*, Z_2 \subset X \cap X^*$.*

Now we can prove a performance guarantee of algorithm A_1.

Theorem 1. *For every graph $G = (V, E)$ and for any disjoint nonempty subsets $Z_1, Z_2 \subset V$ the following inequality holds:*

$$\rho(G, M_1) \le 2\rho(G, M^*),$$

where $M^* \in \mathcal{M}_2(V)$ is an optimal solution to problem **SSGC$_2$** *on the graph G and $M_1 \in \mathcal{M}_2(V)$ is the solution returned by algorithm* A_1.

Proof. Let $M^* = M(X^*, Y^*)$ and d_{min} be the minimum vertex degree in the graph $D = D(G, M^*)$. By Lemma 3, among all graphs constructed by algorithm $\mathbf{A_1}$ at step 1.1 there is the cluster graph $M = M(X, Y)$ satisfying the conditions 1 and 2 of Lemma 3. By condition 1, $|X \cap Y^*| \cup |Y \cap X^*| \leq d_{min}$.

Consider the performance of procedure $\mathbf{LS}(M, X, Y, Z_1, Z_2)$ on the graph $M = M(X, Y)$.

Local search procedure $\mathbf{LS}$ starts with $X_0 = X$ and $Y_0 = Y$. At every iteration k either $\mathbf{LS}$ moves some vertex $u_k \in V \setminus (Z_1 \cup Z_2)$ to another cluster, or no vertex is moved and $\mathbf{LS}$ finishes.

Consider in detail iteration $t + 1$ such that

– at every iteration $k = 1, \ldots, t$ procedure $\mathbf{LS}$ selects some vertex

$$u_k \in (X \cap Y^*) \cup (Y \cap X^*);$$

– at iteration $t + 1$ either procedure $\mathbf{LS}$ selects some vertex

$$u_{t+1} \in ((X \cap X^*) \cup (Y \cap Y^*)) \setminus (Z_1 \cup Z_2),$$

or iteration $t + 1$ is the last iteration of $\mathbf{LS}$.

Let us introduce the following quantities:

$$\alpha_{t+1}(u) = \begin{cases} (X_t \cap X^*)_u^- - (X_t \cap X^*)_u^+ + (Y_t \cap Y^*)_u^+ - (Y_t \cap Y^*)_u^- & \text{for } u \in X_t \cap Y^* \\ (Y_t \cap Y^*)_u^- - (Y_t \cap Y^*)_u^+ + (X_t \cap X^*)_u^+ - (X_t \cap X^*)_u^- & \text{for } u \in Y_t \cap X^*. \end{cases}$$

Consider the cluster graph $M_t = M(X_t, Y_t)$. By Lemma 2,

$$\rho(G, M_t) - \rho(G, M^*) = \sum_{u \in X_t \cap Y^*} \alpha_{t+1}(u) + \sum_{u \in Y_t \cap X^*} \alpha_{t+1}(u).$$

Put $r = |X_t \cap Y^*| + |Y_t \cap X^*|$. Since at all iterations preceding iteration $t + 1$ only vertices from the set $(X \cap Y^*) \cup (Y \cap X^*)$ were moved, then

$$r = |X_t \cap Y^*| + |Y_t \cap X^*| \leq d_{min}. \tag{1}$$

Hence

$$\rho(G, M_t) - \rho(G, M^*) \leq r \max\{\alpha_{t+1}(u) : u \in (X_t \cap Y^*) \cup (Y_t \cap X^*)\}. \tag{2}$$

Note that at iteration $t + 1$ for every vertex $u \in (X_t \cap Y^*) \cup (Y_t \cap X^*)$ the following inequality holds:

$$\alpha_{t+1}(u) \leq \frac{n}{2}. \tag{3}$$

The proof of this inequality is similar to the proof of inequality (5) in [5].

Denote by L the graph returned by procedure $\mathbf{LS}(M, X, Y, Z_1, Z_2)$. Using (1), (2), (3), and Lemma 1 we obtain

$$\rho(G, L) - \rho(G, M^*) \leq \rho(G, M_t) - \rho(G, M^*) \leq$$
$$r \max\{\alpha_{t+1}(u) : u \in (X_t \cap Y^*) \cup (Y_t \cap X^*)\} \leq r\frac{n}{2} \leq d_{min}\frac{n}{2} \leq \rho(G, M^*).$$

Thus, $\rho(G, L) \leq 2\rho(G, M^*)$.

The graph L is constructed among all graphs L_u, $\overline{L}_u$, $\overline{\overline{L}}_u$ at step 1.2 of algorithm $\mathbf{A_1}$. Performance guarantee of algorithm $\mathbf{A_1}$ follows.

Theorem 1 is proved.

It is easy to see that problem $\mathbf{SGC_2}$ is a special case of problem $\mathbf{SSGC_2}$ if $|Z_1| = |Z_2| = 1$. The following theorem is the direct corollary of Theorem 1.

Theorem 2. *For every graph $G = (V, E)$ and for any subset $Z = \{z_1, z_2\} \subset V$ the following inequality holds:*

$$\rho(G, M_1) \leq 2\rho(G, M^*),$$

where $M^* \in \mathcal{M}_2(V)$ is an optimal solution to problem $\mathbf{SGC_2}$ on the graph G and $M_1 \in \mathcal{M}_2(V)$ is the solution returned by algorithm $\mathbf{A_1}$.

References

1. Bair, E.: Semi-supervised clustering methods. Wiley Interdisc. Rev. Comput. Stat. **5**(5), 349–361 (2013)
2. Bansal, N., Blum, A., Chawla, S.: Correlation clustering. Mach. Learn. **56**, 89–113 (2004)
3. Chapelle, O., Schölkopf, B., Zein, A.: Semi-Supervised Learning. MIT Press, Cambridge (2006)
4. Coleman, T., Saunderson, J., Wirth, A.: A local-search 2-approximation for 2-correlation-clustering. In: Halperin, D., Mehlhorn, K. (eds.) ESA 2008. LNCS, vol. 5193, pp. 308–319. Springer, Heidelberg (2008). https://doi.org/10.1007/978-3-540-87744-8_26
5. Il'ev, V., Il'eva, S., Morshinin, A.: A 2-approximation algorithm for the graph 2-clustering problem. In: Khachay, M., Kochetov, Y., Pardalos, P. (eds.) MOTOR 2019. LNCS, vol. 11548, pp. 295–308. Springer, Cham (2019). https://doi.org/10.1007/978-3-030-22629-9_21
6. Shamir, R., Sharan, R., Tsur, D.: Cluster graph modification problems. Discrete Appl. Math. **144**(1–2), 173–182 (2004)

Exact Algorithm for the One-Dimensional Quadratic Euclidean Cardinality-Weighted 2-Clustering with Given Center Problem

Vladimir Khandeev[1,2] and Anna Panasenko[1,2(✉)]

[1] Sobolev Institute of Mathematics, 4 Koptyug Avenue, 630090 Novosibirsk, Russia
{khandeev,a.v.panasenko}@math.nsc.ru
[2] Novosibirsk State University, 2 Pirogova Street, 630090 Novosibirsk, Russia

Abstract. We consider a strongly NP-hard problem of clustering a finite set of points in Euclidean space into two clusters. In this problem, we find a partition of the input set minimizing the sum over both clusters of the weighted intracluster sums of the squared distances between the elements of the clusters and their centers. The weight factors are the cardinalities of the corresponding clusters and the centers are defined as follows. The center of the first cluster is unknown and determined as the centroid, while the center of the other one is given as input (is the origin without loss of generality). In this paper, we present a polynomial-time exact algorithm for the one-dimensional case of the problem.

Keywords: Euclidean space · Minimum sum-of-squares · Weighted clustering · NP-hard problem · One-dimensional case · Exact algorithm · Polynomial-time

1 Introduction

The subject of this study is one strongly NP-hard cardinality-weighted 2-clustering problem of a finite set of points in Euclidean space. Our goal is to substantiate an exact polynomial-time algorithm for the one-dimensional case of the problem.

The motivation of our research consists of two parts. The first part is the strong NP-hardness of the general case of the considered problem. The important question is whether the one-dimensional case of the strongly NP-hard problem is polynomial-time solvable or not? One can find some examples of the algorithmic results for the one-dimensional cases of the clustering problems in [1–3]. The second part of our motivation is the problem importance for some applications, for example, in Data Analysis and Data mining [4,5]. It is known that the efficient cluster approximation algorithms are the main mathematical tools in the applied field of testing hypotheses about the data structure.

Y. Kochetov et al. (Eds.): MOTOR 2020, CCIS 1275, pp. 30–35, 2020.
https://doi.org/10.1007/978-3-030-58657-7_4

The rest of the paper is organized as follows. Section 2 contains the problem formulation and known results. In the same section, we announce our new result. We describe the structure of the optimal solution in Sect. 3. The exact algorithm is presented in Sect. 4. Also in Sect. 4, we substantiate the time complexity of our algorithm.

2   Problem Statement, Known and Obtained Results

Everywhere below $\mathbb{R}$ denotes the set of real numbers, $\|\cdot\|$ denotes the Euclidean norm, and $\langle\cdot,\cdot\rangle$ denotes the scalar product. In this paper we consider the following problem.

Problem 1 (Cardinality-weighted variance-based 2-clustering with given center). Given an N-element set $\mathcal{Y}$ of points in $\mathbb{R}^d$ and a positive integer number M. Find a partition of $\mathcal{Y}$ into two non-empty clusters $\mathcal{C}$ and $\mathcal{Y} \setminus \mathcal{C}$ such that

$$f(\mathcal{C}) = |\mathcal{C}| \sum_{y \in \mathcal{C}} \|y - \overline{y}(\mathcal{C})\|^2 + |\mathcal{Y} \setminus \mathcal{C}| \sum_{y \in \mathcal{Y} \setminus \mathcal{C}} \|y\|^2 \to \min, \tag{1}$$

where $\overline{y}(\mathcal{C}) = \frac{1}{|\mathcal{C}|} \sum_{y \in \mathcal{C}} y$ is the centroid of $\mathcal{C}$, subject to constraint $|\mathcal{C}| = M$.

Due to the limited size of the paper, we omit the examples of the applied problems. An interested reader can find them in [6].

Problem 1 has been studied since 2015 and a number of results have already been proposed. First of all, it was proved that this problem is strongly NP-hard [7,8]. Let us recall that some algorithmic results were obtained for the particular case of Problem 1 when $2M = N$ (see, for example, [9] and the references cited therein). One can easily check that in this case the optimal clusters are separated by a hyperplane. It is known that the construction of optimal separating surfaces (i.e. optimal classifiers) is important for Pattern recognition and Machine learning [10,11].

Further, in [12], an exact pseudopolynomial algorithm was constructed for the case of integer components of the input points and fixed dimension of the space. An approximation scheme that implements an FPTAS in the case of the fixed space dimension was proposed in [13]. In [14], the modification of the FPTAS was constructed. It improves the previous algorithm, implements an FPTAS in the same case and remains polynomial (implements a PTAS) for instances of dimension $\mathcal{O}(\log n)$. An approximation algorithm that allows one to find a 2-approximate solution to the problem in $\mathcal{O}\left(dN^2\right)$ time was constructed in [15]. In [16], a randomized algorithm was constructed. The conditions were found under which the algorithm is asymptotically exact and runs in $\mathcal{O}(dN^2)$ time. In [6], an approximation algorithm that implements a PTAS was constructed.

In this paper, we present an exact algorithm for the one-dimensional case of Problem 1. The time complexity of the proposed algorithm is $\mathcal{O}(N \log N)$.

3   The Structure of the Optimal Solution

In this section, we prove the statement which is necessary for substantiation of our algorithm. The proof of the following well-known lemma is presented in many publications (see, for example, [17]).

Lemma 1. *For an arbitrary point $x \in \mathbb{R}^d$ and a finite set $\mathcal{Z} \subset \mathbb{R}^d$, it is true that*

$$\sum_{y \in \mathcal{Z}} \|y - x\|^2 = \sum_{y \in \mathcal{Z}} \|y - \overline{y}(\mathcal{Z})\|^2 + |\mathcal{Z}| \cdot \|x - \overline{y}(\mathcal{Z})\|^2 .$$

Let $C \subseteq \mathcal{Y}$, $|C| = M$ and $x \in \mathbb{R}^d$. Let us denote:

$$S(C, x) = M \sum_{y \in C} \|y - x\|^2 + (N - M) \sum_{y \in \mathcal{Y} \setminus C} \|y\|^2 .$$

Everywhere below $d = 1$. Let C^* be an optimal solution of Problem 1.

Lemma 2. *1) Let $M \geq \frac{N}{2}$. If $x_a, x_b \in C^*$, $x_k \in \mathcal{Y}$ and $x_a < x_k < x_b$ then $x_k \in C^*$.*
2) Let $M \leq \frac{N}{2}$. If $x_a, x_b \in \mathcal{Y} \setminus C^*$, $x_k \in \mathcal{Y}$ and $x_a < x_k < x_b$ then $x_k \in \mathcal{Y} \setminus C^*$.

Proof. 1) Suppose that there are $x_a, x_b \in C^*$ and $x_k \in \mathcal{Y} \setminus C^*$ such that $x_a < x_k < x_b$. Let us denote $C_1 = (C^* \setminus \{x_b\}) \cup \{x_k\}$, $C_2 = (C^* \setminus \{x_a\}) \cup \{x_k\}$. Then

$$f(C^*) = S(C^*, \overline{y}(C^*)) = S(C_1, \overline{y}(C^*)) +$$
$$+ M\|x_b - \overline{y}(C^*)\|^2 - M\|x_k - \overline{y}(C^*)\|^2 - (N - M)\|x_b\|^2 + (N - M)\|x_k\|^2 .$$

By Lemma 1, $S(C_1, \overline{y}(C^*)) = f(C_1) + M^2\|\overline{y}(C^*) - \overline{y}(C_1)\|^2 = f(C_1) + \|x_b - x_k\|^2$.
So,

$$f(C^*) = f(C_1) + M^2\|\overline{y}(C^*) - \overline{y}(C_1)\|^2 + M\|x_b - \overline{y}(C^*)\|^2 +$$
$$+ (N - M)\|x_k\|^2 - M\|x_k - \overline{y}(C^*)\|^2 - (N - M)\|x_b\|^2 \geq$$
$$\geq f(C^*) + M^2\|\overline{y}(C^*) - \overline{y}(C_1)\|^2 + M\|x_b - \overline{y}(C^*)\|^2 +$$
$$+ (N - M)\|x_k\|^2 - M\|x_k - \overline{y}(C^*)\|^2 - (N - M)\|x_b\|^2 =$$
$$= f(C^*) + \|x_b - x_k\|^2 + (2M - N)(\|x_b\|^2 - \|x_k\|^2) + 2M\langle x_k - x_b, \overline{y}(C^*)\rangle .$$

Then, since $d = 1$, we have:

$$(x_b - x_k)^2 + (2M - N)(x_b^2 - x_k^2) + 2M(x_k - x_b)\overline{y}(C^*) \leq 0 , \tag{2}$$

$$(x_b - x_k)((2M - N + 1)x_b + (2M - N - 1)x_k - 2M\overline{y}(C^*)) \leq 0 .$$

Since $x_b > x_k$, then

$$2M\overline{y}(C^*) \geq (2M - N + 1)x_b + (2M - N - 1)x_k > (2M - N)(x_b + x_k). \tag{3}$$

Similarly, working with C_2, we have

$$(x_a - x_k)((2M - N + 1)x_a + (2M - N - 1)x_k - 2M\bar{y}(C^*)) \leq 0 .$$

But $x_a < x_k$, so

$$2M\bar{y}(C^*) \leq (2M - N + 1)x_a + (2M - N - 1)x_k < (2M - N)(x_a + x_k) . \quad (4)$$

Inequalities (3) and (4) imply

$$(2M - N)x_b < (2M - N)x_a .$$

If $2M - N > 0$ then $x_b < x_a$. It is a contradiction.

If $2M - N = 0$ then inequalities (3) and (4) imply $2M\bar{y}(C^*) = 0$. Substituting it in (2), we get $(x_b - x_k)^2 \leq 0$. It is a contradiction.

2) The case $M \leq \frac{N}{2}$ is treated similarly.   □

Remark 1. If $N = 2M$ then one of the following statements holds:

1) $x < y$ for all $x \in C^*$, $y \in \mathcal{Y} \setminus C^*$; 2) $x > y$ for all $x \in C^*$, $y \in \mathcal{Y} \setminus C^*$.

4   Exact Algorithm

We present an exact algorithm for the one-dimensional case of Problem 1 in this section. The main idea of this algorithm can be described as follows. First of all the algorithm sorts the input set in ascending order. If the desired cardinality $M \geq N/2$, the algorithm forms the sequence of $(N - M + 1)$ sets which consist of M consecutive points. In the other case, the algorithm forms the sequence of $(M + 1)$ sets which are the complements to the sets of $N - M$ consecutive points. In the end, the algorithm chooses (as an output) one of the constructed sets with the minimal value of the objective function.

Let us define some notations for the following algorithm and for cases of the desired cardinality value. Reorder points in $\mathcal{Y} = \{x_1, \ldots, x_N\}$ so that $x_i < x_{i+1}$ for $i \in \{1, \ldots, N - 1\}$.

1. $M \geq \frac{N}{2}$. Let us denote: $C_i = \{x_i, x_{i+1}, \ldots, x_{i+M-1}\}$ for $i \in \{1, \ldots, N - M + 1\}$. Then we can notice that

$$f(C_{k+1}) = f(C_k) +$$
$$+(x_{k+M} - x_k)((2M - N + 1)x_{k+M} + (2M - N - 1)x_k - 2M\bar{y}(C_{k+1})).(5)$$

2. $M \leq \frac{N}{2}$. Let us denote: $B_i = \{x_i, x_{i+1}, \ldots, x_{i+N-M-1}\}$ for $i \in \{1, \ldots, M+1\}$, $C_i = \mathcal{Y} \setminus B_i$. Then we can notice that

$$f(C_{k+1}) = f(C_k) + (x_k - x_{k+N-M})((2M - N + 1)x_k +$$
$$+(2M - N - 1)x_{k+N-M} - 2M\bar{y}(C_{k+1})) . \quad (6)$$

The step-by-step description of the algorithm is as follows.

Algorithm $\mathcal{A}$.
Input: a set $\mathcal{Y}$, a positive integer M.
Step 0. Sort the set $\mathcal{Y}$ in ascending order.
Step 1. If $M \geq \frac{N}{2}$ then go to Step 2, otherwise go to Step 4.
Step 2. Compute $\overline{y}(C_1)$. For all $k \in \{1, \ldots, N - M\}$ compute $\overline{y}(C_{k+1})$ by equation $\overline{y}(C_{k+1}) = \overline{y}(C_k) + \frac{1}{M}(x_{k+M} - x_k)$.
Step 3. Compute $f(C_1)$ by formula (1). For all $k \in \{1, \ldots, N - M\}$ compute $f(C_{k+1})$ by formula (5). Go to Step 6.
Step 4. Compute $\overline{y}(C_1)$. For all $k \in \{1, \ldots, M\}$ compute $\overline{y}(C_{k+1})$ by equation $\overline{y}(C_{k+1}) = \overline{y}(C_k) + \frac{1}{M}(x_k - x_{k+N-M})$.
Step 5. Compute $f(C_1)$ by formula (1). For all $k \in \{1, \ldots, M\}$ compute $f(C_{k+1})$ by formula (6).
Step 6. If $M \geq \frac{N}{2}$ then $k \in \{1, \ldots, N-M+1\}$, otherwise $k \in \{1, \ldots, M+1\}$. Choose as a solution $C_{\mathcal{A}}$ the set C_k with the minimal value $f(C_k)$.
Output: The set $C_{\mathcal{A}}$.

Theorem 1. *Algorithm $\mathcal{A}$ finds the optimal solution of one-dimensional case of Problem 1 in $\mathcal{O}(N \log N)$ time.*

Proof. Algorithm $\mathcal{A}$ finds the optimal solution by Lemma 2 and the fact that we check each appropriate subset while running the algorithm.

Step 0 of the algorithm sorts $\mathcal{Y}$, so it runs in $\mathcal{O}(N \log N)$ time. Step 1 of the algorithm requires $\mathcal{O}(1)$ operations. Step 2 (or Step 4) requires $\mathcal{O}(N)$ operations. Step 3 (or Step 5) requires $\mathcal{O}(N)$ operations. Step 6 requires $\mathcal{O}(N - M)$ (or $\mathcal{O}(M)$) operations. So, the total time complexity of Algorithm $\mathcal{A}$ is $\mathcal{O}(N \log N)$. □

Remark 2. If the points of the input set are pre-ordered, then one can find the optimal solution of Problem 1 in $\mathcal{O}(N)$ time.

5   Conclusion

In this paper, we presented an exact polynomial-time algorithm for one-dimensional case of the Euclidean cardinality-weighted 2-clustering problem of a finite set of points. Our algorithm is based on the optimal solution structure that was established. It was proved that the algorithm is almost linear ($\mathcal{O}(N \log N)$).

This is the first algorithmic result for the one-dimensional case of the considered problem. In other words, we have found out that strongly NP-hard Problem 1 can be solved exactly in a polynomial time when the dimension of the space equals 1. Moreover, the proposed algorithm is very effective one.

Acknowledgments. The study presented in Sects. 1 and 2 was supported by the Russian Academy of Science (the Program of basic research), project 0314-2019-0015, and by the Russian Ministry of Science and Education under the 5-100 Excellence Programme. The study presented in other sections was supported by the Russian Foundation for Basic Research, project 19-31-90031.

References

1. Rao, M.: Cluster analysis and mathematical programming. J. Am. Stat. Assoc. **66**, 622–626 (1971)
2. Grønlund, A., Larsen, K. G., Mathiasen, A., Nielsen, J. S., Schneider, S., Song, M.: Fast exact k-Means, k-medians and bregman divergence clustering in 1D (2017). arXiv:1701.07204
3. Kel'manov, A., Khandeev, V.: Fast and exact algorithms for some np-hard 2-clustering problems in the one-dimensional case. In: van der Aalst, W. et al. (eds.) Analysis of Images, Social Networks and Texts, AIST 2019. Lecture Notes in Computer Science, vol. 11832, pp. 377–387. Springer, Cham (2019). https://doi.org/10.1007/978-3-319-73013-4_30
4. Aggarwal, C.C.: Data Mining: The Textbook. Springer, Heidelberg (2015). https://doi.org/10.1007/978-3-319-14142-8
5. Hastie, T., Tibshirani, R., Friedman, J.: The Elements of Statistical Learning: Data Mining, Inference, and Prediction. Springer, New York (2009). https://doi.org/10.1007/978-0-387-84858-7
6. Panasenko, A.: A PTAS for one cardinality-weighted 2-clustering problem. In: Khachay, M., Kochetov, Y., Pardalos, P. (eds.) MOTOR 2019. LNCS, vol. 11548, pp. 581–592. Springer, Cham (2019). https://doi.org/10.1007/978-3-030-22629-9_41
7. Kel'manov, A.V., Pyatkin, A.V.: NP-hardness of some quadratic euclidean 2-clustering problems. Doklady Math. **92**(2), 634–637 (2015)
8. Kel'manov, A.V., Pyatkin, A.V.: On the complexity of some quadratic euclidean 2-clustering problems. Comput. Math. Math. Phys. **56**(3), 491–497 (2016)
9. Kel'manov, A.V., Khandeev, V.I.: Fully polynomial-time approximation scheme for a special case of a quadratic euclidean 2-clustering problem. Comput. Math. Math. Phys. **56**(2), 334–341 (2016)
10. Bishop, C.M.: Pattern Recognition and Machine Learning. Springer, New York (2006)
11. James, G., Witten, D., Hastie, T., Tibshirani, R.: An Introduction to Statistical Learning. Springer, New York (2013). https://doi.org/10.1007/978-1-4614-7138-7
12. Kel'manov, A.V., Motkova, A.V.: Exact pseudopolynomial algorithms for a balanced 2-clustering problem. J. Appl. Ind. Math. **10**(3), 349–355 (2016)
13. Kel'manov, A., Motkova, A.: A fully polynomial-time approximation scheme for a special case of a balanced 2-clustering problem. In: Kochetov, Y. et al. (eds.) Discrete Optimization and Operations Research (DOOR 2016), LNCS, vol. 9869, pp. 182–192. Springer, Cham (2016). https://doi.org/10.1007/978-3-319-44914-2_15
14. Kel'manov, A., Motkova, A., Shenmaier, V.: An approximation scheme for a weighted two-cluster partition problem. In: van der Aalst, W. et al. (eds.) Analysis of Images, Social Networks and Texts (AIST 2017), LNCS, vol. 10716, pp. 323–333. Springer, Cham (2018). https://doi.org/10.1007/978-3-319-73013-4_30
15. Kel'manov, A.V., Motkova, A.V.: Polynomial-time approximation algorithm for the problem of cardinality-weighted variance-based 2-clustering with a given center. Comp. Math. Math. Phys. **58**(1), 130–136 (2018)
16. Kel'manov, A., Khandeev, V., Panasenko, A.: Randomized algorithms for some clustering problems. In: Eremeev, A. et al. (eds.) Optimization Problems and Their Applications (OPTA 2018), CCIS, vol. 871, pp. 109–119. Springer, Cham (2018). https://doi.org/10.1007/978-3-319-93800-4_9
17. Kel'manov, A.V., Romanchenko, S.M.: An approximation algorithm for solving a problem of search for a vector subset. J. Appl. Ind. Math. **6**(1), 90–96 (2012)

Bilevel Models for Investment Policy in Resource-Rich Regions

Sergey Lavlinskii[1,2,3(✉)], Artem Panin[1,2], and Alexander Plyasunov[1,2]

[1] Sobolev Institute of Mathematics, Novosibirsk, Russia
{lavlin,apljas}@math.nsc.ru, aapanin1988@gmail.com
[2] Novosibirsk State University, Novosibirsk, Russia
[3] Zabaikalsky State University, Chita, Russia

Abstract. This article continues the research of the authors into cooperation between public and private investors in the natural resource sector. This work aims to analyze the partnership mechanisms in terms of efficiency, using the game-theoretical Stackelberg model. Such mechanisms determine the investment policy of the state and play an important role in addressing a whole range of issues related to the strategic management of the natural resource sector in Russia. For bilevel mathematical programming problems, the computational complexity will be evaluated and effective solution algorithms based on metaheuristics and allowing solving large-dimensional problems will be developed. This opens up the possibility of a practical study on the real data of the properties of Stackelberg equilibrium, which determines the design of the mechanism for forming investment policies. The simulation results will allow not only to assess the impact of various factors on the effectiveness of the generated subsoil development program but also to formulate the basic principles that should guide the state in the management process.

Keywords: Stackelberg game · Bilevel mathematical programming problems · Subsoil development program · Probabilistic local search algorithm

1 Introduction

The development and evaluation of mechanisms for stimulating private investment presents an as-yet unresolved problem for the Russian government. The established practice of making this kind of decisions in subsoil resource management tends to operate with political arguments and most unsophisticated effectiveness evaluations, which are derived from analysis of technological projects and current raw materials prices [1–3].

This problem cannot be solved separately from the general problems of strategic planning, the core of which lies with the goal of forming a program of development of the mineral raw materials base (MRB) [4–6]. This program would set a framework for decision-making on many issues, e.g., the follows.

© Springer Nature Switzerland AG 2020
Y. Kochetov et al. (Eds.): MOTOR 2020, CCIS 1275, pp. 36–50, 2020.
https://doi.org/10.1007/978-3-030-58657-7_5

What production infrastructure do we need to facilitate spatial development and attract investors? Can we spend additional money from the state budget to help investors when it comes to infrastructural or environmental projects?

How can we help the investor overcome the barriers posed by the lack of necessary infrastructure and by the high costs of environmental protection, which are so typical of most of Siberian and Far-Eastern regions of Russia? What kind of mechanism should we employ to stimulate private investment? If we want this mechanism to unite the various measures of government investment policy and lay a foundation for a program of development of regional natural resources?

These problems are at the center of attention in this work. The aim of this article is to work out a model that could lay a foundation for a practical methodology to generate an MRB development program. To this end, we propose to use the apparatus of bilevel mathematical programming [7] and thus take into account the features of the hierarchy of interactions between the government and the private investor in the mineral raw materials sector. This approach allows us to find a compromise between the interests of the state budget and those of the private investor and generate a natural resources development program that should be effective in terms of sustainable development prospects.

The first section of the article presents the problem statement and formulates a model. The second one focuses on analyzing the computational complexity of the model and on building effective solution algorithms by means of random local search. The third section presents the results of numerical experiments, which make it possible to study the properties of the Stackelberg equilibrium using real data and determine the principles of investment policy formation. The fourth section discusses the results obtained and formulates recommendations for subsoil resource management.

2   Mathematical Models

Here, we consider a model of cooperation between the government and the private investor in the mineral raw materials sector. This model is a generalization of two models, which were considered by the authors in [8,9].

The first one is the classical model of public-private partnership [10–12]. In this model, the investor coordinates with the government a list of infrastructural projects that open for him an opportunity of realizing the desired mineral resource development projects and then implements the coordinated infrastructural projects at his own expense. The government compensates for his expenses when it begins to receive taxes from the private investor's mineral resource extraction operations.

The second model has been in practical use in Russia for a while. This model suggests that on a frontier territory, the government can help the investor build the infrastructure and conduct some of the necessary environmental activities [13–15]. Thus levying some of the issues that arise from the territorial linkage of development projects, the government encourages the arrival of the investor.

In the generalized cooperation model, the government uses an "all-in-one" investment policy by taking on the responsibility for a part of the infrastructural and environmental projects. The investor also builds the infrastructure, and the corresponding expenses are compensated by the government with a time lag. The aim of the government is to develop the territory and obtain the maximum possible share of the natural resource rent in the form of tax payments.

The investor seeks to maximize his net present value, i.e., an overall effectiveness estimate of his participation in the MRB development program, which commensurates his expenses and revenues, respectively, incurred and obtained at different times during the forecasting period. The key role here belongs to the mechanism of compensating the investor's expenses related to the infrastructural projects.

In the first case, the investor claims compensation of his expenses regardless of the overall outcomes of the MRB development program (model A). Thus, the government builds a schedule of payments within its budget constraints in order to compensate for the infrastructural expenses of the investor with a discount factor. The second scheme of the mutual settlements builds upon coordinated estimation of the investor's integral effect from his participation in the joint (i.e., implemented together with the government) MRB development program. The estimation takes into account the investor's infrastructural expenses and the government's compensation payments, which guarantee that the investor's resulting net present value is positive (model B).

Thus, the input data of the investment policy model are as follows:

- a set of industrial projects implemented by the private investor to open mineral deposits;
- a set of infrastructural projects, which can be implemented both by the private investor and by the government;
- a list of environmental projects necessary to compensate for environmental losses due to the implementation of the industrial projects; a part of the environmental projects can be implemented by the government.

The output of the model is the key investment policy parameters, which define the compensation schedule and the investor incentivation (i.e., expense sharing) mechanism. Formally, these data fully defines the MRB development program and the lists of infrastructural and environmental projects implemented by the government and the private investor, respectively.

A formal description of the model can be presented as follows. We use the following notation:

T is a planning horizon; T_0 is a compensation lag; I is a set of investment projects; J is a set of infrastructure development projects; K is a set of environmental projects;

Investment project i in year t:

CFP_i^t is the cashflow (the difference between the incomes and expenses of all kinds, taking into account a transaction costs, constructive borrowed from [3]);

EPP_i^t is the environmental damage from the implementation of the project;

DBP_i^t is the government revenue from the implementation of the project.

Infrastructure development project j in year t:

ZI_j^t is the costs of implementation of the project;
EPI_j^t is the environmental damage from the implementation of the project;
VDI_j^t is the government revenue from local economic development as a result of the implementation of the project.

Environmental project k in year t: ZE_k^t is the costs of implementation of the project.

The matrices μ and ν define the relationship between the projects, where μ_{ij} is a coherence indicator for the infrastructure and investment projects, $i \in I$, $j \in J$, and ν_{ij} is a coherence indicator for the environmental and investment projects, $i \in I$, $k \in K$:

$$\mu_{ij} = \begin{cases} 1, \text{ if the implementation of investment project } i \\ \quad \text{requires the implementation of infrastructure development project } j, \\ 0 \text{ otherwise;} \end{cases}$$

$$\nu_{ik} = \begin{cases} 1, \text{ if the implementation of investment project } i \\ \quad \text{requires the implementation of environmental project } k, \\ 0 \text{ otherwise.} \end{cases}$$

The discounts of the government and the investor:
DG is the discount of the government; DI is the discount of the investor;
The budget constraints:
b_t^G is the government budget in year t; b_t^O is the investor budget in year t.
We use the following integer variables:

$$\bar{x}_j = \begin{cases} 1, \text{ if the government is prepared to launch infrastructure development project } j \\ \quad \text{(the government has included it into the budget expenses),} \\ 0 \text{ otherwise;} \end{cases}$$

$$x_j = \begin{cases} 1, \text{ if the government launches infrastructure development project } j, \\ 0 \text{ otherwise;} \end{cases}$$

$$\bar{y}_k = \begin{cases} 1, \text{ if the government is prepared to launch environmental project } k \\ \quad \text{(the government has included it into the budget expenses),} \\ 0 \text{ otherwise;} \end{cases}$$

$$y_k = \begin{cases} 1, \text{ if the government launches environmental project } k \\ \quad \text{as agreed with the investor,} \\ 0 \text{ otherwise;} \end{cases}$$

$$v_j = \begin{cases} 1, \text{ if the investor launches infrastructure development project } j, \\ 0 \text{ otherwise;} \end{cases}$$

$$z_i = \begin{cases} 1, \text{ if the investor launches investment project } i, \\ 0 \text{ otherwise;} \end{cases}$$

$$u_k = \begin{cases} 1, \text{ if the investor launches environmental project } k, \\ 0 \text{ otherwise.} \end{cases}$$

W_t, $\bar{W}_t$ is the schedule of compensation payments for infrastructure development in year t, which was proposed by the government and used by the investor. The government problem $\widetilde{\mathcal{PS}}$ can be formulated as follows:

$$\sum_{t \in T} \Big(\sum_{i \in I} (DBP_i^t - EPP_i^t) z_i + \sum_{j \in J} (VDI_j^t - EPI_j^t)(x_j + v_j)$$

$$- \sum_{j \in J} ZI_j^t x_j - \sum_{k \in K} ZE_k^t y_k - W_t \Big)/(1 + DG)^t \to \max_{x,y,W,v,u,z} \quad (1)$$

subject to:

$$\sum_{1 \le t \le \omega} \Big(\sum_{j \in J} ZI_j^t \bar{x}_j + \sum_{k \in K} ZE_k^t \bar{y}_k + \bar{W}_t \Big) \le \sum_{1 \le t \le \omega} b_t^G; \omega \in T; \quad (2)$$

$$\bar{W}_t \ge 0; t \in T; \quad (3)$$

$$\bar{W}_t = 0; 0 \le t \le T_0; \quad (4)$$

$$(x, y, W, z, u, v) \in \mathcal{F}^*(\bar{x}, \bar{y}, \bar{W}). \quad (5)$$

The set $\mathcal{F}^*$ is a set of optimal solutions of the following low-level parametric investor problem $\widetilde{\mathcal{PI}}(\bar{x}, \bar{y}, \bar{W})$:

$$\sum_{t \in T} \Big(\sum_{i \in I} CFP_i^t z_i - \sum_{k \in K} ZE_k^t u_k - \sum_{j \in J} ZI_j^t v_j + W_t \Big)/(1 + DI)^t \to \max_{x,y,W,z,u,v} \quad (6)$$

subject to:

$$\sum_{t \in T} \Big(W_t - \sum_{j \in J} ZI_j^t v_j \Big)/(1 + DI)^t \ge 0; \quad (7)$$

$$\sum_{1 \le t \le \omega} \Big(\sum_{k \in K} ZE_k^t u_k + \sum_{j \in J} ZI_j^t v_j - \sum_{i \in I} CFP_i^t z_i - W_t \Big) \le \sum_{1 \le t \le \omega} b_t^O; \omega \in T; \quad (8)$$

$$x_j + v_j \ge \mu_{ij} z_i; i \in I, j \in J; \quad (9)$$

$$x_j + v_j \le 1; j \in J; \quad (10)$$

$$y_k + u_k \ge \nu_{ik} z_i; i \in I, k \in K; \quad (11)$$

$$y_k + u_k \le 1; k \in K; \quad (12)$$

$$\sum_{i \in I} \nu_{ik} z_i \ge y_k + u_k; k \in K; \quad (13)$$

$$\sum_{t \in T} \Big(\sum_{i \in I} (DBP_i^t - EPP_i^t) z_i - W_t \Big)/(1 + DG)^t \ge 0; \quad (14)$$

$$x_j \leq \bar{x}_j; j \in J; \tag{15}$$

$$y_k \leq \bar{y}_k; k \in K; \tag{16}$$

$$W_t \leq \bar{W}_t; t \in T; \tag{17}$$

$$x_j, y_k, v_j, z_i, u_k \in \{0,1\}; i \in I, k \in K, j \in J. \tag{18}$$

There are mixed integer linear programming problems at each level. In the formulated model, the investor maximizes his NPV and the government sets its aim on obtaining the highest possible budget revenues, taking into account the costs of infrastructure and environmental protection and a cost estimate for environmental losses from the MRB program. The government starts the infrastructure compensation payments to the investor after a lapse of T_0 years (e.g., since the time of receipt of the first tax payments from the investor) (3), (4). The schedule of the compensation payments should ensure: (i) for the government, a balance between the budget revenues and the compensation payments to the investor (14), and (ii) for the investor, a compensation of his infrastructure expenses with a discount factor (7).

Constraints (9)–(13) formalize the relationships between the industrial, infrastructural, and environmental projects. Each infrastructural and environmental project can only be launched by one of the partners and must be necessary for the realization of some industrial project. An infrastructural or environmental project can likewise be assigned to the government only under the condition that the government has put the respective project onto its list (15), (16). The model output provides the key investment policy parameters: x, y, W, v, u, z, which define the investor incentivization (expense sharing) mechanism and the long-term effective MRB development program.

Problem (1)–(18) describes model A and the cooperation mechanism whereby the investor has low trust in the government, i.e., does not expect the latter to fairly compensate for his infrastructural expenses. Constraint (7) formalizes the first mechanism of compensation payments, which arises from unconditional reclamation of the incurred infrastructural expenses, regardless of the overall outcome of the MRB development program. If the partners have high trust in each other, the second scheme of mutual settlements can take place (model B), which builds upon coordinated estimation of the investor's integral effect in the joint (with the government) MRB development program. This scheme is formalized in problem (1)–(6), (8)–(18).

3   Computational Complexity and Solution Algorithm

We recall the definition of the first level of the polynomial hierarchy of complexity classes of decision problems. The first level consists of classes P, NP and co-NP. The class P contains problems solvable in polynomial time on deterministic Turing machines. The class NP is defined as the class of problems solvable in polynomial time on nondeterministic Turing machines. The third basic class co-NP consists of decision problems whose complements belong to NP. These classes are also denoted as Δ_1^P, Σ_1^P, and Π_1^P, respectively. The second level of the polynomial hierarchy is defined by deterministic and nondeterministic Turing machines with oracle [16]. It is said that the decision problem belongs to class Δ_2^P if there exists a deterministic Turing machine with an oracle that recognizes its in polynomial time, using as oracle some language from class NP. Similarly, the decision problem belongs to class Σ_2^P if there exists a nondeterministic Turing machine with an oracle that recognizes its in polynomial time, using as oracle some language from class NP.

The paper showed that the public-private partnership problem with static budget distribution (without carry-over to next year and to the investor) is Σ_2^P-hard. Based on the ideas of the proof of this fact, we obtain the following statement.

Theorem 1. *The problem (1)–(6), (8)–(18) is Σ_2^P-hard.*

Proof. Consider the Subset-Sum-Interval problem [18]. There are positive integers $q_i, i \in \{1, ..., k\}$, R, and r, where r does not exceed k. It is required to determine whether there exists an S such that $R \leq S < R + 2^r$ and for any $I \subseteq \{1, ..., k\}$ it holds $\sum_{i \in I} q_i \neq S$. It is known that the Subset-Sum-Interval problem is Σ_2^P-hard [18].

We construct the next input of the government problem. Let there be $k + 2^r + 2$ production projects and $R + 2^r - 1$ ecological projects. Suppose that no infrastructure projects are required to implement production projects. Planning Horizon $T = T_0 = 3$. For the first k production projects $CFP_i^1 = 0$, $CFP_i^2 = -q_i$, and $CFP_i^3 = 2q_i$. Suppose that $CFP_{k+1}^2 = -1/2$, $CFP_{k+1}^3 = 1$, $DBP_{k+1}^3 = \Delta$, $CFP_{k+2}^3 = DBP_{k+2}^3 = 2\Delta$, where $\Delta = (R + 2^r + 1)^2$, and $CFP_i^1 = -1$, $CFP_i^3 = R + 2^r + 1$, $k + 3 \leq i \leq k + 2^r + 2$. All other parameters of production projects will be set equal to zero. All production projects, with the exception of the $(k+2)$th, do not require the implementation of ecological projects. The production project $(k + 2)$ requires the implementation of all ecological projects. $ZE_j^1 = ZE_j^2 = 1$, for any ecological project j. All other parameters of ecological projects are equal to zero. The government's budget in any year is equal $R + 2^r - 1$. The investor's budget in the first year is equal 2^r, in the second years it is $R + 2^r - 1$, in the third year it is equal to zero.

Obviously, in the optimal solution, a production project $(k + 2)$ is being implemented. For this, due to the limited budgetary opportunities of the investor in the first year, the government must implement S ecological projects, where $R \leq S < R + 2^r$. The investor has to implement the remaining projects and then he will spend the remainder of the budget in the first year on the production projects $\{k + 3, ..., k + 2^r + 2\}$. After that, the investor in the second year has exactly S left from the budget, which he can spend on the first $k + 1$ production projects. Obviously, if there is $I \subseteq \{1, ..., k\}$ such that $\sum_{i \in I} q_i = S$, then the investor will not implement the project $(k + 1)$. Note that the $(k+1)$th project is very beneficial to the government. This means that the government will select S ($R \leq S < R + 2^r$) in such a way that for any $I \subseteq \{1, ..., k\}$ it will be carried out $\sum_{i \in I} q_i \neq S$, if possible. The theorem is proved.

Corollary 1. *The problem (1)–(18) is Σ_2^P-hard.*

Also in [17], an algorithm for solving the public-private partnership problem with static budget distribution is proposed. We modify this algorithm to solve our problem. The first two steps are similar to the original algorithm. Key differences are in the third step. We describe the algorithm scheme.

Step 1: Compute the upper bound UB by solving the government's problem with constraints of the investor's problem.

Step 2: Let *iter* be the number of iteration of the algorithm on step 2. Find a feasible solution using the following procedure:

Step 2.1: Solve the investor's problem with constraints of the government's problem and additional constraint on the value of the objective function of the government: value $\geq UB/iter$.

Step 2.2: In the previous step, we obtain the values of the government's variables. Solve the investor's problem to get the real objective function value. If the real objective function value is very different from optimal value of the investor's problem with constraints of the government's problem and additional constraint then iter:= iter - 1 and repeat the step 2.1.

Step 3: We apply steps 3.1 and 3.2 a given number of times to the solution obtained in the previous step:

Step 3.1: For a fixed value of $\overline{W}$, a specified number of times randomly change the value of the government's Boolean variables. Take the best.

Step 3.2: For a fixed values of the government's Boolean variables, a specified number of times randomly change the value of $\overline{W}$. Take the best.

Note that all auxiliary problems and the investor's problem are solved by CPLEX software. To solve the examples described in the next chapter, the following values of the algorithm parameters were a posteriori selected. In the step 2, *iter* is 30. The step 3 is limited to 2 hours. At steps 3.1 and 3.2, 100 repetitions are performed.

4 Numerical Experiment

The database of model (1)–(18) builds upon special forecasting models, which describe in detail the processes of realization of all the three types of projects [17]. The actual data describe a fragment of the Zabaykalsky Krai MRB, which consists of 50 deposits of polymetallic ores. The experiment considers the implementation of 50 environmental and 10 infrastructural projects (railroad, powerlines, autoroads), combined in such a way that the realization of the entire infrastructural and environmental program would enable the launching of all the MRB development (i.e., industrial) projects.

The numerical experiment technique builds upon analysis of the changes in the properties of solutions of (1)–(18) under varying parameters of the model. These properties include: the values of the objective functions of the government and the investor; the number of implemented infrastructural and industrial projects; the expense sharing proportions; the share of rent received by the government in the form of taxes; etc. This list allows for a meaningful economic interpretation of the implications of a chosen investment policy and helps identify the expected tendencies of change in effectiveness evaluations based on sustainable development criteria.

The following figures present the results of the calculations that studied the reaction of solutions of models A and B to changes in the key model parameters, i.e., the discounts of the investor and the government.

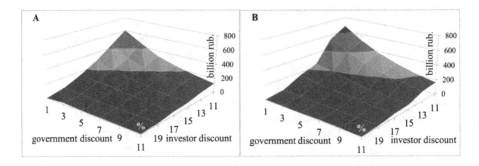

Fig. 1. The government objective function and the partner discounts

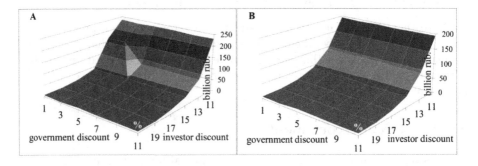

Fig. 2. The investor objective function and the partner discounts

Figure 1 shows the dependence of the government's objective function on the discounts of the MRB development stakeholders. Both surfaces reach their highest values at small discounts, consistent with the fact that under the conditions of a good investment climate, the government finds effective both investment policies, generated by models A and B, respectively. If the conditions worsen (i.e., the discounts increase), the effectiveness of the interaction between the government and the investor drops, predictably, to almost zero in both models. Thus, the problem of policy choice comes to the fore: What policy will provide the best results in the range of high discounts for the majority of resource-rich regions in Russia?

The optimal strategy for a small-discount investor is to claim unconditional compensation of all his infrastructure expenses (model A, Fig. 2). In contrast to Investor in model B, whose functional depends on his discount only, Investor in model A reduces the volume of his infrastructure building operations if the government begins to raise its discount.

Which model is preferable from the viewpoint of the functional (i.e., the decision effectiveness indicator) for the government and the investor? And under what conditions?

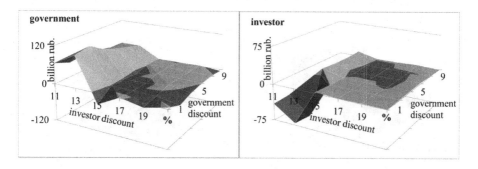

Fig. 3. Difference between the values of the objective functions in models B and A

The answers to these questions are contained in Fig. 3, which presents the difference between the functionals in models B and A. The light-colored part of the surfaces corresponds to the case where model B is preferable in terms of the functional, within these parameter ranges. A meaningful interpretation of Fig. 3 enables the government to choose a strategy that would underpin its investment policy under given conditions.

Thus, in resource-rich regions with a good investment climate, which induces a small investor discount, the government should consider using model B. Under worse investment conditions (high inflation, volatile exchange rates, growing transaction costs, etc.), which force the investor to take decisions with higher discounts, the government should use model B and a high subsoil owner discount.

A small-discount investor should consider the option with unconditional reclamation of his infrastructure expenses. At high investor discounts, model B becomes preferable if the government chooses its investment policy accordingly. This policy builds upon choosing a discount that defines the volume of government investment into the infrastructure and ensures "hitting" the light zone of the surface in Fig. 3.

Which model is preferable from the point of view of the government costs?

Figure 4 shows a relationship between the government costs on compensation payments to the investor in the different models. Here, model B proves to be more effective for the government.

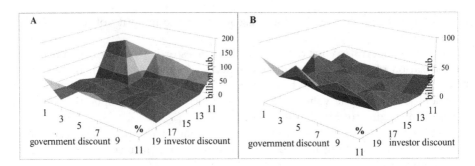

Fig. 4. Government expenses on the compensation payments

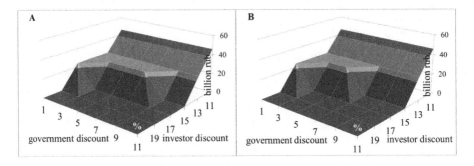

Fig. 5. Government expenses on the infrastructure projects

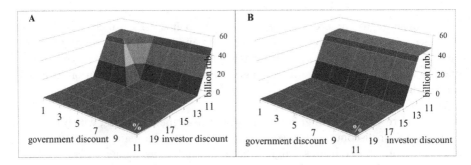

Fig. 6. Investor expenses on the infrastructure projects

Figures 5 and 6 show the dependence between the volumes of the government and investor infrastructure investments on their discounts. Model B gives a greater volume of infrastructure building operations to a small-discount investor. Under adverse conditions, infrastructure is built in both models mostly by the government, and the volume of these operations narrows down with the growing investor discount. As a result, model B is also more preferable in terms of the share of government investment in the infrastructure projects (Fig. 7).

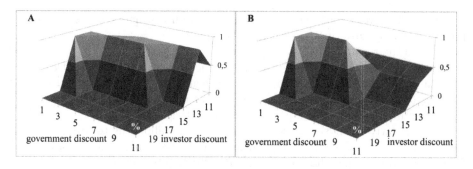

Fig. 7. Share of the government in infrastructure investments

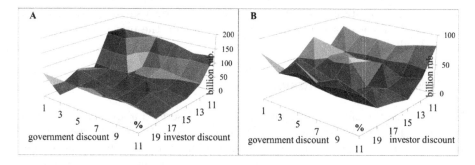

Fig. 8. Total government expenses

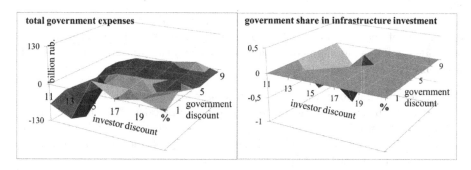

Fig. 9. Difference between the solutions of models B and A

The total government costs, including the expenses on investments and compensations, are shown in Fig. 8. Figure 9 fixes the parameter ranges within which model B is more preferable than A in total costs, which are negative and are marked with dark color in the figure. This figure means that the government costs in model B can be made lower than in A by choosing an appropriate investment policy. At low investor discounts, this happens automatically; under worse investment conditions, the government must choose a high discount, which corresponds to the dark part of the surface.

As for the government's share in infrastructure investments, model B is also preferable for the government (Fig. 9, right panel).

How do the results of this article, [8] and [9] compare?

Substantially, the models differ in the key mechanism for building infrastructure. In [8], infrastructure projects are implemented by the government. In [9], an investor builds infrastructure, its costs are compensated with some lag. The infrastructure is built by both partners in this article.

A comparative analysis of the calculation results allows us to draw the following conclusions. The model [8] provides the highest values of the objective function of stakeholders, however, it requires the highest government spending. The classical model of public-private partnership [9] minimizes budgetary costs but does not provide a sufficient level of profitability today. Models A, B occupy an intermediate position, realizing a compromise of budget savings and efficiency. The choice in favor of a particular model depends on the prevailing conditions of a particular region.

5   Results and Discussion

The bilevel mathematical programming models described above can serve as a foundation for a practical methodology to form a complex of investment policy measures in a resource-rich region. The algorithms proposed in this work may help solve problems of high dimension and formulate real strategic plans for building industrial infrastructure, which encourage the arrival of the private investor.

The numerical experiments conducted on the actual data reveal the practical significance of the proposed tools. Based on the results of the experiments, we can draw the following main conclusions to underpin the process of management in the mineral raw materials sector.

1. In regions with a favorable investment climate and mature institutions, which together ensure a small discount of the potential investor, both models maintain an acceptable effectiveness level for the government. Under the same conditions, the investor should consider a strategy of unconditional reclamation of his infrastructure expenses.
2. If the conditions worsen (the investor discount increases), the government must use model B and a high subsoil owner discount. This discount defines the government investment policy and should be chosen in such a way that model B becomes preferable for the investor as well.
3. Given a budget deficit, the government should consider model B. This model would enable it not only reduce the volume of compensation payments but also cut the total costs incurred by it, which include, apart from the payments to the investor, the government's own expenses on infrastructure.

Thus, the main goal of the government on a frontier territory rich in natural resources when it comes to investment policy formation is to create the conditions for model B to realize. The key condition is a high level of mutual trust between the government and the investor, which enables them to use a mutual settlement scheme based on coordinated estimation of the investor's integral effect in the partnership-based MRB development program. If the parties achieve such a level of trust, then the proposed mathematical tools will allow the formation of a long-term effective investment policy.

Acknowledgements. This work was financially supported by the Russian Foundation for Basic Research (projects numbers 20-010-00151 and 19-410-240003).

References

1. Glazyrina, I.P., Kalgina, I.S., Lavlinskii, S.M.: Problems in the development of the mineral and raw-material base of Russia's Far East and prospects for the modernization of the region's economy in the framework of Russian-Chinese cooperation. Reg. Res. Russ. **3**(4), 405–413 (2013). https://doi.org/10.1134/S2079970514010055
2. Glazyrina, I.P., Lavlinskii, S.M., Kalgina, I.S.: Public-private partnership in the mineral resources complex of Zabaikalskii krai: problems and prospects. Geogr. Natural Res. **35**(4), 359–364 (2014). https://doi.org/10.1134/S1875372814040088
3. Glazyrina, I., Lavlinskii, S.: Transaction costs and problems in the development of the mineral and raw-material base of the resource region. J New Econ. Assoc. New Econ. Assoc. **38**(2), 121–143 (2018)
4. Weisbrod, G., Lynch, T., Meyer, M.: Extending monetary values to broader performance and impact measures: transportation applications and lessons for other fields. Eval. Program Plann. **32**, 332–341 (2009)
5. Lakshmanan, T.R.: The broader economic consequences of transport infrastructure investments. J. Transp. Geogr. **19**(1), 1–12 (2011)
6. Mackie, P., Worsley, T., Eliasson, J.: Transport appraisal revisited. Res. Trans. Econ. **47**, 3–18 (2014)
7. Dempe, S.J.: Foundations of Bilevel Programming. Kluwer Academ Publishers, Dordrecht (2002)
8. Lavlinskii, S.M., Panin, A.A., Plyasunov, A.V.: A two-level planning model for Public-Private Partnership. Autom. Remote Control **76**(11), 1976–1987 (2015). https://doi.org/10.1134/S0005117915110077
9. Lavlinskii, S., Panin, A., Plyasunov, A.V.: Stackelberg model and public-private partnerships in the natural resources sector of Russia. In: Khachay, M., Kochetov, Y., Pardalos, P. (eds.) MOTOR 2019. LNCS, vol. 11548, pp. 158–171. Springer, Cham (2019). https://doi.org/10.1007/978-3-030-22629-9_12
10. Reznichenko, N.V.: Public-private partnership models. Bull. St. Petersburg Univ., Series 8: Manage. **4**, 58–83 (2010). (in Russian)
11. Quiggin, J.: Risk, PPPs and the public sector comparator. Aust. Acc. Rev. **14**(33), 51–61 (2004)
12. Grimsey, D., Levis, M.K.: Public Private Partnerships: The Worldwide Revolution in Infrastructure Provision and Project Finance. Edward Elgar, Cheltenham (2004)
13. Lavlinskii, S.M.: Public-private partnership in a natural resource region: ecological problems, models, and prospects. Studi. Russ. Econ. Dev. **21**(1), 71–79 (2010). https://doi.org/10.1134/S1075700710010089
14. Lavlinskii, S.M., Panin, A.A., Plyasunov, A.V.: Comparison of models of planning public-private partnership. J. Appl. Ind. Math. **10**(3), 356–369 (2016). https://doi.org/10.1134/S1990478916030066

15. Lavlinskii, S., Panin, A., Pliasunov, A.: Public-private partnership models with tax incentives: numerical analysis of solutions. CCIS **871**, 220–234 (2018). https://doi.org/10.1007/978-3-319-93800-4-18
16. Ausiello, G., Crescenzi, P., Gambosi, G., Kann, V., Marchetti- Spaccamela, A., Protasi, M.: Complexity and Approximation: Combinatorial Optimization Problems and Their Approximability Properties. Springer, Berlin (1999). https://doi.org/10.1007/978-3-642-58412-1
17. Lavlinskii, S., Panin, A., Plyasunov, A.: The Stackelberg model in territotial planning. Autom. Remote Control **80**(2), 286–296 (2019)
18. Eggermont, C., Woeginger, G.J.: Motion planning with pulley, rope, and baskets. In: Proceedings of the 29th International Symposium on Theoretical Aspects of Computer Science (STACS2012), Leibniz International Proceedings in Informatics, vol. 14, Wadern, Germany, pp. 374–383 (2012)

One Problem of the Sum of Weighted Convolution Differences Minimization, Induced by the Quasiperiodic Sequence Recognition Problem

Sergey Khamidullin and Liudmila Mikhailova[✉]

Sobolev Institute of Mathematics, 4 Koptyug Avenue, 630090 Novosibirsk, Russia
{kham,mikh}@math.nsc.ru

Abstract. We consider an unexplored discrete optimization problem of summing the elements of two numerical sequences. One of them belongs to the given set (alphabet) of sequences, while another one is given. We have to minimize the sum of M terms (M is unknown), each of them being the difference between the unweighted auto-convolution of the first sequence stretched to some length and the weighted convolution of this stretched sequence with the subsequence of the second one. We show that this problem is equivalent to the problem of recognizing a quasiperiodic sequence as a sequence induced by some sequence U from the given alphabet.

We have constructed the algorithm which finds the exact solution to this problem in polynomial time. The numerical simulation demonstrates that this algorithm can be used to solve modeled applied problems of noise-proof processing of quasiperiodic signals.

Keywords: Discrete optimization problem · Minimization · Weighted convolutions' difference · Recognition · Quasiperiodic · Polynomial-time solvability

1 Introduction

We study an unexplored discrete optimization problem of summing the elements of two numerical sequences. The research goal is to prove the polynomial-time solvability of the problem and construct an algorithm guaranteeing the solution optimality. The research is motivated by the absence of efficient (polynomial-time) algorithms solving this problem with theoretical guarantees of quality (accuracy and complexity).

The problem under consideration is relevant for natural objects noise-resistant monitoring in the case of quasiperiodic repeatability of their typical

The study was supported by the Russian Foundation for Basic Research, projects 19-07-00397 and 19-01-00308, by the Russian Academy of Science (the Program of basic research), project 0314-2019-0015.

state in the presence of non-linear temporal fluctuations. That is, the distance between two consecutive repetitions lies in the given interval, and a typical state allows some variations from one repetition to another. Namely, it is relevant for applied problems when we need to identify (recognize) either the object itself or the state of the object among the set of admissible ones in addition to detecting these typical repetitions.

This type of state repeatability is typical, first of all, for bio-medical problems (for example, problems of analysis and recognition of ECG signals). For illustration, we give an example of modeled ECG-like quasi-periodic signal processing.

2   Problem Formulation and Related Problems

The discrete optimization problem under consideration is

Problem 1. *Given:* a numerical sequence $Y = (y_1, \ldots, y_N)$, a collection $W = \{U^{(1)}, \ldots, U^{(K)} \mid U^{(k)} = (u_1^{(k)}, \ldots, u_{q_k}^{(k)}) \in \Re^{q_k}, \ k = 1, \ldots, K\}$, positive integers $T_{\max}$ and ℓ. *Find:* a numerical sequence $U = (u_1, \ldots, u_{q(U)}) \in W$; a collection $\mathcal{M} = \{n_1, \ldots n_m, \ldots\}$ of indices of the sequence Y, a collection $\mathcal{P} = \{p_1, \ldots, p_m, \ldots\}$ of positive integers; a collection $\mathcal{J} = \{J^{(1)}, \ldots, J^{(m)}, \ldots\}$ of contraction mappings, where $J^{(m)} : \{1, \ldots, p_m\} \longrightarrow \{1, \ldots, q(U)\}$; and the size M of these collections; which minimize the objective function

$$F(U, \mathcal{M}, \mathcal{P}, \mathcal{J}) = \sum_{m=1}^{M} \sum_{i=1}^{p_m} \{u_{J^{(m)}(i)}^2 - 2y_{n_m+i-1}u_{J^{(m)}(i)}\}, \tag{1}$$

under the constraints

$$\begin{aligned}
q(U) \le p_m \le \ell \le T_{\max} \le N, \quad m = 1, \ldots, M, \\
p_{m-1} \le n_m - n_{m-1} \le T_{\max}, \quad m = 2, \ldots, M, \\
p_M \le N - n_M + 1,
\end{aligned} \tag{2}$$

on the elements of the collections $\mathcal{M}$, $\mathcal{P}$, and under the constraints

$$\begin{aligned}
J^{(m)}(1) = 1, \quad J^{(m)}(p_m) = q(U), \\
0 \le J^{(m)}(i) - J^{(m)}(i-1) \le 1, \quad i = 2, \ldots, p_m,
\end{aligned} \quad m = 1, \ldots, M, \tag{3}$$

on the constraction mappings.

Problem 1 is a problem of optimal (in the sense of the minimum of (1)) summation of elements of two numerical sequences. One of these two sequences— Y—is given; another one—U—belongs to the given set of sequences. If we rewrite (1) as follows

$$F(U, \mathcal{M}, \mathcal{P}, \mathcal{J}) = \sum_{m=1}^{M} \Big\{ \sum_{i=1}^{p_m} u_{J^{(m)}(i)}^2 - 2 \sum_{i=1}^{p_m} y_{n_m+i-1}u_{J^{(m)}(i)} \Big\}$$

we can see that Problem 1 is a minimization problem for the sum of weighted convolutions differences. Indeed, for every $m = 1, \ldots, M$, the first expression in the curly brackets is the unweighted autoconvolution of the sequence obtained as some nonlinear extension of the sequence U (by repetitions of its elements), while the second sequence is the weighted convolution of this extended sequence and some subsequence from Y of the same length p_m.

The source of Problem 1 is a problem of simultaneous choice of $U \in W$ and approximation of Y by $X \in \mathcal{X}(U)$ according to the criterion of minimizing the sum of squared distances between the elements of Y and X, i.e., the problem

$$\|Y - X\|^2 \longrightarrow \min_{U, \mathcal{X}(U)} . \tag{4}$$

Here $\mathcal{X}(U)$, $U \in W$, is the set of all permissible approximating sequences engendered by U. Every element $X = (x_1, \ldots, x_N) \in \mathcal{X}(U)$ is uniquely defined by the collections $\mathcal{M}$, $\mathcal{P}$, and $\mathcal{J}$ satisfying (2), (3) according to the rule

$$x_n = \sum_{m=1}^{M} u_{J^{(m)}(n-n_m+1)}, \quad n = 1, \ldots, N, \tag{5}$$

where $u_{J^{(m)}(i)} = 0$, $m = 1, \ldots, M$, if $i < 0$ or $i > p_m$; i.e. $X = X(U, \mathcal{M}, \mathcal{P}, \mathcal{J})$. It means that the problem (4) is equivalent to the problem

$$\|Y - X\|^2 = \|Y - X(U, \mathcal{M}, \mathcal{P}, \mathcal{J})\|^2 \longrightarrow \min_{U, \mathcal{M}, \mathcal{P}, \mathcal{J}} . \tag{6}$$

The expression on the right-hand side of (5) is the sum of M extended sequences U; the following formula is valid for duplication multiplicities of its elements

$$k_t^{(m)} = \left| \left\{ i \,|\, J^{(m)}(i) = t, \ i \in \{1, \ldots, p_m\} \right\} \right|, \quad t = 1, \ldots, q(U),$$

at that $p_m = k_1^{(m)} + \ldots + k_{q(U)}^{(m)}$, $m = 1, \ldots, M$. Thus, the sequence X includes M extended repetitions of U. The index value $n = n_m$, $n_m \in \mathcal{M}$, defines the initial number of the m-th repetition; the value $p = p_m$, $p_m \in \mathcal{P}$, is its length; the mapping $J = J^{(m)}$, $J^{(m)} \in \mathcal{J}$, determines the multiplicities of duplications for elements from U.

It is easy to see that the total quantity of possible solutions to Problem 1 coincides with the size of the set $\mathcal{X} = \bigcup_{U \in W} \mathcal{X}(U)$ and except the trivial case when $q_1 = \ldots = q_K = T_{\max}$, we have the lower bound

$$|\mathcal{X}| = \sum_{U \in W} |\mathcal{X}(U)| \geq K 2^{\lfloor \frac{N - q_{\max} + 1}{q_{\max} + 1} \rfloor},$$

where $q_{\max} = \max_{U \in W} q(U)$ and K is the alphabet size. It means that if $q_{\max}$ is bounded by some constant (which is common in applications), the size of $\mathcal{X}$ grows exponentially with increasing N. Despite this exponential growth, the algorithm below provides an optimal solution in polynomial time.

Finally, by transforming $\|Y - X\|^2$ with (5), we have

$$\sum_{n=1}^{N} (x_n - y_n)^2 = \sum_{n=1}^{N} y_n^2 + \sum_{m=1}^{M} \sum_{i=1}^{p_m} \{ u_{J^{(m)}(i)}^2 - 2 y_{n_m + i - 1} u_{J^{(m)}(i)} \}.$$

The first term on the right-hand side of this equation is constant and doesn't depend on variables of Problem 1; the second term coincides with the objective function (1) of Problem 1. Therefore, the problem (6), as well as the problem (4), is equivalent to optimization Problem 1.

Problem 1 is a generalization of previously studied recognition problems. A particular case of Problem 1, where $q_1 = \ldots q_K = q$, $p_m = q$, $J^{(m)}(i) = i$, $m = 1, \ldots, M$, has been examined in [1]. Its modification, where M is a part of input data, has been considered in [2]. In the problems studied in [1] and [2] we also have to recognize a quasiperiodic sequence, but all repetitions in it are identical. In these papers, the algorithms that allow obtaining optimal solutions in time $\mathcal{O}(KT_{\max}N)$ and $\mathcal{O}(KMT_{\max}N)$, respectively, have been presented. Another known particular case, where $K = 1$, has been studied in [3]. In this case, the dictionary contains only one sequence, so it can be treated as an approximation problem solely. The exact $\mathcal{O}(T_{\max}^3 N)$-time algorithm solving this problem has been presented in [3].

The algorithm for solving Problem 1 seems to be a suitable tool to solve applied problems of recognition and analysis of signals that have a quasiperiodic structure in the form of fluctuating signal sample repetitions. Such problems are relevant for various applications dealing with processing quasiperiodic pulse signals received from natural sources: biomedical, geophysical, etc.

3   Problem Solution and Numerical Simulation

The main mathematical result of this paper is the following.

Theorem 1. *There exists an algorithm that finds an exact solution to Problem 1 in time $\mathcal{O}(KT_{\max}^3 N)$.*

The proof of this theorem is constructive. Specifically, we construct an algorithm and show that this algorithm provides an exact solution to Problem 1. The algorithm is based on solving a family of the following auxiliary problems.

Problem 2 [3]. *Given:* numeric sequences $Y = (y_1, \ldots, y_N)$, $U = (u_1, \ldots, u_{q(U)})$, and positive integers $T_{\max}$, ℓ. *Find:* a collection $\mathcal{M} = \{n_1, \ldots n_m, \ldots\}$ of indices of the sequence Y, a collection $\mathcal{P} = \{p_1, \ldots, p_m, \ldots\}$ of positive integers, a collection $\mathcal{J} = \{J^{(1)}, \ldots, J^{(m)}, \ldots\}$ of contraction mappings, where $J^{(m)} : \{1, \ldots, p_m\} \longrightarrow \{1, \ldots, q(U)\}$; and the size M of these collections; which minimize the objective function

$$G(\mathcal{M}, \mathcal{P}, \mathcal{J}) = F(\bullet \,|\, U),$$

under the constraints (2), on the elements of the collections $\mathcal{M}$ and $\mathcal{P}$, and under the constraints (3) on the contraction mappings. Here the notation $F(\bullet \,|\, U)$ means that we consider F as a function of three arguments, while U is fixed.

The algorithm that finds an exact solution to auxiliary Problem 2 in time $\mathcal{O}(T_{\max}^3 N)$ has been presented in [3].

Remark 1. If $T_{\max}$ is a part of input data and K (the size of the alphabet) is a fixed parameter, then the running time of the algorithm is $\mathcal{O}(N^4)$, since $T_{\max} \leq N$; thus, the algorithm solving Problem 1 is polynomial-time.

We present an example of processing a modeled sequence (time series) that can be interpreted as quasi-periodic sequence of fluctuating ECG-like sequences (Fig. 1) pulses in the additive noise presence. In fact, from the mathematical point of view, it doesn't matter which sequences are included in the alphabet. The main reason for choosing exactly ECG-like signal is our desire to illustrate potential applicability of the algorithm for biomedical applications.

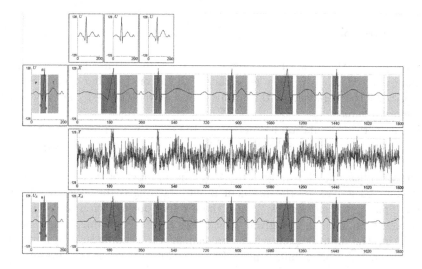

Fig. 1. Example of processing an ECG-like pulse train

In Fig. 1, the alphabet depicted in the top row includes three examples of ECG-pulses. The typical shapes of these pulses, as well as their characteristic sections and significant points, were identified by experts in medicine, for example, see [4,5]. In Fig. 1, these sections are marked by coloring. Below the alphabet, on the left, you can see one of the sequences from the alphabet—sequence U. There is a program-generated sequence to the right of it. This sequence is a quasiperiodic one engendered by fluctuating repetitions of U. The sequence Y is depicted in the third row from the top. It is the element-wise sum of the modeled sequence and the sequence of independent identically distributed Gaussian random variables with zero mathematical expectation. It should be mentioned that only Y and W belong to the input data of the algorithm. The sequence U from the alphabet and the modeled sequence X (in the second row of the figure) are given for illustration only. This data are not available.

The bottom row of the figure represents the result of algorithm operation, namely, the sequences U_A (on the left) and X_A (on the right). Here U_A is the recognition result; the components of the sequence X_A are recovered using (5) and four collections obtained as the algorithm output. The example is computed for $K = 3$, $q(U) = 203$, $U \in W$, $T_{\max} = 370$, $\ell = 370$, $N = 1800$, the maximum amplitude pulse value is 128, and the noise level $\sigma = 35$.

The numerical simulation example shows that the algorithm presented allows processing data in the form of quasiperiodic sequences of fluctuating pulses with quite acceptable quality. Firstly, the unobservable sequence U and the sequence U_A obtained by the algorithm coincide (recognition is carried out correctly). Secondly, a visual comparison of two graphs (of the unobserved sequence X and the recovered sequence X_A) shows only insignificant deviations of one graph from another and almost exact coincidence of the marked sections.

4   Conclusion

We have proved that one of the unexplored discrete optimization problems is polynomially solvable. We have constructed the algorithm that guarantees the optimality of the solution to the problem and have obtained the polynomial complexity estimate.

The numerical simulation has demonstrated that the proposed algorithm can serve as a suitable tool to solve the problems of noise-resistant recognition and analysis of quasiperiodic pulse sequences.

The modification of Problem 1, where the number of convolutions to be summed up is a part of the problem input, remains to be studied. Of considerable mathematical interest is also the discrete optimization problem when the sequence alphabet is not given, that is, we have to recognize a sequence U engendering the input sequence Y as an element of an infinite set of numerical sequences having a fixed finite length. Investigating these problems presents the nearest perspective.

References

1. Kel'manov, A., Khamidullin, S., Okol'nishnikova, L.: Recognition of a quasiperiodic sequence containing identical subsequences-fragments. Patt. Recogn. Image Anal. **14**(1), 72–83 (2004)
2. Kel'manov, A., Khamidullin, S.: Recognizing a quasiperiodic sequence composed of a given number of identical subsequences. Patt. Recogn. Image Anal. **10**(1), 127–142 (2000)
3. Kel'manov, A., Khamidullin, S., Mikhailova, L., Ruzankin, P.: Polynomial-time solvability of one optimization problem induced by processing and analyzing quasiperiodic ECG and PPG signals. In: Jaćimović, M., Khachay, M., Malkova, V., Posypkin, M. (eds.) OPTIMA 2019. CCIS, vol. 1145, pp. 88–101. Springer, Cham (2020). https://doi.org/10.1007/978-3-030-38603-0_7
4. Rajni, R., Kaur, I.: Electrocardiogram signal analysis – an overview. Int. J. Comput. Appl. **84**(7), 22–25 (2013)
5. Al-Ani, M.S.: ECG waveform classification based on P-QRS-T wave recognition. UHD J. Sci. Technol. **2**(2), 7–14 (2018)

Stability Analysis for Pricing

Artem A. Panin[(⊠)] and Alexander V. Plyasunov

Sobolev Institute of Mathematics, Novosibirsk, Russia
aapanin1988@gmail.com, apljas@math.nsc.ru

Abstract. This article is dedicated to finding stable solutions to change input data on the example of pricing problems. In other words, we investigate stability analysis problems based on pricing problems.

Initial pricing problems can be described as the following Stackelberg game. There are a company and its potential clients. First, the company sets prices at own facilities for a homogeneous product. After that, each client chooses the facility in which the minimum of his costs is achieved. The cost consists of purchase and transportation prices. At the same time, clients can make a purchase only if their budget allows it. The goal is to establish prices at which the maximum profit of the company is achieved. In the generalized problem of competitive pricing, two companies compete with each other for the client demand. They set prices sequentially. Clients are also the last to decide.

For the pricing of one company, we discuss the computational complexity and algorithm solution of the stability analysis problem for three different pricing strategies. We also look at the competitive pricing problem with uniform pricing when the same price is set at all facilities. In conclusion, we discuss the relationship between the computational complexity of stability analysis problems and initial problems.

Keywords: Stability analysis · Pricing · Bilevel and three-level problems · Computational complexity

1 Introduction

When solving application problems, it is often necessary to choose a solution that is acceptable not only for the current source data but also remains acceptable when changing this data within a sufficiently wide range. In recent years, a new direction of research in this area has arisen, which is based on the idea of transforming the formulation [1,2]. For a given set of input data of the problem to the maximum, instead of maximizing income, in the new formulation, we will maximize the region of the input data of the problem close to the selected example, for which there is a solution that leads to income not less than the specified threshold.

The research was supported by Russian Foundation for Basic Research (project No. 19-410-240003) (chapter 1,2) and the program of fundamental scientific researches of the SB RAS (project No. 0314-2019-0014) (chapter 3).

This article proposes the new implementation of this idea for the bilevel pricing problems based on the idea of the stability radius (proposed in the works of V.K. Leontiev and E. N. Gordeev [6, 9–11]). The type of stability they examined eventually led to four more new types of stability. These five types of stability are formulated in terms of the behavior of the set of optimal solutions to the problem and are independent of specific solutions. Therefore, they are types of stability of optimization problems. The optimization problem may be:

1) strongly stable [8, 10];
2) strongly quasi-stable [5, 8];
3) stable [4, 8];
4) quasi-stable [4, 8];
5) invariable [5, 8].

In addition to the stability of optimization problems, the stability of a given optimal solution was studied in [4, 8]. There are an optimization problem P (with the criterion of maximization), some instance X. We denote by $F^*(X)$ $(F(X))$ the set of optimal solutions (the set of feasible solutions) of the problem P for input X. Let $\Delta(\rho) = \{\delta : ||\delta|| \leq \rho\}$ ($\Delta^=(\rho) = \{\delta : ||\delta|| = \rho\}$) be the set of variances of the instance X, where $\rho > 0$. The optimal solution $Y^* \in F^*(X)$ is called stable if the set $\Gamma^P(X, Y^*) = \{\rho > 0 : \forall \delta \in \Delta(\rho) [Y^* \in F^*(X + \delta)]\}$ is not empty. The value $\sup \Gamma^P(X, Y^*)$ is called the stability radius of the optimal solution $Y^* \in F^*(X)$.

The concept of stability, which is studied in the paper, is obtained by relaxing the condition that solution $Y^* \in F^*(X)$ remains optimal when the instance X is varied, and replacing it with the condition that the solution remains feasible when the instance X is varied and the value of the objective function on it is not less the specified threshold V. Denote objective function of P as $f_P(X, Y)$, where Y is an arbitrary feasible solution. The feasible solution $Y \in F(X)$ is called stable with respect to threshold V if the set $\Gamma^P(X, Y, V) = \{\rho \geq 0 : \forall \delta \in \Delta(\rho) [Y \in F(X + \delta) \& f_P(X + \delta, Y) \geq V]\}$ is not empty. The value $\rho(X, Y, V) = \sup \Gamma^P(X, Y, V)$ is called the stability radius with respect to threshold V of the feasible solution $Y \in F(X)$.

Definition 1. *The stability analysis problem for input X and threshold V is generally formulated as follows:*

$$\rho(X, Y, V) \to \max_{Y},$$

that is, we need to find the feasible solution $\widetilde{Y}$ that is stable with respect to threshold V and has a maximum stability radius $\rho(X, \widetilde{Y}, V)$.

In the definition of the set $\Gamma^P(X, Y, V)$, we replace the set of variations $\Delta(\rho)$ with the set $\Delta^=(\rho)$.

Definition 2. *The simplified stability analysis problem for input X and threshold V is generally formulated as follows:*

$$\rho(X, Y, V) \to \max_{Y},$$

that is, we need to find the feasible solution $\widetilde{Y}$ that is stable with respect to the set $\Delta^=(\rho)$ of variances and threshold V and has a maximum stability radius $\rho(X, \widetilde{Y}, V)$.

For one-level problems, this definition is sufficient. For bilevel problems, the bottleneck is the concept of a feasible solution. The variables of the bilevel problem are divided into two groups (Y_u, Y_l), where Y_u is the upper-level variables, but Y_l are the lower-level variables. Let us denote by $F_l^*(Y_u)$ the set of optimal solutions of the lower-level problem and by $F(X)\mid_u= \{Y_u : \exists Y_l \in F_l^*(Y_u) \& (Y_u, Y_l) \in F(X)\}$ the projection of the feasible domain $F(X)$ of the two-level problem P onto the variables of the upper level. The vector $Y_u \in F(X)\mid_u$ is called stable with respect to threshold V if the set

$$\Gamma^P(X, Y_u, V) = \{\rho \geq 0 : \forall \delta \in \Delta(\rho) \exists Y_l(\delta) \in F_l^*(Y_u)$$

$$[(Y_u, Y_l(\delta)) \in F(X + \delta) \& f_P(X + \delta, Y) \geq V]\}$$

is not empty. The value $\rho(X, Y_u, V) = \sup \Gamma^P(X, Y_u, V)$ is called the stability radius with respect to threshold V of the vector $Y_u \in F(X)\mid_u$. These and subsequent definitions can be used in both optimistic and pessimistic cases.

Definition 3. *The stability analysis of the bilevel problem P for input X and threshold V is generally formulated as follows:*

$$\rho(X, Y_u, V) \rightarrow \max_{Y_u \in F(X)\mid_u},$$

that is, we need to find the vector $\widetilde{Y}_u \in F(X)\mid_u$ that is stable with respect to threshold V and has a maximum stability radius $\rho(X, \widetilde{Y}_u, V)$.

In the definition of the set $\Gamma^P(X, Y, V)$, we replace the set of variations $\Delta(\rho)$ with the set $\Delta^=(\rho)$ and replace the universal quantifier before δ with the existential quantifier and get the following definition

Definition 4. *The simplified stability analysis of the bilevel problem P for input X and threshold V is generally formulated as follows:*

$$\rho(X, Y_u, V) \rightarrow \max_{Y_u \in F(X)\mid_u},$$

that is, we need to find the vector $\widetilde{Y}_u \in F(X)\mid_u$ that is stable with respect to the set $\Delta^=(\rho)$ of variances and threshold V and has a maximum stability radius $\rho(X, \widetilde{Y}_u, V)$.

Consider the following Stackelberg game. There are a company and its potential clients. First the company prices at own facilities for a homogeneous product. After that, each client chooses the facility in which the minimum of its costs is achieved. The cost consists of purchase and transportation prices. At the same time, clients can make a purchase only if their budget allows it. The goal is to establish prices at which the maximum profit of the company is achieved. This problem is considered in [13,14].

Hereinafter, by variances of instance, we mean the change in the client's budget. This is acceptable since transportation costs or other language distances are often computable with great accuracy. We also note that for pricing problems, the choice of pricing strategy is important. We discuss the computational complexity and algorithm solution of the stability analysis problem for three different pricing strategies. We restrict ourselves to considering the following three strategies: uniform pricing, mill pricing, and discriminatory pricing [7]. Under uniform pricing, the company sets a product's price. The mill pricing implies the assignment of its price at each facility. Customer rights are even further violated in discriminatory pricing when each facility has own price for each customer.

One of the main motivations of this article is the question of changing the complexity status when the initial problem is formulated as a stability analysis problem. If the original problem is polynomially solvable, will the stability analysis problem also be polynomially solvable?

In the next section, we consider the computational complexity of bilevel pricing problems. The third section contains results on exact algorithms for competitive pricing.

2   Bilevel Models of Pricing

First of all, we formulate the problems of pricing. Then we proceed to reformulate it in terms of the stability analysis problem under Definitions 1(3) and 2(4). Next, we will see that for the pricing problems under consideration in the absence of competition, algorithms for solving the stability analysis problem under both definitions are identical. In other words, such problems are equivalent.

We introduce the following notation:

$I = \{1, ..., m\}$ is the set of facilities;
$J = \{1, ..., n\}$ is the set of clients;
$c_{ij} \in Z^+$ is the non-negative transportation cost of a product from the facility i to the client j;
$b_j \in Z^+$ is the non-negative budget of the client j.

To identify the company's product price and the allocation of clients to facilities, we use the following variables:

p is the non-negative product price;
$$x_{ij} = \begin{cases} 1, & \text{if the client } j \text{ is served from the facility } i, \\ 0 & \text{otherwise}; \end{cases}$$

Then the uniform pricing problem can be written as a bilevel model of quadratic mixed-integer programming:

$$\sum_{i \in I} \sum_{j \in J} p x_{ij} \to \max_{p \geq 0, x}$$

where x is an optimal solution of the lower-level (client's) problem:

$$\sum_{j \in J} \sum_{i \in I} (b_j - c_{ij} - p)x_{ij} \to \max_{x_{ij} \in \{0,1\}; i \in I, j \in J}$$

$$\sum_{i \in I} x_{ij} \leq 1; j \in J;$$

If we add an integer constraint on the price, we get the uniform integer pricing problem. To obtain a model for the problem of mill (discriminatory) pricing, it is enough to change the variable p to the variable p_i (p_{ij}) where p_i is the product price in the facility i (the product price in the facility i for the client j).

As the norm of the variance of instance, we consider the norm $|| \cdot ||_{min}$ that is the minimum budget deviation. In other words, we use the following metric: $r(b, \bar{b}) = \min_{j \in J}\{|b_j - \bar{b}_j|\}$.

2.1 Equivalence of Definitions 1 and 2 for the Bilevel Pricing

We formulate the stability analysis problem for the uniform, mill, and discriminatory pricing problem:

$$\rho \to \max_{\rho, p \geq 0}$$

provided that for all values of the variable $\bar{\rho}$ from the segment $[0, \rho]$ exists x such as:

$$\sum_{i \in I} \sum_{j \in J} px_{ij} \geq V;$$

where x is an optimal solution of the lower-level (clients) problem:

$$\sum_{j \in J} \sum_{i \in I} (b_j - \bar{\rho} - c_{ij} - p)x_{ij} \to \max_{x_{ij} \in \{0,1\}; i \in I, j \in J}$$

$$\sum_{i \in I} x_{ij} \leq 1; j \in J.$$

We call this problem the problem of finding a stable price (stability analysis problem) with uniform pricing and the norm $||\cdot||_{min}$. Here $||\cdot||_{min} = \rho$. Similarly, we can write the problem of finding a stable price with mill or discriminatory pricing and integer constraint.

We fix the price p. Let's consider how the set of serviced clients will change with decreasing ρ. Since transportation costs remain unchanged and client budgets are growing, clients will be served at the same facilities that were previously served. At the same time, the set of serviced clients can only increase due to the growth of budgets. This means that the company's income will not decrease. Thus we can rewrite the problem of finding a stable price with uniform pricing and the norm $|| \cdot ||_{min}$ as follows:

$$\rho \to \max_{\rho, p \geq 0, x}$$

$$\sum_{i\in I}\sum_{j\in J} px_{ij} \geq V;$$

where x is an optimal solution of the lower-level (client's) problem:

$$\sum_{j\in J}\sum_{i\in I}(b_j - \rho - c_{ij} - p)x_{ij} \rightarrow \max_{x_{ij}\in\{0,1\};i\in I, j\in J}$$

$$\sum_{i\in I} x_{ij} \leq 1; j \in J.$$

But it is the simplified stability analysis problem. Analogically, it is for the mill or discriminatory pricing and integer constraint. Thus, for bilevel pricing, Definitions 1 and 2 are equivalent. Therefore, in the future, we will not subdivide problems according to two different definitions of stability analysis.

2.2   Algorithms Solution for the Minimum Budget Deviation

To solve the stability analysis problem for the uniform pricing, we first offer an algorithm for solving the problem of uniform pricing. As a consequence of the theorem on necessary optimality conditions [14], the optimal price is $b_j - \min_{i\in I} c_{ij}$ for some $j \in J$. Then by looking at all clients, we can find the best solution in time $mn + n\log n$. For the discriminatory pricing, the optimal price p_{ij} is equal to $b_j - c_{ij}$. Then the discriminatory pricing problem is solvable in time mn. The mill pricing problem is NP-hard [13]. But if we fix the Boolean variables x_{ij}, then we get the linear programming problem. Thus, the mill pricing problem is solvable for $(m + 1)^n$ of calls to the algorithm for solving the linear programming problem.

The optimal value of ρ does not exceed the largest budget. Then, for solving the problem of finding a stable price with uniform, mill, or discriminatory integer pricing, we apply the following algorithm:

Algorithm 1

Iterate through the binary search algorithm all values of ρ from the integer segment $[0, \max_{j\in J, i\in I}\{b_j - c_{ij}\}]$ and solve the pricing problem with uniform, mill or discriminatory pricing where $b_j := b_j - \rho$. The goal is to find the maximum value of ρ, provided that the optimal income of the company is not less V.

Obviously, if we remove the requirement of integer prices, then Algorithm 1 will become useless.

We propose a more efficient algorithm for solving the problem of finding a stable price with uniform pricing:

Algorithm 2

Define $\overline{b_j}$ as $b_j - \min_{i \in I} c_{ij}$ and $b_j(\rho)$ as $\overline{b_j} - \rho$. As a consequence of the theorem on the necessary optimality conditions [14], the optimal price is $b_j - \min_{i \in I} c_{ij}$ for some $j \in J$. Define k_j as $|\{k \in J : \overline{b_k} \geq \overline{b_j}\}|$. The optimal objective function value f is equal $b_j(\rho)k_j$ for some client j. But $f \geq V$. Then we have:

$$b_j(\rho)k_j = (\overline{b_j} - \rho)k_j \geq V$$

$$\Updownarrow$$

$$\rho \geq \overline{b_j} - \frac{V}{k_j}$$

$$\Downarrow$$

$$\rho = \max_{j \in J}\{b_j - \min_{i \in I} c_{ij} - \frac{V}{k_j}\}.$$

For the integer price, we have:

$$\rho = \max_{j \in J}\{b_j - \min_{i \in I} c_{ij} - \lceil \frac{V}{k_j} \rceil\}.$$

The complexity of Algorithm 1 applied to uniform pricing is $logB(mn + n \log n)$ where $B = \max_{j \in J, i \in I}\{b_j - c_{ij}\}$. The complexity of Algorithm 2 is $mn + n \log n$ that is equal to the complexity of the solution algorithm of the uniform pricing problem and less than the complexity of Algorithm 1 at $logB$ times.

For mill pricing, we offer the following enumeration algorithm:

Algorithm 3

We apply the idea of solving the mill pricing problem. Fix the Boolean variables x_{ij}. Then we have the following problem [13,14]:

$$\rho \to \max_{\rho,p \geq 0}$$

$$\sum_{i \in I} \sum_{j \in J} p_i x_{ij} \geq V;$$

$$\sum_{i \in I} (b_j - \rho - c_{ij} - p_i)x_{ij} \geq 0, j \in J;$$

$$\sum_{i \in I} (c_{ij} + p_i)x_{ij} \leq c_{kj} + p_k, k \in I, j \in J.$$

Note that this problem is a linear programming problem. Then, by sequentially sorting all the Boolean variables and solving the linear programming problem, we will find the optimal value of ρ. For the integer pricing, we have to take $\lfloor \rho \rfloor$.

It turns out that the solution algorithm of the mill pricing problem and Algorithm 3 have similar complexity.

In conclusion, we present a solution algorithm for discriminatory pricing:

Algorithm 4

Obviously, the optimal price p_{ij} is equal $\max\{0, b_j(\rho)\}$. Then the optimal objective function value f is $\sum_{j \in J} \max\{0, b_j(\rho)\}$. This leads to the following inequality:

$$f = \sum_{j \in J} \max\{0, b_j(\rho)\} \geq V.$$

Assume that the maximum is reached on the right side on some set of clients $\widetilde{J}$. To calculate the optimal $\widetilde{J}$ we offer the following procedure. Sort clients by increasing this value $b_j(\rho)$. Then we can easily calculate at what value of ρ the first client leaves the set $\widetilde{J}$, the second client, and so on. So for each set $\widetilde{J}$ we have:

$$\rho = \max_{\widetilde{J}} \min\{\rho(\widetilde{J}), \frac{\sum_{j \in \widetilde{J}}(b_j - \min_{i \in I}) - V}{|\widetilde{J}|}\},$$

where $\rho(\widetilde{J})$ is the maximum possible value of ρ for the selected set $\widetilde{J}$. As for the mill pricing for the integer pricing, we have to take $\lfloor \rho \rfloor$.

The complexity of Algorithm 4 is $m(n + 1) + n \log n$.

2.3   Results of the Chapter

The main result of this section is algorithms for solving the stability analysis problems of uniform, mill, and discriminatory pricing. The complexity of these algorithms shows that the complexity of stability analysis problems is comparable to the original pricing problems. In particular, polynomial solvability is preserved. Looking ahead, for competitive pricing, we cannot get such a result.

Another equally important result is the equivalence of Definitions 1 and 2 for the bilevel pricing. As will be shown in the next chapter for the three-level problems of competitive pricing, it is not true.

3   Three-Level Models of Competitive Pricing

Suppose now that two companies compete with each other for the client demand. The first company is the leader. She sets prices first. After that, the second company (the follower) reacts to this with own prices. We restrict ourselves to considering only the uniform pricing strategy. Also note that only if the prices are integer, the problem has an optimal solution.

We introduce the following notations:

$I_L = \{1, ..., m_L\}$ is the set of facilities of the leader company;
$I_F = \{1, ..., m_F\}$ is the set of facilities of the follower company.

Then the leader problem is described as follows:

$$\sum_{i \in I_L} \sum_{j \in J} p x_{ij} \to \max_{x,p,q}$$

$$p \in N;$$

where (q,x) is an optimal solution of the second-level (follower's) problem:

$$\sum_{i \in I_F} \sum_{j \in J} q x_{ij} \to \max_{x,q}$$

$$q \in N;$$

where x is an optimal solution of the lower-level (client's) problem:

$$\sum_{i \in I_L} \sum_{j \in J} [x_{ij}(b_j - c_{ij} - p)] + \sum_{i \in I_F} \sum_{j \in J} [x_{ij}(b_j - c_{ij} - q)] \to \max_x$$

$$\sum_{i \in I_L \cup I_F} x_{ij} \leq 1, \quad j \in J;$$

$$x_{ij} \in \{0, 1\}, i \in I_L \cup I_F, j \in J.$$

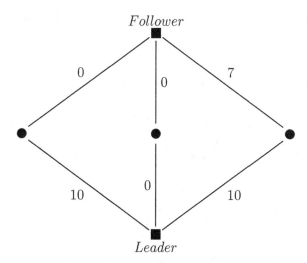

Fig. 1. An example of the three-level pricing problem.

It is known that this problem is polynomially solvable [12]. We already understood that the uniform pricing problem and its stability analysis analog are polynomially solvable too. Can the follower become a serious problem when developing an algorithm for solving the stability analysis problem of competitive pricing? It turns out so. Let's try to understand it. The stability analysis problem under Definition 1 can be described as follows:

$$\rho \to \max_{\rho, p \in N}$$

provided that for all values of the variable $\bar{\rho}$ from the segment $[0, \rho]$ exists q, x such as:

$$\sum_{i \in I_L} \sum_{j \in J} p x_{ij} \geq V;$$

where (q, x) is an optimal solution of the second-level (follower's) problem:

$$\sum_{i \in I_F} \sum_{j \in J} q x_{ij} \to \max_{x, q \in N}$$

where x is an optimal solution of the lower-level (client's) problem:

$$\sum_{i \in I_L} \sum_{j \in J} [x_{ij}(b_j - \bar{\rho} - c_{ij} - p)] + \sum_{i \in I_F} \sum_{j \in J} [x_{ij}(b_j - \bar{\rho} - c_{ij} - q)] \to \max_x$$

$$\sum_{i \in I_L \cup I_F} x_{ij} \leq 1, \quad j \in J;$$

$$x_{ij} \in \{0, 1\}, i \in I_L \cup I_F, j \in J.$$

To complete the picture, we also introduce the stability analysis problem under Definition 2:

$$\rho \to \max_{\rho, p \in N}$$

$$\sum_{i \in I_L} \sum_{j \in J} p x_{ij} \geq V;$$

where (q, x) is an optimal solution of the second-level (follower's) problem:

$$\sum_{i \in I_F} \sum_{j \in J} q x_{ij} \to \max_{x, q \in N}$$

where x is an optimal solution of the lower-level (client's) problem:

$$\sum_{i \in I_L} \sum_{j \in J} [x_{ij}(b_j - \rho - c_{ij} - p)] + \sum_{i \in I_F} \sum_{j \in J} [x_{ij}(b_j - \rho - c_{ij} - q)] \to \max_x$$

$$\sum_{i \in I_L \cup I_F} x_{ij} \leq 1, \quad j \in J;$$

$$x_{ij} \in \{0, 1\}, i \in I_L \cup I_F, j \in J.$$

We show that these two problems are not equivalent. To do this, consider the following example:

Example 1: There are one leader's facility, one follower's facility, and three clients. All clients have the same budget equal to 11. Transportation costs to the leader facility are distributed as follows: from the first client, the cost is 10; from the second client, the cost is 0; from the third client, it is equal 10. Transportation costs to the follower facility are equal to 0, 0, and 7 respectively (Fig. 1). V is equal to 5. Let ρ is 1. Then there is only one feasible solution to the stability analysis problem in which $p = 5$. Is this solution feasible for $\rho = 0$? It is not so because the follower, in this case, will change his decision to serve only the first client and will serve all clients.

Hence for competitive pricing problems, Definitions 1 and 2 are not equivalent. It also follows from the example that we cannot guarantee that if the budget changes, the follower does not decide to change his price and thereby intercept some clients from the leader. Therefore, we cannot use the idea of constructing a solution algorithm (for example, Algorithm 2) that has been successfully applied for bilevel problems.

What can we offer to solve these problems? For Definition 2, we can use the idea of Algorithm 1. That is, to maximize ρ, we can use the binary search:

Algorithm 5

Iterate through the binary search algorithm all values of ρ from the integer segment $[0, \max_{j \in J, i \in I_L}\{b_j - c_{ij}\}]$ and solve the three-level competitive pricing problem [12] where $b_j := b_j - \rho$. The goal is to find the maximum value of ρ, provided that the optimal income of the leader is not less V.

It is clear that for Definition 2 such the algorithm is correct. It is also clear that for Definition 1 it cannot be used. In the case of the stability problem under Definition 1, we need to have a feasible solution that would give the required income for all budget changes from 0 to ρ. We cannot guarantee this for sure due to the unpredictability of the behavior of the follower. This leads us to the idea of enumerating not only the value of ρ and enumerating all the feasible solutions that provide the required income. We get the following algorithm:

Algorithm 6

View from left to right all values of ρ from the segment $[0, \max_{j \in J, i \in I_L}\{b_j - c_{ij}\}]$ while it is possible to present the desired feasible solution for all budget changes from 0 to ρ. In the first step, we look through all the leader's prices and remember those at which, after solving the follower's problem, the leader achieves the required income. In the future, we will only look at the selected prices and discard those for which the condition for achieving the required income greater than or equal to V is not fulfilled.

Algorithm 5 is pseudopolynomial and Algorithm 6 is exponential. Although the original problem is polynomially solvable. This leads us to the following conclusion.

4   Conclusion

In this work, we examined several problems of stability analysis corresponding to previously studied pricing problems. For the simplest pricing problems, it was found that the computational complexity of stability analysis problems does not differ much from the computational complexity of the initial problems. At the same time, for the problems of stability analysis corresponding to the problems of competitive pricing, the complexity of the developed algorithms significantly exceeds the complexity of algorithms for solving the initial problems. Therefore, the following question arises: Can the problem of stability analysis for some metrics fall into a higher complexity class, i.e. to be higher in the polynomial hierarchy than the original problem? For example, if the original problem is polynomially solvable, can the stability analysis problem turn out to be NP-hard?

References

1. Carrizosa, E., Nickel, S.: Robust facility location. Math. Methods Oper. Res. **58**(2), 331–349 (2003)
2. Carrizosa, E., Ushakov, A., Vasilyev, I.: Threshold robustness in discrete facility location problems: a bi-objective approach. Optim. Lett. **9**(7), 1297–1314 (2015). https://doi.org/10.1007/s11590-015-0892-5
3. Davydov, I., Kochetov, Yu., Plyasunov, A.: On the complexity of the $(r \mid p)$-centroid problem in the plane. TOP **22**(2), 614–623 (2014)
4. Emelichev, V., Podkopaev, D.: Quantitative stability analysis for vector problems of 0–1 programming. Discrete Optim. **7**(1–2), 48–63 (2010)
5. Emelichev, V.A., Podkopaev, D.P.: Stability and regularization of vector problems of integer linear programming. Diskretn. Anal. Issled. Oper. Ser. 2 **8**(1), 47–69 (2001)
6. Gordeev, E.N., Leont'ev, V.K.: A general approach to the study of the stability of solutions in discrete optimization problems. Zh. Vychisl. Mat. Mat. Fiz. **36**(1), 66–72 (1996); Comput. Math. Math. Phys. **36**(1), 53–58 (1996)
7. Hanjoul, P., Hansen, P., Peeters, D., Thisse, J.-F.: Uncapacitated plant location under alternative spatial price policies. Mark. Sci. **36**, 41–57 (1990)
8. Kuzmin, K.G.: A united approach to finding the stability radii in a multicriteria problem of a maximum cut. Diskretn. Anal. Issled. Oper. **22**(5), 30–51 (2015); Appl. J. Ind. Math. **9**(4), 527–539 (2015)
9. Leont'ev, V.K.: Stability of the travelling salesman problem. Comput. Math. Math. Phys. **15**(5), 199–213 (1975)
10. Leont'ev, V.K., Gordeev, E.N.: Qualitative analysis of trajectory problems. Kibernetika **5**, 82–90 (1986)
11. Leont'ev, V.K., Gordeev, E.N.: Stability in bottleneck problems. Comput. Math. Math. Phys. **20**(4), 275–280 (1980)
12. Plyasunov, A.V., Panin, A.A.: On three-level problem of competitive pricing. numerical computations: theory and algorithms. In: Proceedings of 2nd International Conference, Pizzo Calabro, Italy, 19–25 June 2016, pp. 050006-1–050006-5. AIP Publ., Melville (2016)

13. Plyasunov, A.V., Panin, A.A.: The pricing problem. I: exact and approximate algorithms. J. Appl. Ind. Math. **7**(2), 241–251 (2013)
14. Plyasunov, A.V., Panin, A.A.: The pricing problem. I: computational complexity. J. Appl. Ind. Math. **7**(3), 420–430 (2013)

Easy NP-hardness Proofs of Some Subset Choice Problems

Artem V. Pyatkin[1,2(✉)] 🆔

[1] Sobolev Institute of Mathematics, Koptyug Avenue, 4, Novosibirsk 630090, Russia
[2] Novosibirsk State University, Pirogova Street, 2, Novosibirsk 630090, Russia
artempyatkin@gmail.com

Abstract. We consider the following subset choice problems: given a
family of Euclidean vectors, find a subset having the largest a) norm of
the sum of its elements; b) square of the norm of the sum of its elements
divided by the cardinality of the subset. The NP-hardness of these problems was proved in two papers about ten years ago by reduction of 3-SAT
problem. However, that proofs were very tedious and hard to read. In
the current paper much easier and natural proofs are presented.

Keywords: Euclidean space · Subset choice · Clustering ·
2-partition · Strong np-hardness

1 Introduction

This paper deals with well-known vector subset choice problems that are induced
by data analysis and pattern recognition problems. A typical problem in data
analysis requires finding in a set of data a subset of the most similar elements
where the similarity is defined according to some criterion. The cardinality of
the sought subset could be known or unknown in advance. One of the possible
criteria is minimum of the sum of squared deviations. This criterion arises, in particular, in a noise-proof data analysis where the aim is to detect informationally
significant fragments in noisy datasets, to estimate them, and to classify them
afterwards [8,12]. The problem of finding a subset of vectors with the longest
sum has applications in the pattern recognition (finding a correct direction to a
certain object) [25].

Although these problems are known to be NP-hard both in the case of known
(given as a part of input) cardinality of a sought subset [3,8] and in the case of
unknown one [14,15,22], the latter proofs are much more complicated and hard
to read (see the discussion in the next section). In this paper we suggest much
more easy and natural NP-hardness proofs for the case of unknown size of the

The research was supported by the program of fundamental scientific researches of
the SB RAS, project 0314-2019-0014, by the Russian Foundation for Basic Research,
project 19-01-00308, and by the Top-5-100 Program of the Ministry of Education and
Science of the Russian Federation.

Y. Kochetov et al. (Eds.): MOTOR 2020, CCIS 1275, pp. 70–79, 2020.
https://doi.org/10.1007/978-3-030-58657-7_8

sought set. We believe that the new proofs can be helpful for analyzing related problems with the unknown cardinalities of the sought subset.

The paper is organized as follows. In the next section the mathematical formulation of the problems are given and the motivation of the research and some related results are discussed. In Sect. 3 the main results of the paper are presented. Section 4 concludes the paper.

2   Problem Formulation, Motivation and Related Results

The problem of noise-proof data analysis in noisy data sets [8,12,15] is as follows. Each record of the data is a vector representing a set of measured characteristics of an object transmitted via a noisy channel. The object can be either in an active or in a passive state. In the passive state all characteristics are 0, while in the active state all measured characteristics are stabile and at least one of them must be non-zero. The noise has a d-dimensional normal distribution with zero mean and an arbitrary dispersion. The goal is to determine the moments when the object was in the active state and to evaluate the measured characteristics.

As it was shown in [8,12,15], this problem can be reduced to the following optimization problem.

Problem 1.   Given a set of vectors $\mathcal{Y} = \{y_1, \ldots, y_N\}$ in d-dimensional Euclidean space, find a non-empty subset $\mathcal{C} \subseteq \mathcal{Y}$ maximizing

$$h(\mathcal{C}) := \frac{\|\sum_{x \in \mathcal{C}} x\|^2}{|\mathcal{C}|}.$$

Everywhere in the paper the norm is Euclidean, unless otherwise stated. A version of Problem 1 with an additional restriction on the cardinality of the sought set $\mathcal{C}$ is referred to as

Problem 2.   Given a set of vectors $\mathcal{Y} = \{y_1, \ldots, y_N\}$ in d-dimensional Euclidean space and a positive integer M, find a subset $\mathcal{C} \subseteq \mathcal{Y}$ of cardinality M maximizing $h(\mathcal{C})$.

The following two subset choice problems are very close in formulation to these ones.

Problem 3.   Given a set of vectors $\mathcal{Y} = \{y_1, \ldots, y_N\}$ in d-dimensional Euclidean space, find a non-empty subset $\mathcal{C} \subseteq \mathcal{Y}$ maximizing $\|\sum_{x \in \mathcal{C}} x\|$.

Problem 4.   Given a set of vectors $\mathcal{Y} = \{y_1, \ldots, y_N\}$ in d-dimensional Euclidean space and a positive integer M, find a subset $\mathcal{C} \subseteq \mathcal{Y}$ of cardinality M minimizing $\sum_{x \in \mathcal{C}} \|x - \overline{x}\|^2$ where $\overline{x} = (\sum_{x \in \mathcal{C}} x)/|\mathcal{C}|$ is the centroid of the set $\mathcal{C}$.

Note that the variant of Problem 3 with a given cardinality of the subset $\mathcal{C}$ is equivalent to Problem 2, while the variant of Problem 4 without the restriction on the cardinality of $\mathcal{C}$ is trivial (every subset of cadrinality 1 is an optimal solution).

Problem 3 has a following interpretation [25]. Each vector is a measurement result of a direction to some interesting object. Each measurement result has an additive error having a normal distribution, and there are some redundant vectors in the set (related to other objects or reflections). The goal is to delete the redundant vectors and find the correct direction. This can be done by finding a subset of vectors having the longest sum.

If the dimension of the space d is fixed then all Problems 1–4 are polynomially solvable. Namely, Problems 1 and 2 can be solved [9] in time $O(dN^{2d+2})$; Problem 3 is a particular case of Shaped Partition problem [11], which yields an $O(N^d)$ algorithm for it; a better algorithm of complexity $O(dN^{d-1} \log N)$ is presented in [25]. Problem 4 can be solved [1] in time $O(dN^{d+1})$. The universal algorithm solving Problems 1–4 in time $O(dN^{d+1})$ using Voronoi diagrams can be found in [24]. Note that this algorithm indeed can solve any vector subset choice problem satisfying one of the following two locality properties:

– For every input there is a point x^* such that the optimal solution consists of the set of M closest to x^* points of $\mathcal{Y}$.
– For every input there is a vector y^* such that the optimal solution consists of the set of M vectors of $\mathcal{Y}$ having minimum scalar products with y^*.

If the dimension of the space d is a part of input then all four problems mentioned above are NP-hard in a strong sense. Moreover, for Problems 2 and 3 an inapproximability bound $(16/17)^{1/p}$ was proved in [26] for an arbitrary norm l_p where $p \in [1, \infty)$.

There are a lot of approximation results for these problems. Let us mention randomized algorithms finding $(1 + \varepsilon)$-approximate solution for Problems 2 and 3 of complexity $O(d^{3/2} N \log \log N / (2\varepsilon - \varepsilon^2)^{(d-1)/2})$ in [10] and of complexity $O(d^{O(1)} N (1 + 2/\varepsilon)^d)$ with probability $1 - 1/e$ in [26]. For Problem 4 a $(1 + \varepsilon)$-approximation algorithm of complexity $O(N^2 (M/\varepsilon)^d)$ was suggested in [19] and a PTAS of complexity $O(dN^{1+2/\varepsilon} (9/\varepsilon)^{3/\varepsilon})$ was constructed in [23]. For Problem 1 a $(1+\varepsilon)$-approximation algorithm of complexity $O(Nd(d+\log N)(\sqrt{(d-1)/\varepsilon}+1)^{d-1})$ can be found in [15].

The NP-hardness of Problem 2 (i. e. in case of known—given as a part of input—cardinality of a sought subset) was proved in [3,8]. The proof uses a natural reduction from the classical NP-hard Clique problem. In this reduction, each vector corresponds to a vertex of a graph and a subset $\mathcal{C}$ is optimal if and only if the corresponding subset of vertices induces a clique in the graph. This proof is so natural that the similar idea was used later, in particular, for proving NP-hardness of Problem 4 in [16], of Maximum Diversity problem in [5] and of 1-Mean and 1-Median 2-Clustering Problem in [18].

The NP-hardness of Problem 1 was proved in [14,15]. It uses quite complicated reduction of 3-SAT problem, where several vectors correspond to each clause and to each variable, and some irrational numbers (square roots) are used in their coordinates (and thus, additional arguments justifying the possibility of rational approximation become necessary). The NP-hardness of Problem 3 was proved in [22] also by reduction of 3-SAT; although there are no irrational numbers, the reduction still remains complicated and the proof is hard to follow.

These reductions are highly inconvenient and hard to generalize. So, many other vector choice or clustering problems with unknown cardinality of the sought set stay open (see, for example, [18]). In this paper we present an easy and natural NP-hardness proof for Problems 1 and 3 with almost the same reduction of Exact Cover by 3-Sets problem.

Let us mention some other problems that are related to Problems 1–4. Make use of the following well-known folklore identities (the proofs can be found, for instance, in [15,17]):

$$\sum_{y \in \mathcal{Y}} \|y\|^2 - \frac{\|\sum_{x \in \mathcal{C}} x\|^2}{|\mathcal{C}|} = \sum_{y \in \mathcal{C}} \|y - \overline{y}\|^2 + \sum_{y \in \mathcal{Y} \setminus \mathcal{C}} \|y\|^2$$

$$= \frac{\sum_{y \in \mathcal{C}} \sum_{z \in \mathcal{C}} \|y - z\|^2}{2|\mathcal{C}|} + \sum_{y \in \mathcal{Y} \setminus \mathcal{C}} \|y\|^2. \tag{1}$$

Since the sum of the squared norms of all vectors from $\mathcal{Y}$ does not depend on $\mathcal{C}$, Problems 1 and 2 are equivalent to minimization of the function

$$\sum_{y \in \mathcal{C}} \|y - \overline{y}\|^2 + \sum_{y \in \mathcal{Y} \setminus \mathcal{C}} \|y\|^2,$$

that can be treated as a minimum sum of squares 2-clustering where the center of one cluster is known. This problem is very close to a classical MSSC (minimum sum of squares clustering) problem also known as k-means [2,6,20,21], but not equivalent to it. Note that in such equivalent formulations these problems admit polynomial 2-approximation algorithms of complexity $O(dN^2)$ both for known [4] and unknown [13] cardinality of the sought set (cluster with an unknown center). As far as we know, no polynomial approximation algorithm with a guaranteed exactness bound is known for Problem 1.

3   Main Results

In this section we present the new NP-hardness proofs for Problems 1 and 3.

3.1   NP-hardness of Problem 1

Let us rewrite Problem 1 in the equivalent (due to (1)) form of the decision problem.

Problem 5.   Given a set of vectors $\mathcal{Y} = \{y_1, \dots, y_N\}$ in d-dimensional Euclidean space and a number $K > 0$, is there a non-empty subset $\mathcal{C} \subseteq \mathcal{Y}$ such that

$$f(\mathcal{C}) := \frac{1}{2|\mathcal{C}|} \sum_{x \in \mathcal{C}} \sum_{y \in \mathcal{C}} \|x - y\|^2 + \sum_{z \in \mathcal{Y} \setminus \mathcal{C}} \|z\|^2 \leq K?$$

We need the following well-known NP-hard [7] version of the Exact Cover by 3-Sets problem where each element lies in at most 3 subsets.

Problem 6 (X3C3). Given a family $E = \{e_1, \ldots, e_m\}$ of 3-element subsets of the set $V = \{v_1, \ldots, v_n\}$ where $n = 3q$ such that every $v \in V$ meets in at most 3 subsets from E, find out whether there exist a subfamily $E_0 = \{e_{i_1}, \ldots, e_{i_q}\}$ covering the set V, i. e. such that $V = \cup_{j=1}^{q} e_{i_j}$.

The main result of this subsection is the following theorem.

Theorem 1. *Problem 1 is NP-hard in a strong sense.*

Proof. Consider an arbitrary instance of X3C3 problem and reduce it to an instance of Problem 5 in the following way. Put $N = m, d = 3n + 1$ and $K = 18a^2(m-1) + m - q$ where a is a positive integer such that $a^2 > m(m-q)/6$. Each vector $y_i \in \mathcal{Y}$ corresponds to a set $e_i \in E$. For every $i \in \{1, \ldots, n\}$ refer to the coordinates $3i$, $3i - 1$, $3i - 2$ of a vector $y \in \mathcal{Y}$ as *i-th coordinate triple*. Denote by $y_i(j)$ the j-th coordinate of y_i. If $v_i \notin e_j$ then the i-th triple of the vector y_j contains zeroes: $y_j(3i - 2) = y_j(3i - 1) = y_j(3i) = 0$. Otherwise, let $k = |\{l < j \mid v_i \in e_l\}|$ be the number of subsets from E with lesser indices than j containing the element v_i. Since each v_i lies in at most 3 subsets from E, we have $k \in \{0, 1, 2\}$. Put

$$y_j(3i - 2) = 2a, \; y_j(3i - 1) = y_j(3i) = -a, \; \text{if } k = 0;$$
$$y_j(3i - 1) = 2a, \; y_j(3i - 2) = y_j(3i) = -a, \; \text{if } k = 1;$$
$$y_j(3i) = 2a, \; y_j(3i - 2) = y_j(3i - 1) = -a, \; \text{if } k = 2.$$

Also, put $y_j(3n + 1) = 1$ for all $j \in \{1, \ldots, m\}$.

For example, if $E = \{(v_1, v_2, v_3), (v_1, v_3, v_4), (v_1, v_5, v_6), (v_2, v_3, v_5), (v_4, v_5, v_6)\}$ then the family $\mathcal{Y}$ contains the following five vectors of dimension 19:

$$y_1 = (2a, -a, -a \mid 2a, -a, -a \mid 2a, -a, -a \mid 0, 0, 0 \mid 0, 0, 0 \mid 0, 0, 0 \mid 1);$$

$$y_2 = (-a, 2a, -a \mid 0, 0, 0 \mid -a, 2a, -a \mid 2a, -a, -a \mid 0, 0, 0 \mid 0, 0, 0 \mid 1);$$

$$y_3 = (-a, -a, 2a \mid 0, 0, 0 \mid 0, 0, 0 \mid 0, 0, 0 \mid 2a, -a, -a \mid 2a, -a, -a \mid 1);$$

$$y_4 = (0, 0, 0 \mid -a, 2a, -a \mid -a, -a, 2a \mid 0, 0, 0 \mid -a, 2a, -a \mid 0, 0, 0 \mid 1);$$

$$y_5 = (0, 0, 0 \mid 0, 0, 0 \mid 0, 0, 0 \mid -a, 2a, -a \mid -a, -a, 2a \mid -a, 2a, -a \mid 1).$$

For the convenience, different coordinate triples are separated by the vertical lines.

Note that $\|y_i\|^2 = 18a^2 + 1$ for all i and also

$$\|y_i - y_j\|^2 = \begin{cases} 36a^2, & \text{if } e_i \cap e_j = \emptyset; \\ 42a^2, & \text{if } |e_i \cap e_j| = 1; \\ 48a^2, & \text{if } |e_i \cap e_j| = 2; \\ 54a^2, & \text{if } |e_i \cap e_j| = 3 \end{cases}$$

for every $i \neq j$.

Assume first that an exact cover E_0 exists. Put $\mathcal{C} = \{y_j \mid e_j \in E_0\}$. Then

$$f(\mathcal{C}) = \frac{q(q-1)36a^2}{2q} + (m-q)(18a^2+1) = 18a^2(m-1) + m - q = K,$$

as required.

Assume now that there is a subset $\mathcal{C}$ of size $t > 0$ such that $f(\mathcal{C}) \le K$. Note that each coordinate triple can be non-zero in at most 3 vectors from $\mathcal{C}$. For each $k \in \{0, 1, 2, 3\}$ denote by a_k the number of coordinate triples that are non-zero in exactly k vectors from $\mathcal{C}$ and estimate the contributions of such triples into the first addend of $f(\mathcal{C})$. Note that $a_0 + a_1 + a_2 + a_3 = n = 3q$. Clearly, the contribution of a_0 zero triples is 0. If a triple is non-zero in one vector from $\mathcal{C}$ then it contributes

$$\frac{(t-1)(4a^2 + a^2 + a^2)}{t},$$

and the total contribution of such triples is

$$\frac{6a^2 a_1(t-1)}{t}. \tag{2}$$

If a triple is non-zero in two vectors from $\mathcal{C}$ then it contributes

$$\frac{2(t-2)(4a^2 + a^2 + a^2) + (9a^2 + 9a^2)}{t};$$

so, the total contribution of such triples is

$$\frac{6a^2 a_2(2t-1)}{t}. \tag{3}$$

Finally, the total contribution of triples that are non-zero in three vectors from $\mathcal{C}$ is

$$\frac{(3(t-3)6a^2 + 3 \cdot 18a^2)a_3}{t} = 18a^2 a_3. \tag{4}$$

Since $|e_j| = 3$ for all j, we have $a_1 + 2a_2 + 3a_3 = 3t$. Using (2)–(4), estimate the objective function

$$f(\mathcal{C}) = \frac{6a^2}{t}((t-1)a_1 + (2t-1)a_2 + 3ta_3) + (m-t)(18a^2+1)$$

$$= \frac{6a^2}{t}(3t^2 - a_1 - a_2) + (m-t)(18a^2+1) = 18ma^2 + m - t - \frac{6a^2}{t}(a_1 + a_2)$$

$$= K + 18a^2 - \frac{6a^2}{t}(a_1 + a_2) + q - t.$$

If $t < q$ then $f(\mathcal{C}) > K$ since $a_1 + a_2 \le 3t$.

Assume now that $t \ge q$ and $a_2 + a_3 \ge 1$. Then $a_1 + a_2 = 3t - a_2 - 2a_3 \le 3t - 1$ and since $t \le m$ we obtain

$$f(\mathcal{C}) = K + 18a^2 - \frac{6a^2}{t}(a_1 + a_2) + q - t$$

$$\geq K + 18a^2 - \frac{6a^2(3t-1)}{t} + q - t \geq K + \frac{6a^2}{m} + q - m > K$$

by the choice of a.

    Therefore, $t \geq q$ and $a_2 = a_3 = 0$. But then $a_1 = 3t$ and $a_0 + a_1 = 3q$, i. e. $a_0 = 0$ and $t = q$. Hence, the set $E_0 = \{e_j \mid y_j \in \mathcal{C}\}$ induces an exact cover. $\square$

3.2   NP-hardness of Problem 3

Since the norm is always non-negative, maximizing it is the same as maximizing its square, which is much more convenient. So, the decision version of Problem 3 is equivalent to the following

Problem 7.   Given a set of vectors $\mathcal{Y} = \{y_1, \ldots, y_N\}$ in d-dimensional Euclidean space and a number K, is there a non-empty subset $\mathcal{C} \subseteq \mathcal{Y}$ such that

$$g(\mathcal{C}) := \|\sum_{x \in \mathcal{C}} x\|^2 \geq K?$$

    In order to prove its NP-hardness we first need to show that X3C3 problem remains NP-complete for 3-uniform family of subsets (i. e. if each $v_i \in V$ lies in exactly 3 subsets from E). We refer to this variant of X3C3 problem as X3CE3 problem.

Proposition 1. *The* X3CE3 *problem is NP-complete.*

Proof. Consider an arbitrary instance of X3C3 problem. We may assume that each v_i lies in at least 2 subsets (if some v_i lies in a unique subset then this subset must always be in E_0 and the instance can be simplified). Denote by α and β the number of elements lying in 3 and 2 subsets from E respectively. Since $3\alpha + 2\beta = 3m$, there must be $\beta = 3\gamma$. Enumerate the elements of V so that $v_1, \ldots, v_{3\gamma}$ would lie in two subsets from E. Construct an instance of X3CE3 problem by adding to V a set of new elements $U = \{u_i \mid i = 1, \ldots, 3\gamma\}$ and by adding to E the subsets $\{v_{3i-2}, u_{3i-2}, u_{3i-1}\}, \{v_{3i-1}, u_{3i-2}, u_{3i}\}, \{v_{3i}, u_{3i-1}, u_{3i}\}$, and $\{u_{3i-2}, u_{3i-1}, u_{3i}\}$ for all $i = 1, \ldots, \gamma$. Clearly, no exact cover (a subfamily E_0) in the constructed instance can contain a subset that intersects both with U and V. Therefore, the constructed instance of X3CE3 problem has an exact cover if and only if the initial instance of X3C3 problem has one. $\square$

Theorem 2. *Problem 3 is NP-hard in a strong sense.*

Proof. Consider an arbitrary instance of X3CE3 problem. Note that $m = n = 3q$. Reduce it to an instance of Problem 7 as follows. Put $N = n, d = 3n+1$ and $K = 6a^2n + 4q^2$ where a is a positive integer such that $a^2 > (n^2 - 4q^2)/6$, and construct the set of vectors $\mathcal{Y}$ in exactly the same way as in proof of Theorem 1.

    In an evident way, each $\mathcal{C} \subseteq \mathcal{Y}$ corresponds to a subfamily $E(\mathcal{C}) \subseteq E$. Put $u(\mathcal{C}) = \sum_{y \in \mathcal{C}} y$. Since $g(\mathcal{C}) = \|u(\mathcal{C})\|^2$, the contribution of the i-th coordinate triple into the objective function $g(\mathcal{C})$ is $6a^2$ if 1 or 2 vectors corresponding to subsets containing v_i lies in $E(\mathcal{C})$, and the contribution is 0 otherwise.

If there is an exact cover E_0 in X3CE3 problem then let $\mathcal{C}$ contain all $n - q = 2q$ vectors corresponding to the elements from $E \setminus E_0$. Since each element of V lies in exactly 2 subsets from $E \setminus E_0$, we have $g(\mathcal{C}) = 6a^2 n + 4q^2 = K$.

Suppose now that there exists a subset $\mathcal{C} \subseteq \mathcal{Y}$ of cardinality $t > 0$ such that $g(\mathcal{C}) \geq K$. As in the proof of Theorem 1, for each $k \in \{0, 1, 2, 3\}$ denote by a_k the number of coordinate triples that are non-zero in exactly k vectors from $\mathcal{C}$. We have $a_0 + a_1 + a_2 + a_3 = n = 3q$ and $a_1 + 2a_2 + 3a_3 = 3t$. It follows form the arguments above that $g(\mathcal{C}) = 6a^2(a_1 + a_2) + t^2$.

If $t < 2q$ then $g(\mathcal{C}) < K$ since $a_1 + a_2 \leq n$.

If $t > 2q$ then $0 < 3t - 6q = a_3 - a_1 - 2a_0 \leq a_3$ and thus $a_3 \geq 1$ implying $a_1 + a_2 \leq n - 1$. Therefore,

$$g(\mathcal{C}) \leq 6a^2(n-1) + n^2 = K - 6a^2 + n^2 - 4q^2 < K$$

by the choice of a.

Hence, $t = 2q$ and $a_1 + a_2 = n$, which implies $a_0 = a_1 = a_3 = 0$ and $a_2 = 3q$. This means that each element $v_i \in V$ lies exactly in 2 subsets from $E(\mathcal{C})$. But then the subfamily $E_0 = E \setminus E(\mathcal{C})$ induces an exact cover in X3CE3 problem.□

4 Conclusions

In this paper we have presented two new NP-hardness proofs for the subset choice problems with unknown cardinalities of the sought subsets. Namely, the problems of finding a subset with the longest sum and a subset with the maximum squared norm of the sum normalized by the size of the subset were considered. These problems find their applications in the areas of data analysis and pattern recognition. Namely, the first problem can be used for finding a correct direction to a certain object, and the second one arises in problem of detection an informationally significant fragment in a noisy data.

The suggested new NP-hardness proofs use an easy and natural reduction from Ecact Cover by 3-Sets problem. We believe that new natural reductions could be helpful for proving NP-hardness of related problems with unknown cardinalities of the sought subsets.

Acknowledgement. The author is grateful to the unknown referees for their valuable comments.

References

1. Aggarwal, A., Imai, H., Katoh, N., Suri, S.: Finding k points with minimum diameter and related problems. J. Algorithms **12**(1), 38–56 (1991)
2. Aloise, D., Deshpande, A., Hansen, P., Popat, P.: NP-hardness of Euclidean sum-of-squares clustering. Mach. Learn. **75**(2), 245–248 (2009)
3. Baburin, A.E., Gimadi, E.K., Glebov, N.I., Pyatkin, A.V.: The problem of finding a subset of vectors with the maximum total weight. J. Appl. Ind. Math. **2**(1), 32–38 (2008). https://doi.org/10.1134/S1990478908010043

4. Dolgushev, A.V., Kel'manov, A.V.: An approximation algorithm for solving a problem of cluster analysis. J. Appl. Ind. Math. **5**(4), 551–558 (2011)
5. Eremeev, A.V., Kel'manov, A.V., Kovalyov, M.Y., Pyatkin, A.V.: Maximum diversity problem with squared euclidean distance. In: Khachay, M., Kochetov, Y., Pardalos, P. (eds.) MOTOR 2019. LNCS, vol. 11548, pp. 541–551. Springer, Cham (2019). https://doi.org/10.1007/978-3-030-22629-9_38
6. Fisher, W.D.: On grouping for maximum homogeneity. J. Am. Stat. Assoc. **53**(284), 789–798 (1958)
7. Garey, M.R., Johnson, D.S.: Computers and Intractability. The Guide to the Theory of NP-Completeness. W. H. Freeman and Company, San Francisco (1979)
8. Gimadi, E.K., Kel'manov, A.V., Kel'manova, M.A., Khamidullin, S.A.: A posteriori detection of a quasi periodic fragment in numerical sequences with given number of recurrences. Sib. Zh. Ind. Mat. **9**(1), 55–74 (2006). (in Russian)
9. Gimadi, E.K., Pyatkin, A.V., Rykov, I.A.: On polynomial solvability of some problems of a vector subset choice in a Euclidean space of fixed dimension. J. Appl. Ind. Math. **4**(1), 48–53 (2010)
10. Gimadi, E., Rykov, I.A.: Efficient randomized algorithm for a vector subset problem. In: Kochetov, Y., Khachay, M., Beresnev, V., Nurminski, E., Pardalos, P. (eds.) DOOR 2016. LNCS, vol. 9869, pp. 148–158. Springer, Cham (2016). https://doi.org/10.1007/978-3-319-44914-2_12
11. Hwang, F.K., Onn, S., Rothblum, U.G.: A polynomial time algorithm for shaped partition problems. SIAM J. Optim. **10**, 70–81 (1999)
12. Kel'manov, A.V., Khamidullin, S.A.: Posterior detection of a given number of identical subsequences in a quasi-periodic sequence. Comput. Math. Math. Phys. **41**, 762–774 (2001)
13. Kel'manov, A.V., Khandeev, V.I.: A 2-approximation polynomial algorithm for a clustering problem. J. Appl. Ind. Math. **7**(4), 515–521 (2013). https://doi.org/10.1134/S1990478913040066
14. Kelmanov, A.V., Pyatkin, A.V.: On the complexity of a search for a subset of "similar" vectors. Doklady Math. **78**(1), 574–575 (2008)
15. Kel'manov, A.V., Pyatkin, A.V.: On a version of the problem of choosing a vector subset. J. Appl. Ind. Math. **3**(4), 447–455 (2009)
16. Kel'manov, A.V., Pyatkin, A.V.: NP-Completeness of some problems of choosing a vector subset. J. Appl. Ind. Math. **5**(3), 352–357 (2011)
17. Kel'manov, A.V., Pyatkin, A.V.: On the complexity of some quadratic euclidean 2-clustering problems. Comput. Math. Math. Phys. **56**(3), 491–497 (2016)
18. Kel'manov, A.V., Pyatkin, A.V., Khandeev, V.I.: NP-hardness of quadratic euclidean 1-mean and 1-median 2-clustering problem with constraints on the cluster sizes. Doklady Math. **100**(3), 545–548 (2019). https://doi.org/10.1134/S1064562419060127
19. Kelmanov, A.V., Romanchenko, S.M.: An FPTAS for a vector subset search problem. J. Appl. Ind. Math. **8**(3), 329–336 (2014). https://doi.org/10.1134/S1990478914030041
20. MacQueen, J.: Some methods for classification and analysis of multivariate observations. In: Proceedings of 5-th Berkeley Symposium on Mathematics, Statistics and Probability, vol. 1, pp. 281–297 (1967)
21. Mahajan, M., Nimbhorkar, P., Varadarajan, K.: The planar k-means problem is NP-hard. Theor. Comput. Sci. **442**, 13–21 (2012)
22. Pyatkin, A.V.: On complexity of a choice problem of the vector subset with the maximum sum length. J. Appl. Ind. Math. **4**(4), 549–552 (2010)

23. Shenmaier, V.V.: An approximation scheme for a problem of search for a vector subset. J. Appl. Ind. Math. **6**(3), 381–386 (2012)
24. Shenmaier, V.V.: Solving some vector subset problems by Voronoi diagrams. J. Appl. Ind. Math. **10**(4), 560–566 (2016). https://doi.org/10.1134/S199047891604013X
25. Shenmaier, V.V.: An exact algorithm for finding a vector subset with the longest sum. J. Appl. Ind. Math. **11**(4), 584–593 (2017). https://doi.org/10.1134/S1990478917040160
26. Shenmaier, V.V.: Complexity and approximation of finding the longest vector sum. Comput. Math. Math. Phys. **58**(6), 850–857 (2018). https://doi.org/10.1134/S0965542518060131

Sensitive Instances of the Cutting Stock Problem

Artem V. Ripatti[(⊠)][iD] and Vadim M. Kartak[iD]

Ufa State Aviation Technical University, Karl Marx str. 12, 450008 Ufa, Russia
ripatti@inbox.ru, kvmail@mail.ru

Abstract. We consider the well-known cutting stock problem (CSP). The gap of a CSP instance is the difference between its optimal function value and optimal value of its continuous relaxation. For most instances of CSP the gap is less than 1 and the maximal known gap $6/5 = 1.2$ was found by Rietz and Dempe [11]. Their method is based on constructing instances with large gaps from so-called sensitive instances with some additional constraints, which are hard to fulfill. We adapt our method presented in [15] to search for sensitive instances with required properties and construct a CSP instance with gap $77/64 = 1.203125$. We also present several instances with large gaps much smaller than previously known.

Keywords: Cutting Stock Problem · Integer Round Up Property · Integrality gap · Sensitive instances

1 Introduction

In the classical formulation, the cutting stock problem (CSP) is stated as follows: there are infinite pieces of stock material of fixed length L. We have to produce $m \in \mathbb{N}$ groups of pieces of different lengths $l_1, \cdots, l_m$ and demanded quantities $b_1, \cdots, b_m$ by cutting initial pieces of stock material in such a way that the number of used initial pieces is minimized.

The cutting stock problem is one of the earliest problems that have been studied through methods of operational research [6]. This problem has many real-world applications, especially in industries where high-value material is being cut [3] (steel industry, paper industry). No exact algorithm is known that solves practical problem instances optimally, so there are lots of heuristic approaches. The number of publications about this problem increases each year, so we refer the reader to bibliography [18] and the most recent survey [2].

Throughout this paper we abbreviate an instance of CSP as $E := (L, l, b)$. The total number of pieces is $n = \sum_{i=1}^{m} b_i$. W.l.o.g., we assume that all numbers in the input data are positive integers and $L \geq l_1 > \cdots > l_m > 0$.

The classical approach for solving CSP is based on the formulation by Gilmore and Gomory [5]. Any subset of pieces (called a *pattern*) is formalized as

---

Supported by RFBR, project 19-07-00895.

Y. Kochetov et al. (Eds.): MOTOR 2020, CCIS 1275, pp. 80–87, 2020.
https://doi.org/10.1007/978-3-030-58657-7_9

a vector $a = (a_1, \cdots, a_m)^\top \in \mathbb{Z}_+^m$ where $a_i \in \mathbb{Z}_+$ denotes the number of pieces i in the pattern a. A pattern a of E is *feasible* if $a^\top l \leq L$. So, we can define the set of all feasible patterns $P^f(L, l) = \{a \in \mathbb{Z}_+^m \mid a^\top l \leq L\}$. For a given set of patterns $P = \{a^1, \cdots, a^r\}$, let $A(P)$ be the $(n \times r)$-matrix whose columns are given by the patterns a^i. Then the CSP can be formulated as follows:

$$z(E) := \sum_{i=1}^{r} x_i \to \min \text{ subject to } A(P^f(L, l))x = b, x \in \mathbb{Z}_+^r.$$

The common approximate solution approach involves considering *the continuous relaxation* of CSP

$$z_C(E) := \sum_{i=1}^{r} x_i^C \to \min \text{ subject to } A(P^f(L, l))x^C = b, x^C \in \mathbb{R}_+^r.$$

Here $z(E)$ and $z_C(E)$ are called *the optimal function values* for the instance E. The difference $\Delta(E) = z(E) - z_C(E)$ is called *the gap* of instance E. Practical experience and numerous computations have shown that for most instances the gap is very small. An instance E has *the integer round up property* (IRUP) if $\Delta(E) < 1$. Otherwise, E is called a non-IRUP instance. This notation was introduced by Baum and Trotter [1].

Subsequently, the largest known gap was increased. In 1986 Marcotte constructed the first known non-IRUP instance with the gap of exactly 1 [9]. Fieldhouse found an instance with gap $31/30 \approx 1.033333$ in 1990 [4]. In 1991 Schiethauer and Terno slightly improved this result to $137/132 \approx 1.037879$ [16]. Rietz, Scheithauer and Terno subsequently constructed non-IRUP instances with gaps $10/9 \approx 1.111111$ and $7/6 \approx 1.166666$ in 1998 and 2000 respectively [12,13] (both papers were published in 2002). Finally, Rietz constructed an instance with gap $6/5 = 1.2$ and published it in his PhD thesis in 2003 [10] and a slightly smaller instance with the same gap together with Dempe in 2008 [11].

The MIRUP (modified IRUP) conjecture [17] states that $\Delta(E) < 2$ for all CSP instances E, but it is still open. More investigations about non-IRUP instances can be found in [7,8,14].

The main idea of our paper is to connect our algorithm for enumeration of instances published in [15] together with ideas of Rietz and Dempe [11] in aim to construct CSP instances with the gap larger than currently known.

The paper has the following structure. In Sect. 2, we describe the construction of Rietz and Dempe, in Sect. 3, we describe our enumeration algorithm. In Sect. 4, we present the computational results and, finally, we draw a conclusion in Sect. 5.

2 Preliminaries

The construction principles of Rietz and Dempe are based on the instance

$$E_0(p, q) = (33 + 3p + q, (21 + p + q, 19 + p + q, 15 + p + q, 10 + p, 9 + p, 7 + p, 6 + p, 4 + p)^\top, b_0),$$

where p and q are positive integers, $b_0 = (1, 1, 1, 1, 1, 2, 1, 1)^\top$, and the following theorem:

Theorem 1 (Rietz and Dempe). *Consider an instance $E = (L, l, b)$ of CSP with the following properties: $l_1 > l_2 > \ldots > l_{m-1} > 2l_m$ and $l_m \leq L/4$. Moreover, assume that this instance is sensitive, i.e. its optimal function value increases if b_m is increased by 1. Then, there are integers p and q such that instance $E' = E \oplus E_0(p, q)$ has gap $\Delta(E') = 1 + \Delta(E)$.*

Here $\oplus$ means a composition of instances. Let $E_1 = (L_1, l_1, b_1)$ and $E_2 = (L_2, l_2, b_2)$ denote two instances of CSP having n_1 and n_2 pieces respectively and with $L_1 = L_2$. The composed instance $E := E_1 \oplus E_2$ of CSP consists of the task of cutting all the $n_1 + n_2$ pieces of lengths from the both vectors l_1 and l_2 and with demands according to both vectors b_1 and b_2. In case when L_1 and L_2 are different, they can be multiplied by one common multiplier (together with piece lengths) to adjust the stock material lengths of both instances. For example, the instances $(2, (1)^\top, (1)^\top)$ and $(5, (2)^\top, (2)^\top)$ can be composed into the new instance $(2, (1)^\top, (1)^\top) \oplus (5, (2)^\top, (2)^\top) = (10, (5, 4)^\top, (1, 2)^\top)$.

Note that $b_m = 0$ is possible in Theorem 1, this means that the maximal possible trimloss in a cutting pattern used in an optimal solution is smaller than half of the length of the shortest piece.

Searching for sensitive instances with properties described in Theorem 1 is a very difficult task. An example of a suitable instance mentioned by Rietz and Dempe in their paper is the following:

$$E_{ST'} = (132, (44, 33, 12)^\top, (2, 3, 5)^\top).$$

Indeed, this instance is sensitive, because its optimal function value $z(E_{ST'}) = 2$ increases to 3 when we insert an additional piece of length 12. Also, $l_1 > l_2 > 2l_3$ and $l_3 < L/4$. $\Delta(E_{ST'}) = 17/132$, so by Theorem 1 there are integers p and q such that $\Delta(E_0(p, q) \oplus E_{ST'}) = 149/132 \approx 1.128787$. Namely, the instance $E_1 = E_0(p, q) \oplus E_{ST'}$ for $p = 74$ and $q = 669$ is the following:

$$E_1 = (924, (764, 762, 758, 308, 231, 84, 83, 81, 80, 78)^\top, (1, 1, 1, 2, 3, 6, 1, 2, 1, 1)^\top).$$

3   Enumeration Algorithm

Consider an instance $E = (L, l, b)$. If L and l are fixed, then the matrix of patterns $A(P^f(L, l))$ is fixed too. We will consider vector b as a vector of variables. Setting $l = (L - l_m, L - l_m - 1, \ldots, 2l_m + 2, 2l_m + 1, l_m)$, where $l_m \leq L/4$, we ensure that the most of required properties of Theorem 1 are satisfied, and now we have to ensure that E is sensitive.

We will enumerate all sensitive instances with a fixed objective function value. Namely, let $S_k(L, l)$ be the set of all patterns b such that $z((L, l, b)) = k$ and b corresponds to a sensitive instance (L, l, b).

Consider the set of *inextensible* feasible patterns $P_*^f(E) = \{a \in \mathbb{Z}_+^m \mid a^\top l \leq L \wedge a^\top l + l_1 > L\}$. Obviously, $S_0(L, l) = \{\mathbf{0}\}$, and $S_1(L, l) = P_*^f(L, l)$. Now we will build the set $S_{i+1}(L, l)$ from $S_i(L, l)$ by adding vectors from $P^f_*(E)$ and considering only those patterns which lead to sensitive instances.

To transform the set $S_i(L, l)$ into the set $S_{i+1}(L, l)$ we need a data structure called a "map", which contains a set of pairs <key, value> (all keys are pairwise distinct) and allows us to make the following operations: insert a pair, find a value by a key (or determine that there is no pair with this key), modify a value by a key and return the list of all pairs. The algorithm is the following:

```
1 create an empty map A
2 for all s ∈ S_i(L, l)
3     for all a ∈ P^f_*(L, l)
4         x ← (s_1 + a_1, ..., s_{m-1} + a_{m-1})
5         y ← s_m + a_m
6         if A has no key x, then
7             insert into A the pair (x, y)
8         else A[x] ← max(A[x], y)
9 S_{i+1}(L, l) = {(x_1, ..., x_{m-1}, y) | (x, y) ∈ A}
```

To find a sensitive instance with maximum gap with fixed L, l and k we generate $S_k(L, l)$ and then simply calculate $\Delta(E)$ over all $E = (L, l, s)$, $s \in S_k(L, l)$.

4 Results

We implemented our algorithm as a C++ program using CPLEX 12.7. The program was run on an Intel Core i7-5820K 4.2 GHz machine with 6 cores and 32 Gb RAM.

Results for the runs where $l = (L - l_m, L - l_m - 1, ..., 2l_m + 1, l_m)$ are presented in Table 1 and Table 2. Maximum gaps greater than 0.1 are marked in bold, and the maximal gap in every column is underlined.

Several sensitive instances with large gaps found during the search are presented in Table 3. Here E_1, E_2 and E_3 correspond to some maximum gaps presented in Table 1 and Table 2. For instance E_4 we continued the search up to $L = 250$ setting $l = (\lfloor L/2 \rfloor, \lfloor L/2 \rfloor - 1, ..., 2l_m + 1, l_m)$. The gap 0.1875 is the maximal over all considered instances with $k \leq 4$.

The instance E_5 is built from E_4 and a non-IRUP instance

$$E_T(t) = (3t, (t + 4, t + 3, t, t - 2, t - 6)^\top, (1, 1, 2, 1, 1)^\top)$$

for some integer t. E_6 is a combination of E_4 and some pieces from two copies of $E_T(t)$ with different values of t.

Using Theorem 1, we constructed a series of non-IRUP instances $E'_1, ..., E'_6$ from the sensitive instances $E_1, ..., E_6$. They are presented in Table 4. In Table 5 we compare our instances with the previously known ones considering the number of piece types m.

Table 1. Maximum gaps for sensitive instances with fixed L, l_m and $k \leq 4$

$L \backslash l_m$	2	3	4	5	6	7
8	0.000000					
9	0.000000					
10	0.100000					
11	0.000000					
12	0.083333	0.000000				
13	0.000000	0.000000				
14	0.071429	0.000000				
15	0.083333	0.100000				
16	0.062500	0.100000	0.000000			
17	0.058824	0.083333	0.000000			
18	0.100000	0.083333	0.000000			
19	0.075000	0.083333	0.000000			
20	0.068182	0.071429	0.100000	0.000000		
21	0.066667	**0.119048**	0.100000	0.000000		
22	0.078947	0.100000	0.100000	0.000000		
23	0.066667	0.093750	0.083333	0.000000		
24	0.083333	**0.129630**	0.083333	0.000000	0.000000	
25	0.060606	0.100000	0.083333	0.100000	0.000000	
26	0.078125	0.083333	0.083333	0.100000	0.000000	
27	0.069444	**0.111111**	**0.119048**	0.100000	0.000000	
28	0.071429	0.100000	**0.119048**	0.100000	0.000000	0.000000
29	0.064815	0.087500	**0.113636**	0.083333	0.000000	0.000000
30	0.076389	**0.125000**	**0.145833**	0.083333	0.100000	0.000000
31		0.097222	**0.129630**	0.083333	0.100000	0.000000
32		0.100000	**0.127907**	0.083333	0.100000	0.000000
33		**0.102564**	**0.106061**	**0.119048**	0.100000	0.000000
34		0.096154	**0.129630**	**0.119048**	0.100000	0.000000
35		0.092857	**0.111111**	**0.125000**	0.083333	0.100000
36		**0.106061**	**0.133333**	**0.138889**	0.083333	0.100000
37			0.105263	**0.145833**	0.083333	0.100000
38			0.125000	**0.131579**	0.083333	0.100000
39			0.128788	**0.153333**	**0.119048**	0.100000
40			0.130435	**0.138889**	**0.119048**	0.100000
41			0.105263	**0.136364**	**0.125000**	0.083333
42			0.125000	**0.136364**	**0.142857**	0.083333
43				**0.133333**	**0.138889**	0.083333
44				**0.136364**	**0.156250**	0.083333
45				0.130952	**0.161458**	0.119048
46				**0.133333**	0.149123	0.119048
47				**0.136364**	0.144068	0.125000
48				**0.136364**	0.156863	0.142857
49					0.136364	0.142857
50					0.148148	0.140000
51					0.141026	**0.166667**

Table 2. Maximum gaps for sensitive instances with fixed L, l_m and $k \leq 4$

L	$l_m = 7$	L	$l_m = 8$	L	$l_m = 9$	L	$l_m = 10$
45	**0.119048**	51	**0.119048**	57	**0.119048**	63	**0.119048**
46	**0.119048**	52	**0.119048**	58	**0.119048**	64	**0.119048**
47	**0.125000**	53	**0.125000**	59	**0.125000**	65	**0.125000**
48	**0.142857**	54	**0.142857**	60	**0.142857**	66	**0.142857**
49	**0.142857**	55	**0.142857**	61	**0.142857**	67	**0.142857**
50	**0.140000**	56	**0.142857**	62	**0.142857**	68	**0.142857**
51	<u>**0.166667**</u>	57	**0.149123**	63	**0.150794**	69	**0.150794**
52	**0.150000**	58	<u>**0.171875**</u>	64	**0.149123**	70	**0.150794**
53	**0.160000**	59	**0.167969**	65	<u>**0.175000**</u>	71	**0.149123**
54	**0.154762**	60	**0.166667**	66	**0.166667**	72	<u>**0.177083**</u>
55	**0.151515**	61	**0.153333**	67	**0.171875**	73	**0.171875**
56	**0.145833**	62	**0.161765**	68	**0.160000**	74	**0.175000**
57	<u>**0.166667**</u>	63	**0.166667**	69	**0.172043**	75	**0.166667**
58	**0.156863**	64	**0.161765**	70	**0.166667**	76	**0.171875**

Table 3. Sensitive instances with required properties and large gaps

E_i	$z(E_i)$	$\Delta(E_i)$
$E_1 = (30, (14, 13, 10, 4)^\top, (1, 1, 2, 2)^\top)$	2	7/48 0.145833
$E_2 = (51, (23, 22, 19, 17, 16, 7)^\top, (2, 1, 1, 1, 1, 3)^\top)$	3	1/6 0.166667
$E_3 = (72, (32, 31, 28, 25, 24, 22, 10)^\top, (2, 1, 1, 1, 2, 2, 3)^\top)$	4	17/96 0.177083
$E_4 = (183, (81, 79, 65, 64, 61, 59, 55, 25)^\top, (1, 1, 2, 1, 2, 1, 1, 4)^\top)$	4	3/16 0.187500
$E_5 = (1281, (567, 553, 455, 448, 430, 427, 425, 413, 385, 175)^\top, $ $(1, 1, 2, 1, 2, 1, 1, 2, 1, 4)^\top)$	5	19/96 0.197917
$E_6 = (1281, (567, 553, 455, 448, 431, 430, 427, 425, 421, 413, 385, 175)^\top, $ $(1, 1, 2, 1, 2, 1, 2, 1, 1, 2, 1, 4)^\top)$	6	13/64 0.203125

Table 4. Non-IRUP instances with large gaps

$E_i' = E_0(p, q) \oplus E_i$	$z(E_i')$	$\Delta(E_i')$
$E_1' = (300, (228, 226, 222, 140, 130, 100, 40, 39, 37, 36, 34)^\top, $ $(1, 1, 1, 1, 1, 2, 3, 1, 2, 1, 1)^\top)$	6	55/48 1.145833
$E_2' = (510, (378, 376, 372, 230, 220, 190, 170, 160, 70, 69, 67, 66, 64)^\top, $ $(1, 1, 1, 2, 1, 1, 1, 1, 4, 1, 2, 1, 1)^\top)$	7	7/6 1.166667
$E_3' = (720, (528, 526, 522, 320, 310, 280, 250, 240, 220, 100, $ $99, 97, 96, 94)^\top, (1, 1, 1, 2, 1, 1, 1, 2, 2, 4, 1, 2, 1, 1)^\top)$	8	113/96 1.177083
$E_4' = (1830, (1338, 1336, 1332, 810, 790, 650, 640, 610, 590, 550, 250, $ $249, 247, 246, 244)^\top, (1, 1, 1, 1, 1, 2, 1, 2, 1, 1, 5, 1, 2, 1, 1)^\top)$	8	19/16 1.187500
$E_5' = (12810, (9318, 9316, 9312, 5670, 5530, 4550, 4480, 4300, $ $4270, 4250, 4130, 3850, 1750, 1749, 1747, 1746, 1744)^\top, $ $(1, 1, 1, 1, 1, 2, 1, 2, 1, 1, 2, 1, 5, 1, 2, 1, 1)^\top)$	9	115/96 1.197917
$E_6' = (12810, (9318, 9316, 9312, 5670, 5530, 4550, 4480, 4310, 4300, $ $4270, 4250, 4210, 4130, 3850, 1750, 1749, 1747, 1746, 1744)^\top, $ $(1, 1, 1, 1, 1, 2, 1, 2, 1, 2, 1, 1, 2, 1, 5, 1, 2, 1, 1)^\top)$	10	77/64 1.203125

Table 5. The number of piece types in old and new non-IRUP instances

m	Old		New	
3	137/132	1.0378787		
4				
5	16/15	1.0666667		
6	38/35	1.0857143		
7	11/10	1.1000000		
8	10/9	1.1111111		
9				
10	149/132	1.1287879		
11			55/48	1.1458333
12				
13			7/6	1.1666667
14	51/44	1.1590909	113/96	1.1770833
15			19/16	1.1875000
16	7/6	1.1666667		
17			115/96	1.1979167
18	13/11	1.1818182		
19			77/64	1.2031250
$\vdots$				
28	6/5	1.2000000		

5   Conclusion

We have combined the construction of Rietz and Dempe and our enumeration
algorithm for searching for sensitive instances. We have found a lot of sensitive instances with large gaps. This allowed us to construct a lot of non-IRUP
instances with gap, say, greater than 1.17. We also constructed a non-IRUP
instance with gap 1.203125 which is greater than the previously known world
record 1.2. Also the non-IRUP instances with large gaps that we found are
smaller than the previously known ones.

Producing instances with large gaps using our search method requires a lot
of computational resources, so we do not expect that it will handle the MIRUP
conjecture directly. But the instances we found may provide the hints about
improved constructions. In the future research we are going to improve our technique of combining instances (using which we produced E_5 and E_6) and construct
new instances with much larger gaps.

Acknowledgements. The authors would like to thank the anonymous referees for
their valuable remarks.

References

1. Baum, S., Trotter Jr., L.: Integer rounding for polymatroid and branching optimization problems. SIAM J. Algebraic Discrete Methods **2**(4), 416–425 (1981)
2. Delorme, M., Iori, M., Martello, S.: Bin packing and cutting stock problems: mathematical models and exact algorithms. Eur. J. Oper. Res. **255**(1), 1–20 (2016)
3. Dyckhoff, H., Kruse, H.J., Abel, D., Gal, T.: Trim loss and related problems. Omega **13**(1), 59–72 (1985)
4. Fieldhouse, M.: The duality gap in trim problems. SICUP Bull. **5**(4), 4–5 (1990)
5. Gilmore, P., Gomory, R.: A linear programming approach to the cutting-stock problem. Oper. Res. **9**(6), 849–859 (1961)
6. Kantorovich, L.V.: Mathematical methods of organizing and planning production. Manage. Sci. **6**(4), 366–422 (1960)
7. Kartak, V.M., Ripatti, A.V.: Large proper gaps in bin packing and dual bin packing problems. J. Global Optim. **74**(3), 467–476 (2018). https://doi.org/10.1007/s10898-018-0696-0
8. Kartak, V.M., Ripatti, A.V., Scheithauer, G., Kurz, S.: Minimal proper non-IRUP instances of the one-dimensional cutting stock problem. Discrete Appl. Math. **187**(Complete), 120–129 (2015)
9. Marcotte, O.: An instance of the cutting stock problem for which the rounding property does not hold. Oper. Res. Lett. **4**(5), 239–243 (1986)
10. Rietz, J.: Untersuchungen zu MIRUP für Vektorpackprobleme. Ph.D. thesis, Technischen Universität Bergakademie Freiberg (2003)
11. Rietz, J., Dempe, S.: Large gaps in one-dimensional cutting stock problems. Discrete Appl. Math. **156**(10), 1929–1935 (2008)
12. Rietz, J., Scheithauer, G., Terno, J.: Families of non-IRUP instances of the one-dimensional cutting stock problem. Discrete Appl. Math. **121**(1), 229–245 (2002)
13. Rietz, J., Scheithauer, G., Terno, J.: Tighter bounds for the gap and non-IRUP constructions in the one-dimensional cutting stock problem. Optimization **51**(6), 927–963 (2002)
14. Ripatti, A.V., Kartak, V.M.: Bounds for non-IRUP instances of cutting stock problem with minimal capacity. In: Bykadorov, I., Strusevich, V., Tchemisova, T. (eds.) MOTOR 2019. CCIS, vol. 1090, pp. 79–85. Springer, Cham (2019). https://doi.org/10.1007/978-3-030-33394-2_7
15. Ripatti, A.V., Kartak, V.M.: Constructing an instance of the cutting stock problem of minimum size which does not possess the integer round-up property. J. Appl. Ind. Math. **14**(1), 196–204 (2020). https://doi.org/10.1134/S1990478920010184
16. Scheithauer, G., Terno, J.: About the gap between the optimal values of the integer and continuous relaxation one-dimensional cutting stock problem. In: Gaul, W., Bachem, A., Habenicht, W., Runge, W., Stahl, W.W. (eds.) Operations Research Proceedings. Operations Research Proceedings 1991, vol. 1991. Springer, Heidelberg (1992). https://doi.org/10.1007/978-3-642-46773-8_111
17. Scheithauer, G., Terno, J.: The modified integer round-up property of the one-dimensional cutting stock problem. Eur. J. Oper. Res. **84**(3), 562–571 (1995)
18. Sweeney, P.E., Paternoster, E.R.: Cutting and packing problems: a categorized, application-oriented research bibliography. J. Oper. Res. Soc. **43**(7), 691–706 (1992)

Some Estimates on the Discretization of Geometric Center-Based Problems in High Dimensions

Vladimir Shenmaier$^{(\boxtimes)}$ (ID)

Sobolev Institute of Mathematics, Novosibirsk, Russia
`shenmaier@mail.ru`

Abstract. We consider the following concept. A set C in multidimensional real space is said to be a $(1+\varepsilon)$-collection for a set X if C contains a $(1+\varepsilon)$-approximation of every point of space with respect to the Euclidean distances to all the elements of X. A $(1+\varepsilon)$-collection allows to find approximate solutions of any geometric center-based problem where it is required to choose points in space (centers) minimizing a continuity-type function which depends on the distances from the input points to the centers. In fact, it gives a universal reduction of such problems to their discrete versions where all the centers must belong to a prescribed set of points. As was shown recently, for every fixed $\varepsilon > 0$ and any finite set in high-dimensional space, there exists a $(1+\varepsilon)$-collection which consists of a polynomial number of points and can be constructed by a polynomial-time algorithm. We slightly improve this algorithm and supplement it with a lower bound for the cardinality of $(1+\varepsilon)$-collections in the worst case. Also, we show the non-existence of polynomial $(1+\varepsilon)$-collections for some sets of points in the case of ℓ_∞ distances.

Keywords: Geometric clustering · Continuous Facility Location · Approximate centers · Euclidean space · High dimensions

1 Introduction

We prove some geometric properties of finite sets of points in high-dimensional real space which may be useful for developing approximation algorithms for data analysis and optimization problems.

Our interest is the following concept. A set C in space $\mathbb{R}^d$ is said to be a $(1 + \varepsilon)$-*collection for a set* $X \subseteq \mathbb{R}^d$ if, for every point $p \in \mathbb{R}^d$, the set C contains a point p' such that the Euclidean distance from p' to each element of X is at most $1 + \varepsilon$ times of that from p. One may ask: what is the minimum cardinality of a $(1 + \varepsilon)$-collection for a given set of n points in any-dimensional space in the worst case? In this paper, we show that, for any fixed $\varepsilon \in (0, 1]$, this cardinality is at most $\Theta\big(n^{\lceil \frac{1}{\varepsilon} \log \frac{2}{\varepsilon} \rceil}\big)$ and is at least $\Theta\big(n^{\lfloor \frac{1}{16\varepsilon}+1 \rfloor}\big)$.

The concept of a $(1 + \varepsilon)$-collection is closely related to the question of the discretization of geometric center-based problems. In these problems, we are

© Springer Nature Switzerland AG 2020
Y. Kochetov et al. (Eds.): MOTOR 2020, CCIS 1275, pp. 88–101, 2020.
https://doi.org/10.1007/978-3-030-58657-7_10

given an n-element set of points in $\mathbb{R}^d$ and the goal is to choose a given number of new points (centers) in space to minimize some objective function which depends on the distances between the input points and the chosen centers. One of widespread ways for finding approximate solutions of such problems is generating a set of candidate centers which contains approximations of the optimal points with respect to the given objective function [5,6,8,10,12,13,15,16,19]. A $(1+\varepsilon)$-collection is a set of points which contains approximations of all the points of space with respect to the distances to the input points. Thereby, it contains approximate centers at once for all the objective functions continuously depending on the distances between the input points and the centers. In fact, a $(1+\varepsilon)$-collection gives a reduction of the original instance of a geometric center-based problem to its discrete version in which all the centers are restricted to be selected from a prescribed finite set of points.

The only thing we require for the objective function is the natural "continuity-type" property: small relative changes of the distances between the input points and the centers must give a bounded relative change of the objective function value. It may be formalized as follows:

Definition. *Let $\|.\|$ denote the Euclidean norm and f be a non-negative function defined for each finite set $X \subset \mathbb{R}^d$ and every tuple $c_1, \ldots, c_k \in \mathbb{R}^d$. Then f is called a continuity-type function if there exists a mapping $\mu : [1, \infty) \to [1, \infty)$ such that, for each $\varepsilon > 0$ and any tuples $c_i, c_i' \in \mathbb{R}^d$ satisfying the inequalities $\|x - c_i'\| \leq (1 + \varepsilon)\|x - c_i\|$, $x \in X$, $i = 1, \ldots, k$, we have*

$$f(X; c_1', \ldots, c_k') \leq \mu(1 + \varepsilon) f(X; c_1, \ldots, c_k).$$

For example, the objective functions of the Euclidean k-Median [4,5,7], Euclidean k-Center [2,5,14], Continuous Facility Location [15], and Smallest m-Enclosing Ball [1,17–19] problems are continuity-type with $\mu(1 + \varepsilon) = 1 + \varepsilon$. Those of the k-Means [6,12,13] and m-Variance [3,9,16] problems are continuity-type with $\mu(1 + \varepsilon) = (1 + \varepsilon)^2$. Note that continuity-type functions are not required to be continuous but, if we want to find approximate solutions close to optimal, the preferred case is that when $\mu(1 + \varepsilon) \to 1$ as $\varepsilon \to 0$.

On the practical side, using such a universal instrument as $(1+\varepsilon)$-collections may be actual in the cases when the known fast methods of generating candidate centers for geometric center-based problems are not applicable or do not give desired approximation guarantees:

- The objective function has a more complicated dependence on the distances between the input points and the centers than the sum or the maximum of these distances or of their squares. In general, it may be an arbitrary continuity-type function of the point-to-center distance matrix.

- The centers we find are required to cover not all the input points but a given number of them, which is typical for the clustering and facility location problems with outliers or penalties. It breaks the standard techniques based on random sampling, e.g., if the total number of input points we need to cover allows to be arbitrarily small and, therefore, any constant number of random samples may "miss" good clusters.

- The input points are served by the desired centers in a more complicated manner than in usual clustering models, e.g., if the service areas of the centers are allowed to overlap by a given number of input points, each input point is required to be served by a given number of centers, each center has its own capacity, service radius, and unit distance costs which depend both on the center and the demand points it serves, etc.

- The problem has one or multiple objectives, possibly not specified explicitly, and an oracle is given which, for any two tuples of centers, answers which one is better. In this case, if the objectives are known to be continuity-type, then enumerating all the tuples of elements of a $(1 + \varepsilon)$-collection for the input set provides a $\mu(1 + \varepsilon)$-approximate solution of the problem.

Besides the considerations mentioned above, the concept of a $(1+\varepsilon)$-collection seems to be theoretically interesting. It is formulated independently of any optimization problems and can be easily extended to any metrics. The cardinality of $(1 + \varepsilon)$-collections describes how the given metrics is convenient for the discretization of space in terms of the distances to given n points.

Related Work. The concept of an α-collection was introduced in [20], where it was suggested an algorithm which, given an n-element set X in any-dimensional real space and any $\varepsilon \in (0, 1]$, constructs an $N(n, \varepsilon)$-element $(1 + \varepsilon)$-collection for the set X with $N(n, \varepsilon) = \mathcal{O}\left(\left(\frac{n}{\varepsilon}\right)^{\frac{2}{\varepsilon}} \log \frac{2}{\varepsilon}\right)$ (here and everywhere, "log" means the logarithm to the base 2). As a corollary, it was described a reduction of the general geometric center-based problem to its discrete version with the same continuity-type objective function:

Geometric Center-Based Problem. Given an n-element set X in space $\mathbb{R}^d$ and an integer $k \geq 1$. Find a tuple $c_1, \ldots, c_k \in \mathbb{R}^d$ to minimize the value of $f(X; c_1, \ldots, c_k)$.

Discrete Center-Based Problem. Given an n-element set $X \subset \mathbb{R}^d$, an ℓ-element set $Y \subset \mathbb{R}^d$, and an integer $k \geq 1$. Find a tuple $c_1, \ldots, c_k \in Y$ to minimize the value of $f(X; c_1, \ldots, c_k)$.

Fact 1. [20] *Suppose that $\beta \geq 1$ and there exists an algorithm which computes a β-approximate solution of the Discrete Center-Based problem with a continuity-type function f in time $T(n, \ell, k, d)$. Then there exists an algorithm which, given $\varepsilon \in (0, 1]$, computes a $\beta\mu(1 + \varepsilon)$-approximate solution of the Geometric Center-Based problem with the function f in time $\mathcal{O}\left(N(n, \varepsilon) d\right) + T\left(n, N(n, \varepsilon), k, d\right)$.*

In particular, it follows the constant-factor approximability of the Geometric k-Median problem, in which we need to choose k centers in high-dimensional Euclidean space to minimize the total distance from the input points to nearest centers. Another known application of $(1 + \varepsilon)$-collections is approximation algorithms for the following k-clustering problems:

Problem 1. Given points $x_1, \ldots, x_n$ in space $\mathbb{R}^d$, integers $k, m \geq 1$, unit distance costs $f_{ij} \geq 0$, and powers $\alpha_{ij} \in [0, \alpha]$, $i = 1, \ldots, k$, $j = 1, \ldots, n$, where α is some parameter. Find disjoint subsets $S_1, \ldots, S_k \subseteq \{1, \ldots, n\}$ with the property $|S_1 \cup \ldots \cup S_k| = m$ and select a tuple $c_1, \ldots, c_k \in \mathbb{R}^d$ to minimize the value of

$$\sum_{i=1}^{k} \sum_{j \in S_i} f_{ij} \|x_j - c_i\|^{\alpha_{ij}}.$$

Problem 2. The same as in Problem 1 except that, instead of the condition $|S_1 \cup \ldots \cup S_k| = m$, each subset S_i is required to have its own given cardinality m_i, $i = 1, \ldots, k$.

Fact 2. [20] *If the values of k and α are fixed, Problems 1 and 2 admit polynomial-time approximation schemes PTAS computing $(1 + \varepsilon)^\alpha$-approximate solutions of these problems in time $\mathcal{O}\big(N(n, \varepsilon)^k n k d\big)$ and $\mathcal{O}\big(N(n, \varepsilon)^k (n^3 + n k d)\big)$ respectively.*

Our Contributions. We slightly improve the upper bound for the minimum cardinality of $(1 + \varepsilon)$-collections for a set of n points and obtain the first lower bound for this cardinality in the worst case, i.e., the value

$$C(n, \varepsilon) = \max_{|X|=n} \min \big\{ |C| \,\big|\, C \text{ is a } (1 + \varepsilon)\text{-collection for } X \big\}$$

is estimated. We prove that, in high-dimensional Euclidean space, this value is at most $\Theta\big((\frac{n}{\varepsilon} \log \frac{2}{\varepsilon})^{\lceil \frac{1}{\varepsilon} \log \frac{2}{\varepsilon} \rceil}\big)$ and is at least $n^{\lfloor \frac{1}{16\varepsilon}+1 \rfloor} \varepsilon^{\lfloor \frac{1}{16\varepsilon} \rfloor}$ for each $\varepsilon \in (0, 1]$, thereby, it lies between $n^{\Theta(\frac{1}{\varepsilon} \log \frac{2}{\varepsilon})}$ and $n^{\Theta(\frac{1}{\varepsilon})}$ if ε is fixed.

Both bounds are obtained constructively. To justify the upper bound, we describe an algorithm which computes a $(1 + \varepsilon)$-collection of the required cardinality for any given n-element set in time proportional to the cardinality of the output. The suggested algorithm is a modification of that from [20] and is based on the "affine hull" technique developed in [16]. The main idea of the algorithm is that we approximate an arbitrary point of space by grids in the affine hulls of small subsets of the input points. The differences of the modified algorithm from that from [20] are an improved construction of the approximating grids and an optimized set of values for the parameters of the algorithm.

To get the lower bound, we present an n-element set of points such that every $(1+\varepsilon)$-collection for this set contains at least the declared number of points. Note that both upper and lower bounds are tight at least for $\varepsilon = 1$.

In contrast to the obtained estimates for Euclidean $(1 + \varepsilon)$-collections, we show that the value of $C(n, \varepsilon)$ is not polynomial in high-dimensional ℓ_∞ space. To state it, we construct a set of n points such that every $(1 + \varepsilon)$-collection for this set in the ℓ_∞ metrics contains at least $2^{\lfloor n/2 \rfloor}$ elements if $\varepsilon \in (0, 1)$.

2 An Upper Bound

In this section, we prove that, for each fixed ε and every set of n points in any-dimensional Euclidean space, there exists and can be constructed in polynomial time a $(1 + \varepsilon)$-collection which consists of $\mathcal{O}\big(n^{\lceil \frac{1}{\varepsilon} \log \frac{2}{\varepsilon} \rceil}\big)$ elements.

Definition. *Given points $p, p' \in \mathbb{R}^d$, a set $X \subseteq \mathbb{R}^d$, and a real number $\alpha \geq 1$, we say that p' is an α-approximation of p with respect to X if $\|x - p'\| \leq \alpha\|x - p\|$ for all $x \in X$.*

Definition. *Given sets $X, C \subseteq \mathbb{R}^d$ and a real number $\alpha \geq 1$, we say that C is an α-collection for X if C contains α-approximations of all the points of space with respect to X.*

Example. Every finite set $X \subset \mathbb{R}^d$ is a 2-collection for itself. Indeed, let p be any point in $\mathbb{R}^d$ and p' be a point of X nearest to p. Then, by the triangle inequality and the choice of p', we have $\|x - p'\| \leq \|x - p\| + \|p - p'\| \leq 2\|x - p\|$ for all $x \in X$. So p' is a 2-approximation of p with respect to X.

Theorem 1. *For any n-element set $X \subset \mathbb{R}^d$ and each $\varepsilon \in (0, 1]$, there exists a $(1+\varepsilon)$-collection for X which consists of $N(n, \varepsilon) = \mathcal{O}\big((\frac{n}{\varepsilon} \log \frac{2}{\varepsilon})^{\lceil \frac{1}{\varepsilon} \log \frac{2}{\varepsilon} \rceil}\big)$ elements and can be constructed in time $\mathcal{O}\big(N(n, \varepsilon)\, d\big)$.*

Proof. As mentioned above, every finite set of points is a 2-collection for itself. Therefore, the theorem holds if $\varepsilon = 1$. Further, we will assume that $\varepsilon < 1$.

First, describe some geometric constructions underlying both the algorithm from [20] and its modification we suggest in this paper for computing a required $(1 + \varepsilon)$-collection. Suppose that $\delta \in (0, \varepsilon)$ and O is an arbitrary point in $\mathbb{R}^d$. Define the following sequences $(x_t)_{t \geq 1}$ and $(y_t)_{t \geq 1}$ depending on O and δ:

set $x_1 = y_1$ to be a point of X nearest to O;

for $t \geq 2$, consider the ball B_t consisting of the points $x \in \mathbb{R}^d$ such that $\|x - y_{t-1}\| \geq (1 + \delta)\|x - O\|$;

if the set $X \cap B_t$ is empty, finish the sequences (x_t) and (y_t); otherwise, define x_t as any point from $X \cap B_t$ and let y_t be the orthogonal projection of the point O into the affine hull of the set $\{x_1, \ldots, x_t\}$.

Denote by T the length of the constructed sequences (x_t) and (y_t).

Lemma 1. [20] *If $2 \leq t \leq T$, then the vector $y_t - y_{t-1}$ is orthogonal to the affine hull of the set $\{x_1, \ldots, x_{t-1}\}$.*

Lemma 2. [20] *If $2 \leq t \leq T$, then the vectors $x_2 - x_1, \ldots, x_t - x_1$ are linearly independent, $y_t \neq y_{t-1}$, and the vectors $e_{t-1} = \dfrac{y_t - y_{t-1}}{\|y_t - y_{t-1}\|}$ can be computed by the Gram-Schmidt process for orthonormalising the set $x_2 - x_1, \ldots, x_t - x_1$.*

Lemma 3. [20] *If $2 \leq t \leq T$ and $r_t = \|y_t - O\|$, then $r_t \leq \left(\dfrac{1}{1+\delta}\right)^{t-1} r_1$.*

Proposition 1. [20] *Let $t = \min\{T, T(\delta)\}$, where $T(\delta) = \left\lceil \dfrac{\log \frac{1}{\delta}}{\log(1 + \delta)} \right\rceil + 1$. Then the point y_t is a $(1 + \delta)$-approximation of the point O with respect to X.*

Lemma 4. [20] *If $T \geq 2$ and $dist_1 = \dfrac{\|x_2 - x_1\|}{1 + \delta}$, then $dist_1 \delta \leq r_1 \leq dist_1$.*

Lemma 5. [20] *If* $2 \leq t \leq T$, *then* $\|y_t - y_{t-1}\| \leq \|y_{t-1} - O\| \leq \|y_1 - O\|$.

Proposition 2. *If* $1 \leq t \leq T$, *then the point* y_t *belongs to the hyperrectangle*

$$Box_t(r_1), \text{ where } Box_t(r) = \left\{ y_1 + \sum_{i=1}^{t-1} \alpha_i e_i \,\middle|\, 0 \leq \alpha_i \leq \frac{r}{(1+\delta)^{i-1}} \right\}.$$

Proof. This statement directly follows from Lemmas 1–3 and 5. □

Idea of the Algorithm. Propositions 1, 2 and Lemmas 2, 4 give an idea how to approximate any unknown point $O \in \mathbb{R}^d$. We can enumerate all the tuples $x_1, \ldots, x_t \in X$ for all $t \leq T(\delta)$ and, for each of these tuples, construct the vectors $e_1, \ldots, e_{t-1}$ defined in Lemma 2. Next, to approximate the point O, we approximate the point y_t. For this aim, we construct a set R_1 containing approximations of the unknown value of r_1 by using the bounds established in Lemma 4. Then, based on Proposition 2, we consider grids in the hyperrectangles $Box_t(r)$, $r \in R_1$. Note that no information about the point O is used, so the constructed set will contain approximations of all the points of space.

The described idea can be implemented in the following form:

Algorithm $\mathcal{A}$.

Select real parameters $\delta \in (0, \varepsilon)$, $h \in (0, \delta)$, and an integer parameter $I \geq 1$.

Step 1. Include to the output set all the elements of X.

Step 2. Enumerate all the tuples $x_1, \ldots, x_t \in X$, $2 \leq t \leq T(\delta)$, such that the vectors $x_2 - x_1, \ldots, x_t - x_1$ are linearly independent and, for each of these tuples, perform Steps 3–5.

Step 3. Execute the Gram-Schmidt process for orthonormalising the set of vectors $x_2 - x_1, \ldots, x_t - x_1$ and obtain the orthonormal vectors $e_1, \ldots, e_{t-1}$.

Step 4. Construct the set R_1 of the numbers $\dfrac{\|x_2 - x_1\|}{1+\delta} \delta^{i/I}$, $i = 0, \ldots, I$.

Step 5. For each value $r \in R_1$, include to the output set the nodes of the grid

$$Grid(x_1, \ldots, x_t; r, h) = \left\{ x_1 + \sum_{i=1}^{t-1} e_i rh(0.5 + \alpha_i) \,\middle|\, \alpha_i = 0, \ldots, \left\lfloor \frac{1}{h(1+\delta)^{i-1}} \right\rfloor \right\}.$$

To justify this algorithm and to estimate the size of its output, we need the following statements.

Lemma 6. *If* $2 \leq t \leq T(\delta)$, *then the union of the* $(t-1)$-*dimensional hyper-cubes centered at the nodes of* $Grid(x_1, \ldots, x_t; r, h)$ *with side* rh *contains the hiperrectangle* $Box_t(r)$ *and is contained in the hiperrectangle* $Box_t(r + rh/\delta)$.

Proof. Given $i = 1, \ldots, t-1$, denote by b_i and g_i the maximum values of the i-th coordinate of the elements of $Box_t(r)$ and $Grid(x_1, \ldots, x_t; r, h)$ respectively in the coordinate system centered at the point x_1 with basis $e_1, \ldots, e_{t-1}$:

$$b_i = \frac{r}{(1+\delta)^{i-1}} \text{ and } g_i = 0.5rh + rh \left\lfloor \frac{1}{h(1+\delta)^{i-1}} \right\rfloor.$$

Note that $g_i - 0.5rh \geq b_i - rh$, so $g_i + 0.5rh \geq b_i$. It follows the first statement.

Next, note that $(1 + \delta)^{i-1} \leq (1 + \delta)^{t-2} < 1/\delta$ by the choice of t and the definition of $T(\delta)$. On the other hand, we have $g_i - 0.5rh \leq b_i$. So

$$g_i + 0.5rh \leq b_i + rh = \frac{r + rh(1 + \delta)^{i-1}}{(1 + \delta)^{i-1}} < \frac{r + rh/\delta}{(1 + \delta)^{i-1}},$$

which follows the second statement. Lemma 6 is proved. □

Proposition 3. *The output set of Algorithm $\mathcal{A}$ is a $(1 + \delta + \delta_+)$-collection for the set X, where $\delta_+ = \dfrac{h\sqrt{T(\delta) - 1}}{2\delta^{1/I}}$.*

Proof. Suppose that O is an arbitrary point in space $\mathbb{R}^d$ and consider the sequences $(x_t)_{t \geq 1}$ and $(y_t)_{t \geq 1}$ defined for this point. By Proposition 1, there exists a number $t \leq T(\delta)$ such that the point y_t is a $(1 + \delta)$-approximation of the point O with respect to X. If $t = 1$, then the required approximation is the point y_1, so it is included to the output set at Step 1. Suppose that $t \geq 2$.

In this case, by Lemma 2, the vectors $x_2 - x_1, \ldots, x_t - x_1$ are linearly independent, so the tuple $x_1, \ldots, x_t$ is listed at Step 2. Lemma 4 implies that there exists a number $r \in R_1$ such that $r \geq r_1 \geq r\delta^{1/I}$, where $r_1 = \|x_1 - O\|$. Let z be a node of $Grid(x_1, \ldots, x_t; r, h)$ nearest to y_t. By Proposition 2, the point y_t belongs to the hyperrectangle $Box_t(r_1) \subseteq Box_t(r)$. On the other hand, by Lemma 6, the hyperrectangle $Box_t(r)$ is covered by the union of the hypercubes centered at the nodes of $Grid(x_1, \ldots, x_t; r, h)$ with side rh. So the distance between z and y_t is at most $\dfrac{rh\sqrt{t - 1}}{2}$. Then, by the triangle inequality, we have

$$\|x - z\| \leq \|x - y_t\| + \|y_t - z\| \leq (1 + \delta)\|x - O\| + \frac{rh\sqrt{t - 1}}{2}$$

for any $x \in X$. But $r \leq r_1/\delta^{1/I} \leq \|x - O\|/\delta^{1/I}$ by the choice of x_1. Therefore, we have $\|x - z\| \leq \left(1 + \delta + \dfrac{h\sqrt{t - 1}}{2\delta^{1/I}}\right)\|x - O\|$. Proposition 3 is proved. □

Proposition 4. *The output set of Algorithm $\mathcal{A}$ consists of $\mathcal{O}(n^{T(\delta)}\ell(\delta, h, I))$ points, where $\ell(\delta, h, I) = (1/h + 1/\delta)^{T(\delta)-1}(1 + \delta)^{-(T(\delta)-1)(T(\delta)-2)/2}I$. The running time of the algorithm is $\mathcal{O}(n^{T(\delta)}\ell(\delta, h, I)d)$.*

Proof. The number of all the tuples $x_1, \ldots, x_t \in X$, where $t \leq T(\delta)$, is $\mathcal{O}(n^{T(\delta)})$. The set R_1 consists of $I + 1$ elements. Next, to estimate the cardinality of the set $Grid(x_1, \ldots, x_t; r, h)$, we can assume that $t = T(\delta)$ since this set grows with increasing t. Then, according to Lemma 6 and the fact that the volume of the hyperrectangle $Box_t(r + rh/\delta)$ is $(r + rh/\delta)^{t-1}(1 + \delta)^{-(t-1)(t-2)/2}$, each set $Grid(x_1, \ldots, x_t; r, h)$ contains at most

$$(1/h + 1/\delta)^{t-1}(1 + \delta)^{-(t-1)(t-2)/2} = \ell(\delta, h, I)/I$$

nodes. Thus, the total number of such nodes is $\mathcal{O}(n^{T(\delta)}\ell(\delta, h, I))$.

Estimate the running time of Algorithm $\mathcal{A}$. For each tuple $x_1, \ldots, x_t \in X$, the vectors $e_1, \ldots, e_{t-1}$ can be constructed in time $\mathcal{O}(t^2 d)$. The vector operations in space $\mathbb{R}^d$ take time $\mathcal{O}(d)$. On the other hand, by the definition of $T(\delta)$, we have $(1 + \delta)^{t-2} < 1/\delta$, so $\ell(\delta, h, I) > (1/\delta)^{(t-1)/2} = \Omega(t^2)$ if $\delta < 1$. Therefore, the algorithm runs in time

$$\mathcal{O}\big(n^{T(\delta)}\big(T(\delta)^2 d + \ell(\delta, h, I)\, d\big)\big) = \mathcal{O}\big(n^{T(\delta)} \ell(\delta, h, I)\, d\big).$$

Proposition 4 is proved. □

To prove Theorem 1, it remains to choose appropriate values of the parameters δ, h, I. Let $\delta = 0.87\varepsilon$, $I = T(\delta) - 1$, and $h = \dfrac{0.26\varepsilon\, \delta^{1/I}}{\sqrt{T(\delta) - 1}}$. Then $\delta_+ \le 0.13\varepsilon$ and, by Proposition 3, Algorithm $\mathcal{A}$ outputs a $(1 + \varepsilon)$-collection for the set X.

Based on Proposition 4, estimate the cardinality of the constructed set and the running time of the algorithm. We have

$$T(\delta) = \zeta + 1, \text{ where } \zeta = \left\lceil \frac{\log \frac{1}{0.87\varepsilon}}{\log(1 + 0.87\varepsilon)} \right\rceil,$$

and

$$\ell(\delta, h, I) = \left(\frac{\sqrt{\zeta}}{0.26\varepsilon\, (0.87\varepsilon)^{1/\zeta}} + \frac{1}{0.87\varepsilon} \right)^{\zeta} (1 + 0.87\varepsilon)^{-\zeta(\zeta-1)/2} \zeta.$$

The values of $T(\delta)$ and $\ell(\delta, h, I)$ can be estimated as follows. Consider the functions

$$a(\varepsilon) = \frac{\zeta + 1}{\lceil \frac{1}{\varepsilon} \log \frac{2}{\varepsilon} \rceil} \text{ and } b(\varepsilon) = \frac{\ell(\delta, h, I)}{\left(\frac{1}{\varepsilon} \log \frac{2}{\varepsilon} \right)^{\lceil \frac{1}{\varepsilon} \log \frac{2}{\varepsilon} \rceil}}.$$

By using asymptotic properties of these functions for small ε and computer calculations of the values of $a(.)$ and $b(.)$ on a grid with sufficiently small step, we obtain that $a(\varepsilon) \le 1$ and $b(\varepsilon) < 37$ for all $\varepsilon \in (0, 1)$. It follows that

$$T(\delta) \le \left\lceil \tfrac{1}{\varepsilon} \log \tfrac{2}{\varepsilon} \right\rceil \text{ and } \ell(\delta, h, I) < 37 \left(\tfrac{1}{\varepsilon} \log \tfrac{2}{\varepsilon} \right)^{\lceil \frac{1}{\varepsilon} \log \frac{2}{\varepsilon} \rceil}.$$

Thus, by Proposition 4, the number of points in the output of Algorithm $\mathcal{A}$ is $N(n, \varepsilon) = \mathcal{O}\big(\big(\tfrac{n}{\varepsilon} \log \tfrac{2}{\varepsilon}\big)^{\lceil \frac{1}{\varepsilon} \log \frac{2}{\varepsilon} \rceil}\big)$ and the algorithm runs in time $\mathcal{O}\big(N(n, \varepsilon)\, d\big)$. Theorem 1 is proved. □

Remark 1. The proposed algorithm for constructing $(1 + \varepsilon)$-collections is a modification of that from [20]. The differences of the new algorithm are a more optimal form of the grids we use to approximate the points y_t at Step 5 and a better way to fill the chosen form with the grid nodes. Another factor which reduces the cardinality of the output $(1 + \varepsilon)$-collections is using a more optimal set of values for the parameters of the algorithm.

Remark 2. The obtained upper bound for the cardinality of $(1 + \varepsilon)$-collections is less than that suggested in [20] both in the expression under the $\mathcal{O}$-notation and in the hidden constant in this notation: ≈ 37 vs. ≈ 4400.

3 A Lower Bound

In this section, we prove that, for any fixed ε, the minimum cardinality of a $(1+\varepsilon)$-collection for a given set of n points in high-dimensional Euclidean space is $\Omega\big(n^{\lfloor \frac{1}{16\varepsilon}+1\rfloor}\big)$ in the worst case.

Theorem 2. *For each $\varepsilon > 0$ and every positive integer $n \geq \frac{1}{8\varepsilon}$, there exists an n-element set $X \subset \mathbb{R}^n$ such that any $(1+\varepsilon)$-collection for this set consists of at least $n^{\lfloor \frac{1}{16\varepsilon}+1\rfloor}\varepsilon^{\lfloor\frac{1}{16\varepsilon}\rfloor}$ elements.*

Proof. Define X as the set of the n-dimensional unit vectors e_i, $i = 1,\ldots,n$, where $e_i(i) = 1$ and $e_i(j) = 0$ for $j \neq i$. Note that every $(1+\varepsilon)$-collection for a finite set of points contains this set itself. At the same time, we have $\lfloor\frac{1}{16\varepsilon}\rfloor = 0$ if $\varepsilon > 1/16$. Therefore, the theorem holds in this case. Further, we will assume that $\varepsilon \leq 1/16$.

Make some notations. Given a positive integer k, denote by $\mathcal{S}_k$ the family of k-element multisets consisting of elements of X, i.e.,

$$\mathcal{S}_k = \{\,(x_1,\ldots,x_k)\,|\,x_1 \in X,\ldots,x_k \in X\,\}.$$

Given a multiset $S = (x_1,\ldots,x_k)$, denote by $c(S)$ its mean: $c(S) = \frac{1}{k}\sum_{i=1}^{k} x_i$.

Given a point $x \in X$ and a multiset $S = (x_1,\ldots,x_k)$, denote by $m(x,S)$ the multiplicity of x in S, i.e., the number of indices i for which $x_i = x$. Finally, we say that multisets $S_1, S_2 \in \mathcal{S}_k$ *differ by t elements* if $\sum_{x\in X}|m(x,S_1) - m(x,S_2)| = t$.

Fact 3 (e.g., see [11,16]). *For every multiset $S = (x_1,\ldots,x_k)$, the following holds:* $\sum_{i=1}^{k}\|x_i - c(S)\|^2 = \frac{1}{2k}\sum_{i=1}^{k}\sum_{j=1}^{k}\|x_i - x_j\|^2.$

Fact 4 (e.g., see [11,13]). *For every multiset $S = (x_1,\ldots,x_k)$ and any point $y \in \mathbb{R}^n$, the following holds:* $\sum_{i=1}^{k}\|x_i - y\|^2 = \sum_{i=1}^{k}\|x_i - c(S)\|^2 + k\,\|y - c(S)\|^2.$

Lemma 7. *Suppose that $k \geq 2$, $S \in \mathcal{S}_k$, $y \in \mathbb{R}^n$, and $\delta = \|y - c(S)\|$. Then there exists a point x in the multiset S with the property $\|x - y\| \geq \alpha(\delta,k)\|x - c(S)\|$, where $\alpha(\delta,k) = \sqrt{1 + \delta^2\dfrac{k}{k-1}}.$*

Proof. By using Fact 3, it can be proved that the sum of the squared distances from the elements of S to the point $c(S)$ is at most $k - 1$. Therefore, by Fact 4, the sum of the squared distances from these elements to the point y is at least $1 + \delta^2\dfrac{k}{k-1}$ times of that to $c(S)$. It follows that the set S contains an element x such that $\|x - y\|^2$ is at least $1 + \delta^2\dfrac{k}{k-1}$ times of $\|x - c(S)\|^2$. Lemma 7 is proved. $\qquad\square$

Lemma 8. *Given integers k and t, where $0 \leq t < k \leq n$, there exists a subfamily $\mathcal{F}_{k,t} \subseteq \mathcal{S}_k$ such that any different multisets $S_1, S_2 \in \mathcal{F}_{k,t}$ differ by at least $2t + 2$ elements and $\mathcal{F}_{k,t}$ consists of at least $\dfrac{n^{k-t}}{2^{2t}(k-1+t)^{k-1-t}k}$ multisets.*

Proof. First, make some combinatorial observations. Denote by $M_{n,k}$ the number of k-element multisets whose elements belong to an n-element set. It is well-known (e.g., see [21]) that $M_{n,k} = \dbinom{n-1+k}{k}$. Choose an arbitrary multiset $S \in \mathcal{S}_k$. It is easy to see that there exist at most $M_{k,t}$ ways to remove t (possibly repeating) elements from S. At the same time, given a multiset S', there exist exactly $M_{n,t}$ ways to add t (possibly repeating) elements of X to S'. So the set $N_t(S)$ of multisets in $\mathcal{S}_k$ which differ from S by at most $2t$ elements consists of at most $M_{k,t}M_{n,t}$ elements.

Based on this fact, define the following simple algorithm for constructing a required subfamily $\mathcal{F}_{k,t}$. Initially, put $\mathcal{S} = \mathcal{S}_k$; then, while $\mathcal{S}$ is non-empty, add to $\mathcal{F}_{k,t}$ an arbitrary multiset $S \in \mathcal{S}$ and remove from $\mathcal{S}$ all the elements of $N_t(S)$. Since the family $\mathcal{S}_k$ consists of $M_{n,k}$ multisets, we have

$$|\mathcal{F}_{k,t}| \geq \frac{M_{n,k}}{M_{k,t}M_{n,t}} = \frac{(n-1+k)!\,(t!)^2}{(k-1+t)!\,(n-1+t)!\,k}.$$

Taking into account that $2^{2t}(t!)^2 \geq (2t)!$, the obtained expression is at least

$$\frac{(n-1+k)!}{(n-1+t)!} \frac{(2t)!}{2^{2t}(k-1+t)!\,k} \geq \frac{n^{k-t}}{2^{2t}(k-1+t)^{k-1-t}k}.$$

Lemma 8 is proved. $\qquad\qquad\square$

Lemma 9. *For every $k = 1, \ldots, n$, each $\left(1 + \dfrac{1}{8k}\right)$-collection for the set X contains at least $\dfrac{n^{\lfloor k/2 \rfloor + 1}}{(8k)^{k/2}}$ elements.*

Proof. For $k = 1$, the statement follows from the fact that any $(1+\varepsilon)$-collection for a finite set of points contains this set itself. Let $k \geq 2$.

Lemma 7 implies that, for every multiset $S \in \mathcal{S}_k$ and any $\delta > 0$, all the $\alpha(\delta, k)$-approximations of the point $c(S)$ with respect to the set X belong to the δ-neighborhood of $c(S)$. So every $\alpha(\delta, k)$-collection for the set X contains at least one element in each of these δ-neighborhoods.

On the other hand, it is easy to see that $\|c(S_1) - c(S_2)\| \geq \dfrac{\sqrt{2t+2}}{k}$ for every $t < k$ and any different multisets S_1, S_2 from the family $\mathcal{F}_{k,t}$ defined in Lemma 8. Therefore, if $\delta < \dfrac{\sqrt{t+1}}{\sqrt{2}\,k}$, then the δ-neighborhoods of the points $c(S)$ for the multisets $S \in \mathcal{F}_{k,t}$ are disjoint. By the above observations and Lemma 8, it follows that every $\alpha(\delta, k)$-collection for the set X contains at least βn^{k-t} elements, where $\beta = \dfrac{1}{2^{2t}(k-1+t)^{k-1-t}k}$.

Put $t = \lceil k/2 \rceil - 1$ and $\delta = \sqrt{(k-1)\left(\dfrac{t+1}{2k^3} + \dfrac{(t+1)^2}{16k^5}\right)}$. Then we have

$\delta < \dfrac{\sqrt{t+1}}{\sqrt{2}\,k}$ and $\alpha(\delta,k) = \sqrt{1 + \dfrac{t+1}{2k^2} + \dfrac{(t+1)^2}{16k^4}} = 1 + \dfrac{t+1}{4k^2} \geq 1 + \dfrac{1}{8k}$. It

follows that every $\left(1 + \dfrac{1}{8k}\right)$-collection for the set X is an $\alpha(\delta,k)$-collection, so it contains at least $\beta n^{\lfloor k/2 \rfloor + 1}$ elements.

Estimate the value of β. Since $2t \leq k-1$, $k-1+t \leq 3k/2 - 3/2$, and $k-1-t \leq k/2$, we have $\beta \geq \dfrac{1}{2^{k-1}(3k/2 - 3/2)^{k/2}k} = \dfrac{1}{2^{k/2}(3k-3)^{k/2}k/2}$. Next, it can be proved that $(3k-3)^{k/2}k/2 \leq (4k)^{k/2}$ for all $k \geq 1$, so $\beta \geq \dfrac{1}{(8k)^{k/2}}$. Lemma 9 is proved. $\qquad\square$

To finish the proof of the theorem, select $k = 2\left\lfloor \dfrac{1}{16\varepsilon} \right\rfloor$. In this case, we have

$\varepsilon \leq \dfrac{1}{8k}$ and $2 \leq k \leq n$ since $\varepsilon \leq \dfrac{1}{16}$ and $n \geq \dfrac{1}{8\varepsilon}$. So every $(1 + \varepsilon)$-collection

for the set X is also a $\left(1 + \dfrac{1}{8k}\right)$-collection and, by Lemma 9, contains at least

$\dfrac{n^{\lfloor k/2 \rfloor + 1}}{(8k)^{k/2}} = n^{\lfloor \frac{1}{16\varepsilon} + 1 \rfloor}\left(\dfrac{1}{16\lfloor \frac{1}{16\varepsilon} \rfloor}\right)^{\lfloor \frac{1}{16\varepsilon} \rfloor} \geq n^{\lfloor \frac{1}{16\varepsilon} + 1 \rfloor}\varepsilon^{\lfloor \frac{1}{16\varepsilon} \rfloor}$ elements. Theorem 2 is

proved. $\qquad\square$

4   The Case of ℓ_∞ Distances

In contrast to the obtained estimations of the worst-case cardinality of minimum $(1 + \varepsilon)$-collections in Euclidean space, we show that this cardinality is not polynomial when the distances between points are defined by the ℓ_∞ norm:

$$\|v\|_\infty = \max_{i=1,\ldots,d} |v(i)|.$$

First, formulate the concept of an α-collection in an arbitrary metric space. Let $\mathcal{M}$ be any set and $dist$ be any metrics on this set.

Definition. *Given points $p, p' \in \mathcal{M}$, a set $X \subseteq \mathcal{M}$, and a real number $\alpha \geq 1$, we say that p' is an α-approximation of p with respect to X in the metrics $dist$ if $dist(x, p') \leq \alpha\, dist(x, p)$ for all $x \in X$.*

Definition. *Given sets $X, C \subseteq \mathcal{M}$, and a real number $\alpha \geq 1$, we say that C is an α-collection for X in the metric space $(\mathcal{M}, dist)$ if C contains α-approximations of all the points of $\mathcal{M}$ with respect to X in the metrics $dist$.*

By using the observations in Example from Sect. 2, it is easy to prove that every finite set is a 2-collection for itself in any metric space. The following theorem shows that, in the case of ℓ_∞ distances, some sets of points do not admit polynomial-cardinality α-collections if $\alpha < 2$:

Theorem 3. *For each $\varepsilon \in (0,1)$ and every integer $n \geq 4$, there exists an n-element set $X \subset \mathbb{R}^{\lfloor n/2 \rfloor}$ such that any $(1 + \varepsilon)$-collection for this set in space $(\mathbb{R}^{\lfloor n/2 \rfloor}, \ell_\infty)$ consists of at least $2^{\lfloor n/2 \rfloor}$ elements.*

Proof. Without loss of generality, we can assume that n is even: if n is odd we will construct the desired set of cardinality $n - 1$ and add the zero vector.

Let $d = n/2$ and define X as the set of the d-dimensional unit vectors $\pm e_i$, $i = 1, \ldots, d$, where $e_i(i) = 1$ and $e_i(j) = 0$ for $j \neq i$. Next, given a vector $v \in \{1, -1\}^d$, define the set $S_v = \{v(1) e_1, \ldots, v(d) e_d\}$.

It is easy to see that, for every $v \in \{1, -1\}^d$, the point $v/2$ has the property $\|x - v/2\|_\infty = 1/2$ for all $x \in S_v$. Moreover, the following statement holds:

Lemma 10. *For each $\varepsilon \in (0,1)$ and every $v \in \{-1, 1\}^d$, any $(1 + \varepsilon)$-collection for the set X in space $(\mathbb{R}^d, \ell_\infty)$ contains a point $x_v \in \mathbb{R}^d$ such that each coordinate $x_v(i)$, $i = 1, \ldots, d$, lies strictly between 0 and $v(i)$.*

Proof. Indeed, any $(1+\varepsilon)$-collection for the set X contains a point $x_v \in \mathbb{R}^d$ which is a $(1 + \varepsilon)$-approximation of the point $v/2$ with respect to X. Since $1 + \varepsilon < 2$, it follows that $\|v(i) e_i - x_v\|_\infty < 2 \|v(i) e_i - v/2\|_\infty = 1$ for each $i = 1, \ldots, d$. Therefore, for each i, the coordinate $x_v(i)$ lies strictly between 0 and $2v(i)$. On the other hand, since $d \geq 2$, the set S_v contains at least one point y for which $y(i) = 0$. This yields that $|x_v(i)| \leq \|y - x_v\|_\infty < 2 \|y - v/2\|_\infty = 1$. So the coordinate $x_v(i)$ lies strictly between 0 and $v(i)$. Lemma 10 is proved. $\qquad\square$

It remains to note that, for every different vectors $v, u \in \{1, -1\}^d$, there exists at least one coordinate $i \in \{1, \ldots, d\}$ such that $v(i) = -u(i)$. It follows that the points x_v, x_u defined in Lemma 10 for the vectors v and u differ. Then, by Lemma 10, any $(1+\varepsilon)$-collection for the set X in space $(\mathbb{R}^d, \ell_\infty)$ contains at least 2^d elements. Theorem 3 is proved. $\qquad\square$

5   Conclusion

We study the concept of a $(1 + \varepsilon)$-collection, which is closely related to the question of the polynomial discretization of geometric center-based problems. Our main result is an upper and a lower bounds for the minimum cardinality of $(1+\varepsilon)$-collections for a given set of n points in high-dimensional Euclidean space in the worst case. We prove that this cardinality is at most $\Theta\big(n^{\lceil \frac{1}{\varepsilon} \log \frac{2}{\varepsilon} \rceil}\big)$ and is at least $\Theta\big(n^{\lfloor \frac{1}{16\varepsilon} + 1 \rfloor}\big)$ for any fixed $\varepsilon \in (0, 1]$. In contrast, it turned out that, in the case of ℓ_∞ distances, there exist sets of points which do not admit polynomial-cardinality $(1 + \varepsilon)$-collections. An interesting open question is the situation in high-dimensional ℓ_1 space. Another question is an asymptotically exact bound for the worst-case cardinality of minimum Euclidean $(1 + \varepsilon)$-collections: is it closer to $n^{\Theta(\frac{1}{\varepsilon} \log \frac{2}{\varepsilon})}$ or to $n^{\Theta(\frac{1}{\varepsilon})}$?

Acknowledgments. The work was supported by the program of fundamental scientific researches of the SB RAS, project 0314-2019-0014.

References

1. Agarwal, P.K., Har-Peled, S., Varadarajan, K.R.: Geometric approximation via coresets. Comb. Comput. Geom. MSRI **52**, 1–30 (2005)
2. Agarwal, P.K., Procopiuc, C.M.: Exact and approximation algorithms for clustering. Algorithmica **33**(2), 201–226 (2002). https://doi.org/10.1007/s00453-001-0110-y
3. Aggarwal, A., Imai, H., Katoh, N., Suri, S.: Finding k points with minimum diameter and related problems. J. Algorithms **12**(1), 38–56 (1991)
4. Arora, S., Raghavan, P., Rao, S.: Approximation schemes for Euclidean k-medians and related problems. In: Proceedings of the 30th ACM Symposium on Theory of Computing (STOC 1998), pp. 106–113 (1998)
5. Bădoiu, M., Har-Peled, S., Indyk, P.: Approximate clustering via core-sets. In: Proceedings of the 34th ACM Symposium on Theory of Computing (STOC 2002), pp. 250–257 (2002)
6. Bhattacharya, A., Jaiswal, R., Kumar, A.: Faster algorithms for the constrained k-means problem. Theory Comput. Syst. **62**(1), 93–115 (2018). https://doi.org/10.1007/s00224-017-9820-7
7. Chen, K.: On k-median clustering in high dimensions. In: Proceedings of the 17th ACM-SIAM Symposium on Discrete Algorithms (SODA 2006), pp. 1177–1185 (2006)
8. Ding, H., Xu, J.: A unified framework for clustering constrained data without locality property. In: Proceedings of the 26th ACM-SIAM Symposium on Discrete Algorithms (SODA 2015), pp. 1471–1490 (2015)
9. Eppstein, D., Erickson, J.: Iterated nearest neighbors and finding minimal polytopes. Discrete Comp. Geom. **11**(3), 321–350 (1994)
10. Har-Peled, S., Mazumdar, S.: On coresets for k-means and k-median clustering. In: Proceedings of the 36th ACM Symposium on Theory of Computing (STOC 2004), pp. 291–300 (2004)
11. Inaba, M., Katoh, N., Imai, H.: Applications of weighted Voronoi diagrams and randomization to variance-based k-clustering. In: Proceedings of the 10th ACM Symposium on Computational Geometry, pp. 332–339 (1994)
12. Jaiswal, R., Kumar, A., Sen, S.: A simple D^2-sampling based PTAS for k-means and other clustering problems. Algorithmica **70**(1), 22–46 (2014). https://doi.org/10.1007/s00453-013-9833-9
13. Kumar, A., Sabharwal, Y., Sen, S.: Linear-time approximation schemes for clustering problems in any dimensions. J. ACM **57**(2), 1–32 (2010)
14. Kumar, P., Mitchell, J.S.B., Yıldırım, E.A.: Approximate minimum enclosing balls in high dimensions using core-sets. J. Exp. Algorithmics. **8**(1.1), 1–29 (2003)
15. Meira, A.A., Miyazawa, F.K., Pedrosa, L.L.C.: Clustering through continuous facility location problems. Theor. Comp. Sci. **657**(B), 137–145 (2017)
16. Shenmaier, V.V.: An approximation scheme for a problem of search for a vector subset. J. Appl. Ind. Math. **6**(3), 381–386 (2012). https://doi.org/10.1134/S1990478912030131
17. Shenmaier, V.V.: The problem of a minimal ball enclosing k points. J. Appl. Industr. Math. **7**(3), 444–448 (2013). https://doi.org/10.1134/S1990478913030186
18. Shenmaier, V.V.: Computational complexity and approximation for a generalization of the Euclidean problem on the Chebyshev center. Doklady Math. **87**(3), 342–344 (2013). https://doi.org/10.7868/S0869565213170052

19. Shenmaier, V.V.: Complexity and approximation of the smallest k-enclosing ball problem. Eur. J. Comb. **48**, 81–87 (2015). https://doi.org/10.1016/j.ejc.2015.02.011
20. Shenmaier, V.V.: A structural theorem for center-based clustering in high-dimensional Euclidean space. In: Nicosia, G., Pardalos, P., Umeton, R., Giuffrida, G., Sciacca, V. (eds.) LOD 2019. LNCS, vol. 11943, pp. 284–295. Springer, Cham (2019). https://doi.org/10.1007/978-3-030-37599-7_24
21. Wikipedia: The Free Encyclopedia. Multiset [e-resource]. https://en.wikipedia.org/wiki/Multiset

Mathematical Programming

Gradient-Free Methods with Inexact Oracle for Convex-Concave Stochastic Saddle-Point Problem

Aleksandr Beznosikov[1,2(✉)], Abdurakhmon Sadiev[1(✉)],
and Alexander Gasnikov[1,2,3,4(✉)]

[1] Moscow Institute of Physics and Technology, Dolgoprudny, Russia
beznosikov.an@phystech.edu, sadiev1998@mail.ru, gasnikov@yandex.ru
[2] Sirius University of Science and Technology, Krasnoyarsk, Russia
[3] Institute for Information Transmission Problems RAS, Moscow, Russia
[4] Caucasus Mathematical Center, Adyghe State University, Maykop, Russia

Abstract. In the paper, we generalize the approach Gasnikov et al. 2017, which allows to solve (stochastic) convex optimization problems with an inexact gradient-free oracle, to the convex-concave saddle-point problem. The proposed approach works, at least, like the best existing approaches. But for a special set-up (simplex type constraints and closeness of Lipschitz constants in 1 and 2 norms) our approach reduces $n/\log n$ times the required number of oracle calls (function calculations). Our method uses a stochastic approximation of the gradient via finite differences. In this case, the function must be specified not only on the optimization set itself, but in a certain neighbourhood of it. In the second part of the paper, we analyze the case when such an assumption cannot be made, we propose a general approach on how to modernize the method to solve this problem, and also we apply this approach to particular cases ofsomeclassical sets.

Keywords: Zeroth-order optimization · Saddle-point problem · Stochastic optimization

1 Introduction

In the last decade in the ML community, a big interest cause different applications of Generative Adversarial Networks (GANs) [10], which reduce the ML problem to the saddle-point problem, and the application of gradient-free methods for Reinforcement Learning problems [17]. Neural networks become rather

The research of A. Beznosikov was partially supported by RFBR, project number 19-31-51001. The research of A. Gasnikov was partially supported by RFBR, project number 18-29-03071 mk and was partially supported by the Ministry of Science and Higher Education of the Russian Federation (Goszadaniye) no 075-00337-20-03.

Y. Kochetov et al. (Eds.): MOTOR 2020, CCIS 1275, pp. 105–119, 2020.
https://doi.org/10.1007/978-3-030-58657-7_11

popular in Reinforcement Learning [13]. Thus, there is an interest in gradient-free methods for saddle-point problems

$$\min_{x \in \mathcal{X}} \max_{y \in \mathcal{Y}} \varphi(x, y). \tag{1}$$

One of the natural approach for this class of problems is to construct a stochastic approximation of a gradient via finite differences. In this case, it is natural to expect that the complexity of the problem (1) in terms of the number of function calculations is $\sim n$ times large in comparison with the complexity in terms of number of gradient calculations, where $n = \dim \mathcal{X} + \dim \mathcal{Y}$. Is it possible to obtain better result? In this paper, we show that this factor can be reduced in some situation to a much smaller factor $\log n$.

We use the technique, developed in [8,9] for stochastic gradient-free non-smooth convex optimization problems (gradient-free version of mirror descent [2]) to propose a stochastic gradient-free version of saddle-point variant of mirror descent [2] for non-smooth convex-concave saddle-point problems.

The concept of using such an oracle with finite differences is not new (see [5,16]). For such an oracle, it is necessary that the function is defined in some neighbourhood of the initial set of optimization, since when we calculate the finite difference, we make some small step from the point, and this step can lead us outside the set. As far as we know, in all previous works, the authors proceed from the fact that such an assumption is fulfilled or does not mention it at all. We raise the question of what we can do when the function is defined only on the given set due to some properties of the problem.

1.1 Our Contributions

In this paper, we present a new method called zeroth-order Saddle-Point Algorithm (zoSPA) for solving a convex-concave saddle-point problem (1). Our algorithm uses a zeroth-order biased oracle with stochastic and bounded deterministic noise. We show that if the noise $\sim \varepsilon$ (accuracy of the solution), then the number of iterations necessary to obtain $\varepsilon-$solution on set with diameter $\Omega \subset \mathbb{R}^n$ is $\mathcal{O}\left(\frac{M^2 \Omega^2}{\varepsilon^2} n\right)$ or $\mathcal{O}\left(\frac{M^2 \Omega^2}{\varepsilon^2} \log n\right)$ (depends on the optimization set, for example, for a simplex, the second option with $\log n$ holds), where M^2 is a bound of the second moment of the gradient together with stochastic noise (see below, (3)).

In the second part of the paper, we analyze the structure of an admissible set. We give a general approach on how to work in the case when we are forbidden to go beyond the initial optimization set. Briefly, it is to consider the "reduced" set and work on it.

Next, we show how our algorithm works in practice for various saddle-point problems and compare it with full-gradient mirror descent.

One can find the proofs together and additional numerical experiments in the full version of this paper available on arXiv [4].

2   Notation and Definitions

We use $\langle x, y \rangle \overset{\text{def}}{=} \sum_{i=1}^{n} x_i y_i$ to define inner product of $x, y \in \mathbb{R}^n$ where x_i is the i-th component of x in the standard basis in $\mathbb{R}^n$. Hence we get the definition of ℓ_2-norm in $\mathbb{R}^n$ in the following way $\|x\|_2 \overset{\text{def}}{=} \sqrt{\langle x, x \rangle}$. We define ℓ_p-norms as $\|x\|_p \overset{\text{def}}{=} \left(\sum_{i=1}^{n} |x_i|^p \right)^{1/p}$ for $p \in (1, \infty)$ and for $p = \infty$ we use $\|x\|_\infty \overset{\text{def}}{=} \max_{1 \le i \le n} |x_i|$. The dual norm $\|\cdot\|_q$ for the norm $\|\cdot\|_p$ is defined in the following way: $\|y\|_q \overset{\text{def}}{=} \max \{ \langle x, y \rangle \mid \|x\|_p \le 1 \}$. Operator $\mathbb{E}[\cdot]$ is full mathematical expectation and operator $\mathbb{E}_\xi[\cdot]$ express conditional mathematical expectation.

Definition 1 (M-Lipschitz continuity). *Function $f(x)$ is M-Lipschitz continuous in $X \subseteq \mathbb{R}^n$ with $M > 0$ w.r.t. norm $\|\cdot\|$ when*

$$|f(x) - f(y)| \le M\|x - y\|, \quad \forall \, x, y \in X.$$

Definition 2 (μ-strong convexity). *Function $f(x)$ is μ-strongly convex w.r.t. norm $\|\cdot\|$ on $X \subseteq \mathbb{R}^n$ when it is continuously differentiable and there is a constant $\mu > 0$ such that the following inequality holds:*

$$f(y) \ge f(x) + \langle \nabla f(x), y - x \rangle + \frac{\mu}{2}\|y - x\|^2, \quad \forall \, x, y \in X.$$

Definition 3 (Prox-function). *Function $d(z) : \mathcal{Z} \to \mathbb{R}$ is called prox-function if $d(z)$ is 1-strongly convex w.r.t. $\|\cdot\|$-norm and differentiable on $\mathcal{Z}$ function.*

Definition 4 (Bregman divergence). *Let $d(z) : \mathcal{Z} \to \mathbb{R}$ is prox-function. For any two points $z, w \in \mathcal{Z}$ we define Bregman divergence $V_z(w)$ associated with $d(z)$ as follows:*

$$V_z(w) = d(z) - d(w) - \langle \nabla d(w), z - w \rangle.$$

We denote the Bregman-diameter $\Omega_{\mathcal{Z}}$ of $\mathcal{Z}$ w.r.t. $V_{z_1}(z_2)$ as $\Omega_{\mathcal{Z}} \overset{\text{def}}{=} \max\{ \sqrt{2 V_{z_1}(z_2)} \mid z_1, z_2 \in \mathcal{Z} \}$.

Definition 5 (Prox-operator). *Let $V_z(w)$ Bregman divergence. For all $x \in \mathcal{Z}$ define prox-operator of ξ:*

$$\text{prox}_x(\xi) = \arg\min_{y \in \mathcal{Z}} \left(V_x(y) + \langle \xi, y \rangle \right).$$

3   Main Result

3.1   Non-smooth Saddle-Point Problem

We consider the saddle-point problem (1), where $\varphi(\cdot, y)$ is convex function defined on compact convex set $\mathcal{X} \subset \mathbb{R}^{n_x}$, $\varphi(x, \cdot)$ is concave function defined on compact convex set $\mathcal{Y} \subset \mathbb{R}^{n_y}$.

We call an inexact stochastic zeroth-order oracle $\tilde{\varphi}(x, y, \xi)$ at each iteration. Our model corresponds to the case when the oracle gives an inexact noisy function value. We have stochastic unbiased noise, depending on the random variable ξ and biased deterministic noise. One can write it the following way:

$$\tilde{\varphi}(x, y, \xi) = \varphi(x, y, \xi) + \delta(x, y),$$
$$\mathbb{E}_\xi[\tilde{\varphi}(x, y, \xi)] = \bar{\varphi}(x, y), \quad \mathbb{E}_\xi[\varphi(x, y, \xi)] = \varphi(x, y), \tag{2}$$

where random variable ξ is responsible for unbiased stochastic noise and $\delta(x, y)$ – for deterministic noise.

We assume that exists such positive constant M that for all $x, y \in \mathcal{X} \times \mathcal{Y}$ we have

$$\|\nabla\varphi(x, y, \xi)\|_2 \leq M(\xi), \quad \mathbb{E}[M^2(\xi)] = M^2. \tag{3}$$

By $\nabla\varphi(x, y, \xi)$ we mean a block vector consisting of two vectors $\nabla_x\varphi(x, y, \xi)$ and $\nabla_y\varphi(x, y, \xi)$. One can prove that $\varphi(x, y, \xi)$ is $M(\xi)$-Lipschitz w.r.t. norm $\|\cdot\|_2$ and that $\|\nabla\varphi(x, y)\|_2 \leq M$.

Also the following assumptions are satisfied:

$$|\tilde{\varphi}(x, y, \xi) - \varphi(x, y, \xi)| = |\delta(x, y)| \leq \Delta. \tag{4}$$

For convenience, we denote $\mathcal{Z} = \mathcal{X} \times \mathcal{Y}$ and then $z \in \mathcal{Z}$ means $z \overset{\text{def}}{=} (x, y)$, where $x \in \mathcal{X}$, $y \in \mathcal{Y}$. When we use $\varphi(z)$, we mean $\varphi(z) = \varphi(x, y)$, and $\varphi(z, \xi) = \varphi(x, y, \xi)$.

For $\mathbf{e} \in \mathcal{RS}_2^n(1)$ (a random vector uniformly distributed on the Euclidean unit sphere) and some constant τ let $\tilde{\varphi}(z + \tau\mathbf{e}, \xi) \overset{\text{def}}{=} \tilde{\varphi}(x + \tau\mathbf{e}_x, y + \tau\mathbf{e}_y, \xi)$, where $\mathbf{e}_x$ is the first part of $\mathbf{e}$ size of dimension $n_x \overset{\text{def}}{=} \dim(x)$, and $\mathbf{e}_y$ is the second part of dimension $n_y \overset{\text{def}}{=} \dim(y)$. And $n \overset{\text{def}}{=} n_x + n_y$. Then define estimation of the gradient through the difference of functions:

$$g(z, \xi, \mathbf{e}) = \frac{n\left(\tilde{\varphi}(z + \tau\mathbf{e}, \xi) - \tilde{\varphi}(z - \tau\mathbf{e}, \xi)\right)}{2\tau} \begin{pmatrix} \mathbf{e}_x \\ -\mathbf{e}_y \end{pmatrix}. \tag{5}$$

$g(z, \xi, \mathbf{e})$ is a block vector consisting of two vectors.

Next we define an important object for further theoretical discussion – a smoothed version of the function $\tilde{\varphi}$ (see [15, 16]).

Definition 6. *Function $\hat{\varphi}(x, y) = \hat{\varphi}(z)$ defines on set $\mathcal{X} \times \mathcal{Y}$ satisfies:*

$$\hat{\varphi}(z) = \mathbb{E}_{\mathbf{e}}\left[\varphi(z + \tau\mathbf{e})\right].$$

Note that we introduce a smoothed version of the function only for proof; in the algorithm, we use only the zero-order oracle (5). Now we are ready to present our algorithm:

Algorithm 1. Zeroth-Order Saddle-Point Algorithm (zoSPA)

Input: Iteration limit N.

Let $z_1 = \underset{z \in \mathcal{Z}}{\operatorname{argmin}}\, d(z)$.

for $k = 1, 2, \ldots, N$ **do**

    Sample $\mathbf{e}_k, \xi_k$ independently.

    Initialize γ_k.

    $z_{k+1} = \operatorname{prox}_{z_k}(\gamma_k g(z_k, \xi_k, \mathbf{e}_k))$.

end for

Output: $\bar{z}_N$,

---

where

$$\bar{z}_N = \frac{1}{\Gamma_N}\left(\sum_{k=1}^{N} \gamma_k z_k\right), \quad \Gamma_N = \sum_{k=1}^{N} \gamma_k. \tag{6}$$

In Algorithm 1, we use the step γ_k. In fact, we can take this step as a constant, independent of the iteration number k (see Theorem 1).

Note that we work only with norms $\|\cdot\|_p$, where p is from 1 to 2 (q is from 2 to ∞). In the rest of the paper, including the main theorems, we assume that p is from 1 to 2.

Lemma 1 (see Lemma 2 from [3]). *For $g(z, \xi, \mathbf{e})$ defined in (5) the following inequalitie holds:*

$$\mathbb{E}\left[\|g(z, \xi, \mathbf{e})\|_q^2\right] \leq 2\left(cnM^2 + \frac{n^2 \Delta^2}{\tau^2}\right) a_q^2,$$

where c is some positive constant (independent of n) and a_q^2 is determined by $\sqrt{\mathbb{E}[\|e\|_q^4]} \leq a_q^2$ and the following statement is true

$$a_q^2 = \min\{2q - 1, 32 \log n - 8\} n^{\frac{2}{q}-1}, \quad \forall n \geq 3. \tag{7}$$

Note that in the case with $p = 2$, $q = 2$ we have $a_q = 1$, this follows not from (7), but from the simplest estimate. And from (7) we get that with $p = 1$, $q = \infty$ – $a_q = \mathcal{O}(\log n/n)$ (see also Lemma 4 from [16]).

Lemma 2 (see Lemma 8 from [16]). *Let $\mathbf{e}$ be from $\mathcal{RS}_2^n(1)$. Then function $\hat{\varphi}(z, \xi)$ is convex-concave and satisfies :*

$$\sup_{z \in \mathcal{Z}} |\hat{\varphi}(z) - \varphi(z)| \leq \tau M + \Delta.$$

Lemma 3 (see Lemma 10 from [16] and Lemma 2 from [3]). *It holds that*

$$\tilde{\nabla}\hat{\varphi}(z) = \mathbb{E}_{\mathbf{e}}\left[\frac{n\left(\varphi(z + \tau \mathbf{e}) - \varphi(z - \tau \mathbf{e})\right)}{2\tau}\begin{pmatrix} \mathbf{e}_x \\ -\mathbf{e}_y \end{pmatrix}\right],$$

$$\|\mathbb{E}_{\mathbf{e}}[g(z, \mathbf{e})] - \tilde{\nabla}\hat{\varphi}(z)\|_q \leq \frac{\Delta n a_q}{\tau},$$

where

$$g(z, \mathbf{e}) = \mathbb{E}_\xi \left[g(z, \xi, \mathbf{e}) \right]$$
$$= \frac{n \left(\tilde{\varphi}(z + \tau \mathbf{e}) - \tilde{\varphi}(z - \tau \mathbf{e}) \right)}{2\tau} \begin{pmatrix} \mathbf{e}_x \\ -\mathbf{e}_y \end{pmatrix}.$$

Hereinafter, by $\tilde{\nabla}\hat{\varphi}(z)$ we mean a block vector consisting of two vectors $\nabla_x \hat{\varphi}(x, y)$ and $-\nabla_y \hat{\varphi}(x, y)$.

Lemma 4 (see Lemma 5.3.2 from [2]). *Define $\Delta_k \overset{def}{=} g(z_k, \xi_k, \mathbf{e}_k) - \tilde{\nabla}\hat{\varphi}(z_k)$. Let $D(u) \overset{def}{=} \sum_{k=1}^{N} \gamma_k \langle \Delta_k, u - z_k \rangle$. Then we have*

$$\mathbb{E}\left[\max_{u \in \mathcal{Z}} D(u) \right] \leq \Omega^2 + \frac{\Delta \Omega n a_q}{\tau} \sum_{k=1}^{N} \gamma_k + M_{all}^2 \sum_{k=1}^{N} \gamma_k^2,$$

where $M_{all}^2 \overset{def}{=} 2 \left(cnM^2 + \frac{n^2 \Delta^2}{\tau^2} \right) a_q^2$ is from Lemma 1.

Theorem 1. *Let problem (1) with function $\varphi(x, y)$ be solved using Algorithm 1 with the oracle $g(z_k, \xi_k, \mathbf{e}_k)$ from (5). Assume, that the function $\varphi(x, y)$ and its inexact modification $\tilde{\varphi}(x, y)$ satisfy the conditions (2), (3), (4). Denote by N the number of iterations. Let step in Algorithm 1 $\gamma_k = \frac{\Omega}{M_{all}\sqrt{N}}$. Then the rate of convergence is given by the following expression*

$$\mathbb{E}\left[\varepsilon_{sad}(\bar{z}_N) \right] \leq \frac{3M_{all}\Omega}{\sqrt{N}} + \frac{\Delta \Omega n a_q}{\tau} + 2\tau M,$$

where $\bar{z}_N$ is defined in (6), Ω is a diameter of $\mathcal{Z}$, $M_{all}^2 = 2 \left(cnM^2 + \frac{n^2 \Delta^2}{\tau^2} \right) a_q^2$ and

$$\varepsilon_{sad}(\bar{z}_N) = \max_{y' \in \mathcal{Y}} \varphi(\bar{x}_N, y') - \min_{x' \in \mathcal{X}} \varphi(x', \bar{y}_N),$$

$\bar{x}_N, \bar{y}_N$ are defined the same way as $\bar{z}_N$ in (6).

Next we analyze the results.

Corollary 1. *Under the assumptions of the Theorem 1 let ε be accuracy of the solution of the problem (1) obtained using Algorithm 1. Assume that*

$$\tau = \Theta \left(\frac{\varepsilon}{M} \right), \quad \Delta = \mathcal{O} \left(\frac{\varepsilon^2}{M \Omega n a_q} \right), \tag{8}$$

then the number of iterations to find ε-solution

$$N = \mathcal{O} \left(\frac{\Omega^2 M^2 n^{2/q}}{\varepsilon^2} C^2(n, q) \right),$$

where $C(n, q) \overset{def}{=} \min\{2q - 1, 32 \log n - 8\}$.

Consider separately cases with $p = 1$ and $p = 2$.

Note that in the case with $p = 2$, we have that the number of iterations increases n times compared with [2], and in the case with $p = 1$ – just $\log^2 n$ times (Table 1).

Table 1. Summary of convergence estimation for non-smooth case: $p = 2$ and $p = 1$.

$p, (1 \leqslant p \leqslant 2)$	$q, (2 \leqslant q \leqslant \infty)$	N, Number of iterations
$p = 2$	$q = 2$	$\mathcal{O}\left(\frac{\Omega^2 M^2}{\varepsilon^2} n\right)$
$p = 1$	$q = \infty$	$\mathcal{O}\left(\frac{\Omega^2 M^2}{\varepsilon^2} \log^2(n)\right)$

3.2 Admissible Set Analysis

As stated above, in works (see [5,16]), where zeroth-order approximation (5) is used instead of the "honest" gradient, it is important that the function is specified not only on an admissible set, but in a certain neighborhood of it. This is due to the fact that for any point x belonging to the set, the point $x + \tau\mathbf{e}$ can be outside it.

But in some cases we cannot make such an assumption. The function and values of x can have a real physical interpretation. For example, in the case of a probabilistic simplex, the values of x are the distribution of resources or actions. The sum of the probabilities cannot be negative or greater than 1. Moreover, due to implementation or other reasons, we can deal with an oracle that is clearly defined on an admissible set and nowhere else.

In this part of the paper, we outline an approach how to solve the problem raised above and how the quality of the solution changes from this.

Our approach can be briefly described as follows:

- Compress our original set X by $(1 - \alpha)$ times and consider a "reduced" version X^α. Note that the parameter α should not be too small, otherwise the parameter τ must be taken very small. But it's also impossible to take large α, because we compress our set too much and can get a solution far from optimal. This means that the accuracy of the solution ε bounds α: $\alpha \leq h(\varepsilon)$, in turn, α bounds τ: $\tau \leq g(\alpha)$.
- Generate a random direction $\mathbf{e}$ so that for any $x \in X^\alpha$ follows $x + \tau\mathbf{e} \in X$.
- Solve the problem on "reduced" set with $\varepsilon/2$-accuracy. The α parameter must be selected so that we find ε-solution of the original problem.

In practice, this can be implemented as follows: 1) do as described in the previous paragraph, or 2) work on the original set X, but if $x_k + \tau\mathbf{e}$ is outside X, then project x_k onto the set X^α. We provide a theoretical analysis only for the method that always works on X^α.

Next, we analyze cases of different sets. General analysis scheme:

- Present a way to "reduce" the original set.
- Suggest a random direction $\mathbf{e}$ generation strategy.
- Estimate the minimum distance between X^α and X in ℓ_2-norm. This is the border of τ, since $\|\mathbf{e}\|_2$.
- Evaluate the α parameter so that the $\varepsilon/2$-solution of the "reduced" problem does not differ by more than $\varepsilon/2$ from the ε-solution of the original problem.

The first case of set is a **probability simplex**:

$$\triangle_n = \left\{ \sum_{i=1}^{n} x_i = 1, \quad x_i \geq 0, \quad i \in 1 \ldots n \right\}.$$

Consider the hyperplane

$$\mathcal{H} = \left\{ \sum_{i=1}^{n} x_i = 1 \right\},$$

in which the simplex lies. Note that if we take the directions $\mathbf{e}$ that lies in $\mathcal{H}$, then for any x lying on this hyperplane, $x + \tau \mathbf{e}$ will also lie on it. Therefore, we generate the direction $\mathbf{e}$ randomly on the hyperplane. Note that $\mathcal{H}$ is a subspace of $\mathbf{R}^n$ with size dim$\mathcal{H} = n - 1$. One can check that the set of vectors from $\mathbf{R}^n$

$$\mathbf{v} = \begin{pmatrix} \mathbf{v}_1 = 1/\sqrt{2}(1, -1, 0, 0, \ldots 0), \\ \mathbf{v}_2 = 1/\sqrt{6}(1, 1, -2, 0, \ldots 0), \\ \mathbf{v}_3 = 1/\sqrt{12}(1, 1, 1, -3, \ldots 0), \\ \ldots \\ \mathbf{v}_k = 1/\sqrt{k+k^2}(1, \ldots 1, -k, \ldots, 0), \\ \ldots \\ \mathbf{v}_{n-1} = 1/\sqrt{n-1+(n-1)^2}(1, \ldots, 1, -n+1) \end{pmatrix},$$

is an orthonormal basis of $\mathcal{H}$. Then generating the vectors $\tilde{\mathbf{e}}$ uniformly on the euclidean sphere $\mathcal{RS}_2^{n-1}(1)$ and computing $\mathbf{e}$ by the following formula:

$$\mathbf{e} = \tilde{e}_1 \mathbf{v}_1 + \tilde{e}_2 \mathbf{v}_2 + \ldots + \tilde{e}_k \mathbf{v}_k + \ldots \tilde{e}_{n-1} \mathbf{v}_{n-1}, \tag{9}$$

we have what is required. With such a vector $\mathbf{e}$, we always remain on the hyperplane, but we can go beyond the simplex. This happens if and only if for some i, $x_i + \tau e_i < 0$. To avoid this, we consider a "reduced" simplex for some positive constant α:

$$\triangle_n^\alpha = \left\{ \sum_{i=1}^{n} x_i = 1, \quad x_i \geq \alpha, \quad i \in 1 \ldots n \right\}.$$

One can see that for any $x \in \triangle_n^\alpha$, for any $\mathbf{e}$ from (9) and $\tau < \alpha$ follows that $x + \tau \mathbf{e} \in \triangle_n$, because $|e_i| \leq 1$ and then $x_i + \tau e_i \geq \alpha - \tau \geq 0$.

The last question to be discussed is the accuracy of the solution that we obtain on a "reduced" set. Consider the following lemma (this lemma does not apply to the problem (1), for it we prove later):

Lemma 5. *Suppose the function $f(x)$ is M-Lipschitz w.r.t. norm $\|\cdot\|_2$. Consider the problem of minimizing $f(x)$ not on original set X, but on the "reduced" set X_α. Let we find x_k solution with $\varepsilon/2$-accuracy on $f(x)$. Then we found $(\varepsilon/2 + rM)$-solution of original problem, where*

$$r = \max_{x \in X} \left\| x - \operatorname*{argmin}_{\hat{x} \in X^\alpha} \|x - \hat{x}\|_2 \right\|_2.$$

It is not necessary to search for the closest point to each x and find r. It's enough to find one that is "pretty" close and find some upper bound of r. Then it remains to find a rule, which each point x from X associated with some point $\hat{x}$ from X_α and estimate the maximum distance $\max_X \|\hat{x} - x\|_2$. For any simplex point, consider the following rule:

$$\hat{x}_i = \frac{(x_i + 2\alpha)}{(1 + 2\alpha n)}, \qquad i = 1, \ldots n.$$

One can easy to see, that for $\alpha \leq 1/2n$:

$$\sum_{i=1}^n \hat{x}_i = 1, \qquad \hat{x}_i \leq \alpha, \qquad i = 1, \ldots n.$$

It means that $\hat{x} \in X_\alpha$. The distance $\|\hat{x} - x\|_2$:

$$\|\hat{x} - x\|_2 = \sqrt{\sum_{i=1}^n (\hat{x}_i - x_i)^2} = \frac{2\alpha n}{1 + 2\alpha n} \sqrt{\sum_{i=1}^n \left(\frac{1}{n} - x_i\right)^2}.$$

$\sqrt{\sum_{i=1}^n \left(\frac{1}{n} - x_i\right)^2}$ is a distance to the center of the simplex. It can be bounded by the radius of the circumscribed sphere $R = \sqrt{\frac{n-1}{n}} \leq 1$. Then

$$\|\hat{x} - x\|_2 \leq \frac{2\alpha n}{1 + 2\alpha n} \leq 2\alpha n. \tag{10}$$

(10) together with Lemma 5 gives that $f(x_k) - f(x^*) \leq \frac{\varepsilon}{2} + 2\alpha n M$. Then by taking $\alpha = \varepsilon/4nM$ (or less), we find ε-solution of the original problem. And it takes $\tau \leq \alpha = \varepsilon/4nM$.

The second case is a **positive orthant**:

$$\perp_n = \{x_i \geq 0, \quad i \in 1 \ldots n\}.$$

We propose to consider a "reduced" set of the following form:

$$\perp_n^\alpha = \{y_i \geq \alpha, \quad i \in 1 \ldots n\}.$$

One can note that for all i the minimum of the expression $y_i + \tau e_i$ is equal to $\alpha - \tau$, because $e_i \geq -1$ and $y_i \geq \alpha$. Therefore, it is necessary that $\alpha - \tau \geq 0$. It means that for any $e \in RS_2^n(1)$, for the vector $y + \tau e$ the following expression is valid:

$$y_i + \tau e_i \geq 0, \quad i \in 1 \ldots n.$$

The projection onto $\perp_n^\alpha$ is carried out as well as onto $\perp_n$: if $x_i < \alpha$ then $x_i \to \alpha$.

Then let find r in Lemma 5 for orthant. Let for any $x \in \perp_n$ define $\hat{x}$ in the following way:

$$\hat{x}_i = \begin{cases} \alpha, & x_i < \alpha, \\ x_i, & x_i \geq \alpha, \end{cases} \qquad i = 1, \ldots n.$$

One can see that $\hat{x}_i \in \perp_n^\alpha$ and

$$\|\hat{x} - x\|_2 = \sqrt{\sum_{i=1}^{n} (\hat{x}_i - x_i)^2} \leq \sqrt{\sum_{i=1}^{n} \alpha^2} = \alpha\sqrt{n}.$$

By Lemma 5 we have that $f(x_k) - f(x^*) \leq \frac{\varepsilon}{2} + \alpha\sqrt{n}M$. Then by taking $\alpha = \varepsilon/2\sqrt{n}M$ (or less), we find ε-solution of the original problem. And it takes $\tau \leq \alpha = \varepsilon/2\sqrt{n}M$.

The above reasoning can easily be generalized to an arbitrary orthant:

$$\tilde{\perp}_n = \{b_i x_i \geq 0, \quad b_i = \pm 1, \quad i \in 1 \ldots n\}.$$

The third case is a **ball in p-norm** for $p \in [1; 2]$:

$$\mathcal{B}_p^n(a, R) = \{\|x - a\|_p \leq R\},$$

where a is a center of ball, R – its radii. We propose reducing a ball and solving the problem on the "reduced" ball $\mathcal{B}_p^n(a, R(1-\alpha))$. We need the following lemma:

Lemma 6. *Consider two concentric spheres in p norm, where $p \in [1; 2]$, $\alpha \in (0; 1)$:*

$$\mathcal{S}_p^n(a, R) = \{\|x - a\|_p = R\}, \quad \mathcal{S}_p^n(a, R(1 - \alpha)) = \{\|y - a\|_p = R(1 - \alpha)\}.$$

Then the minimum distance between these spheres in the second norm

$$m = \frac{\alpha R}{n^{1/p - 1/2}}.$$

Using the lemma, one can see that for any $x \in \mathcal{B}_n^\alpha(a, R(1-\alpha))$, $\tau \leq \alpha R/n^{1/p-1/2}$ and for any $\mathbf{e} \in \mathcal{RS}_2^n(1)$, $x + \tau\mathbf{e} \in \mathcal{B}_n(a, R)$.

Then let find r in Lemma 5 for ball. Let for any x define $\hat{x}$ in the following way:

$$\hat{x}_i = a + (1 - \alpha)(x_i - a), \qquad i = 1, \ldots n.$$

One can see that $\hat{x}_i$ is in the "reduced" ball and

$$\|\hat{x} - x\|_2 = \sqrt{\sum_{i=1}^{n} (\hat{x}_i - x_i)^2} = \sqrt{\sum_{i=1}^{n} (\alpha(x_i - a))^2} = \alpha \sqrt{\sum_{i=1}^{n} (x_i - a)^2} \leq \alpha \sum_{i=1}^{n} |x_i - a|.$$

By Holder inequality:

$$\|\hat{x} - x\|_2 \leq \alpha \sum_{i=1}^{n} |x_i - a| \leq \alpha n^{\frac{1}{q}} \left(\sum_{i=1}^{n} |x_i - a|^p \right)^{\frac{1}{p}} = \alpha n^{\frac{1}{q}} R.$$

By Lemma 5 we have that $f(x_k) - f(x^*) \leq \frac{\varepsilon}{2} + \alpha n^{1/q} RM$. Then by taking $\alpha = \varepsilon/2n^{1/q} RM$ (or less), we find ε-solution of the original problem. And it takes $\tau \leq \alpha R/n^{1/p-1/2} = \varepsilon/2M\sqrt{n}$.

The fourth case is a **product of sets** $\mathcal{Z} = \mathcal{X} \times \mathcal{Y}$. We define the "reduced" set Z^α as

$$Z^\alpha = X^\alpha \times Y^\alpha,$$

We need to find how the parameter α and τ depend on the parameters α_x, τ_x and α_y, τ_y for the corresponding sets X and Y, i.e. we have bounds: $\alpha_x \leq h_x(\varepsilon)$, $\alpha_y \leq h_y(\varepsilon)$ and $\tau_x \leq g_x(\alpha_x) \leq g_x(h_x(\varepsilon))$, $\tau_y \leq g_y(\alpha_y) \leq g_y(h_y(\varepsilon))$. Obviously, the functions g, h are monotonically increasing for positive arguments. This follows from the physical meaning of τ and α.

Further we are ready to present an analogue of Lemma 5, only for the saddle-point problem.

Lemma 7. *Suppose the function $\varphi(x,y)$ inthe saddle-point problem is M-Lipschitz. Let we find $(\tilde{x}, \tilde{y})$ solution on X^α and Y^α with $\varepsilon/2$-accuracy. Then we found $(\varepsilon/2 + (r_x + r_y)M)$-solution of the original problem, where r_x and r_y we define in the following way:*

$$r_x = \max_{x \in X} \left\| x - \operatorname*{argmin}_{\hat{x} \in X^\alpha} \|x - \hat{x}\|_2 \right\|_2,$$

$$r_y = \max_{y \in Y} \left\| y - \operatorname*{argmin}_{\hat{y} \in Y^\alpha} \|y - \hat{y}\|_2 \right\|_2.$$

In the previous cases we found the upper bound $\alpha_x \leq h_x(\varepsilon)$ from the condition that $r_x M \leq \varepsilon/2$. Now let's take $\tilde{\alpha}_x$ and $\tilde{\alpha}_y$ so that $r_x M \leq \varepsilon/4$ and $r_y M \leq \varepsilon/4$. For this we need $\tilde{\alpha}_x \leq h_x(\varepsilon/2)$, $\tilde{\alpha}_y \leq h_y(\varepsilon/2)$. It means that if we take $\alpha = \min(\tilde{\alpha}_x, \tilde{\alpha}_y)$, then $(r_x + r_y)M \leq \varepsilon/2$ for such α. For a simplex, an orthant and a ball the function h is linear, therefore the formula turns into a simpler expression: $\alpha = \min(\alpha_x, \alpha_y)/2$.

For the new parameter $\alpha = \min(\tilde{\alpha}_x, \tilde{\alpha}_y)$, we find $\tilde{\tau}_x = g_x(\alpha) = g_x(\min(\tilde{\alpha}_x, \tilde{\alpha}_y))$ and $\tilde{\tau}_y = g_y(\alpha) = g_y(\min(\tilde{\alpha}_x, \tilde{\alpha}_y))$. Then for any $x \in X^\alpha$, $\mathbf{e}_x \in \mathcal{RS}_2^{\dim X}(1)$, $y \in Y^\alpha$, $\mathbf{e}_y \in \mathcal{RS}_2^{\dim Y}(1)$, $x + \tilde{\tau}_x \mathbf{e}_x \in X$ and $y + \tilde{\tau}_y \mathbf{e}_y \in Y$. Hence, it is easy to see that for $\tau = \min(\tilde{\tau}_x, \tilde{\tau}_y)$ and the vector $\tilde{\mathbf{e}}_x$ of the first $\dim X$ components of $\mathbf{e} \in \mathcal{RS}_2^{\dim X + \dim Y}(1)$ and for the vector $\tilde{\mathbf{e}}_y$ of the remaining $\dim Y$ components, for any $x \in X^\alpha$, $y \in Y^\alpha$ it is true that $x + \tau\tilde{\mathbf{e}}_x \in X$ and $y + \tau\tilde{\mathbf{e}}_y \in Y$. We get $\tau = \min(\tilde{\tau}_x, \tilde{\tau}_y)$. In the previous cases that we analyzed (simplex, orthant and ball), the function g and h are linear therefore the formula turns into a simpler expression: $\tau = \min(\alpha_x, \alpha_y) \cdot \min(\tau_x/\alpha_x, \tau_y/\alpha_y)/2$.

Summarize the results of this part of the paper in Table 2.

One can note that in (8) τ is independent of n. According to Table 2, we need to take into account the dependence on n. In Table 3, we present the constraints on τ and Δ so that Corollary 1 remains satisfied. We consider three cases when

Table 2. Summary of the part 3.2

Set	α of "reduced" set	Bound of τ	e
probability simplex	$\frac{\varepsilon}{4nM}$	$\frac{\varepsilon}{4nM}$	see (9)
positive orthant	$\frac{\varepsilon}{2\sqrt{n}M}$	$\frac{\varepsilon}{2\sqrt{n}M}$	$\mathcal{RS}_2^n(1)$
ball in p-norm	$\frac{\varepsilon}{2n^{1/q}RM}$	$\frac{\varepsilon}{2\sqrt{n}M}$	$\mathcal{RS}_2^n(1)$
$X^\alpha \times Y^\alpha$	$\frac{\min(\alpha_x,\alpha_y)}{2}$	$\frac{\min(\alpha_x,\alpha_y)\cdot\min(\tau_x/\alpha_x,\tau_y/\alpha_y)}{2}$	$\mathcal{RS}_2^n(1)$

both sets X and Y are simplexes, orthants and balls with the same dimension $n/2$.

The second column of Table 3 means whether the functions are defined not only on the set itself, but also in some neighbourhood of it.

Table 3. τ and Δ in Corollary 1 in different cases

Set	Neigh-d?	τ	Δ
Probability simplex	✓	$\Theta\left(\frac{\varepsilon}{M}\right)$	$\mathcal{O}\left(\frac{\varepsilon^2}{M\Omega na_q}\right)$
	✗	$\Theta\left(\frac{\varepsilon}{Mn}\right)$ and $\leq \frac{\varepsilon}{4nM}$	$\mathcal{O}\left(\frac{\varepsilon^2}{M\Omega n^2 a_q}\right)$
Positive orthant	✓	$\Theta\left(\frac{\varepsilon}{M}\right)$	$\mathcal{O}\left(\frac{\varepsilon^2}{M\Omega na_q}\right)$
	✗	$\Theta\left(\frac{\varepsilon}{M\sqrt{n}}\right)$ and $\leq \frac{\varepsilon}{\sqrt{8n}M}$	$\mathcal{O}\left(\frac{\varepsilon^2}{M\Omega n^{3/2} a_q}\right)$
Ball in p-norm	✓	$\Theta\left(\frac{\varepsilon}{M}\right)$	$\mathcal{O}\left(\frac{\varepsilon^2}{M\Omega na_q}\right)$
	✗	$\Theta\left(\frac{\varepsilon}{M\sqrt{n}}\right)$ and $\leq \frac{\varepsilon}{\sqrt{8n}M}$	$\mathcal{O}\left(\frac{\varepsilon^2}{M\Omega n^{3/2} a_q}\right)$

4 Numerical Experiments

In a series of our experiments, we compare zeroth-order Algorithm 1 (zoSPA) proposed in this paper with Mirror-Descent algorithm from [2] which uses a first-order oracle.

We consider the classical saddle-point problem on a probability simplex:

$$\min_{x\in\Delta_n} \max_{y\in\Delta_k} \left[y^T C x\right], \tag{11}$$

This problem has many different applications and interpretations, one of the main ones is a matrix game (see Part 5 in [2]), i.e. the element c_{ij} of the matrix are interpreted as a winning, provided that player X has chosen the ith strategy and player Y has chosen the jth strategy, the task of one of the players is to maximize the gain, and the opponent's task – to minimize.

We briefly describe how the step of algorithm should look for this case. The prox-function is $d(x) = \sum_{i=1}^{n} x_i \log x_i$ (entropy) and $V_x(y) = \sum_{i=1}^{n} x_i \log x_i/y_i$ (KL divergence). The result of the proximal operator is $u = \mathrm{prox}_{z_k}(\gamma_k g(z_k, \xi_k^{\pm}, \mathbf{e}_k)) = z_k \exp(-\gamma_k g(z_k, \xi_k^{\pm}, \mathbf{e}_k))$, by this entry we mean: $u_i = [z_k]_i \exp(-\gamma_k [g(z_k, \xi_k^{\pm}, \mathbf{e}_k)]_i)$. Using the Bregman projection onto the simplex in following way $P(x) = x/\|x\|_1$, we have

$$[x_{k+1}]_i = \frac{[x_k]_i \exp(-\gamma_k [g_x(z_k, \xi_k^{\pm}, \mathbf{e}_k)]_i)}{\sum\limits_{j=1}^{n} [x_k]_j \exp(-\gamma_k [g_x(z_k, \xi_k^{\pm}, \mathbf{e}_k)]_j)},$$

$$[y_{k+1}]_i = \frac{[y_k]_i \exp(\gamma_k [g_y(z_k, \xi_k^{\pm}, \mathbf{e}_k)]_i)}{\sum\limits_{j=1}^{n} [x_k]_j \exp(\gamma_k [g_y(z_k, \xi_k^{\pm}, \mathbf{e}_k)]_j)},$$

where under g_x, g_y we mean parts of g which are responsible for x and for y. From theoretical results one can see that in our case, the same step must be used in Algorithm 1 and Mirror Descent from [2], because $n^{1/q} = 1$ for $q = \infty$.

In the first part of the experiment, we take matrix 200×200. All elements of the matrix are generated from the uniform distribution from 0 to 1. Next, we select one row of the matrix and generate its elements from the uniform from 5 to 10. Finally, we take one element from this row and generate it uniformly from 1 to 5. Then we take the same matrix, but now at each iteration we add to elements of the matrix a normal noise with zero expectation and variance of 10, 20, 30, 40% of the value of the matrix element. The results of the experiment is on Fig. 1.

According to the results of the experiments, one can see that for the considered problems, the methods with the same step work either as described in the theory (slower n times or $\log n$ times) or generally the same as the full-gradient method.

5   Possible Generalizations

In this paper we consider non-smooth cases. Our results can be generalized for the case of strongly convex functions by using restart technique (see for example [7]). It seems that one can do it analogously.[1] Generalization of the results of

---

[1] To say in more details this can be done analogously for deterministic set up. As for stochastic set up we need to improve the estimates in this paper by changing the Bregman diameters of the considered convex sets Ω by Bregman divergence between starting point and solution. This requires more accurate calculations (like in [11]) and doesn't include in this paper. Note that all the constants, that characterized smoothness, stochasticity and strong convexity in all the estimates in this paper can be determine on the intersection of considered convex sets and Bregman balls around the solution of a radii equals to (up to a logarithmic factors) the Bregman divergence between the starting point and the solution.

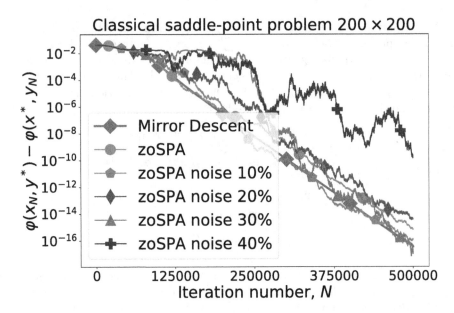

Fig. 1. zoSPA with 0–40% noise and Mirror Descent applied to solve saddle-problem (11).

[6,11,18] and [1,14] for the gradient-free saddle-point set-up is more challenging. Also, based on combinations of ideas from [1,12] it'd be interesting to develop a mixed method with a gradient oracle for x (outer minimization) and a gradient-free oracle for y (inner maximization).

References

1. Alkousa, M., Dvinskikh, D., Stonyakin, F., Gasnikov, A., Kovalev, D.: Accelerated methods for composite non-bilinear saddle point problem. arXiv preprint arXiv:1906.03620 (2019)
2. Ben-Tal, A., Nemirovski, A.: Lectures on Modern Convex Optimization: Analysis, Algorithms, and Engineering Applications. Society for Industrial and Applied Mathematics, Philadelphia (2019)
3. Beznosikov, A., Gorbunov, E., Gasnikov, A.: Derivative-free method for composite optimization with applications to decentralized distributed optimization. arXiv preprint arXiv:1911.10645 (2019)
4. Beznosikov, A., Sadiev, A., Gasnikov, A.: Gradient-free methods for saddle-point problem. arXiv preprint arXiv:2005.05913 (2020)
5. Duchi, J.C., Jordan, M.I., Wainwright, M.J., Wibisono, A.: Optimal rates for zero-order convex optimization: the power of two function evaluations (2013)
6. Dvurechensky, P., Gorbunov, E., Gasnikov, A.: An accelerated directional derivative method for smooth stochastic convex optimization. arXiv preprint arXiv:1804.02394 (2018)
7. Gasnikov, A.: Universal gradient descent. arXiv preprint arXiv:1711.00394 (2017)

8. Gasnikov, A.V., Krymova, E.A., Lagunovskaya, A.A., Usmanova, I.N., Fedorenko, F.A.: Stochastic online optimization. Single-point and multi-point non-linear multi-armed bandits. Convex and strongly-convex case. Autom. Remote Control **78**(2), 224–234 (2017). https://doi.org/10.1134/S0005117917020035

9. Gasnikov, A.V., Lagunovskaya, A.A., Usmanova, I.N., Fedorenko, F.A.: Gradient-free proximal methods with inexact oracle for convex stochastic nonsmooth optimization problems on the simplex. Autom. Remote Control **77**(11), 2018–2034 (2016). https://doi.org/10.1134/S0005117916110114

10. Goodfellow, I.: Nips 2016 tutorial: generative adversarial networks. arXiv preprint arXiv:1701.00160 (2016)

11. Gorbunov, E., Dvurechensky, P., Gasnikov, A.: An accelerated method for derivative-free smooth stochastic convex optimization. arXiv preprint arXiv:1802.09022 (2018)

12. Ivanova, A., et al.: Oracle complexity separation in convex optimization. arXiv preprint arXiv:2002.02706 (2020)

13. Langley, P.: Crafting papers on machine learning. In: Langley, P. (ed.) Proceedings of the 17th International Conference on Machine Learning. (ICML 2000), Stanford, CA, pp. 1207–1216. Morgan Kaufmann (2000)

14. Lin, T., Jin, C., Jordan, M., et al.: Near-optimal algorithms for minimax optimization. arXiv preprint arXiv:2002.02417 (2020)

15. Nesterov, Y., Spokoiny, V.G.: Random gradient-free minimization of convex functions. Found. Comput. Math. **17**(2), 527–566 (2017)

16. Shamir, O.: An optimal algorithm for bandit and zero-order convex optimization with two-point feedback. J. Mach. Learn. Res. **18**(52), 1–11 (2017)

17. Sutton, R.S., Barto, A.G.: Reinforcement Learning: An Introduction. MIT Press, Cambridge (2018)

18. Vorontsova, E.A., Gasnikov, A.V., Gorbunov, E.A., Dvurechenskii, P.E.: Accelerated gradient-free optimization methods with a non-Euclidean proximal operator. Autom. Remote Control **80**(8), 1487–1501 (2019)

On Multiple Coverings of Fixed Size Containers with Non-Euclidean Metric by Circles of Two Types

Alexander Kazakov[1] , Anna Lempert[1(✉)] , and Quang Mung Le[2]

[1] Matrosov Institute for System Dynamics and Control Theory of SB RAS,
664033 Irkutsk, Russia
kazakov@icc.ru, lempert.a.a@gmail.com
[2] DaNang Department of Information and Communications,
Danang 550000, Viet Nam
quangmungle2010@gmail.com
http://www.idstu.irk.ru, https://tttt.danang.gov.vn

Abstract. The paper is devoted to the multiple covering problem by circles of two types. The number of circles of each class is given as well as a ratio radii. The circle covering problem is usually studied in the case when the distance between points is Euclidean. We assume that the distance is determined using some particular metric arising in logistics, which, generally speaking, is not Euclidean. The numerical algorithm is suggested and implemented. It based on an optical-geometric approach, which is developed by the authors in recent years and previously used only for circles of an equal radius. The results of a computational experiment are presented and discussed.

Keywords: Circle covering problem · Multiple covering ·
Non-Euclidean metric · Incongruent circles · Optical-geometric
approach · Logistics

1 Introduction

The covering problems are widely used in various technical and economic fields of human activity. Examples of such tasks are locating ATMs, hospitals, artificial Earth satellites, schools, medical ambulance stations, cell towers [3,6,11], wireless sensors [1,2,8].

In general form, this problem is formulated as follows: how to locate geometric objects in a bounded area so that the covered area is completely inside in the union of these objects. Equal circles are often used as covering elements. In most cases, we are talking about the one-fold circle covering problem (CCP), which is considered in a large number of papers (for example, [20,30,32]).

There are other statements of the covering problem, such as the single covering with circles of different radii and the multiple covering by equal circles. In this case, as a rule, the radii ratio obeys the additional restrictions.

© Springer Nature Switzerland AG 2020
Y. Kochetov et al. (Eds.): MOTOR 2020, CCIS 1275, pp. 120–132, 2020.
https://doi.org/10.1007/978-3-030-58657-7_12

The problem of a single covering by unequal circles was first investigated by F. Toth and J. Molnar [31]. They proposed a hypothesis about the lower bound for the covering density. Then, this hypothesis was proven by G. Toth [29]. Florian and Heppes [14] established a sufficient condition for such a covering to be solid in the sense of [31]. Dorninger presented an analytical description for the general case (covering by unequal circles) in such a way that the conjecture can easily be numerically verified and upper and lower limits for the asserted bound can be gained [12].

The multiple covering problem is as interesting and important as the classical CCP. Global navigation systems GPS (USA), Glonass (Russia), Baidu (China) and Galileo (EU) use a multiple covering (at least 3-fold) of the served areas to ensure positioning accuracy. For the multiple covering of a circle by congruent circles on a plane, the first exact results for $k = 2, 3, 4$ were obtained by Blundon [4]. Some analytical results are obtained in the special cases when the covered area is a regular polygon [17, 26, 33]. These results are very important to verify the correctness of approximate results found by numerical methods. Among approximate methods, we can mention the greedy [10], heuristic [9, 18, 20], and combinatorial [15] algorithms.

Note that the most of known results are obtained for the case when a covered set is a subset of the Euclidean space. In the case of a non-Euclidean metric, this problem is relatively poorly studied. Moreover, the problem of multiple covering with unequal circles, apparently, has not been considered yet.

In this paper, we deal with multiple covering by circles of two types with a specific non-Euclidean metric. This metric allows using the time as a measure of the distance [21, 22]. We expand a technique based on the combination of optical-geometric approach [21] and Voronoi diagram [16, 19, 28].

The results of a computational experiment are presented and discussed.

2   Problem Statement and Modeling

Let us consider some bounded field (service area) where it is required to locate a certain number of service facilities in such a way that their service zones, having a given shape, completely cover it. Such statements appear in problems of cell towers or security points placement [3, 13], designing energy-efficient monitoring of distributed objects by wireless sensor networks [2, 27], etc. The most straightforward problem statement of this type assumes that the service areas have the form of circles whose radii are the same, and it is enough to cover each point of the serviced space at least once. As a result, we have the classical circle covering problem (see, introduction). However, various complications and generalizations are possible in connection with applications.

Firstly, the need to take into account terrain features (for example, relief) leads to the fact that service areas cease to be circles. One way to solve this problem is to introduce a specific metric, which, in fact, replaces the physical distance between points by the minimum time it takes to pass the path between them [7, 24].

Secondly, often, it is required that two or more objects service each point of the area. This situation is more typical for security tasks when it is necessary to ensure the correct operation of the system in case of failure of some of the servicing devices due to an accident or sabotage. However, such requirements may also apply to logistic systems (systems with duplication or redundancy).

Thirdly, service areas may be different. A similar requirement arises if we use service objects of various types.

Each of the additional requirements separately was previously considered (see, for example, [2,3]). Moreover, we have already studied models in which two of the three conditions were taken into account simultaneously [22,25]. However, three conditions are simultaneously considered for the first time. For definiteness, we will further talk about logistic systems (and serving logistic centers) and proceed to model designing.

We make a simplifying assumption. Suppose we are given a bounded domain, in which consumers are continuously distributed and there are only two types of logistic centers. Let n and m be a number of logistic centers of the first and second type, respectively, τ_1 and τ_2 be their maximum delivery time and $\tau_2 = \alpha\tau_1, \alpha > 0$. Here, the maximum delivery time is the time for which the goods are delivered to the most distant consumer at the border of the service area of the logistic center means the "radius" of this zone. It is required to locate the centers so that each consumer must be serviced by at least k of them ($k < n + m$), and the parameters τ_1, τ_2 would be minimal.

Note that in logistics, such a statement is quite natural, since the characteristics of the service centers directly affecting the delivery time of goods (such as the area of storage facilities, handling equipment, the capacity of parking lots and garages, etc.) are determined at the design stage.

If we know only the total number of logistic centers $n+m$ and the multiplicity k, then the best placement is one with the shortest average delivery time $\bar\tau = \tau_1(n + \alpha m)/(n + m)$.

Next, we turn to the mathematical formulation of the described problem.

3   Mathematical Formulation

Assume we are given a metric space X, a bounded domain $M \subset X$ with a continuous boundary ∂M, n circles $C_i(O_i, R_1)$ and m circles $C_i(O_i, R_2)$; here $O_i(x_i, y_i)$ is a circle center, R_1 and R_2 are radii. Let $f(x, y) > 0$ be a continuous function, which shows the instantaneous speed of movement at every point of X. The minimum moving time between two points $a, b \in X$ is determined as follows:

$$\rho(a, b) = \min_{\Gamma \in G(a,b)} \int_{\Gamma} \frac{d\Gamma}{f(x, y)} , \tag{1}$$

where $G(a, b)$ is the set of continuous curves, which belong to X and connect two points a and b. It is easy to verify that for the distance determined by formula (1), all metric axioms are satisfied. In logistic problems, in particular, $\rho(a, b)$

determines the minimum time for the delivery of goods between points. Still, it may also have another meaning, for example, determining the geodetic distance. Therefore, to avoid direct association with transportation, we will further use the traditional symbol R to designate the circle radius in metric (1).

It is required to locate the circles to minimize the radii and to cover M at least k times. The last means that every point of M must belong at least k different circles.

In other words, we have the following optimization problem:

$$R_1 \to \min, \tag{2}$$

$$R_2 = \alpha R_1, \alpha \in \mathbb{R}^+, \tag{3}$$

$$\max_{j \in J_k(s)} \omega \rho(s, O_j) \le R_1. \tag{4}$$

Here $\omega = \begin{cases} 1, & i = 1, ..., n \\ 1/\alpha, & i = n+1, ..., n+m \end{cases}$, and $J_k(s)$ is the set of indexes (numbers) of k centers, that locate closer to s than other $n + m - k$ centers:

$$J_k(s) = \left\{ q_j, j = 1, ..., k : \rho(O_{q_j}, p) \le \omega \rho(O_l, p) \; \forall l = \{1, ..., n+m\} \setminus \{q_1, ..., q_j\} \right\}.$$

The objective function (2) minimizes the radius of the covering. Constraint (3) fixes the radii ratio, and (4) guarantees that each point of M belongs to at least k circles.

Note, if $\alpha = 1$, we have the multiple covering of a bounded domain by equal circles with non-Euclidean metric [25].

4   Solution Method

In this section, we propose a numerical method for solving problem (2)–(4), based on traditional principles for our studies. We combine the analogy between the propagation of the light wave and finding the minimum of integral functional (1) and Voronoi diagram technic. This approach is described in more details in [21,23].

The concept of k–th order Voronoi diagrams was introduced by F.L.Toth [28] and earlier was used in studies [16,17,25]. To apply it, at first, we should determine a k–fold Voronoi-Dirichlet region for the case of two types of circles.

For a set of $n + m$ points O_i, the generalized k–fold Voronoi region M_i^k centered at O_i is defined as follows:

$$M_i^k = \left\{ p \in M : \rho(p, O_i) \le \max_{j \in J_k(p)} \lambda \rho(p, O_j) \right\}, i = 1, ..., n+m, k < n+m, \tag{5}$$

where $\lambda = \begin{cases} 1, & i, j = 1, ..., n; \; i, j = n+1, ..., n+m, \\ 1/\alpha, & i = 1, ..., n, j = n+1, ..., n+m, \\ \alpha, & i = n+1, ..., n+m, j = 1, ..., n. \end{cases}$

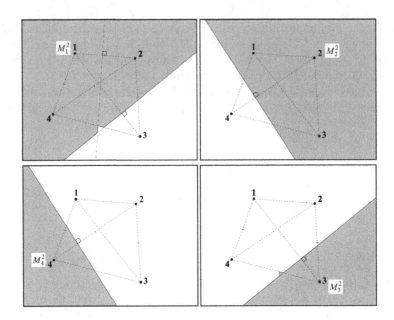

Fig. 1. Double Voronoi-Dirichlet regions

Figure 1 shows double Voronoi-Dirichlet regions (grey color) for the case of four circles (we point out their centers only) where the radii of circles 1 and 2 are equal, but they are three times larger than the radii of circles 3 and 4.

At first, we propose the OCMC (One Covering Minimum Circle) algorithm, which allows finding a circle $C(O^*, R^*)$ centered in O^*, covering region M and having an approximately minimal radius R^*.

The principle of this algorithm is that the randomly generated center of the circle moves in the direction of decreasing the maximum distance from it to the boundary of the covered region. This process finishes when the coordinates of the center stop changing (Fig. 2).

Here and further we cover set M by a uniform rectangular grid with the step h and deal with set M^h approximating M. For brevity, we omit the index h.

Algorithm OCMC

Step 1. Put $R^* = +\infty$, $Iter = 1$.

Step 2. Randomly generate initial coordinates of a point $O(x, y) \in M$.

Step 3. Define the set of nearest points for the point O:

$$\Delta O = \{O(x + \chi, y + \sigma) : \chi, \sigma = \{-h, 0, h\}\}.$$

Step 4. Find a point O_{new}:

$$O_{new} = \arg\min_{p \in \Delta O} \max_{s \in \partial M} \rho(p, s).$$

Step 5. If $\rho(O_{new}, \partial M) \leq \rho(O, \partial M)$, then put $O := O_{new}$ and go to Step 3.

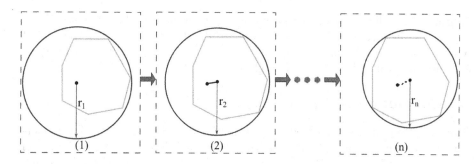

Fig. 2. The principle of OCMC algorithm

Step 6. If $\rho(O, \partial M) < R^*$, then put $O^* := O$, $R^* := \rho(O^*, \partial M)$.
Step 7. The counter *Iter* of an initial solution generations is incremented. If it becomes equal a certain prescribed value, then the algorithm is terminated. Otherwise, go to Step 2.

The general algorithm includes the basic steps: constructing the generalized k-fold Voronoi diagram for the initial set of centers; moving O_i to the point O_i^*, that is the center of the covering circle, which has the minimal radius for each part of the diagram; revising radius ratio and returning to the first step with the new centers. Now we describe the general algorithm in details.

General algorithm

Step 1. Randomly generate initial coordinates of the circles centers $O_i \in M$, $i = 1, ..., n + m$.
Step 2. From $O_i, i = 1, ..., n + m$, we initiate the light waves using the algorithm from [21]. The speed of a light wave emitted from points $O_i, i = 1, ..., n$ is α times less than from $O_i, i = n + 1, ..., n + m$. This allows us to find the time $T_i(x, y), i = 1, ..., n$ which is required to reach $s(x, y)$ by each wave. For every $s(x, y) \in M$ we obtain vector $T(x, y) = T_i(x, y)$.
Step 3. For each $s(x, y)$ we choose k minimal components of vector $T(x, y)$. Thus, we obtain $J_k(s)$ which is the index set of Voronoi domains contained $s(x, y)$.
Step 4. Find k–fold Voronoi domain $M_i^k, i = 1, ..., n + m$ and their boundaries ∂M_i^k.
Step 5. For each $M_i^k, i = 1, ..., n + m$ we find a minimal covering circle $C_i(O_i^*, R_i^*)$ by OCMC algorithm.
Step 6. To ensure full covering of M by circles, we choose the maximum radius $R_1 = \max\limits_{i=1,...,n} R_i$ and $R_2 = \max\limits_{i=n+1,...,n+m} R_i$.
Step 7. Check the inequality $R_2 \geq \alpha R_1$. If it is satisfied, then put $R_1 = R_2/\alpha$, otherwise, put $R_2 = \alpha R_1$.
Step 8. If the value of the founded radius is less than the previous one, we save the current radius and the current set of circles. The counter of an initial solution generations is incremented. If it becomes equal a certain prescribed value, then the algorithm is terminated. Otherwise, go to Step 1.

A drawback of the algorithm is that it does not guarantee a solution that globally minimizes the circles radii. This feature is inherited from the constructing of Voronoi diagram. We use multiple generating of initial positions (Step 1) to increase the probability of finding a global solution.

5   Computational Experiment

The algorithms are implemented in C# using the Visual Studio 2015. The numerical experiment was carried out using the PC of the following configuration: Intel (R) Core i5-3570K (3.4 GHz, 8 GB RAM) and Windows 10 operating system.

Note that in the tables n is a number of big circles, m is a number of small circles, k is multiplicity of the covering, $R_{n,m}^k$ is the best radius of the big circles, $\Delta R_{n,m}^k = \frac{n + \alpha m}{n + m} R$ is the average radius of the covering.

In the figures, the origin is located in the upper left corner, the bold black closed curves are large circles, the thin ones are small circles, the grey dots are the centers of circles, the dashed line lines are the boundary of container M. The number of random generations $Iter = 100$, the grid step $h = 0.001$.

Example 1. This example illustrates how the proposed in the previous section algorithm works in the case of the Euclidean metric $f(x, y) \equiv 1$. The covered set is a square with a side equals to 3, $\alpha = 0.5$, $k = 2, 3, 4$. Table 1 shows the best solutions for 15 circles.

Table 1. The best coverings of a square by 15 circles with Euclidean metric

n	m	$R_{n,m}^2$	$\Delta R_{n,m}^2$	$R_{n,m}^3$	$\Delta R_{n,m}^3$	$R_{n,m}^4$	$\Delta R_{n,m}^4$
14	1	0.27739	0.26814	0.34132	0.32994	0.50000	0.48333
13	2	0.27877	0.26018	0.35184	0.32838	0.50000	0.46667
12	3	0.28985	0.26086	0.35358	0.31823	0.50018	0.45016
11	4	0.30170	0.26147	0.37174	0.32217	0.50028	0.43357
10	5	0.30699	0.25583	0.38873	0.32394	0.50584	0.42154
9	6	0.31457	0.25166	0.40089	0.32071	0.51499	0.41199
8	7	0.32299	0.24763	0.41846	0.32082	0.52389	0.40165
7	8	0.33483	0.24554	0.42964	0.31507	0.53729	0.39401
6	9	0.35668	0.24968	0.46228	0.32359	0.55318	0.38723
5	10	0.37642	0.25094	0.50071	0.33381	0.55607	0.37071
4	11	0.39016	0.24710	0.51579	0.32667	0.58426	0.37003
3	12	0.41236	0.24742	0.54103	0.32462	0.63004	0.37802
2	13	0.44312	0.25110	0.56356	0.31935	0.70711	0.40069
1	14	0.47796	0.25491	0.61036	0.32553	0.70711	0.37712

Table 1 shows that the radii of circles, as one would expect, grow with an increase in the number of small circles and a simultaneous decrease in the number

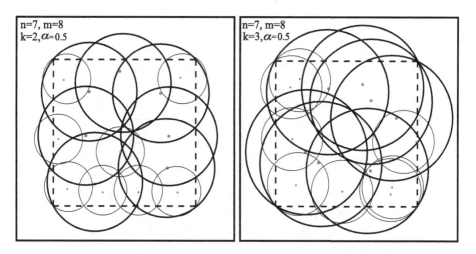

Fig. 3. The best 2-fold (left) and 3-fold (right) coverings with 15 circles

of large ones. Other, more specific laws could not be identified. It is noteworthy that for a 4-fold covering, the radii $R^4_{2,13}$ and $R^4_{1,14}$ are equal.

Average radii behave even less regularly. The best 2, 3-fold coverings consist of 7 large circles and 8 small ones (see Fig. 3), and 4-fold covering contains 4 large circles and 11 small ones. In addition, we note that the average radius of 2-fold covering with circles of two types is always less than the best radius of 2-fold covering with equal ones $R^1_{15} = 0.27012$ (see [25]).

The operating time is $3'20'' \div 4'34''$.

Example 2. Let $f(x,y) = 0.5 + 2x$. It means that instantaneous speed of movement increases linearly along the coordinate x. The covered set M is following

$$M = \{(x,y) : (x - 2.5)^2 + (y - 2.5)^2 \le 4\}.$$

The best solutions for the cases of 2,3,4-fold coverings with 13 circles of two types are shown in Table 2. Here the radii ratio is $1/3$.

Note that in this case the wave fronts also have the form of a circle, as in the Euclidean metric, but the source of the wave (the center of the circle) is displaced (see more in [5]). The apparent size of the covering circles depends on the location of their centers: the closer it to the axis Oy, the smaller it looks (Fig. 4). We emphasize that in the given metric the radii are equal.

Table 2 shows that the radii of circles, as in the previous example, grow with an increase in the number of small circles. The average radii decrease monotonously with an increase in the number of small circles. The best 2, 3-fold coverings consist of 1 large circle and 12 small ones, and 4-fold covering contains 2 large circles and 11 small ones.

Table 2. The best coverings of a circle by 13 circles with the "linear" metric

n	m	$R_{n,m}^2$	$\Delta R_{n,m}^2$	$R_{n,m}^3$	$\Delta R_{n,m}^3$	$R_{n,m}^4$	$\Delta R_{n,m}^4$
12	1	0.27457	0.26049	0.35948	0.34105	0.43305	0.41084
11	2	0.28117	0.25233	0.37712	0.33844	0.45014	0.40397
10	3	0.29531	0.24988	0.39511	0.33433	0.46838	0.39632
9	4	0.30825	0.24502	0.41499	0.32987	0.47750	0.37955
8	5	0.32948	0.24499	0.43755	0.32536	0.48785	0.36276
7	6	0.35322	0.24454	0.45458	0.31471	0.49786	0.34467
6	7	0.37173	0.23829	0.48095	0.30830	0.50551	0.32404
5	8	0.39442	0.23261	0.48902	0.28839	0.51528	0.30388
4	9	0.43990	0.23687	0.49312	0.26553	0.52712	0.28383
3	10	0.46108	0.22463	0.54984	0.26787	0.58098	0.28304
2	11	0.49159	0.21428	0.61712	0.26900	0.63789	0.27805
1	12	0.55774	0.21451	0.69580	0.26762	0.72432	0.27859

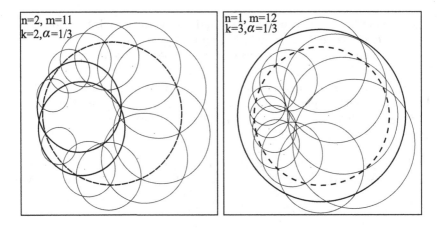

Fig. 4. The best 2-fold (left) and 3-fold (right) coverings of a circle by 13 circles with the "linear" metric

Figure 4 (right) illustrates the interesting 3-fold covering. It splits into 1-fold covering by 1 large circle and 2-fold covering by 12 small ones.

The operating time of the proposed algorithm is $3'11'' \div 4'08''$.

Example 3. Let the covered set M is a polygon with the vertices: (0.5, 1.5); (0.5, 3.5); (1.5, 4.5); (3.5, 4.5); (4.5, 3.5); (4.5, 1.5); (3.5, 0.5); (1.5, 0.5). The instantaneous speed of movement $f(x,y)$ is defined as follows:

$$f(x,y) = \frac{3}{(x-2)^2 + (y-2.5)^2 + 1} + 1.$$

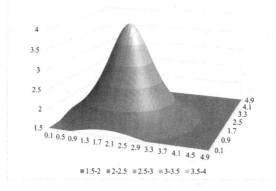

Fig. 5. Level lines of function $f(x, y)$

Figure 5 shows level lines of $f(x, y)$. From the lowest to the highest point, the wave speed increases.

Table 3 shows the best coverings of M by 19 circles of two types for $\alpha = 1/4$. One can see that the radii of the circles, as in the two previous examples, grow with an increasing number of small circles. Moreover, the increase in all cases occurs with acceleration.

The best 2-fold coverings consist of 7 large and 12 small circles, 3-fold coverings includes of 10 large and 9 small circles, and 4-fold covering contains 2 large circles and 17 small ones. Figure 6 shows that the wave fronts differ significantly from the circles, and the covering elements have an oviform shape.

The operating is $4'40'' \div 6'05''$.

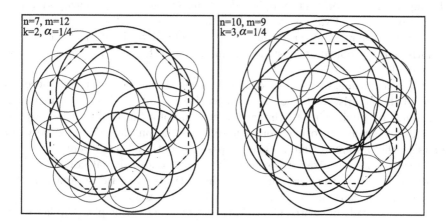

Fig. 6. The best 2-fold (left) and 3-fold (right) coverings of a polygon by 19 circles with the non-Euclidean metric

Table 3. The best multiple coverings in the example 3

n	m	$R_{n,m}^2$	$\Delta R_{n,m}^2$	$R_{n,m}^3$	$\Delta R_{n,m}^3$	$R_{n,m}^4$	$\Delta R_{n,m}^4$
18	1	0.51256	0.49232	0.63925	0.61401	0.78887	0.75773
17	2	0.53152	0.48956	0.65804	0.60609	0.81266	0.74851
16	3	0.55448	0.48882	0.69060	0.60882	0.85300	0.75199
15	4	0.56005	0.47162	0.71822	0.60482	0.86624	0.72947
14	5	0.58186	0.46702	0.74946	0.60154	0.91106	0.73125
13	6	0.60108	0.45872	0.77528	0.59166	0.96564	0.73693
12	7	0.62704	0.45378	0.80655	0.58369	0.96823	0.70069
11	8	0.64791	0.44331	0.82150	0.56208	1.04158	0.71266
10	9	0.68587	0.44221	0.85210	0.54938	1.07594	0.69370
9	10	0.71505	0.43280	0.94860	0.57416	1.11342	0.67391
8	11	0.76475	0.43269	0.98175	0.55546	1.14014	0.64508
7	12	0.80687	0.42467	1.05814	0.55692	1.19896	0.63103
6	13	0.87353	0.42527	1.18404	0.57644	1.27998	0.62315
5	14	0.96564	0.43200	1.25147	0.55987	1.38141	0.61800
4	15	1.05610	0.43078	1.36896	0.55839	1.45417	0.59315
3	16	1.16501	0.42921	1.50528	0.55458	1.62250	0.59776
2	17	1.29896	0.42729	1.67998	0.55262	1.74612	0.57438
1	18	1.47143	0.42594	1.90411	0.55119	2.02766	0.58695

6   Conclusion

The paper considers one of the topical problems for logistic and security systems: optimal placement of various service facilities (sensors, CCTV cameras, logistic centers) with the reservation (duplication). We formulate the subject problem in the form of the problem of constructing an optimal k-fold covering of a bounded set by circles of two types.

At the same time, we use a specific non-Euclidean metric to take into account the local characteristics of the service area (for example, relief). The metric is determined by minimizing the integral functional of a function that defines the speed of movement. In other words, it replaces the physical distance between points by the minimum time it takes to pass the path between them.

To solve the optimization problem, we suggest an original computational algorithm based on the combination of the optical-geometric approach and a new method for constructing generalized multiple Voronoi diagrams.

We have already presented algorithms based on these principles [25]; however, in this case, the procedure for constructing multiple Voronoi diagram is much more complicated. The reason is the presence of various types of elements in the covering, which, in turn, often leads to the non-convexity and the multiply-connection of Voronoi regions.

✱ The algorithm is implemented, and a computational experiment is carried out. It shows that the developed tools effectively solve the problem with the number of objects up to 20. Besides, it turned out that in the best (from the application domain point of view) covering, as a rule, objects of both types are present. This fact is an additional confirmation of the relevance of the study.

Further studies may be associated, firstly, with an increase in the number of types of covering elements; secondly, with an increase in the adequacy of the model, in particular, the use of two-level optimization problems as a mathematical formalization.

Acknowledgements. This work was partially supported by the Russian Foundation for Basic Research, research projects Nos 18-08-00604, 20-010-00724.

References

1. Al-Sultan, K., Hussain, M., Nizami, J.: A genetic algorithm for the set covering problem. J. Operat. Res. Soc. **47**(5), 702–709 (1996)
2. Astrakov, S., Erzin, A., Zalyubovskiy, V.: Sensor networks and covering of plane by discs. J. Appl. Indust. Math. **16**(3), 3–19 (2009)
3. Bánhelyi, B., Palatinus, E., Lévai, B.L.: Optimal circle covering problems and their applications. CEJOR **23**(4), 815–832 (2014). https://doi.org/10.1007/s10100-014-0362-7
4. Blundon, W.: Multiple covering of the plane by circles. Mathematika **4**, 7–16 (1957)
5. Borovskikh, A.: The two-dimensional Eikonal equation. Sib. Math. J. **47**(2), 813–834 (2006)
6. Brusov, V., Piyavskii, S.: A computational algorithm for optimally covering a plane region. Comput. Math. Math. Phys. **11**(2), 17–27 (1971)
7. Bychkov, I.V., Kazakov, A.L., Lempert, A.A., Bukharov, D.S., Stolbov, A.B.: An intelligent management system for the development of a regional transport logistics infrastructure. Automa. Remote Control **77**(2), 332–343 (2016). https://doi.org/10.1134/S0005117916020090
8. Cardei, M., Wu, J., Lu, M.: Improving netwotk lifetime using sensors with adjustible sensing ranges. Int. J. Sens. Netw. **1**(1–2), 41–49 (2006)
9. Ceria, S., Nobili, P., Sassano, A.: A Lagrangian-based heuristic for large scale set covering problems. Math. Progr. **81**, 215–228 (1998)
10. Chvatal, V.: A greedy heuristic for the set covering. Math. Optim. Res. **4**, 233–235 (1979)
11. Das, G., Das, S., Nandy, S., Shina, B.: Efficient algorithm for placing a given number of base station to cover a convex region. J. Parallel Distrib. Comput. **66**, 1353–1358 (2006)
12. Dorninger, D.: Thinnest covering of the Euclidean plane with incongruent circles. Anal. Geom. Metr. Spaces **5**(1), 40–46 (2017)
13. Drezner, Z.: Facility Location: A Survey of Applications and Methods. Springer, New York (1995)
14. Florian, A., Heppes, A.: Solid coverings of the Euclidean plane with incongruent circles. Discret. Comput. Geom. **23**(2), 225–245 (2000)
15. Fujito, T.: On combinatorial approximation of covering 0–1 integer programs and partial set cover. J. Comb. Optim. **8**, 439–452 (2004)

16. Galiev, S., Karpova, M.: Optimization of multiple covering of a bounded set with circles. Comput. Math. Math. Phys. **50**, 721–732 (2010)

17. Galiev, S., Khorkov, A.: Multiple circle coverings of an equilateral triangle, square, and circle. Diskret. Anal. Issled. Oper. **22**, 5–28 (2015)

18. Hall, N., Hochbaum, D.: A fast approximation algorithm for the multicovering problem. Discret. Appl. Math. **15**, 35–40 (1989)

19. Jain, A., Dubes, R.: Algorithms for Clustering Data. Prentice Hall, Upper Saddle River (1988)

20. Johnson, D.: Approximation algorithms for combinatorial problems. J. Comput. System Sci. **9**, 256–278 (1974)

21. Kazakov, A., Lempert, A.: An approach to optimization in transport logistics. Autom. Remote Control **72**(7), 1398–1404 (2011)

22. Kazakov, A., Lempert, A.: On mathematical models for optimization problem of logistics infrastructure. Int. J. Artif. Intell. **13**(1), 200–210 (2015)

23. Kazakov, A., Lempert, A., Le, Q.: An algorithm for packing circles of two types in a fixed size container with non-Euclidean metric. In: Supplementary Proceedings of the 6th International Conference on AIST, vol. 1975, pp. 281–292. CEUR-WS (2017)

24. Lempert, A.A., Kazakov, A.L., Bukharov, D.S.: Mathematical model and program system for solving a problem of logistic objects placement. Autom. Remote Control **76**(8), 1463–1470 (2015). https://doi.org/10.1134/S0005117915080111

25. Lempert, A., Le, Q.: Multiple covering of a closed set on a plane with non-Euclidean metrics. IFAC-PapersOnLine **51**(32), 850–854 (2018)

26. Sriamorn, K.: Multiple lattice packings and coverings of the plane with triangles. Discrete Comput. Geom. **55**, 228–242 (2016)

27. Tabirca, T., Yang, L., Tabirca, S.: Smallest number of sensors for k-covering. Int. J. Comput. Commun. Control **8**(2), 312–319 (2013)

28. Toth, G.: Multiple packing and covering of the plane with circles. Acta Math. Acad. Sci. Hungar. **27**, 135–140 (1976)

29. Toth, G.: Covering the plane with two kinds of circles. Discrete Comput. Geom. **13**, 445–457 (1995)

30. Toth, G.: Thinnest covering of a Circle by Eight, Nine, or Ten congruent circles. Comb. Comput. Geom. **52**, 361–376 (2005)

31. Toth, L.F.: Solid circle-packings and circle-coverings. Studia Sci. Math. Hungar. **3**, 401–409 (1968)

32. Zahn, J.: Black box maximization of circular coverage. J. Res. Natl. Bur. Stand. **66**, 181–216 (1962)

33. Zong, C.: Packing, covering and tiling in two-dimension spaces. Expo. Math. **32**, 297–364 (2012)

Analogues of Switching Subgradient Schemes for Relatively Lipschitz-Continuous Convex Programming Problems

Alexander A. Titov[1,3(✉)] , Fedor S. Stonyakin[1,2] ,
Mohammad S. Alkousa[1,3] , Seydamet S. Ablaev[2] ,
and Alexander V. Gasnikov[1]

[1] Moscow Institute of Physics and Technology, Moscow, Russia
{a.a.titov,mohammad.alkousa}@phystech.edu, gasnikov@yandex.ru
[2] V. I. Vernadsky Crimean Federal University, Simferopol, Russia
fedyor@mail.ru, seydamet.ablaev@yandex.ru
[3] National Research University, Higher School of Economics, Moscow, Russia

Abstract. Recently some specific classes of non-smooth and non-Lipsch-itz convex optimization problems were considered by Yu. Nesterov and H. Lu. We consider convex programming problems with similar smoothness conditions for the objective function and functional constraints. We introduce a new concept of an inexact model and propose some analogues of switching subgradient schemes for convex programming problems for the relatively Lipschitz-continuous objective function and functional constraints. Some class of online convex optimization problems is considered. The proposed methods are optimal in the class of optimization problems with relatively Lipschitz-continuous objective and functional constraints.

Keywords: Convex programming problem · Switching subgradient scheme · Relative lipschitz-continuity · Inexact model · Stochastic mirror descent · Online optimization problem

The research in Sect. 2 was supported by Russian Foundation for Basic Research grant, project number 18-31-20005 mol-a-ved. The research in Sect. 3 and partially in Sect. 6 was supported by Russian Foundation for Basic Research grant, project number 18-31-00219 mol-a. The research in Sect. 5 was supported by the Ministry of Science and Higher Education of the Russian Federation (Goszadaniye) no. 075-00337-20-03. The research in Sect. 4 and Algorithms 6 and 7 (in Appendix) was supported by the grant of the President of Russian Federation for young candidates of sciences (project MK-15.2020.1).

Y. Kochetov et al. (Eds.): MOTOR 2020, CCIS 1275, pp. 133–149, 2020.
https://doi.org/10.1007/978-3-030-58657-7_13

1   Introduction

Different relaxations of the classical smoothness conditions for functions are interesting for a large number of modern applied optimization problems. In particular, in [2] there were proposed conditions of relative smoothness of the objective function, which mean the replacement of the classic Lipschitz condition by the following weaker version

$$f(y) \leq f(x) + \langle \nabla f(x), y - x \rangle + LV_d(y, x), \tag{1}$$

to hold for any x, y from the domain of the objective function f and some $L > 0$; $V_d(y, x)$ represents an analogue of the distance between the points x and y (often called the *Bregman divergence*). Such a distance is widely used in various fields of science, in particular in mathematical optimization. Usually, the *Bregman divergence* is defined on the base of the auxiliary 1-strongly convex and continuously-differentiable function $d : Q \subset \mathbb{R}^n \to \mathbb{R}$ (*distance generating function*) as follows

$$V_d(y, x) = d(y) - d(x) - \langle \nabla d(x), y - x \rangle \quad \forall x, y \in Q, \tag{2}$$

where Q is a convex closed set, $\langle \cdot, \cdot \rangle$ is a scalar product in $\mathbb{R}^n$. In particular, for the Euclidean setting of the problem, we have $d(x) = \frac{1}{2}\|x\|_2^2$ and $V_d(y, x) = d(y - x) = \frac{1}{2}\|y - x\|_2^2$ for arbitrary $x, y \in Q$. However, in many applications, it often becomes necessary to use non-Euclidean norms. Moreover, the considered condition of relative smoothness in [2,16] implies only the convexity (but not strong convexity) of the distance generating function d. As shown in [16], the concept of relative smoothness makes it possible to apply a variant of the gradient method to some problems which were previously being solved only by interior-point methods. In particular, we talk about the well-known problem of construction of an optimal ellipsoid which covers a given set of points. This problem is important in the field of statistics and data analysis.

A similar approach to the Lipschitz property and non-smooth problems was proposed in [17] (see also [24]). This approach is based on an analogue of the Lipschitz condition for the objective function $f : Q \to \mathbb{R}$ with Lipschitz constant $M_f > 0$, which involves replacing the boundedness of the norm of the subgradient, i.e. $\|\nabla f(x)\|_* \leq M_f$, with the so-called *relative Lipschitz condition*

$$\|\nabla f(x)\|_* \leq \frac{M_f \sqrt{2V_d(y, x)}}{\|y - x\|} \quad \forall x, y \in Q, \; y \neq x,$$

where $\|\cdot\|_*$ denotes the conjugate norm, see Sect. 2 below. Moreover, the distance generating function d must not necessarily be strongly convex. In [17] there were proposed deterministic and stochastic Mirror Descent algorithms for optimization problems with convex relatively Lipschitz-continuous objective functionals. Note that some applications of relative Lipschitz-continuity to the well-known classical support vector machine (SVM) problem and to the problem of minimizing the maximum of convex quadratic functions (intersection of m ellipsoids problem in $\mathbb{R}^n$) were discussed in [17].

In this paper we propose a new concept of an inexact model for objective functional and functional constraint. More precisely, we introduce some analogues of the concepts of an inexact oracle [8] and an inexact model [32] for objective functionals. However, unlike [8,32], we do not generalize the smoothness condition. We relax the Lipschitz condition and consider a recently proposed generalization of *relative Lipschitz-continuity* [17,24]. We propose some optimal Mirror Descent methods, in different settings of Relatively Lipschitz-continuous convex optimization problems.

The Mirror Descent method originated in the works of A. Nemirovski and D. Yudin more than 30 years ago [21,22] and was later analyzed in [5]. It can be considered as the non-Euclidean extension of subgradient methods. The method was used in many applications [19,20,31]. Standard subgradient methods employ the Euclidean distance function with a suitable step-size in the projection step. The Mirror Descent extends the standard projected subgradient methods by employing a nonlinear distance function with an optimal step-size in the nonlinear projection step [18]. The Mirror Descent method not only generalizes the standard subgradient descent method, but also achieves a better convergence rate and it is applicable to optimization problems in Banach spaces, while the subgradient descent is not [9]. Also, in some works [4,10,22] there was proposed an extension of the Mirror Descent method for constrained problems.

Also, in recent years, online convex optimization (OCO) has become a leading online learning framework, due to its powerful modeling capability for a lot of problems from diverse domains. OCO plays a key role in solving problems where statistical information is being updated [13,14]. There are many examples of such problems: Internet networks, consumer data sets or financial markets, machine learning applications, such as adaptive routing in networks, dictionary learning, classification and regression (see [33] and references therein). In recent years, methods for solving online optimization problems have been actively developed, in both deterministic and stochastic settings [7,12,15,25]. Among them one can mention the Mirror Descent method for the deterministic setting of the problem [26,30] and for the stochastic setting [1,11,34,35], which allows to solve problems for an arbitrary distance function.

This paper is devoted to Mirror Descent methods for convex programming problems with a relatively Lipschitz-continuous objective function and functional constraints. It consists of an introduction and 5 main sections. In Sect. 2 we consider the problem statement and define the concept of an inexact (δ, ϕ, V)−model for the objective function. Also, we propose some modifications of the Mirror Descent method for the concept of Model Generality. Section 3 is devoted to some special cases of problems with the properties of relative Lipschitz continuity, here we propose two versions of the Mirror Descent method in order to solve the problems under consideration. In Sects. 4 and 5 we consider the stochastic and online (OCO) setting of the optimization problem respectively. In Sect. 6 one can find numerical experiments which demonstrate the efficiency of the proposed methods.

The contribution of the paper can be summarized as follows:

- Continuing the development of Yurii Nesterov's ideas in the direction of the relative smoothness and non-smoothness [24], we introduced the concept of an inexact (δ, ϕ, V)–model of the objective function. For the proposed model we proposed some variants of the well-known Mirror Descent method, which provides an $(\varepsilon + \delta)$–solution of the optimization problem, where ε is the controlled accuracy. There was considered the applicability of the proposed method to the case of the stochastic setting of the considered optimization problem.
- We also considered a special case of the relative Lipschitz condition for the objective function. The proposed Mirror Descent algorithm was specified for the case of such functions. Furthermore, there was introduced one more modification of the algorithm with another approach to the step selection. There was also considered the possibility of applying the proposed methods to the case of several functional constraints.
- We considered an online optimization problem and proposed a modification of the Mirror Descent algorithm for such a case. Moreover, there were conducted some numerical experiments which demonstrate the effectiveness of the proposed methods.

2 Inexact Model for Relative Non-smooth Functionals and Mirror Descent Algorithm

Let $(E, \|\cdot\|)$ be a normed finite-dimensional vector space and E^* be the conjugate space of E with the norm:

$$\|y\|_* = \max_x \{\langle y, x \rangle, \|x\| \leq 1\},$$

where $\langle y, x \rangle$ is the value of the continuous linear functional y at $x \in E$.

Let $Q \subset E$ be a (simple) closed convex set. Consider two subdifferentiable functions $f, g : Q \to \mathbb{R}$. In this paper we consider the following optimization problem

$$f(x) \to \min_{x \in Q, \, g(x) \leq 0}. \tag{3}$$

Let $d : Q \to \mathbb{R}$ be any convex (not necessarily strongly-convex) differentiable function, we will call it *reference function*. Suppose we have a constant $\Theta_0 > 0$, such that $d(x^*) \leq \Theta_0^2$, where x^* is a solution of (3). Note that if there is a set, $X_* \subset Q$, of optimal points for the problem (3), we may assume that

$$\min_{x^* \in X_*} d(x^*) \leq \Theta_0^2.$$

Let us introduce some generalization of the concept of relative Lipschitz continuity [24]. Consider one more auxiliary function $\phi : \mathbb{R} \to \mathbb{R}$, which is strictly increasing and satisfies $\phi(0) = 0$. Clearly, due to the strict monotonicity of $\phi(\cdot)$, there exists the inverse function $\phi^{-1}(\cdot)$.

Definition 1. *Let $\delta > 0$. We say that f and g admit the (δ, ϕ, V)–model at the point $y \in Q$ if*

$$f(x) + \psi_f(y, x) \leq f(y), \quad -\psi_f(y, x) \leq \phi_f^{-1}(V_d(y, x)) + \delta \qquad (4)$$

$$g(x) + \psi_g(y, x) \leq g(y), \quad -\psi_g(y, x) \leq \phi_g^{-1}(V_d(y, x)) + \delta, \qquad (5)$$

where $\psi_f(\cdot, x)$ and $\psi_g(\cdot, x)$ are convex functions for fixed x and $\psi_f(x, x) = \psi_g(x, x) = 0$ for all $x \in Q$.

Let $h > 0$. For problems with a (δ, ϕ, V)–model, the proximal mapping operator (Mirror Descent step) is defined as follows

$$Mirr_h(x, \psi) = \arg\min_{y \in Q} \left\{ \psi(y, x) + \frac{1}{h} V_d(y, x) \right\}.$$

The following lemma describes the main property of this operator.

Lemma 1 (Main Lemma). *Let f be a convex function, which satisfies (4), $h > 0$ and $\tilde{x} = hMirr_h(x, \psi_f)$. Then for any $y \in Q$*

$$h(f(x) - f(y)) \leq -h\psi_f(y, x) \leq \phi_f^*(h) + V_d(y, x) - V_d(y, \tilde{x}) + h\delta,$$

where ϕ_f^* is the conjugate function of ϕ_f.

Proof. From the definition of $\tilde{x}$

$$\tilde{x} = hMirr_h(x, \psi_f) = \arg\min_{y \in Q} \left\{ h\psi_f(y, x) + V_d(y, x) \right\},$$

for any $y \in Q$, we have $h\psi_f(y, x) - h\psi_f(\tilde{x}, x) + \langle \nabla d(\tilde{x}) - \nabla d(x), y - \tilde{x} \rangle \geq 0$. Further, $h(f(x) - f(y)) \leq -h\psi_f(y, x) \leq$

$$\leq -h\psi_f(\tilde{x}, x) + \langle \nabla d(\tilde{x}) - \nabla d(x), y - \tilde{x} \rangle$$
$$= -h\psi_f(\tilde{x}, x) + V_d(y, x) - V_d(y, \tilde{x}) - V_d(\tilde{x}, x) + h\delta$$
$$\leq h\phi_f^{-1}(V_d(\tilde{x}, x)) + V_d(y, x) - V_d(y, \tilde{x}) - V_d(\tilde{x}, x) + h\delta$$
$$\leq \phi_f^*(h) + \phi_f(\phi_f^{-1}(V_d(\tilde{x}, x))) + V_d(y, x) - V_d(y, \tilde{x}) - V_d(\tilde{x}, x) + h\delta$$
$$= \phi_f^*(h) + V_d(\tilde{x}, x) + V_d(y, x) - V_d(y, \tilde{x}) - V_d(\tilde{x}, x) + h\delta$$
$$= \phi_f^*(h) + V_d(y, x) - V_d(y, \tilde{x}) + h\delta.$$

For problem (3) with an inexact (δ, ϕ, V)–model, we consider a Mirror Descent algorithm, listed as Algorithm 1 below. For this proposed algorithm, we will call step k productive if $g(x^k) \leq \varepsilon$, and non-productive if the reverse inequality $g(x^k) > \varepsilon$ holds. Let I and $|I|$ denote the set of indexes of productive steps and their number, respectively. Similarly, we use the notation J and $|J|$ for non-productive steps.

Let x^* denote the exact solution of the problem (3). The next theorem provides the complexity and quality of the proposed Algorithm 1.

Theorem 1 (Modified MDA for Model Generality). *Let f and g be convex functionals, which satisfy (4), (5) respectively and $\varepsilon > 0, \delta > 0$ be fixed positive numbers. Assume that $\Theta_0 > 0$ is a known constant such that $d(x^*) \leq \Theta_0^2$. Then, after the stopping of Algorithm 1, the following inequalities hold:*

$$f(\widehat{x}) - f(x^*) \leq \varepsilon + \delta \quad and \quad g(\widehat{x}) \leq \varepsilon + \delta.$$

Algorithm 1. Modified MDA for (δ, ϕ, V)–model.

Require: $\varepsilon > 0, \delta > 0, h^f > 0, h^g > 0, \Theta_0 : d(x^*) \leq \Theta_0^2$.
1: $x^0 = \arg\min_{x \in Q} d(x)$.
2: $I =: \emptyset$ and $J =: \emptyset$
3: $N \leftarrow 0$
4: **repeat**
5:    **if** $g\left(x^N\right) \leq \varepsilon + \delta$ **then**
6:      $x^{N+1} = Mirr_{h^f}\left(x^N, \psi_f\right),$ "productive step"
7:      $N \rightarrow I$
8:    **else**
9:      $x^{N+1} = Mirr_{h^g}\left(x^N, \psi_g\right),$ "non-productive step"
10:     $N \rightarrow J$
11:    **end if**
12:    $N \leftarrow N + 1$
13: **until** $\Theta_0^2 \leq \varepsilon \left(|J|h^g + |I|h^f\right) - |J|\phi_g^*(h^g) - |I|\phi_f^*(h^f).$
Ensure: $\widehat{x} := \frac{1}{|I|} \sum_{k \in I} x^k.$

Proof. By Lemma 1, we have for all $k \in I$ and $y \in Q$

$$h^f \left(f(x^k) - f(y)\right) \leq \phi_f^*(h^f) + V_d(y, x^k) - V_d(y, x^{k+1}) + h^f \delta. \tag{6}$$

Similarly, for all $k \in J$ and $y \in Q$

$$h^g \left(g(x^k) - g(y)\right) \leq \phi_g^*(h^g) + V_d(y, x^k) - V_d(y, x^{k+1}) + h^g \delta. \tag{7}$$

Summing up these inequalities over productive and non-productive steps, we get

$$\sum_{k \in I} h^f \left(f(x^k) - f(x^*)\right) + \sum_{k \in J} h^g \left(g(x^k) - g(x^*)\right)$$

$$\leq \sum_{k \in I} \phi_f^*(h^f) + \sum_{k \in J} \phi_g^*(h^g) + \sum_{k} \left(V_d(x^*, x^k) - V_d(x^*, x^{k+1})\right) + \sum_{k \in I} h^f \delta + \sum_{k \in J} h^g \delta$$

$$\leq \sum_{k \in I} \phi_f^*(h^f) + \sum_{k \in J} \phi_g^*(h^g) + \Theta_0^2 + \sum_{k \in I} h^f \delta + \sum_{k \in J} h^g \delta.$$

Since for any $k \in J$, $g(x^k) - g(x^*) > \varepsilon + \delta$, we have

$$\sum_{k \in I} h^f \left(f(x^k) - f(x^*) \right) \leq \sum_{k \in I} \phi_f^*(h^f) + \sum_{k \in J} \phi_g^*(h^g) + \Theta_0^2 - \varepsilon \sum_{k \in J} h^g + \sum_{k \in I} h^f \delta$$

$$= |I| \left(\phi_f^*(h^f) + \delta h^f \right) + |J| \phi_g^*(h^g) - |J| h^g \varepsilon + \Theta_0^2 \leq \varepsilon |I| h^f + \delta |I| h^f.$$

So, for $\widehat{x} := \frac{1}{|I|} \sum_{k \in I} x^k$, after the stopping criterion of Algorithm 1 is satisfied, the following inequalities hold

$$f(\widehat{x}) - f(x^*) \leq \varepsilon + \delta \quad \text{and} \quad g(\widehat{x}) \leq \varepsilon + \delta.$$

3   The Case of Relatively Lipschitz-Continuous Functionals

Suppose hereinafter that the objective function f and the constraint g satisfy the so-called relative Lipschitz condition, with constants $M_f > 0$ and $M_g > 0$, i.e. the functions ϕ_f^{-1} and ϕ_g^{-1} from (4) and (5) are modified as follows:

$$\phi_f^{-1} \left(V_d(y, x) \right) = M_f \sqrt{2 V_d(y, x)}, \tag{8}$$

$$\phi_g^{-1} \left(V_d(y, x) \right) = M_g \sqrt{2 V_d(y, x)}. \tag{9}$$

Note that the functions f, g must still satisfy the left inequalities in (4), (5):

$$f(x) + \psi_f(y, x) \leq f(y), \quad -\psi_f(y, x) \leq M_f \sqrt{2 V_d(y, x)} + \delta; \tag{10}$$

$$g(x) + \psi_g(y, x) \leq g(y), \quad -\psi_g(y, x) \leq M_g \sqrt{2 V_d(y, x)} + \delta, \tag{11}$$

For this particular case we say that f and g admit the $(\delta, M_f, V)-$ and (δ, M_g, V)–model at each point $x \in Q$ respectively. The following remark provides an explicit form of ϕ_f, ϕ_g and their conjugate functions ϕ_f^*, ϕ_g^*.

Remark 1. Let $M_f > 0$ and $M_g > 0$. Then functions ϕ_f and ϕ_g which correspond to (8) and (9) are defined as follows:

$$\phi_f(t) = \frac{t^2}{2 M_f^2}, \quad \phi_g(t) = \frac{t^2}{2 M_g^2}.$$

Their conjugate functions have the following form:

$$\phi_f^*(y) = \frac{y^2 M_f^2}{2}, \tag{12}$$

$$\phi_g^*(y) = \frac{y^2 M_g^2}{2}. \tag{13}$$

For the case of a relatively Lipschitz-continuous objective function and constraint, we consider a modification of Algorithm 1, the modified algorithm is listed as Algorithm 2, below. The difference between Algorithms 1 and 2 is represented in the control of productivity and the stopping criterion.

For the proposed Algorithm 2, we have the following theorem, which provides an estimate of its complexity and the quality of the solution of the problem.

Theorem 2. *Let f and g be convex functions, which satisfy (10) and (11) for $M_f > 0$ and $M_g > 0$. Let $\varepsilon > 0, \delta > 0$ be fixed positive numbers. Assume that $\Theta_0 > 0$ is a known constant such that $d(x^*) \leq \Theta_0^2$. Then, after the stopping of Algorithm 2, the following inequalities hold:*

$$f(\widehat{x}) - f(x^*) \leq M_f \varepsilon + \delta \quad and \quad g(\widehat{x}) \leq M_g \varepsilon + \delta.$$

Algorithm 2. Mirror Descent for Relatively Lipschitz-continuous functions, version 1.

Require: $\varepsilon > 0, \delta > 0, M_f > 0, M_g > 0, \Theta_0 : d(x^*) \leq \Theta_0^2$
1: $x^0 = \arg\min_{x \in Q} d(x)$.
2: $I =: \emptyset$
3: $N \leftarrow 0$
4: **repeat**
5:    **if** $g\left(x^N\right) \leq M_g \varepsilon + \delta$ **then**
6:       $h^f = \frac{\varepsilon}{M_f}$,
7:       $x^{N+1} = Mirr_{h^f}\left(x^N, \psi_f\right)$,   "productive step"
8:       $N \to I$
9:    **else**
10:       $h^g = \frac{\varepsilon}{M_g}$,
11:       $x^{N+1} = Mirr_{h^g}\left(x^N, \psi_g\right)$,   "non-productive step"
12:    **end if**
13:    $N \leftarrow N + 1$
14: **until** $N \geq \frac{2\Theta_0^2}{\varepsilon^2}$.
Ensure: $\widehat{x} := \frac{1}{|I|} \sum_{k \in I} x^k$.

Proof. By Lemma 1, we have

$$\sum_{k \in I} h^f \left(f(x^k) - f(x^*)\right) + \sum_{k \in J} h^g \left(g(x^k) - g(x^*)\right) \leq \sum_{k \in I} \phi_f^*(h^f) + \sum_{k \in J} \phi_g^*(h^g)$$
$$+ \Theta_0^2 + \sum_{k \in I} h^f \delta + \sum_{k \in J} h^g \delta$$

Since for any $k \in J$, $g(x^k) - g(x^*) > M_g \varepsilon + \delta$, we have

$$\sum_{k \in I} h^f \left(f(x^k) - f(x^*)\right) \leq \sum_{k \in I} \phi_f^*(h^f) + \sum_{k \in J} \phi_g^*(h^g) + \Theta_0^2 - M_g \varepsilon \sum_{k \in J} h^g + \sum_{k \in I} h^f \delta$$
$$= |I|(\phi_f^*(h^f) + \delta h^f) + |J|\phi_g^*(h^g) - |J|\varepsilon^2 + \Theta_0^2.$$

Taking into account the explicit form of the conjugate functions (12), (13) one can get:

$$\sum_{k\in I} h^f \left(f(x^k) - f(x^*) \right) \leq |I| \left(\frac{M_f^2 (h^f)^2}{2} + \delta h^f \right) + |J| \frac{M_g^2 (h^g)^2}{2} - |J|\varepsilon^2 + \Theta_0^2$$

$$= |I| \left(\frac{\varepsilon^2}{2} + \delta h^f \right) + |J| \frac{\varepsilon^2}{2} - |J|\varepsilon^2 + \Theta_0^2$$

$$\leq M_f \varepsilon |I| h^f + \delta |I| h^f,$$

supposing that the stopping criterion is satisfied.

So, for the output value of the form $\widehat{x} = \frac{1}{|I|} \sum_{k\in I} x^k$, the following inequalities hold:

$$f(\widehat{x}) - f(x^*) \leq M_f \varepsilon + \delta \quad \text{and} \quad g(\widehat{x}) \leq M_g \varepsilon + \delta.$$

Also, for the case of a relatively Lipschitz-continuous objective function and constraint, we consider another modification of Algorithm 1, which is listed as the following Algorithm 3. Note that the difference between Algorithm 2 and Algorithm 3 lies in the choice of steps h^f, h^g and the stopping criterion.

Algorithm 3. Mirror Descent for Relatively Lipschitz-continuous functions, version 2.

Require: $\varepsilon > 0, \delta > 0, M_f > 0, M_g > 0, \Theta_0 : d(x^*) \leq \Theta_0^2.$
1: $x^0 = \arg\min_{x\in Q} d(x).$
2: $I =: \emptyset$ and $J =: \emptyset$
3: $N \leftarrow 0$
4: **repeat**
5:     **if** $g\left(x^N\right) \leq \varepsilon + \delta$ **then**
6:         $h^f = \frac{\varepsilon}{M_f^2},$
7:         $x^{k+1} = Mirr_{h^f}\left(x^N, \psi_f\right),$ "productive step"
8:         $N \rightarrow I$
9:     **else**
10:        $h^g = \frac{\varepsilon}{M_g^2},$
11:        $x^{N+1} = Mirr_{h^g}\left(x^N, \psi_g\right),$ "non-productive step"
12:        $N \rightarrow J$
13:     **end if**
14:     $N \leftarrow N + 1$
15: **until** $\frac{2\Theta_0^2}{\varepsilon^2} \leq \frac{|I|}{M_f^2} + \frac{|J|}{M_g^2}.$
Ensure: $\widehat{x} := \frac{1}{|I|} \sum_{k\in I} x^k.$

By analogy with the proof of Theorem 2 one can obtain the following result concerning the quality of the convergence of the proposed Algorithm 3.

Theorem 3. *Let f and g be convex functions, which satisfy (10) and (11) for $M_f > 0$ and $M_g > 0$. Let $\varepsilon > 0, \delta > 0$ be fixed positive numbers. Assume that $\Theta_0 > 0$ is a known constant such that $d(x^*) \leq \Theta_0^2$. Then, after the stopping of Algorithm 3, the following inequalities hold:*

$$f(\widehat{x}) - f(x^*) \leq \varepsilon + \delta \quad and \quad g(\widehat{x}) \leq \varepsilon + \delta.$$

Moreover, the required number of iterations of Algorithm 3 does not exceed

$$N = \frac{2M^2\Theta_0^2}{\varepsilon^2}, \quad where \ M = \max\{M_f, M_g\}.$$

Remark 2. Clearly, Algorithms 2 and 3 are optimal in terms of the lower bounds [22]. More precisely, let us understand hereinafter the optimality of the Mirror Descent methods as the complexity $O(\frac{1}{\varepsilon^2})$ (it is well-known that this estimate is optimal for Lipschitz-continuous functionals [22]).

Remark 3 (The case of several functional constraints). Let us consider a set of convex functions f and $g_p : Q \to \mathbb{R}$, $p \in [m] \stackrel{\text{def}}{=} \{1, 2, \ldots, m\}$. We will focus on the following constrained optimization problem

$$\min\{f(x) : \ x \in Q \ \text{and} \ g_p(x) \leq 0 \ \text{for all} \ p \in [m]\}. \tag{14}$$

It is clear that instead of a set of functionals $\{g_p(\cdot)\}_{p=1}^m$ we can consider one functional constraint $g : Q \to \mathbb{R}$, such that $g(x) = \max_{p \in [m]}\{g_p(x)\}$. Therefore, by this setting, problem (14) will be equivalent to the problem (3).

Assume that for any $p \in [m]$, the functional g_p satisfies the following condition

$$-\psi_{g_p}(y, x) \leq M_{g_p}\sqrt{2V_d(y, x)} + \delta.$$

For problem (14), we propose a modification of Algorithms 2 and 3 (the modified algorithms are listed as Algorithm 6 and 7 in [29], Appendix A). The idea of the proposed modification allows to save the running time of the algorithms due to consideration of not all functional constraints on non-productive steps.

Remark 4 (Composite Optimization Problems [6,16,23]). Previously proposed methods are applicable to composite optimization problems, specifically

$$\min\{f(x) + r(x) : \ x \in Q, \ g(x) + \eta(x) \leq 0\},$$

where $r, \eta : Q \to \mathbb{R}$ are so-called simple convex functionals (i.e. the proximal mapping operator $Mirr_h(x, \psi)$ is easily computable). For this case, for any $x, y \in Q$, we have

$$\psi_f(y, x) = \langle \nabla f(x), y - x \rangle + r(y) - r(x),$$
$$\psi_g(y, x) = \langle \nabla g(x), y - x \rangle + \eta(y) - \eta(x).$$

4   Stochastic Mirror Descent Algorithm

Let us, in this section, consider the stochastic setting of the problem (3). This means that we can still use the value of the objective function and functional constraints, but instead of their (sub)gradient, we use their stochastic (sub)gradient. Namely, we consider the first-order unbiased oracle that produces $\nabla f(x, \xi)$ and $\nabla g(x, \zeta)$, where ξ and ζ are random vectors and

$$\mathbb{E}[\nabla f(x, \xi)] = \nabla f(x), \quad \mathbb{E}[\nabla g(x, \zeta)] = \nabla g(x).$$

Assume that for each $x, y \in Q$

$$\langle \nabla f(x, \xi), x - y \rangle \leq M_f \sqrt{2V_d(y, x)} \text{ and } \langle \nabla g(x, \zeta), x - y \rangle \leq M_g \sqrt{2V_d(y, x)}, \quad (15)$$

where $M_f, M_g > 0$. Let us consider a proximal mapping operator for f

$$Mirr_h(x, \nabla f(x, \xi)) = \arg\min_{y \in Q} \left\{ \frac{1}{h} V_d(y, x) + \langle \nabla f(x, \xi), y \rangle \right\},$$

and, similarly, we consider a proximal mapping operator for g. The following lemma describes the main property of this operator.

Lemma 2. *Let f be a convex function which satisfies (4), $h > 0, \delta > 0$, ξ be a random vector and $\tilde{x} = Mirr_h(x, \nabla f(x, \xi))$. Then for all $y \in Q$*

$$h(f(x) - f(y)) \leq \phi_f^*(h) + V_d(y, x) - V_d(y, \tilde{x}) + h\langle \nabla f(x, \xi) - \nabla f(x), y - x \rangle + h\delta,$$

where, as earlier, $\phi_f^*(h) = \dfrac{h^2 M_f^2}{2}$.

Suppose $\varepsilon > 0$ is a given positive real number. We say that a (random) point $\hat{x} \in Q$ is an expected ε–solution to the problem (3), in the stochastic setting, if

$$\mathbb{E}[f(\hat{x})] - f(x^*) \leq \varepsilon \text{ and } g(\hat{x}) \leq \varepsilon. \quad (16)$$

In order to solve the stochastic setting of the considered problem (3), we propose the following algorithm.

The following theorem gives information about the efficiency of the proposed Algorithm 4. The proof of this theorem is given in [29], Appendix B.

Algorithm 4. Modified Mirror Descent for the stochastic setting.

Require: $\varepsilon > 0, \delta > 0, h^f > 0, h^g > 0, \Theta_0 : d(x^*) \leq \Theta_0^2.$
1: $x^0 = \arg\min_{x \in Q} d(x).$
2: $I =: \emptyset$ and $J =: \emptyset$
3: $N \leftarrow 0$
4: **repeat**
5:     **if** $g\left(x^N\right) \leq \varepsilon + \delta$ **then**
6:         $x^{N+1} = Mirr_{h^f}\left(x^N, \nabla f(x, \xi^N)\right),$ "productive step"
7:         $N \rightarrow I$
8:     **else**
9:         $x^{N+1} = Mirr_{h^g}\left(x^N, \nabla f(x, \zeta^N)\right),$ "non-productive step"
10:         $N \rightarrow J$
11:     **end if**
12:     $N \leftarrow N+1$
13: **until** $\Theta_0^2 \leq \varepsilon\left(|J|h^g + |I|h^f\right) - |J|\phi_g^*(h^g) - |I|\phi_f^*(h^f).$
Ensure: $\widehat{x} := \frac{1}{|I|}\sum_{k \in I} x^k.$

Theorem 4. *Let f and g be convex functions and (15) hold. Let $\varepsilon > 0, \delta > 0$ be fixed positive numbers. Then, after the stopping of Algorithm 4, the following inequalities hold:*

$$\mathbb{E}[f(\widehat{x})] - f(x^*) \leq \varepsilon + \delta \quad \text{and} \quad g(\widehat{x}) \leq \varepsilon + \delta.$$

Remark 5. It should be noted how the optimality of the proposed method can be understood. With the special assumptions (10)–(11) and choice of h^f, h^g, the complexity of the algorithm is $O(\frac{1}{\varepsilon^2})$, which is optimal in such a class of problems.

5   Online Optimization Problem

In this section we consider the online setting of the optimization problem (3). Namely,

$$\frac{1}{N}\sum_{i=1}^{N} f_i(x) \rightarrow \min_{x \in Q, g(x) \leq 0}, \tag{17}$$

under the assumption that all $f_i : Q \rightarrow \mathbb{R}$ $(i = 1, \ldots, N)$ and g satisfy (10) and (11) with constants $M_i > 0, i = 1, \ldots, N$ and $M_g > 0$.

In order to solve problem (17), we propose an algorithm (listed as Algorithm 5 below). This algorithm produces N productive steps and in each step, the (sub)gradient of exactly one functional of the objectives is calculated. As a result of this algorithm, we get a sequence $\{x^k\}_{k \in I}$ (on productive steps), which can be considered as a solution to problem (17) with accuracy κ (see (18)).

Assume that $M = \max\{M_i, M_g\}, h^f = h^g = h = \frac{\varepsilon}{M}.$

Algorithm 5. Modified Mirror Descent for the online setting.

Require: $\varepsilon > 0, \delta > 0, M > 0, N, \Theta_0 : d(x^*) \leq \Theta_0^2$.
1: $x^0 = \arg\min_{x \in Q} d(x)$.
2: $i := 1, k := 0$
3: set $h = \frac{\varepsilon}{M^2}$
4: **repeat**
5:    **if** $g\left(x^k\right) \leq \varepsilon + \delta$ **then**
6:       $x^{k+1} = Mirr_h\left(x^k, \psi_{f_i}\right),$ "productive step"
7:       $i = i + 1,$
8:       $k = k + 1,$
9:    **else**
10:       $x^{k+1} = Mirr_h\left(x^k, \psi_g\right),$ "non-productive step"
11:       $k = k + 1,$
12:    **end if**
13: **until** $i = N + 1$.
14: Guaranteed accuracy:

$$\kappa = \frac{|J|}{N}\left(-\frac{\varepsilon}{2}\right) + \left(\frac{\varepsilon}{2} + \delta\right) + \frac{M^2\Theta_0^2}{N\varepsilon}. \tag{18}$$

For Algorithm 5, we have the following result.

Theorem 5. *Suppose all $f_i : Q \to \mathbb{R}$ ($i = 1, \ldots, N$) and g satisfy (10) and (11) with constants $M_i > 0, i = 1, \ldots, N$ and $M_g > 0$, Algorithm 5 works exactly N productive steps. Then after the stopping of this Algorithm, the following inequality holds*

$$\frac{1}{N}\sum_{i=1}^{N} f_i(x^k) - \min_{x \in Q} \frac{1}{N}\sum_{i=1}^{N} f_i(x) \leq \kappa,$$

moreover, when the regret is non-negative, there will be no more than $O(N)$ non-productive steps.

The proof of this theorem is given in [29], Appendix C. In particular, note that the proposed method is optimal [13]: if for some $C > 0$, $\kappa \sim \varepsilon \sim \delta = \frac{C}{\sqrt{N}}$, then $|J| \sim O(N)$.

6   Numerical Experiments

To show the practical performance of the proposed Algorithms 2, 3 and their modified versions, which are listed as Algorithm 6 and Algorithm 7 in [29], in the case of many functional constraints, a series of numerical experiments were performed[1], for the well-known *Fermat-Torricelli-Steiner* problem, but with some non-smooth functional constraints.

---

[1] All experiments were implemented in Python 3.4, on a computer fitted with Intel(R) Core(TM) i7-8550U CPU @ 1.80 GHz, 1992 Mhz, 4 Core(s), 8 Logical Processor(s). RAM of the computer is 8 GB.

For a given set $\{P_k = (p_{1k}, p_{2k}, \ldots, p_{nk}); k \in [r]\}$ of r points, in n-dimensional Euclidean space $\mathbb{R}^n$, we need to solve the considered optimization problem (3), where the objective function f is given by

$$f(x) := \frac{1}{r} \sum_{k=1}^{r} \sqrt{(x_1 - p_{1k})^2 + \ldots + (x_n - p_{nk})^2} = \frac{1}{r} \sum_{k=1}^{r} \|x - P_k\|_2. \quad (19)$$

The functional constraint has the following form

$$g(x) = \max_{i \in [m]} \{g_i(x) = \alpha_{i1}x_1 + \alpha_{i2}x_2 + \ldots + \alpha_{in}x_n\}. \quad (20)$$

The coefficients $\alpha_{i1}, \alpha_{i2}, \ldots, \alpha_{in}$, for all $i \in [m]$, in (20) and the coordinates of the points P_k, for all $k \in [r]$, are drawn from the normal (Gaussian) distribution with the location of the mode equaling 1 and the scale parameter equaling 2.

We choose the standard Euclidean norm and the Euclidean distance function in $\mathbb{R}^n$, $\delta = 0$, starting point $x^0 = \left(\frac{1}{\sqrt{n}}, \ldots, \frac{1}{\sqrt{n}}\right) \in \mathbb{R}^n$ and Q is the unit ball in $\mathbb{R}^n$.

We run Algorithms 2, 3 and their modified versions, Algorithms 6 and 7 respectively (see [29]), for $m = 200, n = 500, r = 100$ and different values of $\varepsilon \in \{\frac{1}{2^i} : i = 1, 2, 3, 4, 5\}$. The results of the work of these algorithms are represented in Table 1 below. These results demonstrate the comparison of the number of iterations (Iter.), the running time (in seconds) of each algorithm and the qualities of the solution, produced by these algorithms with respect to the objective function f and the functional constraint g, where we calculate the values of these functions at the output $x^{\text{out}} := \widehat{x}$ of the algorithms. We set $f^{\text{best}} := f(x^{\text{out}})$ and $g^{\text{out}} := g(x^{\text{out}})$.

Table 1. The results of Algorithms 2, 3 and their modified versions Algorithms 6 and 7 respectively, with $m = 200, n = 500, r = 100$ and different values of ε.

	Algorithm 2				Algorithm 6			
$1/\varepsilon$	Iter	Time (sec.)	f^{best}	g^{out}	Iter	Time (sec.)	f^{best}	g^{out}
2	16	5.138	22.327427	2.210041	16	4.883	22.327427	2.210041
4	64	20.911	22.303430	2.016617	64	20.380	22.303430	2.016617
8	256	84.343	22.283362	1.858965	256	79.907	22.283362	2.015076
16	1024	317.991	22.274366	1.199792	1024	317.033	22.273177	1.988190
32	4096	1253.717	22.272859	0.607871	4096	1145.033	22.269038	1.858965
	Algorithm 3				Algorithm 7			
2	167	9.455	22.325994	0.417002	164	7.373	22.325604	0.391461
4	710	39.797	22.305980	0.204158	667	29.954	22.305654	0.188497
8	2910	158.763	22.289320	0.103493	2583	119.055	22.289302	0.088221
16	11613	626.894	22.280893	0.051662	10155	468.649	22.280909	0.045343
32	46380	2511.261	22.277439	0.026000	40149	1723.136	22.277450	0.022639

In general, from the conducted experiments, we can see that Algorithm 2 and its modified version (Algorithm 6) work faster than Algorithms 3 and its modified version (Algorithm 7). But note that Algorithms 3 and 7 guarantee a better quality of the resulting solution to the considered problem, with respect to the objective function f and the functional constraint (20). Also, we can see the efficiency of the modified Algorithm 7, which saves the running time of the algorithm, due to consideration of not all functional constraints on non-productive steps.

7 Conclusion

In the paper, there was introduced the concept of an inexact (δ, ϕ, V)–model of the objective function. There were considered some modifications of the Mirror Descent algorithm, in particular for stochastic and online optimization problems. A significant part of the work was devoted to the research of a special case of relative Lipschitz condition for the objective function and functional constraints. The proposed methods are applicable for a wide class of problems because relative Lipschitz-continuity is an essential generalization of the classical Lipschitz-continuity. However, for relatively Lipschitz-continuous problems, we could not propose adaptive methods like [3,27,28]. Note that Algorithm 3 and its modified version Algorithm 7 (see [29]) are partially adaptive since the resulting number of iterations is not fixed, due to the stopping criterion, although the step-sizes are fixed.

References

1. Alkousa, M.S.: On some stochastic mirror descent methods for constrained online optimization problems. Comput. Res. Model. **11**(2), 205–217 (2019)
2. Bauschke, H.H., Bolte, J., Teboulle, M.: A descent lemma Beyond Lipschitz Gradient continuity: first-order methods revisited and applications. Math. Oper. Res. **42**(2), 330–348 (2017)
3. Bayandina, A., Dvurechensky, P., Gasnikov, A., Stonyakin, F., Titov, A.: Mirror descent and convex optimization problems with non-smooth inequality constraints. In: Giselsson, P., Rantzer, A. (eds.) Large-Scale and Distributed Optimization. LNM, vol. 2227, pp. 181–213. Springer, Cham (2018). https://doi.org/10.1007/978-3-319-97478-1_8
4. Beck, A., Ben-Tal, A., Guttmann-Beck, N., Tetruashvili, L.: The CoMirror algorithm for solving nonsmooth constrained convex problems. Oper. Res. Lett. **38**(6), 493–498 (2010)
5. Beck, A., Teboulle, M.: Mirror descent and nonlinear projected subgradient methods for convex optimization. Oper. Res. Lett. **31**(3), 167–175 (2003)
6. Beck, A., Teboulle, M.: A fast iterative Shrinkage-thresholding algorithm for linear inverse problems. SIAM J. Imaging Sci. **2**(1), 183–202 (2009)
7. Belmega, E.V., Mertikopoulos, P.: Online and stochastic optimization beyond Lipschitz continuity: a Riemannian approach. Published as a Conference Paper at ICLR (2020). https://openreview.net/pdf?id=rkxZyaNtwB

8. Devolder, O., Glineur, F., Nesterov, Yu.: First-order methods of smooth convex optimization with inexact oracle. Math. Program. **146**(1), 37–75 (2013). https://doi.org/10.1007/s10107-013-0677-5

9. Doan, T.T., Bose, S., Nguyen, D.H., Beck, C.L.: Convergence of the iterates in mirror descent methods. IEEE Control Syst. Lett. **3**(1), 114–119 (2019)

10. Gasnikov, A.V.: Modern numerical optimization methods. The method of universal gradient descent (2018). (in Russian). https://arxiv.org/ftp/arxiv/papers/1711/1711.00394.pdf

11. Gasnikov, A.V., Krymova, E.A., Lagunovskaya, A.A., Usmanova, I.N., Fedorenko, F.A.: Stochastic online optimization. Single-point and multi-point non-linear multi-armed bandits. Convex and strongly-convex case. Autom. Remote Control **78**(2), 224–234 (2017). https://doi.org/10.1134/S0005117917020035

12. Hao, Y., Neely, M.J., Xiaohan, W.: Online convex optimization with stochastic constraints. In: Published in NIPS, pp. 1427–1437 (2017)

13. Hazan, E., Kale, S.: Beyond the regret minimization barrier: optimal algorithms for stochastic strongly-convex optimization. JMLR **15**, 2489–2512 (2014)

14. Hazan, E.: Introduction to online convex optimization. Found. Trends Optim. **2**(3–4), 157–325 (2015)

15. Jenatton, R., Huang, J., Archambeau, C.: Adaptive algorithms for online convex optimization with long-term constraints. In: Proceedings of the 33rd International Conference on Machine Learning, PMLR 48, pp. 402–411 (2016)

16. Lu, H., Freund, R.M., Nesterov, Y.: Relatively smooth convex optimization by first-order methods, and applications. SIAM J. Optim. **28**(1), 333–354 (2018)

17. Lu, H.: Relative continuity for Non-Lipschitz nonsmooth convex optimization using stochastic (or deterministic) mirror descent. INFORMS J. Optim. **1**(4), 288–303 (2019)

18. Luong, D.V.N., Parpas, P., Rueckert, D., Rustem, B.: A weighted mirror descent algorithm for nonsmooth convex optimization problem. J. Optim. Theory Appl. **170**(3), 900–915 (2016). https://doi.org/10.1007/s10957-016-0963-5

19. Nazin, A.V., Miller, B.M.: Mirror descent algorithm for homogeneous finite controlled Markov Chains with unknown mean losses. In: IFAC Proceedings, vol. 44(1), pp. 12421–12426 (2011). https://doi.org/10.3182/20110828-6-IT-1002.03450

20. Nazin, A., Anulova, S., Tremba, A.: Application of the mirror descent method to minimize average losses coming by a poisson flow. In: 2014 European Control Conference (ECC), Strasbourg, pp. 2194–2197 (2014)

21. Nemirovskii, A.: Efficient Methods for Large-Scale Convex Optimization Problems. Math. Ekon. (1979). (in Russian)

22. Nemirovsky, A., Yudin, D.: Problem Complexity and Method Efficiency in Optimization. Wiley, New York (1983)

23. Nesterov, Y.: Gradient methods for minimizing composite functions. Math. Program. **140**, 125–161 (2013)

24. Nesterov, Y.: Relative smoothness: new paradigm in convex optimization. In: Conference report, EUSIPCO-2019, A Coruna, Spain, 4 September 2019 (2019). https://www.gtec.udc.es/eusipco2019/wp-content/uploads/2019/10/Relative-Smoothness-New-Paradigm-in-Convex.pdf

25. Orabona, F.: A modern introduction to online learning (2020). https://arxiv.org/pdf/1912.13213.pdf

26. Orabona, F., Crammer, K., Cesa-Bianchi, N.: A generalized online mirror descent with applications to classification and regression. Mach. Learn. **99**(3), 411–435 (2014). https://doi.org/10.1007/s10994-014-5474-8

27. Stonyakin, F.S., Alkousa, M., Stepanov, A.N., Titov, A.A.: Adaptive mirror descent algorithms for convex and strongly convex optimization problems with functional constraints. J. Appl. Ind. Math. **13**(3), 557–574 (2019). https://doi.org/10.1134/S1990478919030165

28. Stonyakin, F.S., Stepanov, A.N., Gasnikov, A.V., Titov, A.A.: Mirror descent for constrained optimization problems with large subgradient values of functional constraints. Comput. Res. Model. **12**(2), 301–317 (2020)

29. Titov, A.A., Stonyakin, F.S., Alkousa, M.S., Ablaev, S.S., Gasnikov, A.V.: Analogues of switching subgradient schemes for relatively Lipschitz-continuous convex programming problems (2020). https://arxiv.org/pdf/2003.09147.pdf

30. Titov, A.A., Stonyakin, F.S., Gasnikov, A.V., Alkousa, M.S.: Mirror descent and constrained online optimization problems. In: Evtushenko, Y., Jaćimović, M., Khachay, M., Kochetov, Y., Malkova, V., Posypkin, M. (eds.) OPTIMA 2018. CCIS, vol. 974, pp. 64–78. Springer, Cham (2019). https://doi.org/10.1007/978-3-030-10934-9_5

31. Tremba, A., Nazin, A.: Extension of a saddle point mirror descent algorithm with application to robust PageRank. In: 52nd IEEE Conference on Decision and Control, 10–13 December, pp. 3691–3696 (2013)

32. Tyurin, A.I., Gasnikov, A.V.: Fast gradient descent method for convex optimization problems with an oracle that generates a (δ, L)-model of a function in a requested point. Comput. Math. Math. Phys. **59**(7), 1137–1150 (2019)

33. Yuan, J., Lamperski, A.: Online convex optimization for cumulative constraints. In: Published in NIPS, pp. 6140–6149 (2018)

34. Yunwen, L., Ding-Xuan, Z.: Convergence of online mirror descent. Appl. Comput. Harmon. Anal. **48**(1), 343–373 (2020)

35. Zhou, Z., Mertikopoulos, P., Bambos, N., Boyd, S.P., Glynn, P.W.: On the convergence of mirror descent beyond stochastic convex programming. SIAM J. Optim. **30**(1), 687–716 (2020)

Constructing Mixed Algorithms
on the Basis of Some Bundle Method

Rashid Yarullin[(✉)] [iD]

Kazan (Volga region) Federal University, Kazan, Russia
yarullinrs@gmail.com

Abstract. In the paper, a method is proposed for minimizing a non-differentiable convex function. This method belongs to a class of bundle methods. In the developed method it is possible to periodically produce discarding all previously constructed cutting planes that form the model of the objective function. These discards are applied when approximation of the epigraph of the objective function is sufficiently good in the a neighborhood of the current iteration point, and the quality of this approximation is estimated by using the model of the objective function. It is proposed an approach for constructing mixed minimization algorithms on the basis of the developed bundle method with involving any relaxation methods. The opportunity to mix the developed bundle method with other methods is provided as follows. In the proposed method during discarding the cutting planes the main iteration points are fixed with the relaxation condition. Any relaxation minimization method can be used to build these points. Moreover, the convergence of all such mixed algorithms will be guaranteed by the convergence of the developed bundle method. It is important to note that the procedures for updating cutting planes introduced in the bundle method will be transferred to mixed algorithms. The convergence of the proposed method is investigated, its properties are discussed, an estimate of the accuracy of the solution and estimation of the complexity of finding an approximate solution are obtained.

Keywords: Nondifferentiable optimization · Mixed algorithms ·
Bundle methods · Cutting planes · Sequence of approximations ·
Convex functions

1 Introduction

Nowadays a lot of different methods have been developed for solving nonlinear programming problems. Each of these optimization methods has its own disadvantages and advantages. In this regard, for solving practical problems these methods are used in a complex manner in order to accelerate the convergence of the optimization process. Namely, at each step to find the next approximation there are opportunities to choose any minimization method among other methods which allows to construct descent direction from the current point faster. The

© Springer Nature Switzerland AG 2020
Y. Kochetov et al. (Eds.): MOTOR 2020, CCIS 1275, pp. 150–163, 2020.
https://doi.org/10.1007/978-3-030-58657-7_14

algorithm that is formed as a result of applying various optimization methods is called mixed (e.g., [1,2]).

In this paper, based on the ideas of [2], an approach is proposed for constructing mixed algorithms on the basis of some proximal bundle method which is characterized by the possibility to periodically discard the cutting planes.

2   Problem Setting

Let $f(x)$ be a convex function defined in an n-dimensional Euclidian space, $\partial f(x)$, $\partial_\epsilon f(x)$ be a subdifferential and an ϵ-subdifferential of the function $f(x)$ at x respectively.

Suppose $f^* = \min\{f(x) : x \in \mathbb{R}^n\}$, $X^* = \{x \in \mathbb{R}^n : f(x) = f^*\} \neq \emptyset$, $X^*(\varepsilon) = \{x \in \mathbb{R}^n : f(x) \leq f^* + \varepsilon\}$, $\varepsilon > 0$, $K = \{0, 1, \dots\}$, $L(y) = \{x \in \mathbb{R}^n : f(x) \leq f(y)\}$, where $y \in \mathbb{R}^n$. Denote by $\lceil \chi \rceil$ the least integer no less than $\chi \in \mathbb{R}^1$. It is assumed that the set $L(y)$ is bounded for any $y \in \mathbb{R}^n$. Fix an arbitrary point $x^* \in X^*$.

It is required to find a point from the set $X^*(\varepsilon)$ with given $\varepsilon > 0$ for a finite number of iterations.

3   Minimization Method

First, consider an auxiliary procedure $\pi = \pi(\bar{x}, \bar{\xi}, \bar{\theta}, \bar{\mu})$ with the following input parameters:

$$\bar{x} \in \mathbb{R}^n, \quad \bar{\xi} > 0, \quad \bar{\theta} \in (0,1), \quad \bar{\mu} > 0.$$

Step 0. Define initial parameters $k = 0$, $x_k = \bar{x}$.

Step 1. Choose a subgradient $s_k \in \partial f(x_k)$. Assign $i = 0$, $s_{k,i} = s_k$, $x_{k,i} = x_k$,

$$\hat{f}_{k,i}(y) = f(x_{k,i}) + \langle s_{k,i}, y - x_{k,i} \rangle. \tag{1}$$

Step 2. Find a point

$$x_{k,i+1} = \arg \min\{\hat{f}_{k,i}(y) + \frac{\bar{\mu}}{2}\|y - x_k\|^2 : y \in \mathbb{R}^n\}. \tag{2}$$

Step 3. Compute a parameter

$$\delta_{k,i} = f(x_k) - \hat{f}_{k,i}(x_{k,i+1}) - \frac{\bar{\mu}}{2}\|x_{k,i+1} - x_k\|^2. \tag{3}$$

Step 4. If the inequality

$$\delta_{k,i} \leq \bar{\xi}, \tag{4}$$

is fulfilled, then the process of finding sequence is stopped, and the point

$$\hat{x} = \arg \min\{f(x_{k,j}) : 0 \leq j \leq i+1\} \tag{5}$$

is a result of the procedure.

Step 5. If the condition

$$f(x_{k,i+1}) \le f(x_k) - \bar{\theta}\delta_{k,i} \tag{6}$$

is fulfilled, then choose a point $x_{k+1} \in \mathbf{R}^n$ according to the inequality

$$f(x_{k+1}) \le f(x_{k,i+1}), \tag{7}$$

fix a number $i_k = i$, increase the value of k by one, and go to Step 1. Otherwise, go to the next step.

Step 6. Choose a subgradient $s_{k,i+1} \in \partial f(x_{k,i+1})$, assign

$$\hat{f}_{k,i+1}(y) = \max\{\hat{f}_{k,i}(y), f(x_{k,i+1}) + \langle s_{k,i+1}, y - x_{k,i+1}\rangle\}, \tag{8}$$

and go to Step 2 with incremented i.

Consider some remarks concerning the procedure π.

Remark 1. For some $k \ge 0$, $i \ge 0$ on the basis of (1), (8) it is not difficult to obtain the equality

$$\hat{f}_{k,i}(y) = \max_{0 \le j \le i}\{f(x_{k,j}) + \langle s_{k,j}, y - x_{k,j}\rangle\}. \tag{9}$$

The function $\hat{f}_{k,i}(y)$ is a model of the convex function $f(x)$. Since the model $\hat{f}_{k,i}(y)$ is the maximum of linear (hence convex) functions, then the function $\hat{f}_{k,i}(y)$ is convex.

One of the main problems arising in the numerical implementation of bundle and cutting methods is the unlimited growth of the count of cutting planes which are used to find iteration points. Currently, several approaches are proposed to discard cutting planes for bundle methods (e. g., [3,5,6]). These approaches are realized according to the aggregation technique of cutting planes proposed in [3] as follows. At the initial step of any bundle method, a storage of cutting planes (called a bundle) is formed and its size is set. Then the overflow of this storage is checked at each step. If the storage of the cutting planes is full, then the procedure is started for discarding the cutting planes in two stages. All inactive cutting planes are discarded at the first stage, and if the first stage does not allow to allocate free spaces in the plane storage, then the second stage is performed. At the second stage any active cutting plane is removed from the storage to free space and one aggregated cutting plane is added which is constructed as a convex combination of active and inactive cutting planes. Note that the application of such an aggregation technique allows approximating the subdifferential of the objective function at the current point and construct some e-subgradient. However, the quality of the approximation of the epigraph of the objective function at the current iteration point is deteriorated after performing the second stage of the procedure for discarding the cutting planes.

A different approach was developed for cutting plane methods for periodically discarding cutting planes in [7–9]. This approach is based on some criteria for estimating the quality of approximating sets formed by cutting planes in a neighborhood of current iteration points. In particular, in [8] the quality of the approximation is estimated by the proximity of the current iteration point to a feasible set of the initial problem, and in [9] the quality is estimated by the assessment of the proximity of the current iteration value to the optimal value. After obtaining sufficiently good approximation sets the proposed approach allows to use update procedures such that it is possible to periodically discard an arbitrary number of any previously constructed cutting planes. Namely, both full and partial updating of approximating sets is permissible. In the case of using partial updating it is possible to leave, for example, only active cutting planes or $n + 1$-last cutting planes.

In this paper, the procedure π is proposed, where cutting planes are discarded based on the approach developed for the cutting plane methods. Namely, at Step 5 of the procedure π there is the possibility of periodically discarding all cutting planes as follows. In the neighborhood of the point $x_{k,i+1}$ the approximation quality of the epigraph of the function $f(x)$ is evaluated by the model $\hat{f}_{k,i}(x)$. If inequality (6) is fulfilled for some $k \geq 0$, $i \geq 0$, then the approximation quality is enough good, and there is a full update of the model of the function $f(x)$ by discarding cutting planes. Otherwise, the model of the convex function $\hat{f}_{k,i}(x)$ is refined and cutting planes are not discarded.

Based on the procedure π the bundle method will be constructed below. Note that at Step 5 of the procedure π during discarding cutting planes basic points x_k, $k \in K$ are determined. In the process of constructing these points can be used any relaxation minimization methods. It is important to note that convergence of such mixed algorithms is guaranteed by the convergence of the proposed bundle method even if the mentioned relaxation methods included in mixed algorithms are heuristic.

Lemma 1. Let $S \subset \mathrm{R}^n$ be a bounded closed set, $\tau \geq 0$. Then the set

$$B(\tau, S) = \bigcup_{v \in S} \{y \in \mathrm{R}^n : \|y - v\| \leq \tau\} \tag{10}$$

is bounded.

Proof. Since the set S is bounded, then there exists a number $\tau' > 0$ such that for any $v \in S$ the inequality

$$\|v\| \leq \tau' \tag{11}$$

is defined. Now suppose that the set $B(\tau, S)$ is not bounded. Then for any $\omega > 0$ there exists a point $y \in B(\tau, S)$ such that $\|y\| > \omega$. Fix any sequence of positive numbers $\{\omega_k\}$, $k \in K$, such that $\omega_k \to +\infty$, $k \in K$. Due to unboundedness of the set $B(\tau, S)$ there is a sequence of points $\{y_k\}$, $k \in K$, such that

$$y_k \in B(\tau, S), \quad \|y_k\| > \omega_k, \quad k \in K. \tag{12}$$

Moreover, in accordance with construction of points $\{y_k\}$, $k \in K$, for each $k \in K$ there exists a point $v_k \in S$ satisfying the condition

$$\|y_k - v_k\| \leq \tau.$$

Hence, from (11), (12) we have

$$\omega_k < \|y_k \pm v_k\| \le \|y_k - v_k\| + \|v_k\| \le \tau + \tau'.$$

The obtained inequality $\omega_k \le \tau + \tau'$ contradicts the assumption $\omega_k \to +\infty$. The lemma is proved.

Lemma 2. *Suppose for some $k \ge 0$, $i \ge 0$ the points x_k, $x_{k,0}$, $x_{k,1}$, ..., $x_{k,i}$ and the model $\hat{f}_{k,i}(y)$ are constructed by the procedure π. Then we obtain*

$$\epsilon_{k,j} = f(x_k) - f(x_{k,j}) - \langle s_{k,j}, x_k - x_{k,j} \rangle \ge 0, \quad 0 \le j \le i, \tag{13}$$

$$\hat{f}_{k,i}(y) = f(x_k) + \max_{0 \le j \le i} \{ \langle s_{k,j}, y - x_k \rangle - \epsilon_{k,j} \}.$$

Proof. Since the function $f(x)$ is convex and $s_{k,j} \in \partial f(x_{k,j})$, $0 \le j \le i$, then using definition of a subgradient it is not difficult to obtain (13). Further, taking account (9) and (13) we have

$$\hat{f}_{k,i}(y) = \max_{0 \le j \le i} \{ f(x_{k,j}) + \langle s_{k,j}, x - x_{k,j} \rangle \pm \epsilon_{k,j} \}$$

$$= \max_{0 \le j \le i} \{ \langle s_{k,j}, y - x_{k,j} \rangle - \epsilon_{k,j} + f(x_k) - \langle s_{k,j}, x_k - x_{k,j} \rangle \}$$

$$= f(x_k) + \max_{0 \le j \le i} \{ \langle s_{k,j}, y - x_k \rangle - \epsilon_{k,j} \}.$$

The lemma is proved.

The following theorem is proved in [3, p. 144].

Theorem 1. *Suppose for some $k \ge 0$, $i \ge 0$ the point $x_{k,i+1}$ is constructed according to (2) by the procedure π. Then*

$$x_{k,i+1} = x_k - \frac{\hat{s}_{k,i}}{\bar{\mu}}, \tag{14}$$

where

$$\hat{s}_{k,i} = \sum_{j=0}^{i} \hat{\alpha}_{k,i}^j s_{k,j}, \tag{15}$$

and the vector $\hat{\alpha}_{k,i} = (\hat{\alpha}_{k,i}^0, \hat{\alpha}_{k,i}^1, \ldots, \hat{\alpha}_{k,i}^i) \in R^{i+1}$ is a solution of the following problem:

$$\min_{\alpha = (\alpha^0, \alpha^1, \ldots, \alpha^i) \in R^{i+1}} \frac{1}{2\bar{\mu}} \| \sum_{j=0}^{i} \alpha^j s_{k,j} \|^2 + \sum_{j=0}^{i} \alpha^j \epsilon_{k,j}, \tag{16}$$

$$s.t. \quad \alpha = (\alpha^0, \alpha^1, \ldots, \alpha^i) \ge 0, \quad \sum_{j=0}^{i} \alpha^j = 1. \tag{17}$$

Moreover, the following expressions

$$\delta_{k,i} = \hat{\epsilon}_{k,i} + \frac{1}{2\bar{\mu}} \| \hat{s}_{k,i} \|^2, \tag{18}$$

$$\hat{s}_{k,i} \in \partial_{\hat{\epsilon}_{k,i}} f(x_k), \tag{19}$$

$$\hat{s}_{k,i} \in \partial \hat{f}_{k,i}(x_{k,i+1}) \tag{20}$$

are valid, where

$$\hat{\epsilon}_{k,i} = \sum_{j=0}^{i} \hat{\alpha}_{k,i}^j \epsilon_{k,j}. \tag{21}$$

From inclusion (19) it follows

Lemma 3. *Suppose the points* $x_k, x_{k,0}, \ldots, x_{k,i+1}$ *and the corresponding subgradients* $s_k, s_{k,0}, \ldots, s_{k,i+1}$ *are constructed for some* $k \geq 0$, $i \geq 0$ *by the proposed procedure* π. *Then for any point* $y \in \mathrm{R}^n$ *the inequality*

$$f(x_k) - f(y) \leq \langle \hat{s}_{k,i}, x_k - y \rangle + \hat{\epsilon}_{k,i} \tag{22}$$

is fulfilled, where $\hat{s}_{k,i}$, $\hat{\epsilon}_{k,i}$ *are defined according to (15), (21) respectively.*

Lemma 4. *Suppose that the stopping criterion (4) is fulfilled for some* $k \geq 0$, $i \geq 0$. *Then the following estimate holds:*

$$f(\hat{x}) - f^* \leq \bar{\rho}\sqrt{2\bar{\mu}\bar{\xi}} + \bar{\xi}, \tag{23}$$

where $\bar{\rho} > 0$ *is the diameter of the set* $L(\bar{x})$.

Proof. Note that the equality $f(x_0) = f(\bar{x})$ is fulfilled in accordance with Step 0 of the procedure π, and from (6), (7) we have $f(x_k) \leq f(\bar{x})$. Consequently, $x_k \in L(\bar{x})$. Moreover, in view of condition (5) the inequality $f(\hat{x}) \leq f(x_k)$ is defined. Hence and from inequality (22) under $y = x^*$ the estimate holds

$$f(\hat{x}) - f^* \leq \|\hat{s}_{k,i}\| \|x_k - x^*\| + \hat{\epsilon}_{k,i}. \tag{24}$$

Further, according to the stopping criterion (4) and equality (18) we obtain

$$\|\hat{s}_{k,i}\| \leq \sqrt{2\bar{\mu}\delta_{k,i}} \leq \sqrt{2\bar{\mu}\bar{\xi}},$$

$$\hat{\epsilon}_{k,i} \leq \delta_{k,i} \leq \bar{\xi}.$$

Hence and from (24), $x^* \in L(\bar{x})$, $x_k \in L(\bar{x})$ it follows the estimate (23). The lemma is proved.

To prove finiteness of the procedure π let's show that values $\delta_{k,i}$, $\|x_{k,i+1} - x_k\|$ are bounded.

Lemma 5. *Suppose that for some* $k \geq 0$, $i \geq 0$ *the points* x_k, $x_{k,i+1}$ *are constructed, the subgradient* s_k *is fixed, the number* $\delta_{k,i}$ *is computed by the procedure* π. *Then the following expressions*

$$\|x_{k,i+1} - x_k\| \leq \frac{2\|s_k\|}{\bar{\mu}}, \tag{25}$$

$$0 \le \delta_{k,i} \le \frac{2\|s_k\|^2}{\bar{\mu}}, \tag{26}$$

$$f(x_k) - \delta_{k,i} = \hat{f}_{k,i}(x_{k,i+1}) + \langle \hat{s}_{k,i}, y - x_{k,i+1} \rangle + \frac{\bar{\mu}}{2}\|y - x_k\|^2 - \frac{\bar{\mu}}{2}\|y - x_{k,i+1}\|^2 \tag{27}$$

are fulfilled, where $y \in R^n$.

Proof. Note that according to (9) for all $j = 0, \ldots, i$, we have

$$\hat{f}_{k,i}(y) \ge f(x_{k,j}) + \langle s_{k,j}, y - x_{k,j} \rangle, \tag{28}$$

where $y \in R^n$, and from Step 1 of the procedure π it follows that $x_{k,0} = x_k$, $s_{k,0} = s_k$. Hence from formula (9) with $y = x_{k,i+1}$, $j = 0$, we obtain

$$f(x_k) - \hat{f}_{k,i}(x_{k,i+1}) \le \langle s_k, x_k - x_{k,i+1} \rangle \le \|s_k\|\|x_k - x_{k,i+1}\|. \tag{29}$$

Moreover, from (3) it follows

$$\frac{\bar{\mu}}{2}\|x_{k,i+1} - x_k\|^2 \le f(x_k) - \hat{f}_{k,i}(x_{k,i+1}). \tag{30}$$

Hence combining inequalities (29), (30) we prove (25).

Further, according to Lemma 2 for all $j = 0, \ldots, i$ we get $\epsilon_{k,j} \ge 0$, therefore, in view of (21) the inequality $\hat{\epsilon}_{k,i} \ge 0$ is determined. Hence and from (18) taking into account $\bar{\mu} > 0$ it follows that $\delta_{k,i} \ge 0$. Moreover, in accordance with (3), (29) we have

$$\delta_{k,i} \le f(x_k) - \hat{f}_{k,i}(x_{k,i+1}) \le \|s_k\|\|x_k - x_{k,i+1}\|.$$

Using the last inequality, (25) and $\delta_{k,i} \ge 0$ expression (26) is obtained.

Let's turn to obtain inequality (27). For any $y \in R^n$ it is determined

$$\|y \pm x_{k,i+1} - x_k\|^2 = \|y - x_{k,i+1}\|^2 + \|x_{k,i+1} - x_k\|^2 + 2\langle y - x_{k,i+1}, x_{k,i+1} - x_k \rangle.$$

Then multiplying the last equality by $\bar{\mu}/2$ and taking into account (14) we get

$$\frac{\bar{\mu}}{2}\|x_{k,i+1} - x_k\|^2 = \frac{\bar{\mu}}{2}\|y - x_k\|^2 - \frac{\bar{\mu}}{2}\|y - x_{k,i+1}\|^2 + \langle \hat{s}_{k,i}, y - x_{k,i+1} \rangle. \tag{31}$$

Moreover, from (3) it follows

$$f(x_k) - \delta_{k,i} = \hat{f}_{k,i}(x_{k,i+1}) + \frac{\bar{\mu}}{2}\|x_{k,i+1} - x_k\|^2.$$

Now substituting $\bar{\mu}/2\|x_{k,i+1} - x_k\|^2$ by (31) in the last equality we obtain (27). The lemma is proved.

Corollary 1. *Suppose that conditions of Lemma 5 are defined, $S \subset R^n$ is bounded closed set satisfying the inclusion*

$$L(\bar{x}) \subset S. \tag{32}$$

Then there exists numbers $\eta = \eta(S) > 0$, $\zeta = \zeta(\eta) > 0$ such that the inequalities

$$\|s_k\| \leq \eta, \tag{33}$$

$$\|x_{k,i+1} - x_k\| \leq \frac{2\eta}{\bar{\mu}}, \tag{34}$$

$$\delta_{k,i} \leq \frac{2\eta^2}{\bar{\mu}}, \tag{35}$$

$$\|s_{k,i+1}\| \leq \zeta \tag{36}$$

are fulfilled.

Proof. Since inclusion (32) is fulfilled according to conditions of the corollary and we have $x_k \in L(\bar{x})$, $s_k \in \partial f(x_k)$ by construction, then in view of boundness of the set S there exists a number $\eta = \eta(S) > 0$ (e. g., [4, p. 121]) such that inequality (33) is determined. Moreover, taking into account inequality (33) from (25), (26) it follows (34), (35).

Further, since the set S is bounded and closed, then according to Lemma 1 the set $B(2\eta/\bar{\mu}, S)$ is bound too. Moreover, from the inclusion $x_k \in L(\bar{x}) \subset S$ and inequality (34) we have $x_{k,i+1} \in B(2\eta/\bar{\mu}, S)$. Therefore, taking into account $s_{k,i+1} \in \partial f(x_{k,i+1})$ there exists a number $\zeta = \zeta(\eta) > 0$ (e. g., [4, p. 121]) such that inequality (36) is determined. The assertion is proved.

Lemma 6. *Suppose that by the proposed procedure π for some $\bar{k} \geq 0$, $\bar{i} \geq 2$ the points $x_{\bar{k}} = x_{\bar{k},0}$,*

$$x_{\bar{k},1}, x_{\bar{k},2}, \ldots, x_{\bar{k},\bar{i}+1} \tag{37}$$

are constructed, the subgradients $s_{\bar{k}} = s_{\bar{k},0}$,

$$s_{\bar{k},1}, s_{\bar{k},2}, \ldots, s_{\bar{k},\bar{i}+1} \tag{38}$$

are chosen, and according to (3) the numbers

$$\delta_{\bar{k},0}, \delta_{\bar{k},1}, \ldots, \delta_{\bar{k},\bar{i}} \tag{39}$$

are computed. Then for each $i = 0, \ldots, \bar{i} - 2$ it is determined that

$$\delta_{\bar{k},i} - \delta_{\bar{k},i+1} \geq \frac{\bar{\mu}(1 - \bar{\theta})^2}{2(\|s_{\bar{k},i+2}\| + \|s_{\bar{k},i+1}\|)^2} \delta_{\bar{k},i+1}^2. \tag{40}$$

Proof. According to Step 5 of the procedure π for each $l = 0, \ldots, \bar{i} - 1$ it is determined

$$f(x_{\bar{k},l+1}) > f(x_{\bar{k}}) - \bar{\theta}\delta_{\bar{k},l}, \tag{41}$$

and in view of equality (14) the vectors

$$\hat{s}_{\bar{k},0}, \hat{s}_{\bar{k},1}, \ldots, \hat{s}_{\bar{k},\bar{i}}$$

correspond to points (37). Choose an arbitrary index i such that $0 \leq i \leq \bar{i} - 2$. Then using definition of a subgradient of a convex function and taking into account (20) we have

$$\hat{f}_{\bar{k},i}(x_{\bar{k},i+1}) \leq \hat{f}_{\bar{k},i}(x_{\bar{k},i+2}) + \langle \hat{s}_{\bar{k},i}, x_{\bar{k},i+1} - x_{\bar{k},i+2} \rangle. \tag{42}$$

Moreover, according to (8) for any $y \in \mathbb{R}^n$ it is defined

$$\hat{f}_{\bar{k},i}(y) \leq \hat{f}_{\bar{k},i+1}(y).$$

Hence under $y = x_{\bar{k},i+2}$ and from (42) it follows that

$$\hat{f}_{\bar{k},i}(x_{\bar{k},i+1}) \leq \hat{f}_{\bar{k},i+1}(x_{\bar{k},i+2}) + \langle \hat{s}_{\bar{k},i}, x_{\bar{k},i+1} - x_{\bar{k},i+2} \rangle,$$

and taking into account for the $(i + 1)$-th element the last inequality has the form

$$\hat{f}_{\bar{k},i}(x_{\bar{k},i+1}) \leq f(x_{\bar{k}}) - \delta_{\bar{k},i+1} - \frac{\bar{\mu}}{2}\|x_{\bar{k},i+2} - x_{\bar{k}}\|^2 + \langle \hat{s}_{\bar{k},i}, x_{\bar{k},i+1} - x_{\bar{k},i+2} \rangle. \tag{43}$$

Now using equality (27) from Lemma 5 under $k = \bar{k}$, $y = x_{\bar{k},i+2}$ it is obtained

$$f(x_{\bar{k}}) - \delta_{\bar{k},i} + \frac{\bar{\mu}}{2}\|x_{\bar{k},i+2} - x_{\bar{k},i+1}\|^2 =$$
$$= \hat{f}_{\bar{k},i}(x_{\bar{k},i+1}) + \langle \hat{s}_{\bar{k},i}, x_{\bar{k},i+2} - x_{\bar{k},i+1} \rangle + \frac{\bar{\mu}}{2}\|x_{\bar{k},i+2} - x_{\bar{k}}\|^2.$$

Hence and from (43) it follows that

$$\frac{\bar{\mu}}{2}\|x_{\bar{k},i+2} - x_{\bar{k},i+1}\|^2 \leq \delta_{\bar{k},i} - \delta_{\bar{k},i+1}. \tag{44}$$

On the other hand, from (3), (9) (for the $(i + 1)$-th element) we get

$$\delta_{\bar{k},i+1} \leq f(x_{\bar{k}}) - f(x_{\bar{k},i+1}) - \langle s_{\bar{k},i+1}, x_{\bar{k},i+2} - x_{\bar{k},i+1} \rangle,$$

and from inequality (41) under $l = i + 1$ it follows that

$$-\bar{\theta}\delta_{\bar{k},i+1} < f(x_{\bar{k},i+2}) - f(x_{\bar{k}}).$$

Now summing the last two inequalities it is determined that

$$(1 - \bar{\theta})\delta_{\bar{k},i+1} \leq f(x_{\bar{k},i+2}) - f(x_{\bar{k},i+1}) - \langle s_{\bar{k},i+1}, x_{\bar{k},i+2} - x_{\bar{k},i+1} \rangle$$
$$\leq (\|s_{\bar{k},i+2}\| + \|s_{\bar{k},i+1}\|)\|x_{\bar{k},i+2} - x_{\bar{k},i+1}\|.$$

Hence and from (44) we obtain (40). The lemma is proved.

Theorem 2. *Let $S \subset \mathbb{R}^n$ be a bounded closed set satisfied condition (32). Then complexity of the procedure π is equal to*

$$\lceil \frac{f(\bar{x}) - f^*}{\bar{\theta}\bar{\xi}} \rceil \lceil 1 + \frac{16\eta^2\zeta^2}{\bar{\mu}^2(1 - \bar{\theta})^2\bar{\xi}^2} \rceil, \tag{45}$$

where $\eta = \eta(S) > 0$, $\zeta = \zeta(\eta) > 0$.

Proof. First, let's estimate the number of iterations of the procedure π by k. Assume that in the procedure π there is a loop in relation to k. In this case, it is constructed a sequence $\{x_k\}$, $k \in K$, such that according to Steps 4, 5 of the procedure π for each $k \in K$ the following conditions hold:

$$\Delta_k > \bar{\xi}, \tag{46}$$

$$f(x_{k+1}) \le f(x_k) - \bar{\theta}\Delta_k, \tag{47}$$

where $\Delta_k = \delta_{k,i_k}$. Now summing the last inequality by k from 0 to $n \ge 0$ we have

$$\sum_{k=0}^{n} \bar{\theta}\Delta_k \le \sum_{k=0}^{n} (f(x_k) - f(x_{k+1})) \le f(x_0) - f^*.$$

Hence under $n \to +\infty$ we obtain $\Delta_k \to 0$ which contradicts condition (46). Consequently, there exists a number $k' \ge 0$ such that the criterion

$$\Delta_{k'} \le \bar{\xi}$$

is fulfilled.

Further, let's consider two cases to estimate the value k'.

1) Suppose that condition (4) is determined under $k = k' = 0$ and $i \ge 0$. Then it is clear that the number of iterations k' does not exceed the value of the first multiplier of valuation (45).

2) Suppose that criterion is fulfilled under $k = k' > 0$ and $i \ge 0$. Then according to Steps 4, 5 of the procedure π and in view of (46), (47) we have

$$\sum_{p=0}^{k'-1} \bar{\theta}\bar{\xi} \le \sum_{p=0}^{k'-1} (f(x_p) - f(x_{p+1})) \le f(x_0) - f^*.$$

Hence taking into account $x_0 = \bar{x}$ (in accordance with Step 0 of the procedure π) it is obtained that

$$k' \le \lceil \frac{f(\bar{x}) - f^*}{\bar{\theta}\bar{\xi}} \rceil. \tag{48}$$

Now let's obtain a complexity of the procedure π in relation to i while k is fixed. Suppose that the point x_k is constructed under some $k \ge 0$ by the procedure π, and there is a loop in relation to i, i. e. for each $i \in K$ conditions (4), (6) are not fulfilled simultaneously. Then there is a sequence $\{\delta_{k,i}\}$, $i \in K$, constructed by the procedure π such that according to Lemma 6 for each $i \in K$ it is determined

$$\frac{\bar{\mu}(1 - \bar{\theta})^2}{2(\|s_{k,i+2}\| + \|s_{k,i+1}\|)^2} \delta_{k,i+1}^2 \le \delta_{k,i} - \delta_{k,i+1}.$$

Hence taking into account (36) from Corollary 1 we get

$$\frac{\bar{\mu}(1 - \bar{\theta})^2}{8\zeta^2} \delta_{k,i+1}^2 \le \delta_{k,i} - \delta_{k,i+1}.$$

After summing the last inequality by i from 0 to $n \geq 0$ we get

$$\sum_{i=0}^{n} \frac{\bar{\mu}(1-\bar{\theta})^2}{8\zeta^2} \delta_{k,i+1}^2 \leq \sum_{i=0}^{n}(\delta_{k,i} - \delta_{k,i+1}) \leq \delta_{k,0}.$$

Hence from $n \to +\infty$ it follows that $\delta_{k,i} \to 0$, $i \in K$. Therefore, there exists a number $i' \in K$ such that the inequality

$$\delta_{k,i'} \leq \bar{\xi}$$

is fulfilled.

To estimate i' consider the following cases.

1) Suppose that it is defined either criterion (4) or condition (6) for some $k \geq 0$, $i' = i \leq 1$. Then i' does not exceed the value of the second multiplier of variable (45).
2) Assume that any condition of (4), (6) is fulfilled for some $k \geq 0$, $i' = i \geq 2$. Then according to Lemma 6, stopping criterion (4) and inequalities (36), (35) from Corollary 1 we get

$$\sum_{j=0}^{i'-2} \frac{\bar{\mu}(1-\bar{\theta})^2}{8\zeta^2} \bar{\xi}^2 \leq \sum_{j=0}^{i'-2}(\delta_{k,j} - \delta_{k,j+1}) \leq \delta_{k,0} \leq \frac{2\eta^2}{\bar{\mu}}.$$

Therefore, the estimate

$$i' \leq \lceil 1 + \frac{16\eta^2\zeta^2}{\bar{\mu}^2(1-\bar{\theta})^2\bar{\xi}^2} \rceil$$

is obtained. Further, taking into account the last estimate and (48) the theorem is proved. Now let's propose a method which permits to find a point allowed to find a point from the set $X^*(\varepsilon)$ under the determined $\varepsilon > 0$ for a finite number of iterations.

Step 0. Assign $t = 0$. Choose a point $z_t \in \mathbf{R}^n$. Determine parameters $\kappa > 0$, $\sigma \in (0,1)$, $\mu > 0$, $\theta \in (0,1)$.
Step 1. Compute $\xi_t = \kappa\sigma^t$.
Step 2. Find a point $z_{t+1} = \pi(z_t, \xi_t, \theta, \mu)$.
Step 3. Increase the value of t by one, and go to Step 1.

Remark 2. According to Steps 0, 4, 5 of the procedure π and Step 2 of the proposed method for each $t \in K$ we obtain

$$f(z_{t+1}) \leq f(z_t). \tag{49}$$

Therefore, the constructed sequence $\{f(z_t)\}$, $k \in K$, is non-increasing.

Theorem 3. *Suppose the sequence $\{z_t\}$, $t \in K$, is constructed by the proposed method. Then for each $t \in K$ it holds*

$$z_t \in L(z_0), \tag{50}$$

$$f(z_{t+1}) - f^* \leq \rho \sqrt{2\mu \xi_t} + \xi_t, \tag{51}$$

where $\rho > 0$ is a diameter of the set $L(z_0)$.

Proof. In accordance with Theorem 2 the procedure π is finite for each $t \in K$, and as already noted in Remark 2 for each $t \in K$ inequality (49) is fulfilled. Consequently, for each $t \in K$ we obtain inclusion (50).

In view of Lemma 4, Step 4 of the procedure π and Step 2 of the proposed method for each $t \in K$ we have

$$f(z_{t+1}) - f^* \leq \varrho_t \sqrt{2\mu \xi_t} + \xi_t,$$

where $\varrho_t > 0$ is a diameter of the set $L(z_t)$. Since for each $t \in K$ inequality (49) is fulfilled, then $L(z_t) \subset L(z_0)$, $t \in K$. Therefore, there is a constant $\rho > 0$ such that estimate (51) is determined for each $t \in K$.

Theorem 4. *Let $\varepsilon > 0$ and $\rho > 0$ be a diameter of the set $L(z_0)$. Then the complexity of the procedure of finding ε-solution by the proposed method is equal to*

$$\lceil 2\log_\sigma \varepsilon - \log_\sigma \kappa - 2\log_\sigma \hat{\rho} \rceil \lceil \frac{(f(z_0) - f^*)\hat{\rho}^2}{\theta \varepsilon^2} \rceil \lceil 1 + \frac{16\eta^2 \zeta^2 \hat{\rho}^4}{\mu^2(1-\theta)^2 \varepsilon^4} \rceil, \tag{52}$$

where $\hat{\rho} = \rho\sqrt{2\mu} + \sqrt{\xi_0}$, $\eta = \eta(L(z_0)) > 0$, $\zeta = \zeta(\eta) > 0$.

Proof. From inequality (51) of Theorem 3 for each $t \in K$ it follows that

$$f(z_{t+1}) - f^* \leq \xi_t^{1/2}(\rho\sqrt{2\mu} + \sqrt{\xi_t}).$$

Since according to Step 1 of the proposed method we have $\xi_t \leq \xi_0$, $\xi_t \to 0$, $t \in K$, then there exists a number $t' \in K$ such that for each $t \geq t'$ the expression

$$f(z_{t+1}) - f^* \leq \xi_t^{1/2}(\rho\sqrt{2\mu} + \sqrt{\xi_t}) \leq \sqrt{\kappa \sigma^t}(\rho\sqrt{2\mu} + \sqrt{\xi_0}) \leq \varepsilon \tag{53}$$

is defined.

If $t' = 0$, then the number of iterations in relations to t does not exceed the first multiplier of value (52). In this connection assume that $t' > 0$. Then from (53) under $t = t'$ it follows

$$t' \leq \lceil 2\log_\sigma \varepsilon - \log_\sigma \kappa - 2\log_\sigma(\rho\sqrt{2\mu} + \sqrt{\xi_0}) \rceil, \tag{54}$$

and for each $p < t'$ the inequality

$$\frac{1}{\xi_p} \leq \frac{\hat{\rho}^2}{\varepsilon^2}. \tag{55}$$

is fulfilled.

Further, since for each $t \in K$ inclusion (50) is determined and $L(z_0)$ is a bounded closed set, then according to Theorem 2 under $S = L(z_0)$ there exists numbers $\eta = \eta(L(z_0))$, $\zeta = \zeta(\eta) > 0$ such that for each $t < t'$ complexity of finding the point z_{t+1} on basis of the point z_t by the procedure π equals

$$\lceil \frac{f(z_t) - f^*}{\theta \xi_t} \rceil \lceil 1 + \frac{16\eta^2 \zeta^2}{\mu^2 (1 - \theta)^2 \xi_t^2} \rceil.$$

Hence and from (54), (55), $f(z_{t+1}) \leq f(z_t)$, $t < t'$ it follows that general complexity of the proposed method equals

$$\sum_{j=0}^{t'-1} \lceil \frac{f(z_j) - f^*}{\theta \xi_j} \rceil \lceil 1 + \frac{16\eta^2 \zeta^2}{\mu^2 (1 - \theta)^2 \xi_j^2} \rceil \leq \sum_{j=0}^{t'-1} \lceil \frac{(f(z_0) - f^*)\hat{\rho}^2}{\theta \varepsilon^2} \rceil \lceil 1 + \frac{16\eta^2 \zeta^2 \hat{\rho}^4}{\mu^2 (1 - \theta)^2 \varepsilon^4} \rceil \leq$$

$$\lceil 2\log_\sigma \varepsilon - \log_\sigma \kappa - 2\log_\sigma(\rho\sqrt{2\mu} + \sqrt{\xi_0}) \rceil \lceil \frac{(f(z_0) - f^*)\hat{\rho}^2}{\theta \varepsilon^2} \rceil \lceil 1 + \frac{16\eta^2 \zeta^2 \hat{\rho}^4}{\mu^2 (1 - \theta)^2 \varepsilon^4} \rceil.$$

The theorem is proved.

4   Conclusion

The bundle method is proposed for minimizing a convex function. To control the count of cutting planes the developed method updates the model of the objective function in case of obtaining good approximation quality of the epigraph in the neighborhood of the current iteration point. Moreover, at the moment of discarding cutting planes there are opportunities to involve any minimization method. The convergence of the proposed method is proved. Estimation of the complexity of finding an ε-solution is equal to $O(\varepsilon^{-6})$.

References

1. Zangwill, W.I.: Nonlinear Programming: A Unified Approach. Prentice-Hall, Englewood Cliffs (1969)
2. Zabotin, I.Y., Yarullin, R.S.: A cutting method and construction of mixed minimization algorithms on its basis. Uchen. Zap. Kazansk. Univ. Ser. Fiz.-Matem. Nauki. **156**(4), 14–24 (2014). (in Russian)
3. Bonnans, J., Gilbert, J., Lemarechal, C., Sagastizabal, C.: Numerical Optimization: Theoretical and Practical Aspects, 2nd edn. Springer, Heidelberg (2003). https://doi.org/10.1007/978-3-662-05078-1
4. Polyak, B.T.: Introduction to Optimization. Nauka, Moscow (1983). [in Russian]
5. de Oliveira, W., Eckstein, J.: A bundle method for exploiting additive structure in difficult optimization problems. Technical report (2015)
6. Kiwiel, K.C.: Efficiency of proximal bundle methods. J. Optim. Theory Appl. **104**, 589–603 (2000)

7. Shulgina, O.N., Yarullin, R.S., Zabotin, I.Ya.: A cutting method with approximation of a constraint region and an epigraph for solving conditional minimization problems. Lobachevskii J. Math. **39**(6), 847–854 (2018). https://doi.org/10.1134/S1995080218060197
8. Zabotin, I.Y., Yarullin, R.S.: One approach to constructing cutting algorithms with dropping of cutting planes. Russ. Math. (Iz. VUZ). **57**(3), 60–64 (2013)
9. Zabotin, I.Ya., Yarullin, R.S.: Cutting-plane method based on epigraph approximation with discarding the cutting planes. Autom. Rem. Control **76**(11), 1966–1975 (2015). https://doi.org/10.1134/S0005117915110065

Global Optimization

Near-Optimal Hyperfast Second-Order Method for Convex Optimization

Dmitry Kamzolov$^{(\boxtimes)}$ (iD)

Moscow Institute of Physics and Technology, Moscow, Russia
kamzolov.dmitry@phystech.edu

Abstract. In this paper, we present a new Hyperfast Second-Order Method with convergence rate $O(N^{-5})$ up to a logarithmic factor for the convex function with Lipshitz 3rd derivative. This method based on two ideas. The first comes from the superfast second-order scheme of Yu. Nesterov (CORE Discussion Paper 2020/07, 2020). It allows implementing the third-order scheme by solving subproblem using only the second-order oracle. This method converges with rate $O(N^{-4})$. The second idea comes from the work of Kamzolov et al. (arXiv:2002.01004). It is the inexact near-optimal third-order method. In this work, we improve its convergence and merge it with the scheme of solving subproblem using only the second-order oracle. As a result, we get convergence rate $O(N^{-5})$ up to a logarithmic factor. This convergence rate is near-optimal and the best known up to this moment.

Keywords: Tensor method · Inexact method · Second-order method · Complexity

1 Introduction

In recent years, it has been actively developing higher-order or tensor methods for convex optimization problems. The primary impulse was the work of Yu. Nesterov [23] about the possibility of the implementation tensor method. He proposed a smart regularization of Taylor approximation that makes subproblem convex and hence implementable. Also Yu. Nesterov proposed accelerated tensor methods [22,23], later A. Gasnikov et al. [4,11,12,18] proposed the near-optimal tensor method via the Monteiro–Svaiter envelope [21] with line-search and got a near-optimal convergence rate up to a logarithmic factor. Starting from 2018–2019 the interest in this topic rises. There are a lot of developments in tensor methods, like tensor methods for Hölder-continuous higher-order derivatives [15, 28], proximal methods [6], tensor methods for minimizing the gradient norm of convex function [9,15], inexact tensor methods [14,19,24], and near-optimal composition of tensor methods for sum of two functions [19]. There are some results about local convergence and convergence for strongly convex functions [7,10,11]. See [10] for more references on applications of tensor method.

The work was funded by RFBR, project number 19-31-27001.

Y. Kochetov et al. (Eds.): MOTOR 2020, CCIS 1275, pp. 167–178, 2020.
https://doi.org/10.1007/978-3-030-58657-7_15

At the very beginning of 2020, Yurii Nesterov proposed a Superfast Second-Order Method [25] that converges with the rate $O(N^{-4})$ for a convex function with Lipshitz third-order derivative. This method uses only second-order information during the iteration, but assume additional smoothness via Lipshitz third-order derivative.[1] Here we should note that for the first-order methods, the worst-case example can't be improved by additional smoothness because it is a specific quadratic function that has all high-order derivatives bounded [24].[2] But for the second-order methods, one can see that the worst-case example does not have Lipshitz third-order derivative. This means that under the additional assumption, classical lower bound $O(N^{-2/7})$ can be beaten, and Nesterov proposes such a method that converges with $O(N^{-4})$ up to a logarithmic factor. The main idea of this method to run the third-order method with an inexact solution of the Taylor approximation subproblem by method from Nesterov with inexact gradients that converges with the linear speed. By inexact gradients, it becomes possible to replace the direct computation of the third derivative by the inexact model that uses only the first-order information. Note that for non-convex problems previously was proved that the additional smoothness might speed up algorithms [1, 3, 14, 26, 29].

In this paper, we propose a Hyperfast Second-Order Method for a convex function with Lipshitz third-order derivative with the convergence rate $O(N^{-5})$ up to a logarithmic factor. For that reason, firstly, we introduce Inexact Near-optimal Accelerated Tensor Method, based on methods from [4, 19] and prove its convergence. Next, we apply Bregman-Distance Gradient Method from [14, 25] to solve Taylor approximation subproblem up to the desired accuracy. This leads us to Hyperfast Second-Order Method and we prove its convergence rate. This method have near-optimal convergence rates for a convex function with Lipshitz third-order derivative and the best known up to this moment.

The paper is organized as follows. In Sect. 2 we formulate problem and introduce some basic facts and notation. In Sect. 3 we propose Inexact Near-optimal Accelerated Tensor Method and prove its convergence rate. In Sect. 4 we propose Hyperfast Second-Order Method and get its convergence speed.

2   Problem Statement and Preliminaries

In what follows, we work in a finite-dimensional linear vector space $E = \mathbb{R}^n$, equipped with a Euclidian norm $\| \cdot \| = \| \cdot \|_2$.

We consider the following convex optimization problem:

$$\min_x f(x), \tag{1}$$

---

[1] Note, that for the first-order methods in non-convex case earlier (see, [5] and references therein) it was shown that additional smoothness assumptions lead to an additional acceleration. In convex case, as far as we know these works of Yu. Nesterov [24, 25] are the first ones where such an idea was developed.

[2] However, there are some results [30] that allow to use tensor acceleration for the first-order schemes. This additional acceleration requires additional assumptions on smoothness. More restrictive ones than limitations of high-order derivatives.

where $f(x)$ is a convex function with Lipschitz p-th derivative, it means that

$$\|D^p f(x) - D^p f(y)\| \leq L_p \|x - y\|. \tag{2}$$

Then Taylor approximation of function $f(x)$ can be written as follows:

$$\Omega_p(f, x; y) = f(x) + \sum_{k=1}^{p} \frac{1}{k!} D^k f(x) [y - x]^k, \ y \in \mathbb{R}^n. \tag{3}$$

By (2) and the standard integration we can get next two inequalities

$$|f(y) - \Omega_p(f, x; y)| \leq \frac{L_p}{(p+1)!} \|y - x\|^{p+1}, \tag{4}$$

$$\|\nabla f(y) - \nabla \Omega_p(f, x; y)\| \leq \frac{L_p}{p!} \|y - x\|^p. \tag{5}$$

3   Inexact Near-Optimal Accelerated Tensor Method

Problem (1) can be solved by tensor methods [23] or its accelerated versions [4,12,18,22]. This methods have next basic step:

$$T_{H_p}(x) = \underset{y}{\operatorname{argmin}} \left\{ \tilde{\Omega}_{p, H_p}(f, x; y) \right\},$$

where

$$\tilde{\Omega}_{p, H_p}(f, x; y) = \Omega_p(f, x; y) + \frac{H_p}{p!} \|y - x\|^{p+1}. \tag{6}$$

For $H_p \geq L_p$ this subproblem is convex and hence implementable.

But what if we can not solve exactly this subproblem. In paper [25] it was introduced Inexact pth-Order Basic Tensor Method (BTMI$_p$) and Inexact pth-Order Accelerated Tensor Method (ATMI$_p$). They have next convergence rates $O(k^{-p})$ and $O(k^{-(p+1)})$, respectively. In this section, we introduce Inexact pth-Order Near-optimal Accelerated Tensor Method (NATMI$_p$) with improved convergence rate $\tilde{O}(k^{-\frac{3p+1}{2}})$, where $\tilde{O}(\cdot)$ means up to logarithmic factor. It is an improvement of Accelerated Taylor Descent from [4] and generalization of Inexact Accelerated Taylor Descent from [19].

Firstly, we introduce the definition of the inexact subproblem solution. Any point from the set

$$\mathcal{N}_{p, H_p}^{\gamma}(x) = \left\{ T \in \mathbb{R}^n \ : \ \|\nabla \tilde{\Omega}_{p, H_p}(f, x; T)\| \leq \gamma \|\nabla f(T)\| \right\} \tag{7}$$

is the inexact subproblem solution, where $\gamma \in [0; 1]$ is an accuracy parameter. $\mathcal{N}_{p, H_p}^0$ is the exact solution of the subproblem.

Next we propose Algorithm 1.

Algorithm 1. Inexact pth-Order Near-optimal Accelerated Tensor Method (NATMI)

1: **Input:** convex function $f : \mathbb{R}^n \to \mathbb{R}$ such that $\nabla^p f$ is L_p-Lipschitz, $H_p = \xi L_p$ where ξ is a scaling parameter, γ is a desired accuracy of the subproblem solution.

2: Set $A_0 = 0, x_0 = y_0$
3: **for** $k = 0$ **to** $k = K - 1$ **do**
4:      Compute a pair $\lambda_{k+1} > 0$ and $y_{k+1} \in \mathbb{R}^n$ such that

$$\frac{1}{2} \leq \lambda_{k+1} \frac{H_p \cdot \|y_{k+1} - \tilde{x}_k\|^{p-1}}{(p-1)!} \leq \frac{p}{p+1},$$

where

$$y_{k+1} \in \mathcal{N}^\gamma_{p,H_p}(\tilde{x}_k) \qquad (8)$$

and

$$a_{k+1} = \frac{\lambda_{k+1} + \sqrt{\lambda_{k+1}^2 + 4\lambda_{k+1} A_k}}{2}, \quad A_{k+1} = A_k + a_{k+1}$$

$$\tilde{x}_k = \frac{A_k}{A_{k+1}} y_k + \frac{a_{k+1}}{A_{k+1}} x_k.$$

5:      Update $x_{k+1} := x_k - a_{k+1} \nabla f(y_{k+1})$
6: **return** y_K

To get the convergence rate of Algorithm 1 we prove additional lemmas. The first lemma gets intermediate inequality to connect theory about inexactness and method's theory.

Lemma 1. *If $y_{k+1} \in \mathcal{N}^\gamma_{p,H_p}(\tilde{x}_k)$, then*

$$\|\nabla \tilde{\Omega}_{p,H_p}(f, \tilde{x}_k; y_{k+1})\| \leq \frac{\gamma}{1-\gamma} \cdot \frac{(p+1)H_p + L_p}{p!} \|y_{k+1} - \tilde{x}_k\|^p. \qquad (9)$$

Proof. From triangle inequality we get

$$\|\nabla f(y_{k+1})\| \leq \|\nabla f(y_{k+1}) - \nabla \Omega_p(f, \tilde{x}_k; y_{k+1})\|$$
$$+ \|\nabla \Omega_p(f, \tilde{x}_k; y_{k+1}) - \nabla \tilde{\Omega}_{p,H_p}(f, \tilde{x}_k; y_{k+1})\| + \|\nabla \tilde{\Omega}_{p,H_p}(f, \tilde{x}_k; y_{k+1})\|$$
$$\overset{(5),(6),(7)}{\leq} \frac{L_p}{p!} \|y_{k+1} - \tilde{x}_k\|^p + \frac{(p+1)H_p}{p!} \|y_{k+1} - \tilde{x}_k\|^p + \gamma \|\nabla f(y_{k+1})\|.$$

Hence,

$$(1 - \gamma)\|\nabla f(y_{k+1})\| \leq \frac{(p+1)H_p + L_p}{p!} \|y_{k+1} - \tilde{x}_k\|^p.$$

And finally from (7) we get

$$\|\nabla \tilde{\Omega}_{p,H_p}(f, \tilde{x}_k; y_{k+1})\| \leq \frac{\gamma}{1-\gamma} \cdot \frac{(p+1)H_p + L_p}{p!} \|y_{k+1} - \tilde{x}_k\|^p.$$

Next lemma plays the crucial role in the prove of the Algorithm 1 convergence. It is the generalization for inexact subproblem of Lemma 3.1 from [4].

Lemma 2. *If $y_{k+1} \in \mathcal{N}_{p,H_p}^{\gamma}(\tilde{x}_k)$, $H_p = \xi L_p$ such that $1 \geq 2\gamma + \frac{1}{\xi(p+1)}$ and*

$$\frac{1}{2} \leq \lambda_{k+1} \frac{H_p \cdot \|y_{k+1} - \tilde{x}_k\|^{p-1}}{(p-1)!} \leq \frac{p}{p+1}, \quad then \tag{10}$$

$$\|y_{k+1} - (\tilde{x}_k - \lambda_{k+1}\nabla f(y_{k+1}))\| \leq \sigma \cdot \|y_{k+1} - \tilde{x}_k\|, \tag{11}$$

$$\sigma \geq \frac{p\xi + 1 - \xi + 2\gamma\xi}{(1-\gamma)2p\xi}, \tag{12}$$

where $\sigma \leq 1$.

Proof. Note, that by definition

$$\nabla\tilde{\Omega}_{p,H_p}(f, \tilde{x}_k; y_{k+1}) = \nabla\Omega_p(f, \tilde{x}_k; y_{k+1})$$
$$+ \frac{H_p(p+1)}{p!}\|y_{k+1} - \tilde{x}_k\|^{p-1}(y_{k+1} - \tilde{x}_k). \tag{13}$$

Hence,

$$y_{k+1} - \tilde{x}_k = \frac{p!}{H_p(p+1)\|y_{k+1} - \tilde{x}_k\|^{p-1}}$$
$$\cdot \left(\nabla\tilde{\Omega}_{p,H_p}(f, \tilde{x}_k; y_{k+1}) - \nabla\Omega_p(f, \tilde{x}_k; y_{k+1})\right). \tag{14}$$

Then, by triangle inequality we get

$$\|y_{k+1} - (\tilde{x}_k - \lambda_{k+1}\nabla f(y_{k+1}))\| = \|\lambda_{k+1}(\nabla f(y_{k+1}) - \nabla\Omega_p(f, \tilde{x}_k; y_{k+1}))$$
$$+ \lambda_{k+1}\nabla\tilde{\Omega}_{p,H_p}(f, \tilde{x}_k; y_{k+1})$$
$$+ \left(y_{k+1} - \tilde{x}_k + \lambda_{k+1}(\nabla\Omega_p(f, \tilde{x}_k; y_{k+1}) - \nabla\tilde{\Omega}_{p,H_p}(f, \tilde{x}_k; y_{k+1}))\right)\|$$
$$\overset{(5),(14)}{\leq} \lambda_{k+1}\frac{L_p}{p!}\|y_{k+1} - \tilde{x}_k\|^p + \lambda_{k+1}\|\nabla\tilde{\Omega}_{p,H_p}(f, \tilde{x}_k; y_{k+1})\|$$
$$+ \left|\lambda_{k+1} - \frac{p!}{H_p \cdot (p+1) \cdot \|y_{k+1} - \tilde{x}_k\|^{p-1}}\right|$$
$$\cdot \|\nabla\tilde{\Omega}_{p,H_p}(f, \tilde{x}_k; y_{k+1}) - \nabla\Omega_p(f, \tilde{x}_k; y_{k+1})\|$$

$$\overset{(9),(13)}{\leq} \|y_{k+1} - \tilde{x}_k\|\left(\lambda_{k+1}\frac{L_p}{p!}\|y_{k+1} - \tilde{x}_k\|^{p-1}\right.$$
$$+ \lambda_{k+1}\frac{\gamma}{1-\gamma} \cdot \frac{(p+1)H_p + L_p}{p!}\|y_{k+1} - \tilde{x}_k\|^{p-1}\right)$$
$$+ \left|\lambda_{k+1} - \frac{p!}{H_p \cdot (p+1) \cdot \|y_{k+1} - \tilde{x}_k\|^{p-1}}\right| \cdot \frac{(p+1)H_p}{p!}\|y_{k+1} - \tilde{x}_k\|^p$$

$$= \|y_{k+1} - \tilde{x}_k\| \left(\frac{\lambda_{k+1}}{p!} \left(L_p + \frac{\gamma}{1-\gamma}((p+1)H_p + L_p) \right) \|y_{k+1} - \tilde{x}_k\|^{p-1} \right)$$

$$+ \|y_{k+1} - \tilde{x}_k\| \left| \frac{\lambda_{k+1}(p+1)H_p}{p!} \|y_{k+1} - \tilde{x}_k\|^{p-1} - 1 \right|$$

$$\overset{(10)}{\leq} \|y_{k+1} - \tilde{x}_k\| \left(\frac{\lambda_{k+1}}{p!} \left(L_p + \frac{\gamma}{1-\gamma}((p+1)H_p + L_p) \right) \|y_{k+1} - \tilde{x}_k\|^{p-1} \right)$$

$$+ \|y_{k+1} - \tilde{x}_k\| \left(1 - \frac{\lambda_{k+1}(p+1)H_p}{p!} \|y_{k+1} - \tilde{x}_k\|^{p-1} \right)$$

$$= \|y_{k+1} - \tilde{x}_k\| \left(1 + \frac{\lambda_{k+1}}{p!} \|y_{k+1} - \tilde{x}_k\|^{p-1} \right.$$

$$\left. \cdot \left(L_p - (p+1)H_p + \frac{\gamma}{1-\gamma}((p+1)H_p + L_p) \right) \right).$$

Hence, by (10) and simple calculations we get

$$\sigma \geq 1 + \frac{1}{2pH_p} \left(L_p - (p+1)H_p + \frac{\gamma}{1-\gamma}((p+1)H_p + L_p) \right)$$

$$= 1 + \frac{1}{2p\xi} \left(1 - (p+1)\xi + \frac{\gamma}{1-\gamma}((p+1)\xi + 1) \right)$$

$$= 1 + \frac{1}{2p\xi} \left(1 - p\xi - \xi + \frac{\gamma p\xi + \gamma\xi + \gamma}{1-\gamma} \right)$$

$$= 1 + \frac{1}{2p\xi} \left(\frac{1 - p\xi - \xi - \gamma + \gamma p\xi + \gamma\xi + \gamma p\xi + \gamma\xi + \gamma}{1-\gamma} \right)$$

$$= 1 + \left(\frac{1 - p\xi - \xi + 2\gamma p\xi + 2\gamma\xi}{(1-\gamma)2p\xi} \right)$$

$$= \frac{p\xi + 1 - \xi + 2\gamma\xi}{(1-\gamma)2p\xi}.$$

Lastly, we prove that $\sigma \leq 1$. For that we need

$$(1-\gamma)2p\xi \geq p\xi + 1 - \xi + 2\gamma\xi$$

$$(p+1)\xi \geq 1 + 2\gamma\xi(1+p)$$

$$\frac{1}{2} - \frac{1}{2\xi(p+1)} \geq \gamma.$$

We have proved the main lemma for the convergence rate theorem, other parts of the proof are the same as [4]. As a result, we get the next theorem.

Theorem 1. *Let f be a convex function whose p^{th} derivative is L_p-Lipschitz and x_* denote a minimizer of f. Then Algorithm 1 converges with rate*

$$f(y_k) - f(x_*) \leq \tilde{O} \left(\frac{H_p R^{p+1}}{k^{\frac{3p+1}{2}}} \right), \tag{15}$$

where

$$R = \|x_0 - x^*\| \tag{16}$$

is the maximal radius of the initial set.

4   Hyperfast Second-Order Method

In recent work [25] it was mentioned that for convex optimization problem (1) with first order oracle (returns gradient) the well-known complexity bound $(L_1 R^2/\varepsilon)^{1/2}$ can not be beaten even if we assume that all $L_p < \infty$. This is because of the structure of the worth case function

$$f_p(x) = |x_1|^{p+1} + |x_2 - x_1|^{p+1} + ... + |x_n - x_{n-1}|^{p+1},$$

where $p = 1$ for first order method. It's obvious that $f_p(x)$ satisfy the condition $L_p < \infty$ for all natural p. So additional smoothness assumptions don't allow to accelerate additionally. The same thing takes place, for example, for $p = 3$. In this case, we also have $L_p < \infty$ for all natural p. But what is about $p = 2$? In this case $L_3 = \infty$. It means that $f_2(x)$ couldn't be the proper worth case function for the second-order method with additional smoothness assumptions. So there appears the following question: Is it possible to improve the bound $(L_2 R^3/\varepsilon)^{2/7}$? At the very beginning of 2020 Yu. Nesterov gave a positive answer. For this purpose, he proposed to use an accelerated third-order method that requires $\tilde{O}\left((L_3 R^4/\varepsilon)^{1/4}\right)$ iterations by using second-order oracle [23]. So all this means that if $L_3 < \infty$, then there are methods that can be much faster than $\tilde{O}\left((L_2 R^3/\varepsilon)^{2/7}\right)$.

In this section, we improve convergence speed and reach near-optimal speed up to logarithmic factor. We consider problem (1) with $p = 3$, hence $L_3 < \infty$. In previous section, we have proved that Algorithm 1 converges. Now we fix the parameters for this method

$$p = 3, \quad \gamma = \frac{1}{2p} = \frac{1}{6}, \quad \xi = \frac{2p}{p+1} = \frac{3}{2}. \tag{17}$$

By (12) we get $\sigma = 0.6$ that is rather close to initial exact $\sigma_0 = 0.5$. For such parameters we get next convergence speed of Algorithm 1 to reach accuracy ε:

$$N_{out} = \tilde{O}\left(\left(\frac{L_3 R^4}{\varepsilon}\right)^{\frac{1}{5}}\right). \tag{18}$$

Note, that at every step of Algorithm 1 we need to solve next subproblem with accuracy $\gamma = 1/6$

$$\operatorname*{argmin}_y \left\{ \langle \nabla f(x_i), y - x_i \rangle + \frac{1}{2} \nabla^2 f(x_i)[y - x_i]^2 \right.$$
$$\left. + \frac{1}{6} D^3 f(x_i)[y - x_i]^3 + \frac{L_3}{4} \|y - x_i\|^4 \right\}. \tag{19}$$

In [14] it was proved, that problem (19) can be solved by Bregman-Distance Gradient Method (BDGM) with linear convergence speed. According to [25] BDGM can be improved to work with inexact gradients of the functions. This made possible to approximate $D^3 f(x)$ by gradients and escape calculations of $D^3 f(x)$ at each step. As a result, in [25] it was proved, that subproblem (19) can be solved up to accuracy $\gamma = 1/6$ with one calculation of Hessian and $O\left(\log\left(\frac{\|\nabla f(x_i)\| + \|\nabla^2 f(x_i)\|}{\varepsilon}\right)\right)$ calculation of gradient.

We use BDGM to solve subproblem from Algorithm 1 and, as a result, we get next Hyperfast Second-Order method as merging NATMI and BDGM.

Algorithm 2. Hyperfast Second-Order Method

1: **Input:** convex function $f : \mathbb{R}^n \to \mathbb{R}$ with L_3-Lipschitz 3rd-order derivative.
2: Set $A_0 = 0, x_0 = y_0$
3: **for** $k = 0$ to $k = K - 1$ **do**
4:    Compute a pair $\lambda_{k+1} > 0$ and $y_{k+1} \in \mathbb{R}^n$ such that

$$\frac{1}{2} \leq \lambda_{k+1} \frac{3L_3 \cdot \|y_{k+1} - \tilde{x}_k\|^2}{4} \leq \frac{3}{4},$$

where $y_{k+1} \in \mathcal{N}_{3,3L_3/2}^{1/6}(\tilde{x}_k)$ solved by Algorithm 3 and

$$a_{k+1} = \frac{\lambda_{k+1} + \sqrt{\lambda_{k+1}^2 + 4\lambda_{k+1}A_k}}{2} , \quad A_{k+1} = A_k + a_{k+1}$$

$$\tilde{x}_k = \frac{A_k}{A_{k+1}} y_k + \frac{a_{k+1}}{A_{k+1}} x_k .$$

5:    Update $x_{k+1} := x_k - a_{k+1}\nabla f(y_{k+1})$
6: **return** y_K

In the Algorithm 3, $\beta_{\rho_k}(z_i, z)$ is a Bregman distance generated by $\rho_k(z)$

$$\beta_{\rho_k}(z_i, z) = \rho_k(z) - \rho_k(z_i) - \langle \nabla \rho_k(z_i), z - z_i \rangle .$$

By $g_{\varphi_k, \tau}(z)$ we take an inexact gradient of the subproblem (19)

$$g_{\varphi_k, \tau}(z) = \nabla f(\tilde{x}_k) + \nabla^2 f(\tilde{x}_k)[z - \tilde{x}_k] + \frac{1}{2}g_{\tilde{x}_k}^\tau(z) + L_3\|z - \tilde{x}_k\|^2(z - \tilde{x}_k) \quad (22)$$

and $g_{\tilde{x}_k}^\tau(z)$ is a inexact approximation of $D^3 f(\tilde{x}_k)[y - \tilde{x}_k]^2$

$$g_{\tilde{x}_k}^\tau(z) = \frac{1}{\tau^2}\left(\nabla f(\tilde{x}_k + \tau(z - \tilde{x}_k)) + \nabla f(\tilde{x}_k - \tau(z - \tilde{x}_k)) - 2\nabla f(\tilde{x}_k)\right). \quad (23)$$

In paper [25] it is proved, that we can choose

$$\delta = O\left(\frac{\varepsilon^{\frac{3}{2}}}{\|\nabla f(\tilde{x}_k)\|_*^{\frac{1}{2}} + \|\nabla^2 f(\tilde{x}_k)\|^{\frac{3}{2}}/L_3^{\frac{1}{2}}}\right),$$

Algorithm 3. Bregman-Distance Gradient Method

1: Set $z_0 = \tilde{x}_k$ and $\tau = \frac{3\delta}{8(2+\sqrt{2})\|\nabla f(\tilde{x}_k)\|}$

2: Set objective function

$$\varphi_k(z) = \langle \nabla f(\tilde{x}_k), z - \tilde{x}_k \rangle + \frac{1}{2}\nabla^2 f(\tilde{x}_k)[z - \tilde{x}_k]^2 + \frac{1}{6}D^3 f(\tilde{x}_k)[z - \tilde{x}_k]^3 + \frac{L_3}{4}\|z - \tilde{x}_k\|^4$$

3: Set feasible set

$$S_k = \left\{ z : \|z - \tilde{x}_k\| \leq 2\left(\frac{2+\sqrt{2}}{L_3}\|\nabla f(\tilde{x}_k)\| \right)^{\frac{1}{3}} \right\} \tag{20}$$

4: Set scaling function

$$\rho_k(z) = \frac{1}{2}\left\langle \nabla^2 f(\tilde{x}_k)(z - \tilde{x}_k), z - \tilde{x}_k \right\rangle + \frac{L_3}{4}\|z - \tilde{x}_k\|^4 \tag{21}$$

5: **for** $k \geq 0$ **do**

6:     Compute the approximate gradient $g_{\varphi_k,\tau}(z_i)$ by (22).

7:     **IF** $\|g_{\varphi_k,\tau}(z_i)\| \leq \frac{1}{6}\|\nabla f(z_i)\| - \delta$, then **STOP**

8:     **ELSE** $z_{i+1} = \underset{z \in S_k}{\operatorname{argmin}} \left\{ \langle g_{\varphi_k,\tau}(z_i), z - z_i \rangle + 2\left(1 + \frac{1}{\sqrt{2}}\right)\beta_{\rho_k}(z_i, z) \right\}$,

9: **return** z_i

then total number of inner iterations equal to

$$T_k(\delta) = O\left(\ln \frac{G+H}{\varepsilon} \right), \tag{24}$$

where G and H are the uniform upper bounds for the norms of the gradients and Hessians computed at the points generated by the main algorithm. Finally, we get next theorem.

Theorem 2. *Let f be a convex function whose third derivative is L_3-Lipschitz and x_* denote a minimizer of f. Then to reach accuracy ε Algorithm 2 with Algorithm 3 for solving subproblem computes*

$$N_1 = \tilde{O}\left(\left(\frac{L_3 R^4}{\varepsilon} \right)^{\frac{1}{5}} \right) \tag{25}$$

Hessians and

$$N_2 = \tilde{O}\left(\left(\frac{L_3 R^4}{\varepsilon} \right)^{\frac{1}{5}} \log\left(\frac{G+H}{\varepsilon} \right) \right) \tag{26}$$

gradients, where G and H are the uniform upper bounds for the norms of the gradients and Hessians computed at the points generated by the main algorithm.

One can generalize this result on uniformly-strongly convex functions by using inverse restart-regularization trick from [13].

So, the main observation of this section is as follows: If $L_3 < \infty$, then we can use this hyperfast[3] second-order algorithm instead of considered in the paper optimal one to make our sliding faster (in convex and uniformly convex cases).

---

[3] Here we use terminology introduced in [25].

5   Conclusion

In this paper, we present Inexact Near-optimal Accelerated Tensor Method and improve its convergence rate. This improvement make it possible to solve the Taylor approximation subproblem by other methods. Next, we propose Hyperfast Second-Order Method and get its convergence speed $O(N^{-5})$ up to logarithmic factor. This method is a combination of Inexact Third-Order Near-Optimal Accelerated Tensor Method with Bregman-Distance Gradient Method for solving inner subproblem. As a result, we prove that our method has near-optimal convergence rates for given problem class and the best known on that moment.

In this paper, we developed near-optimal Hyperfast Second-Order method for sufficiently smooth convex problem in terms of convergence in function. Based on the technique from the work [9], we can also developed near-optimal Hyperfast Second-Order method for sufficiently smooth convex problem in terms of convergence in the norm of the gradient. In particular, based on the work [16] one may show that the complexity of this approach to the dual problem for 1-entropy regularized optimal transport problem will be $\tilde{O}\left(\left((\sqrt{n})^4/\varepsilon\right)^{1/5}\right) \cdot O(n^{2.5}) = O(n^{2.9}\varepsilon^{-1/5})$ a.o., where n is the linear dimension of the transport plan matrix, that could be better than the complexity of accelerated gradient method and accelerated Sinkhorn algorithm $O(n^{2.5}\varepsilon^{-1/2})$ a.o. [8,16]. Note, that the best theoretical bounds for this problem are also far from to be practical ones [2,17,20,27].

Acknowledgements. I would like to thank Alexander Gasnikov, Yurii Nesterov, Pavel Dvurechensky and Cesar Uribe for fruitful discussions.

References

1. Birgin, E.G., Gardenghi, J., Martínez, J.M., Santos, S.A., Toint, L.: Worst-case evaluation complexity for unconstrained nonlinear optimization using high-order regularized models. Math. Program. **163**(1–2), 359–368 (2017)
2. Blanchet, J., Jambulapati, A., Kent, C., Sidford, A.: Towards optimal running times for optimal transport. arXiv preprint arXiv:1810.07717 (2018)
3. Bubeck, S., Jiang, Q., Lee, Y.T., Li, Y., Sidford, A.: Complexity of highly parallel non-smooth convex optimization. In: Advances in Neural Information Processing Systems, pp. 13900–13909 (2019)
4. Bubeck, S., Jiang, Q., Lee, Y.T., Li, Y., Sidford, A.: Near-optimal method for highly smooth convex optimization. In: Conference on Learning Theory, pp. 492–507 (2019)
5. Carmon, Y., Duchi, J., Hinder, O., Sidford, A.: Lower bounds for finding stationary points II: first-order methods. arXiv preprint arXiv:1711.00841 (2017)
6. Doikov, N., Nesterov, Y.: Contracting proximal methods for smooth convex optimization. arXiv preprint arXiv:1912.07972 (2019)
7. Doikov, N., Nesterov, Y.: Local convergence of tensor methods. arXiv preprint arXiv:1912.02516 (2019)
8. Dvurechensky, P., Gasnikov, A., Kroshnin, A.: Computational optimal transport: complexity by accelerated gradient descent is better than by Sinkhorn's algorithm. arXiv preprint arXiv:1802.04367 (2018)

9. Dvurechensky, P., Gasnikov, A., Ostroukhov, P., Uribe, C.A., Ivanova, A.: Near-optimal tensor methods for minimizing the gradient norm of convex function. arXiv preprint arXiv:1912.03381 (2019)
10. Gasnikov, A.: Universal gradient descent. arXiv preprint arXiv:1711.00394 (2017)
11. Gasnikov, A., Dvurechensky, P., Gorbunov, E., Vorontsova, E., Selikhanovych, D., Uribe, C.A.: Optimal tensor methods in smooth convex and uniformly convex optimization. In: Conference on Learning Theory, pp. 1374–1391 (2019)
12. Gasnikov, A., et al.: Near optimal methods for minimizing convex functions with Lipschitz p-th derivatives. In: Conference on Learning Theory, pp. 1392–1393 (2019)
13. Gasnikov, A.V., Kovalev, D.A.: A hypothesis about the rate of global convergence for optimal methods (Newton's type) in smooth convex optimization. Comput. Res. Model. **10**(3), 305–314 (2018)
14. Grapiglia, G.N., Nesterov, Y.: On inexact solution of auxiliary problems in tensor methods for convex optimization. arXiv preprint arXiv:1907.13023 (2019)
15. Grapiglia, G.N., Nesterov, Y.: Tensor methods for minimizing functions with hölder continuous higher-order derivatives. arXiv preprint arXiv:1904.12559 (2019)
16. Guminov, S., Dvurechensky, P., Nazary, T., Gasnikov, A.: Accelerated alternating minimization, accelerated Sinkhorn's algorithm and accelerated iterative Bregman projections. arXiv preprint arXiv:1906.03622 (2019)
17. Jambulapati, A., Sidford, A., Tian, K.: A direct tilde {O}(1/epsilon) iteration parallel algorithm for optimal transport. In: Advances in Neural Information Processing Systems, pp. 11355–11366 (2019)
18. Jiang, B., Wang, H., Zhang, S.: An optimal high-order tensor method for convex optimization. In: Conference on Learning Theory, pp. 1799–1801 (2019)
19. Kamzolov, D., Gasnikov, A., Dvurechensky, P.: On the optimal combination of tensor optimization methods. arXiv preprint arXiv:2002.01004 (2020)
20. Lee, Y.T., Sidford, A.: Solving linear programs with Sqrt (rank) linear system solves. arXiv preprint arXiv:1910.08033 (2019)
21. Monteiro, R.D., Svaiter, B.F.: An accelerated hybrid proximal extragradient method for convex optimization and its implications to second-order methods. SIAM J. Optim. **23**(2), 1092–1125 (2013)
22. Nesterov, Y.: Lectures on Convex Optimization. SOIA, vol. 137. Springer, Cham (2018). https://doi.org/10.1007/978-3-319-91578-4
23. Nesterov, Y.: Implementable tensor methods in unconstrained convex optimization. Math. Program., 1–27 (2019). https://doi.org/10.1007/s10107-019-01449-1
24. Nesterov, Y.: Inexact accelerated high-order proximal-point methods. Technical report, Technical Report CORE Discussion paper 2020, Université catholique de Louvain, Center for Operations Research and Econometrics (2020)
25. Nesterov, Y.: Superfast second-order methods for unconstrained convex optimization. Technical report, Technical Report CORE Discussion paper 2020, Université catholique de Louvain, Center for Operations Research and Econometrics (2020)
26. Nesterov, Y., Polyak, B.T.: Cubic regularization of newton method and its global performance. Math. Program. **108**(1), 177–205 (2006)
27. Quanrud, K.: Approximating optimal transport with linear programs. arXiv preprint arXiv:1810.05957 (2018)

28. Song, C., Ma, Y.: Towards unified acceleration of high-order algorithms under hölder continuity and uniform convexity. arXiv preprint arXiv:1906.00582 (2019)
29. Wang, Z., Zhou, Y., Liang, Y., Lan, G.: Cubic regularization with momentum for nonconvex optimization. In: Proceedings of the Uncertainty in Artificial Intelligence (UAI) Conference (2019)
30. Wilson, A., Mackey, L., Wibisono, A.: Accelerating rescaled gradient descent. arXiv preprint arXiv:1902.08825 (2019)

On a Solving Bilevel D.C.-Convex Optimization Problems

Andrei V. Orlov[(✉)]

Matrosov Institute for System Dynamics and Control Theory of SB of RAS,
Irkutsk, Russia
anor@icc.ru

Abstract. This work addresses the optimistic statement of a bilevel optimization problem with a general d.c. optimization problem at the upper level and a convex optimization problem at the lower level. First, we use the reduction of the bilevel problem to a nonconvex mathematical optimization problem using the well-known Karush-Kuhn-Tucker approach. Then we employ the novel Global Search Theory and Exact Penalty Theory to solve the resulting nonconvex optimization problem. Following this theory, the special method of local search in this problem is constructed. This method takes into account the structure of the problem in question.

Keywords: Bilevel optimization · Optimistic solution ·
KKT-approach · Reduction theorem · Difference of two convex
functions · D.c. optimization · Global Search Theory · Exact Penalty
Theory · Local search

1 Introduction

It is well-known that bilevel optimization is now at the front edge of modern mathematical optimization [1,2]. Bilevel optimization problems (BOPs) represent extreme problems, which – side by side with ordinary constraints such as equalities and inequalities – include a constraint described as an optimization subproblem [1]. BOPs are important theoretically and very prospective in applications. In particular, according to J.-S. Pang [3], a distinguished expert in optimization, the development of methods for solving various problems with hierarchical structure is one of the three challenges faced by optimization theory and methods in the 21st century. Moreover, problems with hierarchical structure arise in investigations of complex control systems, and bilevel optimization is the most popular modeling tool in such systems (see e.g. [2]).

A bilevel optimization problem is not well-posed if the inner (or lower level) problem does not have a unique optimal solution. This situation can be addressed by using the optimistic or pessimistic formulation of the problem. The optimistic approach, when the actions of the lower level might coordinate with the interests of the upper level, is used in most of the investigations since the assumptions that

Y. Kochetov et al. (Eds.): MOTOR 2020, CCIS 1275, pp. 179–191, 2020.
https://doi.org/10.1007/978-3-030-58657-7_16

guarantee the existence of an optimal solution are weaker and reformulations of the problem result in ordinary single-level optimization problems [1].

This paper is focused on one of the classes of bilevel problems with a general d.c. optimization problem (with functions that can be represented as a difference of two convex functions) at the upper level and a general convex optimization problem at the lower level. The task is to find an optimistic solution. Such a class was chosen because it is the most general one which might be solvable by the well-known Karush-Kuhn-Tucker (KKT) approach when the lower level problem has to be replaced by the KKT-conditions (or the equivalent duality conditions) [4–6]. This approach (or the KKT-transformation) leads to a nonconvex single-level mathematical optimization problem with complementarity constraints which is intrinsically irregular. For example, the Mangasarian-Fromovitz constraint qualification is violated at every feasible point (see, e.g., [7]). Moreover, the resulting problem is equivalent to the original one only if global optimal solutions are considered [1,8].

Hence, the principal question here is how we will solve the obtained nonconvex optimization problem.

New Global Optimality Conditions (GOCs) proved in [9,10] by A.S. Strekalovsky for d.c. optimization problem with inequality and equality constraints open a way to develop efficient global search methods for the most general nonconvex optimization problems. Previously, the Global Search Theory (GST) for the canonical nonconvex optimization classes (such as convex maximization, d.c. minimization, and problems with one d.c. constraint) [11,12] allowed to solve some topical problems of Optimization and Operations Research [13–16] including problems with the bilevel structure [17–21].

In particular, our group, under the guidance of A.S. Strekalovsky, has good experience in solving linear bilevel problems with up to 500 variables at each level [17,20], quadratic-linear bilevel problems of dimension up to (150×150) [20], and quadratic bilevel problems of dimension up to (100×100) [21]. Here we intend to generalize our approach for more complicated bilevel problems.

At the same time, as we can see in the available publications, only a few results published so far deal with numerical solutions of high-dimension bilevel problems (for example, up to 100 variables at each level for linear bilevel problems [22]). In most of the cases, authors consider just illustrative examples with the dimension up to 10 (see, e.g. [23,24]) and only the works [25–27] present some results on solving nonlinear bilevel problems of dimension up to 30 at each level (see also the surveys [28,29]).

In contrast to the commonly accepted global optimization methods such as branch-and-bound based techniques, approximation, and diagonal methods, etc. [30–32], the GST developed by A.S. Strekalovsky [9–12] employs a reduction of the nonconvex problem to a family of simpler problems (usually convex) that can be solved by classic convex optimization methods [4–6].

In accordance with the GST, this paper aims at the construction of basic elements of the methods for finding optimistic solutions to the problems under study. Section 2 deals with the reduction of the original bilevel problem to the single-level one, and the obtaining of a d.c. decomposition for all functions from the latter formulation. In Sect. 3 exact penalization and the GOCs in terms of reduced nonconvex problem are presented and discussed. Section 4 is devoted to the description of the Special Local Search Method. Section 5 presents concluding remarks.

2   Problem Statement and Reduction

Consider the following sufficiently general bilevel optimization problem in its optimistic statement. In the optimistic case, according to the theory [1], the minimization at the upper level should be performed with respect to the variables of both levels:

$$(\mathcal{BP}) \quad \left. \begin{array}{l} F(x,\,y) := g_0(x,y) - h_0(x,y) \downarrow \min\limits_{x,y}, \\ x \in X := \{x \in \mathbb{R}^m \mid g_i(x) - h_i(x) \leq 0, \quad i = 1, ..., p\}, \\ y \in Y_*(x) := \operatorname*{Arg\,min}\limits_{y}\{G(x,y) \mid y \in Y(x)\}, \end{array} \right\} \quad (1)$$

where $Y(x) := \{y \in \mathbb{R}^n \mid \varphi_j(x,y) \leq 0, \quad j = 1, ..., q\}$, the functions $g_0(\cdot)$, $h_0(\cdot)$, and $\varphi_j(\cdot)$, $j = 1, ...q$, are convex with respect to the aggregate of x and y on $\mathbb{R}^{m+n}$, the functions $g_i(\cdot)$, $h_i(\cdot)$, $i = 1, ...p$, are convex on $\mathbb{R}^m$, and the function $y \to G(x,y)$ is convex with respect to y on $\mathbb{R}^n$ $\forall x \in X$.

Let us make the following assumptions concerning the problem $(\mathcal{BP})$.

Assume that the set $Y_*(x)$ is a nonempty compact for every fixed $x \in X$. Furthermore, suppose that

$$(\mathcal{H}1) \quad \left. \begin{array}{l} \text{function } F(x,y) \text{ is bounded below on the nonempty set} \\ Z := \{(x,y) \in \mathbb{R}^{m+n} \mid g(x) - h(x) \leq 0_p, \ \varphi(x,y) \leq 0_q\}, \end{array} \right\} \quad (2)$$

where $0_t = (0, ..., 0)^T \in \mathbb{R}^t$; $g(\cdot)$, $h(\cdot)$, and $\varphi(\cdot)$ are appropriate vector-valued functions. Moreover,

$$(\mathcal{H}2) \quad \left. \begin{array}{l} \forall x \in X \text{ function } G(x,y) \text{ is bounded below on the nonempty} \\ \text{set } Y(x), \text{ so that, } \inf\limits_{x}\inf\limits_{y}\{G(x,y) \mid y \in Y(x), \ x \in X\} > -\infty. \end{array} \right\} \quad (3)$$

In addition, assume that the objective function of the problem $(\mathcal{BP})$ satisfies the Lipschitz property [5,6] with respect to both variables. Besides, functions $h_0(x,y)$ and $h_i(x)$, $i = 1, ..., p$, are differentiable with respect to all their variables and these gradients are continuous.

Further, it can be readily seen that the lower level problem of the problem $(\mathcal{BP})$ is convex when $x \in X$ is fixed:

$$(\mathcal{FP}(x)) \quad G(x,y) \downarrow \min\limits_{y}, \quad \varphi_j(x,y) \leq 0, \quad j = 1, ..., q. \quad (4)$$

So that, e.g. when Slater's constraint qualification is fulfilled for the lower level problem at any parameter value $x \in X$:

$$(\mathcal{SCQ}) \quad \exists \bar{y}(x) : \varphi_j(x, \bar{y}(x)) < 0, \quad j = 1, ...q, \quad (5)$$

then the KKT optimality conditions are necessary and sufficient in the problem $(\mathcal{FP}(x))$. Additionally, we have to assume that the functions $G(x, \cdot)$, $\varphi_j(x, \cdot)$, $j = 1, ...q$, are differentiable with respect to y $\forall x \in X$ and these gradients are continuous with respect to both x and y [4–6]. Under these assumptions, the existence of a solution to problem $(\mathcal{BP})$ can be guaranteed [1].

In addition, under these assumptions, the bilevel problem $(\mathcal{BP})$ can be replaced with the following single-level mathematical optimization problem:

$$(\mathcal{P}) \qquad \left.\begin{aligned} F(x, y) &\downarrow \min_{x,y,v}, \quad x \in X, \\ \nabla_y \mathcal{L}(x,y,v) &= 0_n, \quad \varphi(x,y) \le 0_q, \quad v \ge 0_q, \quad \langle v, \varphi(x,y) \rangle = 0, \end{aligned}\right\} \qquad (6)$$

where $\mathcal{L}(x,y,v) = G(x,y) + \langle v, \varphi(x,y) \rangle$ is the normal Lagrange function of problem $(\mathcal{FP}(x))$.

In [8] the equivalence of the problems $(\mathcal{BP})$ and $(\mathcal{P})$ from the viewpoint of searching global solutions is proved. The most important theorem for justification of the approach for finding optimistic solutions to the problem $(\mathcal{BP})$ using solving the problem $(\mathcal{P})$ is the following one.

Theorem 1. [8] *Let the triple* (x^*, y^*, v^*) *be a global solution to the single-level problem* $(\mathcal{P})$–*(6). If the Slater's constraint qualification* $(\mathcal{SCQ})$–*(5) is satisfied* $\forall x \in X$ *for the lower level problem* $(\mathcal{FP}(x))$–*(4), then the pair* (x^*, y^*) *is a global solution to the bilevel problem* $(\mathcal{BP})$–*(1).*

Further, let us study a possibility of finding a global solution to the problem $(\mathcal{P})$ with the help of the new GOCs [9,10] and the GST developed by A.S. Strekalovsky [11,12].

First of all, a solution to the problem $(\mathcal{P})$ exists under the assumptions above. Now let us analyze properties of this problem.

The constraints $\varphi(x,y) \le 0$ and $v \ge 0$ define a convex set S and do not produce additional difficulties to the problem. For simplicity, everywhere further we will use the term "convex constraints" when the constraints describe a convex feasible set in the problem in question, as well as the term "nonconvex constraint" defined similarly.

The constraints $\nabla_y \mathcal{L}(x,y,v) = 0_n$ might be convex or nonconvex depending on properties of the functions $G(\cdot)$ and $\varphi(\cdot)$.

Let function $G(x,y)$ be quadratic, for example,

$$G(x,y) := \frac{1}{2}\langle y, Cy \rangle + \langle d, y \rangle + \langle x, Qy \rangle,$$

where $d \in \mathbb{R}^n$, C is a symmetric and positive semidefinite $(n \times n)$-matrix, and Q is a rectangular $(m \times n)$-matrix. Note that, in that case, the quadratic terms of the form $\frac{1}{2}\langle x, Dx \rangle + \langle c, x \rangle$, where D is symmetric and positive semidefinite too, are not included in the lower level objective function, because for a fixed upper-level variable x these terms are constants and do not affect the structure of the set $Y_*(x)$ [4,6].

Besides, let function $\varphi(x, y)$ be affine, so that, $\varphi(x, y) = Ax + By - b$, where $b \in \mathbb{R}^q$, $A \in \mathbb{R}^{q \times m}$, $B \in \mathbb{R}^{q \times n}$. Then $\nabla_y \mathcal{L}(x, y, v) = Cy + d + x^T Q + v^T B$ and affine constraints $\nabla_y \mathcal{L}(x, y, v) = 0_n$ define a convex set which can be also included in the set S.

If the functions $G(x, y)$ and $\varphi(x, y)$ are more general then the equality constraint $\nabla_y \mathcal{L}(x, y, v) = 0_n$ is becoming nonconvex one.

Further, the constraint $\langle v, \varphi(x, y) \rangle = 0$ is nonconvex in the problem $(\mathcal{P})$–(6) even in the quadratic case above, because it contains the products of components of the variables v and x (as well as components of v and y) which are bilinear structures [13].

To apply the GST for the study of the problem $(\mathcal{P})$, it is necessary to construct explicit representations of all mentioned nonconvex functions in its formulation as differences of two convex functions (d.c. decompositions). If the lower level constraint functions are twice continuously differentiable then, the complementarity constraints are d.c. functions [7].

Let $(\nabla_y \mathcal{L}(x, y, v))_i = g_i(x, y, v) - h_i(x, y, v)$, $i = p+1, ..., p+n$, be the d.c. decompositions of each component of the vector-function $\nabla_y \mathcal{L}(x, y, v)$. At the same time, let $\langle v, \varphi(x, y) \rangle = g_{p+n+1}(x, y, v) - h_{p+n+1}(x, y, v)$. Note that we can build such decompositions where $h_i(x, y, v)$, $i = p+1, ..., p+n+1$, are differentiable with respect to all their variables and these gradients are continuous.

For example, if the function $\varphi(x, y)$ is affine, i.e. $\varphi(x, y) = Ax + By - b$, then the d.c. representation for the function defining the latter constraint in the problem $(\mathcal{P})$–(6) can be obtained by the well-known property of scalar product $(\langle x, y \rangle = \frac{1}{4}\|x + y\|^2 - \frac{1}{4}\|x - y\|^2)$:

$$g_{p+n+1}(x, y, v) = \frac{1}{4}\|v + Ax\|^2 + \frac{1}{4}\|v + By\|^2 - \langle b, v \rangle,$$

$$h_{p+n+1}(x, y, v) = \frac{1}{4}\|v - Ax\|^2 + \frac{1}{4}\|v - By\|^2.$$

So, the problem $(\mathcal{P})$ might be written in the following way:

$$(\mathcal{DCC}) \qquad \left. \begin{array}{c} F(x, \ y) := g_0(x, y) - h_0(x, y) \downarrow \min_{x, y, v}, \\ (x, y, v) \in S := \{(x, y, v) \mid \varphi(x, y) \leq 0_q, \ v \geq 0_q\}, \\ f_i(x) := g_i(x) - h_i(x) \leq 0, \ i \in \{1, ..., p\} =: \mathcal{I}, \\ f_i(x, y, v) := g_i(x, y, v) - h_i(x, y, v) = 0, \\ i \in \{p+1, ..., p+n+1\} =: \mathcal{E}, \end{array} \right\} \qquad (7)$$

where g_i, h_i, $i \in \mathcal{E}$, are convex with respect to the aggregate of all their variables.

Basing on the assumptions above we can conclude that the feasible set $\mathcal{F}$ of the problem $(\mathcal{DCC})$

$$\mathcal{F} := \{(x, y, v) \in S \mid f_i(x) \leq 0, \ i \in \mathcal{I}, \ f_i(x, y, v) = 0, \ i \in \mathcal{E}\},$$

is non-empty and the optimal value $\mathcal{V}(\mathcal{DCC})$ of the problem $(\mathcal{DCC})$ is finite:

$$\mathcal{V}(\mathcal{DCC}) := \inf(F, \mathcal{F}) := \inf_{(x, y, v)} \{F(x, y) \mid (x, y, v) \in \mathcal{F}\} > -\infty.$$

Now we can pass to the characterization of global solutions in this problem (see [9,10]).

3   Exact Penalization and Global Optimality Conditions

First of all, consider the auxiliary problem:

$$(\mathcal{DC}(\sigma)) \quad \Phi_\sigma(x,y,v) := F(x,y) + \sigma W(x,y,v) \downarrow \min_{(x,y,v)}, \quad (x,y,v) \in S, \tag{8}$$

where $\sigma > 0$ is a penalty parameter, and the penalized function $W(\cdot)$ is defined in the following way:

$$W(x,y,v) := \max\{0, f_1(x), \ldots, f_p(x)\} + \sum_{i \in \mathcal{E}} |f_i(x,y,v)|.$$

For a fixed σ this problem belongs to the class of d.c. minimization problems [11,12] with a convex feasible set. In what follows, we show that the objective function of $(\mathcal{DC}(\sigma))$ can be represented as a difference of two convex functions.

It is well-known that if for some σ the triple $(x(\sigma), y(\sigma), v(\sigma))$ is a solution to the problem $(\mathcal{DC}(\sigma))$ (briefly $(x(\sigma), y(\sigma), v(\sigma)) \in \mathrm{Sol}(\mathcal{DC}(\sigma)))$, and $(x(\sigma), y(\sigma), v(\sigma))$ is feasible in the problem $(\mathcal{DCC})$, i.e. $(x(\sigma), y(\sigma), v(\sigma)) \in \mathcal{F}$ and $W[\sigma] := W(x(\sigma), y(\sigma), v(\sigma)) = 0$, then $(x(\sigma), y(\sigma), v(\sigma))$ is a global solution to the problem $(\mathcal{DCC})$ [4–6,9,10].

Also, the following result takes place.

Proposition 1. [1,4–6]   *Suppose that for some fixed $\hat{\sigma} > 0$ the equality $W[\hat{\sigma}] = 0$ holds for the solution $(x(\hat{\sigma}), y(\hat{\sigma}), v(\hat{\sigma}))$ to the problem $(\mathcal{DC}(\hat{\sigma}))$–(8). Then for all values of the parameter $\sigma > \hat{\sigma}$ the function $W[\sigma]$ vanishes, so that the triple $(x(\sigma), y(\sigma), v(\sigma))$ is a solution to $(\mathcal{DCC})$–(7).*

Thus, if the equality $W[\sigma] = 0$ holds, then a solution to the problem $(\mathcal{DC}(\sigma))$ is a solution to the problem $(\mathcal{DCC})$. In addition, this situation remains the same when the value of σ grows.

Hence, the key point for using Exact Penalty Theory here is the existence of a threshold value $\hat{\sigma} > 0$ of the penalty parameter σ for which $W[\sigma] = 0$ $\forall \sigma \geq \hat{\sigma}$. Due to the assumption above that the objective function $F(\cdot)$ of the problem $(\mathcal{DCC})$ satisfies the Lipschitz property [5,6] with respect to both variables, the following assertion holds.

Proposition 2. [5,6,9,10]   *Let the triple (x_*, y_*, v_*) be a global solution to the problem $(\mathcal{DCC})$–(7). Then, there exists $\hat{\sigma} > 0$ such that (x_*, y_*, v_*) is a global solution to the problem $(\mathcal{DC}(\hat{\sigma}))$–(8). Moreover, $\forall \sigma > \hat{\sigma}$ any solution $(x(\sigma), y(\sigma), v(\sigma))$ to the problem $(\mathcal{DC}(\sigma))$–(8) must be feasible in the problem $(\mathcal{DCC})$–(7), i.e. $W[\sigma] = 0$, and, therefore, $(x(\sigma), y(\sigma), v(\sigma))$ is a solution to the problem $(\mathcal{DCC})$–(7), so that $\mathrm{Sol}(\mathcal{DCC}) \subset \mathrm{Sol}(\mathcal{DC}(\sigma))$. The latter inclusion provides the equality*

$$\mathrm{Sol}(\mathcal{DCC}) = \mathrm{Sol}(\mathcal{DC}(\sigma)) \quad \forall \sigma > \hat{\sigma}, \tag{9}$$

so that the problems $(\mathcal{DCC})$–(7) and $(\mathcal{DC}(\sigma))$–(8) turn out to be equivalent (in the sense of (9)).

Therefore, combining Propositions 1 and 2 with Theorem 1, we can conclude that the established connection between the problems $(\mathcal{DC}(\sigma))$ and $(\mathcal{DCC})$ enables us to search for a global solution to the problem $(\mathcal{DC}(\sigma))$ (where $\sigma > \hat{\sigma}$) instead of a solution to the problem $(\mathcal{DCC})$ for finding an optimistic solution to the problem $(\mathcal{BP})$–(1).

So, the existence of the threshold value $\hat{\sigma} > 0$ of a penalty parameter allows us to solve the single problem $(\mathcal{DC}(\sigma))$–(8) (where $\sigma > \hat{\sigma}$ is fixed) instead of solving a sequence of the problems $(\mathcal{DC}(\sigma))$–(8) when the penalty parameter tends to infinity ($\sigma \to +\infty$) (see, for example, [4]).

Before characterizing global solutions in the problem $(\mathcal{DC}(\sigma))$–(8) we need to show that the objective function $\Phi_\sigma(\cdot)$ is a d.c. function, so that it can be represented as a difference of two convex functions. Using the well-known properties: $\max\limits_{i \in \mathcal{I}} f_i(u) = \max\limits_{i \in \mathcal{I}}[g_i(u) + \sum\limits_{j \in \mathcal{I}}^{j \neq i} h_j(u)] - \sum\limits_{i \in \mathcal{I}} h_i(u), \quad \max\{0, f_i(u)\}$
$= \max\{g_i(u), h_i(u)\} - h_i(u)$, and $|f_i(u)| = 2\max\{g_i(u), h_i(u)\} - [g_i(u) + h_i(u)]$
[11,30], it can be readily seen that

$$\Phi_\sigma(x, y, v) \overset{\triangle}{=} F(x,y) + \sigma \max\{0, f_i(x), i \in \mathcal{I}\} \\ + \sigma \sum_{i \in \mathcal{E}} |f_i(x,y,v)| = G_\sigma(x) - H_\sigma(x), \tag{10}$$

where

$$G_\sigma(x,y,v) := g_0(x,y) + \sigma \max\left\{ \sum_{j \in I} h_j(x); \left[g_i(x) + \sum_{j \in I}^{j \neq i} h_j(x) \right], i \in \mathcal{I} \right\} \\ + 2\sigma \sum_{i \in \mathcal{E}} \max\{g_i(x,y,v); h_i(x,y,v)\}, \tag{11}$$

$$H_\sigma(x,y,v) := h_0(x,y) + \sigma\left[\sum_{i \in \mathcal{I}} h_i(x) + \sum_{j \in \mathcal{E}} (g_j(x,y,v) + h_j(x,y,v)) \right]. \tag{12}$$

It is easy to see that $G_\sigma(\cdot)$ and $H_\sigma(\cdot)$ are both convex functions [33,34], so that the function $\Phi_\sigma(\cdot)$ is a d.c. function, as claimed. At the same time, based on the above assumptions, the function $H_\sigma(\cdot)$ is differentiable with respect to all its variables. Moreover, it is obvious that for a feasible (in the problem $(\mathcal{DCC})$) point $(x_*, y_*, v_*) \in S$ we have $W(x_*, y_*, v_*) = 0$, and, therefore, for a number $\zeta := F(x_*, y_*)$ ($\forall \sigma > 0$), we obtain

$$\Phi_\sigma(x_*, y_*, v_*) = F(x_*, y_*) + \sigma W(x_*, y_*, v_*) = F(x_*, y_*) = \zeta. \tag{13}$$

Now we are ready to formulate the necessary GOCs in terms of the problem $(\mathcal{DC}(\sigma))$–(8) that constitute the basis of the Global Search Theory.

Theorem 2. [9,10] *Let a feasible point $(x_*, y_*, v_*) \in \mathcal{F}$, $\zeta := F(x_*, y_*)$ be a (global) solution to the problem $(\mathcal{DCC})$–(7), and a number $\sigma : \sigma \geq \hat{\sigma} > 0$ is selected, where $\hat{\sigma}$ is a threshold value of the penalty parameter, such that $\mathrm{Sol}(\mathcal{DCC}) = \mathrm{Sol}(\mathcal{DC}_\sigma) \; \forall \sigma \geq \hat{\sigma}$.*

Then $\forall (z, u, w, \gamma) \in \mathbb{R}^{m+n+q+1}$, *satisfying the equality*

$$H_\sigma(z, u, w) = \gamma - \zeta, \tag{14}$$

the inequality

$$G_\sigma(x, y, v) - \gamma \geq \langle \nabla H_\sigma(z, u, w), (x, y, v) - (z, u, w) \rangle \quad \forall (x, y, v) \in S \tag{15}$$

takes place. □

The conditions (14)–(15) possess the so-called algorithmic (constructive) property. More precisely, if the GOCs are violated, we can construct a feasible point that will be better than the point in question [9–12]. Indeed, if for some $(\tilde{z}, \tilde{u}, \tilde{w}, \tilde{\gamma})$ from (14) on some level $\zeta := \zeta_k := \Phi_\sigma(x^k, y^k, v^k)$ for the feasible in the problem $(\mathcal{DC}(\sigma))$ point $(\tilde{x}, \tilde{y}, \tilde{v}) \in S$ the inequality (15) is violated:

$$G_\sigma(\tilde{x}, \tilde{y}, \tilde{v}) < \tilde{\gamma} + \langle \nabla H_\sigma(\tilde{z}, \tilde{u}, \tilde{w}), (\tilde{x}, \tilde{y}, \tilde{v}) - (\tilde{z}, \tilde{u}, \tilde{w}) \rangle,$$

then it follows from the convexity of $H_\sigma(\cdot)$ and the equality (14) that

$$\Phi_\sigma(\tilde{x}, \tilde{y}, \tilde{v}) = G_\sigma(\tilde{x}, \tilde{y}, \tilde{v}) - H_\sigma(\tilde{x}, \tilde{y}, \tilde{v})$$
$$< H_\sigma(\tilde{z}, \tilde{u}, \tilde{w}) + \zeta + H_\sigma(\tilde{x}, \tilde{y}, \tilde{v}) - H_\sigma(\tilde{z}, \tilde{u}, \tilde{w}) - H_\sigma(\tilde{x}, \tilde{y}, \tilde{v}) = \Phi_\sigma(x^k, y^k, v^k),$$

or, $\Phi_\sigma(\tilde{x}, \tilde{y}, \tilde{v}) < \Phi_\sigma(x^k, y^k, v^k)$, $(x^k, y^k, v^k) \in \mathcal{F}$, $(\tilde{x}, \tilde{y}, \tilde{v}) \in S$. Therefore, the point (x^k, y^k, v^k) is not a solution to the problem $(\mathcal{DC}(\sigma))$. Moreover, if the triple $(\tilde{x}, \tilde{y}, \tilde{v})$ is also feasible in the problem $(\mathcal{DCC})$, i.e. $W(\tilde{x}, \tilde{y}, \tilde{v}) = 0 = W(x^k, y^k, v^k)$, we obtain the chain $F(x^k, y^k) = \Phi_\sigma(x^k, y^k, v^k) > \Phi_\sigma(\tilde{x}, \tilde{y}, \tilde{v}) = F(\tilde{x}, \tilde{y})$. It means that $(x^k, y^k, v^k) \notin \text{Sol}(\mathcal{DCC})$ and the triple $(\tilde{x}, \tilde{y}, \tilde{v}) \in \mathcal{F}$ is better than the point (x^k, y^k, v^k).

It can be readily seen that Theorem 2 reduces the solution of the nonconvex problem $(\mathcal{DC}(\sigma))$ to study of the family of the convex (linearized) problems

$$(\mathcal{P}_\sigma\mathcal{L}(z, u, w)) \quad \left. \begin{array}{l} \Psi_\sigma(x, y, v) := G_\sigma(x, y, v) \\ \quad - \langle \nabla H_\sigma(z, u, w), (x, y, v) \rangle \downarrow \min_{x, y, v}, \quad (x, y, v) \in S, \end{array} \right\} \tag{16}$$

depending on 4-tuples $(z, u, w, \gamma) \in \mathbb{R}^{m+n+q+1}$ which satisfy the equality (14). Note, the linearization in this problem is implemented with respect to "unified" nonconvexity of the problem $(\mathcal{DC}(\sigma))$ defined by the function $H_\sigma(\cdot)$.

According to our previous experience, varying of the parameters (z, u, w, γ) is convenient to carry out together with a local search. Then, by changing parameters (z, u, w, γ) in (14) for a fixed $\zeta = \zeta_k$ and obtaining approximate solutions $(x(z, u, w, \gamma), y(z, u, w, \gamma), v(z, u, w, \gamma))$ of the linearized problems $(\mathcal{P}_\sigma\mathcal{L}(z, u, w))$, we get a family of starting points to launch a local search procedure.

Additionally, we do not need to go over all (z, u, w, γ) at each level ζ for checking the inequality (15). It is sufficient to prove that the principal inequality (15) is violated at the single 4-tuple $(\tilde{z}, \tilde{u}, \tilde{w}, \tilde{\gamma})$ by the feasible point $(\tilde{x}, \tilde{y}, \tilde{v})$. After that we move to the new level $(x^{k+1}, y^{k+1}, v^{k+1}) := (\tilde{x}, \tilde{y}, \tilde{v})$, $\zeta_{k+1} := \Phi_\sigma(x^{k+1}, y^{k+1}, v^{k+1})$ and vary parameters again.

So, the development of a special Local Search Method that takes into consideration special features of the problem in question is the priority task before constructing a Global Search procedure. According to the GST, such Local Search Method is the principal element of Global Search Algorithms [11–13, 20].

4   Local Search

Note, the problem $(\mathcal{DC}(\sigma))$–(8) belongs to the one of the canonical nonconvex optimization classes namely d.c. minimization when a value of the penalty
parameter $\sigma := \bar{\sigma} > 0$ is fixed. Hence, in order to carry out a local search in
the problem $(\mathcal{DC}(\bar{\sigma}))$ one can apply well-known Special Local Search Method
(SLSM), developed in [11]. This method is very popular in the literature as DC-
algorithm (DCA) [35], and it is based on a consecutive solving the sequence of
problems linearized with respect to the basic nonconvexity (see $(\mathcal{P}_\sigma\mathcal{L}(z, u, w))$–
(16)). It is clear that the linearization is implemented with respect to the function
$H_\sigma(\cdot)$ which accumulates all the nonconvexities of the problems $(\mathcal{DCC})$–(7) and
$(\mathcal{DC}(\sigma))$–(8). But in that case, a question about finding a threshold value of the
penalty parameter (which provide the equality $\mathrm{Sol}(\mathcal{DCC}) = \mathrm{Sol}(\mathcal{DC}(\sigma))$ remains
open, and it should be resolved in advance, before performing a local search.

In this work, we suggest seeking a threshold value of the penalty parameter
at the stage of a local search. Keeping the ideology of linearization, we also use
the recent results concerning steering penalty parameters in nonlinear optimization [36,37]. So, we can present the new Special Penalty Local Search Method
(SPLSM) [38] in terms of the problem $(\mathcal{DC}(\sigma))$–(8).

In accordance with the principles of local search in d.c. minimization [11], let
us organize an iterative process that additionally takes into account a dynamic
update of the penalty parameter.

Let there be given a starting point $(x_0, y_0, v_0) \in S$, an initial value $\sigma^0 > 0$
of the penalty parameter σ. And let at the iteration k we have found the triple
$(x^k, y^k, v^k) \in S$ and the value $\sigma_k \geq \sigma^0$ of the penalty parameter. Introduce the
following notations: $G_k(\cdot) := G_{\sigma_k}(\cdot), \quad H_k(\cdot) := H_{\sigma_k}(\cdot)$.

Now, consider the following linearized problem $(\mathcal{P}_k\mathcal{L}) = (\mathcal{P}_{\sigma_k}\mathcal{L}(x^k, y^k, v^k))$

$$(\mathcal{P}_k\mathcal{L}) \qquad \left. \begin{array}{c} \Psi_k(x, y, v) := G_k(x, y, v) \\ - \langle \nabla H_k(x^k, y^k, v^k), (x, y, v) \rangle \downarrow \min_{x,y,v}, \quad (x, y, v) \in S. \end{array} \right\} \qquad (17)$$

At the same time, taking into account that the penalty function $W(x, y, v)$
is also a d.c. function, we can represent it in the following way with the help of
the decomposition (10)–(12):

$$W(x, y, v) = G_W(x, y, v) - H_W(x, y, v),$$

where

$$G_W(x, y, v) := \frac{1}{\sigma}[G_\sigma(x, y, v) - g_0(x, y)], \quad H_W(x, y, v) := \frac{1}{\sigma}[H_\sigma(x, y, v) - h_0(x, y)].$$

Now, introduce the following auxiliary linearized problem

$$(\mathcal{AP}_W\mathcal{L}_k) \qquad \left. \begin{array}{c} \Psi_W(x, y, v) := G_W(x, y, v) \\ - \langle \nabla H_W(x(\sigma_k), y(\sigma_k), v(\sigma_k)), (x, y, v) \rangle \downarrow \min_{x,y,v}, \\ (x, y, v) \in S. \end{array} \right\} \qquad (18)$$

This problem is also convex and it is related to minimization of the penalty function $W(x, y, v)$. Now, pass to the scheme of the SPLSM.

Let also there be given two scalar parameters $\eta_1, \eta_2 \in]0, 1[$ of the method.

Step 0. Set $k := 0$, $(x^k, y^k, v^k) := (x_0, y_0, v_0)$, $\sigma_k := \sigma^0$.

Step 1. Solve the subproblem $(\mathcal{P}_k \mathcal{L})$ to get $(x(\sigma_k), y(\sigma_k), v(\sigma_k)) \in \text{Sol}(\mathcal{P}_k \mathcal{L})$.

Step 2. If $W(x(\sigma_k), y(\sigma_k), v(\sigma_k)) = 0$ then set $\sigma_+ := \sigma_k$, $(x(\sigma_+), y(\sigma_+), v(\sigma_+)) := (x(\sigma_k), y(\sigma_k), v(\sigma_k))$ and go to **Step 7**.

Step 3. Else (if $W(x(\sigma_k), y(\sigma_k), v(\sigma_k)) > 0$), by solving the linearized problems $(\mathcal{AP}_W \mathcal{L}_k)$ find $(x_W^k, y_W^k, v_W^k) \in \text{Sol}(\mathcal{AP}_W \mathcal{L}_k)$.

Step 4. If $W(x_W^k, y_W^k, v_W^k) = 0$ then solve a few problems $(\mathcal{P}_\sigma \mathcal{L}(x_W^k, y_W^k, v_W^k))$ –(16) (by increasing, if necessary, the value σ_k of a penalty parameter σ), trying to find $\sigma_+ > \sigma_k$ and the triple $(x(\sigma_+), y(\sigma_+), v(\sigma_+)) \in \text{Sol}(\mathcal{P}_{\sigma_+} \mathcal{L}(x_W^k, y_W^k, v_W^k))$, such that $W(x(\sigma_+), y(\sigma_+), v(\sigma_+)) = 0$ and go to **Step 7**.

Step 5. Else, if $W(x_W^k, y_W^k, v_W^k) > 0$, or the value $\sigma_+ > \sigma_k$ such that $W(x(\sigma_+), y(\sigma_+), v(\sigma_+)) = 0$ is not found at the previous step, then find $\sigma_+ > \sigma_k$ satisfying the inequality

$$W(x(\sigma_k), y(\sigma_k), v(\sigma_k)) - W(x(\sigma_+), y(\sigma_+), v(\sigma_+)) \geq \eta_1 [W(x(\sigma_k), y(\sigma_k), v(\sigma_k)) - W(x_W^k, y_W^k, v_W^k)]. \tag{19}$$

Step 6. Increase σ_+, if necessary, to fulfil the inequality

$$\Psi_k(x(\sigma_k), y(\sigma_k), v(\sigma_k)) - \Psi_{\sigma_+}(x(\sigma_+), y(\sigma_+), v(\sigma_+)) \geq \eta_2 \sigma_+ [W(x(\sigma_k), y(\sigma_k), v(\sigma_k)) - W(x(\sigma_+), y(\sigma_+), v(\sigma_+))]. \tag{20}$$

Step 7. $\sigma_{k+1} := \sigma_+$, $(x^{k+1}, y^{k+1}, v^{k+1}) := (x(\sigma_+), y(\sigma_+), v(\sigma_+))$, $k := k + 1$ and loop to **Step 1**. $\qquad\square$

Remark 1. The ideas of additional minimization of the penalty function $W(\cdot)$ and using the parameters η_1 and η_2 were inspired by works [36,37]. In these works also can be found the practical rules on how to select the parameters of the scheme and how to increase the value of a penalty parameter σ.

Remark 2. The presented scheme has a theoretical nature mainly. Its convergence analysis can be found in [38]. In order to apply this scheme in practice, we should take into account the possibility of approximate solving the linearized problems $(\mathcal{P}_k \mathcal{L})$ and $(\mathcal{AP}_W \mathcal{L}_k)$ as well as elaborate the stopping criteria. It is clear that the usage the only obvious criterion $W(x(\sigma_+), y(\sigma_+), v(\sigma_+)) = 0$ (or $W(x(\sigma_+), y(\sigma_+), v(\sigma_+)) \leq \varepsilon$) is not sufficient for the local search goals [38].

Remark 3. The convex linearized problems $(\mathcal{P}_k \mathcal{L})$ and $(\mathcal{AP}_W \mathcal{L}_k)$ are not smooth due to the properties of the functions $G_k(\cdot)$ and $G_W(\cdot)$ respectively. To solve these problems we can use one of the appropriate method of convex non-differentiable optimization [5,39] or reformulate these problems in order to eliminate non-smoothness (see, e.g. [38]).

5   Concluding Remarks

The paper proposes a new approach to solving bilevel optimization problem with a general d.c. optimization problem at the upper level and a convex optimization problem at the lower level. On the one hand, it uses the well-known KKT-approach for reducing the original bilevel problem to the single-level one. On the other hand, the reduced problem is investigated by the new Global Optimality Conditions proved for general nonconvex (d.c.) optimization problems by A.S. Strekalovsky using Exact Penalty Theory.

We described in detail the reduction of the original bilevel problem to a d.c. optimization problem studied the question about an explicit d.c. decomposition of all functions from the formulation of the problem presented the Global Optimality Conditions and Special Penalty Local Search Method in terms of the problem in question.

This paper is the first theoretical stage of scientific research concerning the very difficult problems of bilevel optimization problems in the sufficiently general statement. Our further investigations will be devoted to a building of a field of corresponding test examples, to elaboration and testing of the developed local search scheme as well as constructing and testing a global search method for bilevel problems in question, based on presented theoretical foundations.

References

1. Dempe, S.: Foundations of Bilevel Programming. Kluwer Academic Publishers, Dordrecht (2002)
2. Dempe, S., Kalashnikov, V.V., Perez-Valdes, G.A., Kalashnykova, N.: Bilevel Programming Problems: Theory, Algorithms and Applications to Energy Networks. Springer, Heidelberg (2015). https://doi.org/10.1007/978-3-662-45827-3
3. Pang, J.-S.: Three modeling paradigms in mathematical programming. Math. Program. Ser. B **125**, 297–323 (2010)
4. Bazaraa, M.S., Shetty, C.M.: Nonlinear Programming: Theory and Algorithms. Wiley, New York (1979)
5. Nocedal, J., Wright, S.J.: Numerical Optimization. ORFE. Springer, Heidelberg (2000). https://doi.org/10.1007/b98874
6. Bonnans, J.-F., Gilbert, J.C., Lemarechal, C., Sagastizabal, C.A.: Numerical Optimization: Theoretical and Practical Aspects. Springer, Heidelberg (2006). https://doi.org/10.1007/978-3-540-35447-5
7. Luo, Z.-Q., Pang, J.-S., Ralph, D.: Mathematical Programs with Equilibrium Constraints. Cambridge University Press, Cambridge (1996)
8. Dempe, S., Dutta, J.: Is bilevel programming a special case of a mathematical program with complementarity constraints? Math. Program. Ser. A **131**, 37–48 (2012)
9. Strekalovsky, A.S.: Global optimality conditions and exact penalization. Optim. Lett. **13**(3), 597–615 (2017). https://doi.org/10.1007/s11590-017-1214-x
10. Strekalovsky, A.S.: On a global search in D.C. optimization problems. In: Jaćimović, M., Khachay, M., Malkova, V., Posypkin, M. (eds.) OPTIMA 2019. CCIS, vol. 1145, pp. 222–236. Springer, Cham (2020). https://doi.org/10.1007/978-3-030-38603-0_17

11. Strekalovsky, A.S.: Elements of Nonconvex Optimization. Nauka, Novosibirsk (2003). (in Russian)
12. Strekalovsky, A.S.: On solving optimization problems with hidden nonconvex structures. In: Rassias, T.M., Floudas, C.A., Butenko, S. (eds.) Optimization in Science and Engineering, pp. 465–502. Springer, New York (2014). https://doi.org/10.1007/978-1-4939-0808-0_23
13. Orlov, A.V.: Numerical solution of bilinear programming problems. Comput. Math. Math. Phys. **48**, 225–241 (2008). https://doi.org/10.1134/S0965542508020061
14. Gruzdeva, T.V.: On a continuous approach for the maximum weighted clique problem. J. Glob. Optim. **56**, 971–981 (2013). https://doi.org/10.1007/s10898-012-9885-4
15. Strekalovsky, A.S., Gruzdeva, T.V., Orlov, A.V.: On the problem polyhedral reparability: a numerical solution. Autom. Remote Control **76**, 1803–1816 (2015). https://doi.org/10.1134/S0005117915100082
16. Orlov, A.V., Strekalovsky, A.S., Batbileg, S.: On computational search for Nash equilibrium in hexamatrix games. Optim. Lett. **10**(2), 369–381 (2014). https://doi.org/10.1007/s11590-014-0833-8
17. Gruzdeva, T.V., Petrova, E.G.: Numerical solution of a linear bilevel problem. Comput. Math. Math. Phys. **50**, 1631–1641 (2010). https://doi.org/10.1134/S0965542510100015
18. Orlov, A.: A nonconvex optimization approach to quadratic bilevel problems. In: Battiti, R., Kvasov, D.E., Sergeyev, Y.D. (eds.) LION 2017. LNCS, vol. 10556, pp. 222–234. Springer, Cham (2017). https://doi.org/10.1007/978-3-319-69404-7_16
19. Orlov, A.V.: The global search theory approach to the bilevel pricing problem in telecommunication networks. In: Kalyagin, V.A., Pardalos, P.M., Prokopyev, O., Utkina, I. (eds.) NET 2016. SPMS, vol. 247, pp. 57–73. Springer, Cham (2018). https://doi.org/10.1007/978-3-319-96247-4_5
20. Strekalovsky, A.S., Orlov, A.V.: Linear and Linear-Quadratic Bilevel Optimization Problems. SB RAS, Novosibirsk (2019). (in Russian)
21. Strekalovsky, A.S., Orlov, A.V.: Global search for bilevel optimization with quadratic data. In: Dempe, S., Zemkoho, A. (eds.) Bilevel Optimization: Advances and Next Challenges (2020)
22. Saboia, C.H., Campelo, M., Scheimberg, S.: A computational study of global algorithms for linear bilevel programming. Numer. Algorithms **35**, 155–173 (2004). https://doi.org/10.1023/B:NUMA.0000021760.62160.a4
23. Muu, L.D., Quy, N.V.: A global optimization method for solving convex quadratic bilevel programming problems. J. Glob. Optim. **26**, 199–219 (2003). https://doi.org/10.1023/A:1023047900333
24. Pistikopoulos, E.N., Dua, V., Ryu, J.: Global optimization of bilevel programming problems via parametric programming. In: Floudas, C.A., Pardalos, P.M. (eds.) Frontiers in Global Optimization. NOIA, vol. 74, pp. 457–476. Springer, Boston (2004). https://doi.org/10.1007/978-1-4613-0251-3_25
25. Colson, B., Marcotte, P., Savard, G.: A trust-region method for nonlinear bilevel programming: algorithm and computational experience. Comput. Optim. Appl. **30**, 211–227 (2005). https://doi.org/10.1007/s10589-005-4612-4
26. Etoa Etoa, J.B.: Solving quadratic convex bilevel programming problems using a smoothing method. Appl. Math. Comput. **217**, 6680–6690 (2011)
27. Gumus, Z.H., Floudas, C.A.: Global optimization of nonlinear bilevel programming problems. J. Glob. Optim. **20**, 1–31 (2001). https://doi.org/10.1023/A:1011268113791

28. Dempe, S.: Bilevel programming. In: Audet, C., Hansen, P., Savard, G. (eds.) Essays and Surveys in Global Optimization, pp. 165–193. Springer, Boston (2005). https://doi.org/10.1007/0-387-25570-2_6
29. Colson, B., Marcotte, P., Savard, G.: An overview of bilevel optimization. Ann. Oper. Res. **153**, 235–256 (2007). https://doi.org/10.1007/s10479-007-0176-2
30. Horst, R., Tuy, H.: Global Optimization: Deterministic Approaches. Springer, Heidelberg (1993)
31. Horst, R., Thoai, N.V.: D.C. programming: overview. J. Optim. Theory Appl. **103**(1), 1–43 (1999)
32. Strongin, R.G., Sergeyev, Y.D.: Global Optimization with Non-convex Constraints: Sequential and Parallel Algorithms. NOIA, vol. 45. Springer, Boston (2000). https://doi.org/10.1007/978-1-4615-4677-1
33. Hiriart-Urruty, J.-B., Lemaréchal, C.: Convex Analysis and Minimization Algorithms. GL, vol. 305. Springer, Heidelberg (1993). https://doi.org/10.1007/978-3-662-02796-7
34. Rockafellar, R.T.: Convex Analysis. Princeton University Press, Princeton (1970)
35. Tao, P.D., Souad, L.B.: Algorithms for solving a class of non convex optimization. Methods of subgradients. In: Hiriart-Urruty, J.-B. (ed.) Fermat Days 85, pp. 249–271. Elsevier Science Publishers B.V., North Holland (1986)
36. Byrd, R.H., Nocedal, J., Waltz, R.A.: Steering exact penalty methods for nonlinear programming. Optim. Methods Softw. **23**, 197–213 (2008)
37. Byrd, R.H., Lopez-Calva, G., Nocedal, J.: A line search exact penalty method using steering rules. Math. Program. Ser. A **133**, 39–73 (2012). https://doi.org/10.1007/s10107-010-0408-0
38. Strekalovsky, A.S.: Local search for nonsmooth DC optimization with DC equality and inequality constraints. In: Bagirov, A.M., Gaudioso, M., Karmitsa, N., Mäkelä, M.M., Taheri, S. (eds.) Numerical Nonsmooth Optimization, pp. 229–261. Springer, Cham (2020). https://doi.org/10.1007/978-3-030-34910-3_7
39. Ben-Tal, A., Nemirovski, A.: Non-Euclidean restricted memory level method for large-scale convex optimization. Math. Program. **102**, 407–456 (2005). https://doi.org/10.1007/s10107-004-0553-4

Strongly Convex Optimization
for the Dual Formulation
of Optimal Transport

Nazarii Tupitsa[1,2,4(✉)] ⓘ, Alexander Gasnikov[1,2,4] ⓘ, Pavel Dvurechensky[2,3] ⓘ,
and Sergey Guminov[1,2] ⓘ

[1] Moscow Institute of Physics and Technology, Dolgoprudny, Russia
{tupitsa,sergey.guminov}@phystech.edu
[2] Institute for Information Transmission Problems RAS, Moscow, Russia
gasnikov@yandex.ru
[3] Weierstrass Institute for Applied Analysis and Stochastics, Berlin, Germany
pavel.dvurechensky@wias-berlin.de
[4] National Research University Higher School of Economics,
Moscow, Russia

Abstract. In this paper we experimentally check a hypothesis, that dual problem to discrete entropy regularized optimal transport problem possesses strong convexity on a certain compact set. We present a numerical estimation technique of parameter of strong convexity and show that such an estimate increases the performance of an accelerated alternating minimization algorithm for strongly convex functions applied to the considered problem.

Keywords: Convex optimization · Otimal transport · Sinkhorn's algorithm · Alternating ainimization

1 Introduction

Optimal transport problem has different applications since it allows to define a distance between probability measures including the earth mover's distance [51,62] and Monge-Kantorovich or Wasserstein distance [61]. These distances play an increasing role in different machine learning tasks, such as unsupervised learning [6,11], semi-supervised learning [56], clustering [31], text classification [35], as well as in image retrieval, clustering and classification [13,51,53], statistics [24,49], and other applications [33]. In many of these applications the original optimal distances are substituted by entropically regularized optimal transport problem [13] which gives rise to a so-called Sinkhorn divergence.

A close problem arises in transportation research and consists in recovering a matrix of traffic demands between city districts from the information on population and workplace capacities of each district. As it is shown in [28], a

This research was funded by Russian Science Foundation (project 18-71-10108).

Y. Kochetov et al. (Eds.): MOTOR 2020, CCIS 1275, pp. 192–204, 2020.
https://doi.org/10.1007/978-3-030-58657-7_17

natural model of the district's population dynamics leads to an entropy-linear programming optimization problem for the traffic demand matrix estimation. In this case, the objective function is a sum of an entropy function and a linear function. It is important to note also that the entropy function is multiplied by a regularization parameter γ and the model is close to reality when the regularization parameter is small. The same approach is used in IP traffic matrix estimation [63].

Recent approaches to solving discrete optimal transport problem are based on accelerated primal-dual gradient-based algorithms [21,30] which in some regimes demonstrate better performance than well-known Sinkhorn's algorithm [13,55]. Both these algorithms have complexity polynomially depending on the desired accuracy [3,21,40]. Despite, formally, the dual for the optimal transport problem is not strongly convex, it is strongly convex on any bounded subset of any subspace orthogonal to a one-dimensional subspace. In this paper we suggest and check empirically a hypothesis which helps to increase the rate of convergence for the dual problem to optimal transport. The hypothesis is that dual function demonstrates strong convexity on the orthogonal subspace and Sinkhorn's and other algorithms produce points in this orthogonal subspace meaning that actually the dual problem is strongly convex on the trajectory of the method.

Since we focus mainly on alternating minimization, the related work contains such classical works as [10,48]. AM algorithms have a number of applications in machine learning problems. For example, iteratively reweighted least squares can be seen as an AM algorithm. Other applications include robust regression [41] and sparse recovery [16]. Famous Expectation Maximization (EM) algorithm can also be seen as an AM algorithm [4,42]. Sublinear $O(1/k)$ convergence rate was proved for AM algorithm in [8]. AM-algorithms converge faster in practice in comparison to gradient methods as they are free of the choice of the step-size and are adaptive to the local smoothness of the problem. Besides mentioned above works on AM algorithms, we mention [9,52,58], where non-asymptotic convergence rates for AM algorithms were proposed and their connection with cyclic coordinate descent was discussed, but the analyzed algorithms are not accelerated. Accelerated versions are known for random coordinate descent methods [2,23,26,27,36,37,44,47,54]. These methods use momentum term and block-coordinate steps, rather than full minimization in blocks. A hybrid accelerated random block-coordinate method with exact minimization in the last block and an accelerated alternating minimization algorithm were proposed in [17].

2   Dual Optimal Transport Problem

In this paper we consider the following discrete-discrete entropically regularized optimal transport problem

$$f(X) = \langle C, X \rangle + \gamma \langle X, \ln X \rangle \to \min_{X \in \mathcal{U}(r,c)}, \tag{1}$$

$$\mathcal{U}(r,c) = \{X \in \mathbb{R}_+^{N \times N} : X\mathbf{1} = r, X^T\mathbf{1} = c\},$$

where X is the transportation plan, $\ln X$ is taken elementwise, $C \in \mathbb{R}_+^{N \times N}$ is a given cost matrix, $\mathbf{1} \in \mathbb{R}^N$ is the vector of all ones, $r, c \in S_N(1) := \{s \in \mathbb{R}_+^N : \langle s, \mathbf{1} \rangle = 1\}$ are given discrete measures, and $\langle A, B \rangle$ denotes the Frobenius product of matrices defined as $\langle A, B \rangle = \sum\limits_{i,j=1}^{N} A_{ij} B_{ij}$.

Next, we consider the dual problem for the above optimal transport problem. First, we note that $\mathcal{U}(r, c) \subset Q := \{X \in \mathbb{R}_+^{N \times N} : \mathbf{1}^T X \mathbf{1} = 1\}$ and the entropy $\langle X, \ln X \rangle$ is strongly convex on Q w.r.t 1-norm, meaning that the dual problem has the objective with Lipschitz-continuous gradient [43]. To be more precise, function f is μ-strongly convex on a set Q with respect to norm $\| \cdot \|$ iff

$$f(y) \geqslant f(x) + \langle \nabla f(x), y - x \rangle + \frac{\mu}{2} \|x - y\|^2 \quad \forall x, y \in Q.$$

Further, function f is said to have L-Lipschitz-continuous gradient iff, for all $x, y \in Q$, $\|\nabla f(x) - \nabla f(y)\|_* \leqslant L\|x - y\|$. Here $\| \cdot \|_*$ is the standard conjugate norm for $\| \cdot \|$. The proof that Entropy is 1-strongly convex on the standard simplex w.r.t. to $\| \cdot \|_1$-norm can be found in [43]. The dual problem is constructed as follows

$$\min_{X \in Q \cap \mathcal{U}(r,c)} \langle C, X \rangle + \gamma \langle X, \ln X \rangle \tag{2}$$

$$= \min_{X \in Q} \max_{y, z \in \mathbb{R}^N} \left\{ \langle C, X \rangle + \gamma \langle X, \ln X \rangle + \langle y, X\mathbf{1} - r \rangle + \langle z, X^T\mathbf{1} - c \rangle \right\}$$

$$= \max_{y, z \in \mathbb{R}^N} \left\{ -\langle y, r \rangle - \langle z, c \rangle + \min_{X \in Q} \sum_{i,j=1}^{N} X^{ij} \left(C^{ij} + \gamma \ln X^{ij} + y^i + z^j \right) \right\}.$$

Note that for all i, j and some small ε

$$X^{ij} \left(C^{ij} + \gamma \ln X^{ij} + y^i + z^j \right) < 0$$

for $X^{ij} \in (0, \varepsilon)$ and this quantity approaches 0 as X^{ij} approaches 0. Hence, $X^{ij} > 0$ without loss of generality. Using Lagrange multipliers for the constraint $\mathbf{1}^T X \mathbf{1} = 1$, we obtain the problem

$$\min_{X^{ij} > 0} \max_{\nu} \left\{ \sum_{i,j=1}^{N} \left[X^{ij} \left(C^{ij} + \gamma \ln X^{ij} + y^i + z^j \right) \right] - \nu \left[\sum_{i,j=1}^{N} X^{ij} - 1 \right] \right\}.$$

The solution to this problem is

$$X^{ij} = \frac{\exp\left(-\frac{1}{\gamma} \left(y^i + z^j + C^{ij} \right) - 1 \right)}{\sum_{i,j=1}^{n} \exp\left(-\frac{1}{\gamma} \left(y^i + z^j + C^{ij} \right) - 1 \right)}.$$

With a change of variables $u = -y/\gamma - \frac{1}{2}\mathbf{1}, v = -z/\gamma - \frac{1}{2}\mathbf{1}$ we arrive at the following expression for the dual (minimization) problem

$$\varphi(u, v) = \gamma \left(\ln \left(\mathbf{1}^T B(u, v) \mathbf{1} \right) - \langle u, r \rangle - \langle v, c \rangle \right) \rightarrow \min_{u, v \in \mathbb{R}^N}, \tag{3}$$

where $[B(u,v)]^{ij} = \exp\left(u^i + v^j - \frac{C^{ij}}{\gamma}\right)$. Let us also define

$$\varphi(y,z) = \varphi\left(-y/\gamma - \frac{1}{2}\mathbf{1}, -z/\gamma - \frac{1}{2}\mathbf{1}\right), \tag{4}$$

i.e. $\varphi(y,z)$ is the dual objective before change of variables. Note that the gradient of this function has the form of two blocks

$$\nabla\varphi(y,z) = \begin{pmatrix} r - \dfrac{B\left(-y/\gamma - 1/2, -z/\gamma - 1/2\right)\mathbf{1}}{\mathbf{1}^T B\left(-y/\gamma - 1/2, -z/\gamma - 1/2\right)\mathbf{1}} \\ c - \dfrac{B\left(-y/\gamma - 1/2, -z/\gamma - 1/2\right)^T \mathbf{1}}{\mathbf{1}^T B\left(-y/\gamma - 1/2, -z/\gamma - 1/2\right)\mathbf{1}} \end{pmatrix}. \tag{5}$$

Notably, this dual problem is a smooth minimization problem with the objective having Lipschitz continuous gradient with constant $2/\gamma$ [30]. Unfortunately, generally speaking it is not strongly convex since given a point (u_0, v_0) the value of the objective is the same on the whole line $(u_0 + t\mathbf{1}, v_0 - t\mathbf{1})$ parameterized by t. Yet, this function is strongly convex in the subspace orthogonal to these lines [15]. The goal of this paper is to use this strong convexity to accelerate the accelerated alternating minimization method based on Nesterov extrapolation and alternating minimization.

The variables in the dual problem (3) naturally decompose into two blocks u and v. Moreover, minimization over any one block may be performed analytically.

Lemma 1. *The iterations*

$$u^{k+1} \in \operatorname*{argmin}_{u \in \mathbb{R}^N} \varphi(u, v^k), \ v^{k+1} \in \operatorname*{argmin}_{v \in \mathbb{R}^N} \varphi(u^{k+1}, v),$$

can be written explicitly as

$$u^{k+1} = u^k + \ln r - \ln\left(B\left(u^k, v^k\right)\mathbf{1}\right),$$
$$v^{k+1} = v^k + \ln c - \ln\left(B\left(u^{k+1}, v^k\right)^T \mathbf{1}\right).$$

This lemma implies that an alternating minimization method applied to the dual formulation is a natural algorithm. In fact, this is the celebrated Sinkhorn's algorithm [13,55] in one of its forms [3] listed as Algorithm 1. This algorithm may also be implemented more efficiently as a matrix-scaling algorithm, see [13]. For the reader's convenience, we prove this lemma here.

Algorithm 1. Sinkhorn's Algorithm

Output: x^k
    for $k \geq 1$ **do**
        $u^{k+1} = u^k + \ln r - \ln\left(B\left(u^k, v^k\right)\mathbf{1}\right)$
        $v^{k+1} = v^k$
        $u^{k+2} = u^{k+1}$
        $v^{k+2} = v^{k+1} + \ln c - \ln\left(B\left(u^{k+1}, v^{k+1}\right)^T \mathbf{1}\right)$
    end for

Proof. From optimality conditions, for u to be optimal, it is sufficient to have $\nabla_u \varphi(u, v) = 0$, or

$$r - (\mathbf{1}^T B(u, v^k)\mathbf{1})^{-1} B(u, v^k)\mathbf{1} = 0. \tag{6}$$

Now we check that it is, indeed, the case for $u = u^{k+1}$ from the statement of this lemma. We check that

$$\begin{aligned} B(u^{k+1}, v^k)\mathbf{1} &= \mathrm{diag}(e^{(u^{k+1}-u^k)})B(u^k, v^k)\mathbf{1} \\ &= \mathrm{diag}(e^{\ln r - \ln(B(u^k, v^k)\mathbf{1})})B(u^k, v^k)\mathbf{1} \\ &= \mathrm{diag}(r)\,\mathrm{diag}(B(u^k, v^k)\mathbf{1})^{-1}B(u^k, v^k)\mathbf{1} = \mathrm{diag}(r)\mathbf{1} = r \end{aligned}$$

and the conclusion then follows from the fact that

$$\mathbf{1}^T B(u^{k+1}, v^k)\mathbf{1} = \mathbf{1}^T r = 1.$$

The optimality of v^{k+1} can be proved in the same way.

3   Accelerated Sinkhorn's Algorithm

In this section, we describe accelerated alternating minimization method from [59], which originates from [29, 30, 46], where the latter preprint [30] describes accelerated alternating minimization for non-strongly functions. Our goal is to use the algorithm which has a possibility to use strong convexity. Formally, the dual OT problem (3) is not strongly convex on the whole space. It is strongly convex on any bounded subset of the subspace orthogonal to lines $(u_0 + t\mathbf{1}, v_0 - t\mathbf{1})$. For non-strongly convex problems algorithm (2) has the following sublinear convergence rate $f(x^k) - f(x_*) \leqslant \frac{4nLR^2}{k^2}$. The proof can be found in [30]. The following Algorithm 2 requires the knowledge of the parameter μ of strong convexity. Notice, that this algorithm run with $\mu = 0$ coincides with its modification for non-strongly functions from [30]. But actually, we were able to outperform the algorithm from [30] by estimating a parameter of strong convexity, but only in iterations.

Algorithm 2. Accelerated Alternating Minimization 2

Input: Starting point x_0.
Output: x^k
1: Set $A_0 = 0$, $x^0 = v^0$, $\tau_0 = 1$
2: **for** $k \geqslant 0$ **do**
3:   Set
$$\beta_k = \underset{\beta \in [0,1]}{\operatorname{argmin}} f\left(x^k + \beta(v^k - x^k)\right) \tag{7}$$
4:   Set $y^k = x^k + \beta_k(v^k - x^k)$ {Extrapolation step}
5:   Choose $i_k = \underset{i \in \{1,\dots,n\}}{\operatorname{argmax}} \|\nabla_i f(y^k)\|_2^2$
6:   Set $x^{k+1} = \underset{x \in S_{i_k}(y^k)}{\operatorname{argmin}} f(x)$ {Block minimization}
7:   If L is known choose a_{k+1} s.t. $\frac{a_{k+1}^2}{(A_k+a_{k+1})(\tau_k+\mu a_{k+1})} = \frac{1}{Ln}$
     If L is unknown, find largest a_{k+1} from the equation
$$f(y^k) - \frac{a_{k+1}^2}{2(A_k+a_{k+1})(\tau_k+\mu a_{k+1})}\|\nabla f(y^k)\|_2^2 +$$
$$\frac{\mu \tau_k a_{k+1}}{2(A_k+a_{k+1})(\tau_k+\mu a_{k+1})}\|v^k - y^k\|_2^2 = f(x^{k+1}) \tag{8}$$
8:   Set $A_{k+1} = A_k + a_{k+1}$, $\tau_{k+1} = \tau_k + \mu a_{k+1}$
9:   Set $v^{k+1} = \underset{x \in \mathbb{R}^N}{\operatorname{argmin}} \psi_{k+1}(x)$ {Update momentum term}
10: **end for**

Theoretical justification is given by the following theorem proved in [59].

Theorem 1 [[59] *Theorem 1*]. *After k steps of Algorithm 2 it holds that*

$$f(x^k) - f(x_*) \leqslant nLR^2 \min\left\{\frac{4}{k^2}, \left(1 - \sqrt{\frac{\mu}{nL}}\right)^{k-1}\right\}, \tag{9}$$

where R is an estimate for $\|x_0 - x_*\|$ satisfying $\|x_0 - x_*\| \leqslant R$.

Applying Algorithm 2 to the dual entropy-regularized optimal transport problem (3) with the objective (4), and using the estimate $L = 2/\gamma$ and $R \leqslant \sqrt{n/2}\left(\|C\|_\infty - \frac{\gamma}{2}\ln \min_{i,j}\{r_i, c_j\}\right)$ [30], we obtain the following Corollary.

Corollary 1. *Let the histograms r, c be slightly modified, s.t. $\min_{i,j}\{r_i, c_j\} \geqslant \varepsilon$.
For example, one can set $(\tilde{r}, \tilde{c}) = \left(1 - \frac{\varepsilon}{8}\right)\left((r, c) + \frac{\varepsilon}{n(8-\varepsilon)}(1, 1)\right)$. Let Algorithm 2 be applied to the dual entropy-regularized optimal transport problem (3) with the objective (4). Let this dual problem have μ-strongly convex objective. Then, after k steps of Algorithm 2 it holds that*

$$\varphi(y, z) - \varphi(y_*, z_*) \leqslant \frac{2n}{\gamma}\left(\|C\|_\infty - \frac{\gamma}{2}\ln \varepsilon\right)^2 \min\left\{\frac{4}{k^2}, \left(1 - \sqrt{\frac{\mu\gamma}{4}}\right)^{k-1}\right\}. \tag{10}$$

The specification of Algorithm 2 for the dual entropy regularized optimal transport problem (3) with the objective (4) is listed below as Algorithm 3. Each variable has two blocks that naturally correspond to the variables (y, z) in (4).

Algorithm 3. Accelerated Sinkhorn with Strong Convexity

Input: Starting point x_0.
Output: x^k
1: Set $A_0 = 0$, $x^0 = w^0$, $\tau_0 = 1$
2: **for** $k \geqslant 0$ **do**
3:    Set
$$\beta_k = \operatorname*{argmin}_{\beta \in [0,1]} \varphi \left(x^k + \beta(w^k - x^k) \right) \qquad (11)$$
4:    Set $s^k = x^k + \beta_k(w^k - x^k)$ {Extrapolation step}
5:    Choose $i_k = \operatorname*{argmax}_{i \in \{1,2\}} \|\nabla_i \varphi(s^k)\|_2^2$, where $\nabla \varphi(\cdot)$ is given in (5).
6:    **if** $i_k = 1$ **then**
7:       $x_1^{k+1} = s_1^k + \ln r - \ln \left(B \left(s_1^k, s_2^k \right) \mathbf{1} \right)$, $x_2^{k+1} = s_2^k$
8:    **else**
9:       $x_2^{k+1} = s_2^k + \ln c - \ln \left(B \left(s_1^k, s_2^k \right)^T \mathbf{1} \right)$, $x_1^{k+1} = s_1^k$
10:   **end if**
11:   If L is known choose a_{k+1} s.t. $\frac{a_{k+1}^2}{(A_k + a_{k+1})(\tau_k + \mu a_{k+1})} = \frac{1}{2L}$
      If L is unknown, find largest a_{k+1} from the equation
$$\varphi(s^k) - \frac{a_{k+1}^2}{2(A_k + a_{k+1})(\tau_k + \mu a_{k+1})} \|\nabla \varphi(s^k)\|_2^2 +$$
$$\frac{\mu \tau_k a_{k+1}}{2(A_k + a_{k+1})(\tau_k + \mu a_{k+1})} \|w^k - s^k\|_2^2 = \varphi(x^{k+1}) \qquad (12)$$
12:   Set $A_{k+1} = A_k + a_{k+1}$, $\tau_{k+1} = \tau_k + \mu a_{k+1}$
13:   Set $w^{k+1} = w^k - a_{k+1} \nabla \varphi(s^k)$ {Update momentum term}
14: **end for**

We point out that usually, the goal is to solve the primal OT problem. For simplicity, we consider only dual OT problem since the solution of the primal can be reconstructed via standard primal-dual analysis [5,12,21,22] applied to the discussed methods.

4 Estimating a Parameter of Strong Convexity

We build an initial estimate of strong convexity parameter μ by searching the value $\hat{\mu}$ from $[0, \hat{L}]$ which gives the minimum objective value after 10 iterations. $\hat{L}$ is an upper bound on the parameter of Lipschitz continuity of the gradient.

Dependence of the objective value after 10 iterations on μ is presented on Fig. 1.

Then we restart the algorithm from the best point with $\mu = [2\hat{\mu}, \hat{\mu}, \hat{\mu}/2]$ every 10 iterations.

The significant implementation detail is connected with the accumulation of the momentum term (vector w) by Algorithm 2. If we restart the algorithm

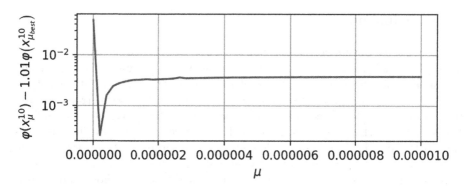

Fig. 1. Empirical dependence of the progress after 10 iterations $h(\mu) = \varphi(x_\mu^{10})$ on the strong convexity parameter μ used in Algorithm 2. The initial value of μ is chosen as a point of minimum of this dependence.

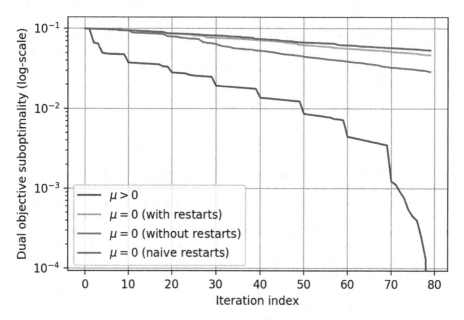

Fig. 2. Performance of Algorithm 2 with the optimal choice of parameter μ on the dual entropy regularized optimal transport problem (3).

naively (with $w^0 = x^0$), we will lose all accumulated information. That is why, we restart the algorithm with w^0 obtained from the last iteration of the previous restart. In order to compare the difference we bring to comparison the case of naive restarts.

As we can see from (Fig. 2), the value of the dual objective decreases faster when one uses the method with positive strong convexity parameter than when one uses the method with $\mu = 0$.

5   Conclusion

In this work we have investigated, how strong convexity can be used to accelerate the accelerated Sinkhorn's algorithm for the dual entropy-regularized optimal transport problem. As we see, the accelerated alternating minimization method in its particular version of accelerated Sinkhorn's algorithm with strong convexity can utilize an estimated value of the strong convexity parameter to converge faster. We underline that it is not clear how one can incorporate this information in the standard Sinkhorn's algorithm to accelerate it. As future work we would like to note the study of automatic strong convexity adaptation procedures like in [25,50], which are now adapted for gradient methods and coordinate descent methods, rather than for alternating minimization methods. Among other extensions, it would be interesting to understand whether restricted strong convexity improves convergence rates of the methods for approximating Wasserstein barycenter [1,14,19,34,38,60] and related distributed optimization methods [18]. Another direction is an application to similar optimization problems, which arise in transportation research in connection to equilibrium in congestion traffic models and traffic demands matrix estimation [7,20] and multimarginal optimal transport [39]. Finally, we use regularization for the OT problem to make the dual problem have Lipshitz gradient. It would be interesting to use universal methods [32,45,57] for the dual OT problem.

References

1. Agueh, M., Carlier, G.: Barycenters in the Wasserstein space. SIAM J. Math. Anal. **43**(2), 904–924 (2011)
2. Allen-Zhu, Z., Qu, Z., Richtarik, P., Yuan, Y.: Even faster accelerated coordinate descent using non-uniform sampling. In: Balcan, M.F., Weinberger, K.Q. (eds.) Proceedings of The 33rd International Conference Machine Learning. Proceedings of Machine Learning Research, vol. 48, pp. 1110–1119. PMLR, New York, New York, USA (2016), http://proceedings.mlr.press/v48/allen-zhuc16.html, arXiv:1512.09103
3. Altschuler, J., Weed, J., Rigollet, P.: Near-linear time approxfimation algorithms for optimal transport via Sinkhorn iteration. In: Guyon, I., et al. (eds.) Advances in Neural Information Processing Systems, vol. 30, pp. 1961–1971. Curran Associates, Inc. (2017). arXiv:1705.09634
4. Andresen, A., Spokoiny, V.: Convergence of an alternating maximization procedure. JMLR **17**(63), 1–53 (2016). http://jmlr.org/papers/v17/15-392.html
5. Anikin, A.S., Gasnikov, A.V., Dvurechensky, P.E., Tyurin, A.I., Chernov, A.V.: Dual approaches to the minimization of strongly convex functionals with a simple structure under affine constraints. Comput. Math. Math. Phys. **57**(8), 1262–1276 (2017). https://doi.org/10.1134/S0965542517080048
6. Arjovsky, M., Chintala, S., Bottou, L.: Wasserstein GAN (2017). arXiv:1701.07875
7. Baimurzina, D.R., et al.: Universal method of searching for equilibria and stochastic equilibria in transportation networks. Comput. Math. Math. Phys. **59**(1), 19–33 (2019). https://doi.org/10.1134/S0965542519010020, arXiv:1701.02473

8. Beck, A.: On the convergence of alternating minimization for convex programming with applications to iteratively reweighted least squares and decomposition schemes. SIAM J. Optim. **25**(1), 185–209 (2015)
9. Beck, A., Tetruashvili, L.: On the convergence of block coordinate descent type methods. SIAM J. Optim. **23**(4), 2037–2060 (2013)
10. Bertsekas, D.P., Tsitsiklis, J.N.: Parallel and Distributed Computation: Numerical Methods, vol. 23. Prentice Hall Englewood Cliffs, Upper Saddle River (1989)
11. Bigot, J., Gouet, R., Klein, T., López, A.: Geodesic PCA in the Wasserstein space by convex PCA. Ann. Inst. H. Poincaré Probab. Statist. **53**(1), 1–26 (2017)
12. Chernov, A., Dvurechensky, P., Gasnikov, A.: Fast primal-dual gradient method for strongly convex minimization problems with linear constraints. In: Kochetov, Y., Khachay, M., Beresnev, V., Nurminski, E., Pardalos, P. (eds.) DOOR 2016. LNCS, vol. 9869, pp. 391–403. Springer, Cham (2016). https://doi.org/10.1007/978-3-319-44914-2_31
13. Cuturi, M.: Sinkhorn distances: lightspeed computation of optimal transport. In: Burges, C.J.C., Bottou, L., Welling, M., Ghahramani, Z., Weinberger, K.Q. (eds.) Advances in Neural Information Processing Systems, vol. 26, pp. 2292–2300. Curran Associates, Inc. (2013)
14. Cuturi, M., Doucet, A.: Fast computation of Wasserstein barycenters. In: Xing, E.P., Jebara, T. (eds.) Proceedings of the the 31st International Conference Machine Learning, vol. 32, pp. 685–693. PMLR, Bejing, China (2014)
15. Cuturi, M., Peyré, G.: A smoothed dual approach for variational Wasserstein problems. SIAM J. Imaging Sci. **9**(1), 320–343 (2016)
16. Daubechies, I., DeVore, R., Fornasier, M., Güntürk, C.S.: Iteratively reweighted least squares minimization for sparse recovery. Commun. Pure Appl. Math. **63**(1), 1–38 (2010)
17. Diakonikolas, J., Orecchia, L.: Alternating randomized block coordinate descent. In: Dy, J., Krause, A. (eds.) Proceedings of the the 35th International Conference Machine Learning, vol. 80, pp. 1224–1232. PMLR, Stockholmsmässan, Stockholm, Sweden (2018). http://proceedings.mlr.press/v80/diakonikolas18a.html
18. Dvinskikh, D., Gorbunov, E., Gasnikov, A., Dvurechensky, P., Uribe, C.A.: On primal and dual approaches for distributed stochastic convex optimization over networks. In: 2019 IEEE 58th Conference on Decision and Control, pp. 7435–7440 (2019). https://doi.org/10.1109/CDC40024.2019.9029798, arXiv:1903.09844
19. Dvurechensky, P., Dvinskikh, D., Gasnikov, A., Uribe, C.A., Nedić, A.: Decentralize and randomize: faster algorithm for Wasserstein barycenters. In: Bengio, S., Wallach, H., Larochelle, H., Grauman, K., Cesa-Bianchi, N., Garnett, R. (eds.) Advances in Neural Information Processing Systems , vol. 31, pp. 10783–10793. Neural Information Processing Systems 2018, Curran Associates, Inc. (2018). arXiv:1806.03915
20. Dvurechensky, P., Gasnikov, A., Gasnikova, E., Matsievsky, S., Rodomanov, A., Usik, I.: Primal-dual method for searching equilibrium in hierarchical congestion population games. In: Supplementary Proceedings of the International Conference on Discrete Optimization and Operations Research and Scientific School (DOOR 2016), Vladivostok, Russia, 19–23 September 2016, pp. 584–595 (2016). arXiv:1606.08988
21. Dvurechensky, P., Gasnikov, A., Kroshnin, A.: Computational optimal transport: complexity by accelerated gradient descent is better than by Sinkhorn's algorithm. In: Dy, J., Krause, A. (eds.) Proceedings of the 35th International Conference on Machine Learning, vol. 80, pp. 1367–1376. PMLR (2018). arXiv:1802.04367

22. Dvurechensky, P., Gasnikov, A., Omelchenko, S., Tiurin, A.: A stable alternative to Sinkhorn's algorithm for regularized optimal transport. In: Kononov, A., et al. (eds.) Mathematical Optimization Theory and Operations Research (MOTOR 2020). Springer, Cham (2020). https://doi.org/10.1007/978-3-030-49988-4_28, arXiv:1706.07622

23. Dvurechensky, P., Gasnikov, A., Tiurin, A.: Randomized similar triangles method: a unifying framework for accelerated randomized optimization methods (coordinate descent, directional search, derivative-free method) (2017). arXiv:1707.08486

24. Ebert, J., Spokoiny, V., Suvorikova, A.: Construction of non-asymptotic confidence sets in 2-Wasserstein space (2017). arXiv:1703.03658

25. Fercoq, O., Qu, Z.: Restarting the accelerated coordinate descent method with a rough strong convexity estimate. Comput. Optim. Appl. **75**(1), 63–91 (2020). https://doi.org/10.1007/s10589-019-00137-2

26. Fercoq, O., Richtárik, P.: Accelerated, parallel, and proximal coordinate descent. SIAM J. Optimiz. **25**(4), 1997–2023 (2015)

27. Gasnikov, A., Dvurechensky, P., Usmanova, I.: On accelerated randomized methods. Proc. Moscow Inst. Phys. Technol. **8**(2), 67–100 (2016), (in Russian). arXiv:1508.02182

28. Gasnikov, A., Gasnikova, E., Mendel, M., Chepurchenko, K.: Evolutionary derivations of entropy model for traffic demand matrix calculation. Matematicheskoe Modelirovanie **28**(4), 111–124 (2016). (in Russian)

29. Guminov, S.V., Nesterov, Y.E., Dvurechensky, P.E., Gasnikov, A.V.: Accelerated primal-dual gradient descent with linesearch for convex, nonconvex, and nonsmooth optimization problems. Doklady Math. **99**(2), 125–128 (2019)

30. Guminov, S., Dvurechensky, P., Tupitsa, N., Gasnikov, A.: Accelerated alternating minimization, accelerated Sinkhorn's algorithm and accelerated iterative bregman projections (2019). arXiv:1906.03622

31. Ho, N., Nguyen, X., Yurochkin, M., Bui, H.H., Huynh, V., Phung, D.: Multilevel clustering via Wasserstein means. In: Precup, D., Teh, Y.W. (eds.) Proceedings of the 34th International Conference on Machine Learning, vol. 70, pp. 1501–1509. PMLR (2017)

32. Kamzolov, D., Dvurechensky, P., Gasnikov, A.V.: Universal intermediate gradient method for convex problems with inexact oracle. Optim. Methods Softw. 1–28 (2020). https://doi.org/10.1080/10556788.2019.1711079, arXiv:1712.06036

33. Kolouri, S., Park, S.R., Thorpe, M., Slepcev, D., Rohde, G.K.: Optimal mass transport: signal processing and machine-learning applications. IEEE Signal Process. Mag. **34**(4), 43–59 (2017)

34. Kroshnin, A., Tupitsa, N., Dvinskikh, D., Dvurechensky, P., Gasnikov, A., Uribe, C.: On the complexity of approximating Wasserstein barycenters. In: Chaudhuri, K., Salakhutdinov, R. (eds.) Proceedings of the 36th Conference on Machine Learning. Proceedings of Machine Learning Research, vol. 97, pp. 3530–3540. PMLR, Long Beach, California, USA (2019). arXiv:1901.08686

35. Kusner, M.J., Sun, Y., Kolkin, N.I., Weinberger, K.Q.: From word embeddings to document distances. In: Proceedings of the the 32nd International Conference Machine Learning (ICML 2015), vol. 37, pp. 957–966. PMLR (2015)

36. Lee, Y.T., Sidford, A.: Efficient accelerated coordinate descent methods and faster algorithms for solving linear systems. In: Proceedings of the 2013 IEEE 54th FOCS (FOCS 2013), pp. 147–156. IEEE Computer Society, Washington, DC, USA (2013). arXiv:1305.1922

37. Lin, Q., Lu, Z., Xiao, L.: An accelerated proximal coordinate gradient method. In: Ghahramani, Z., Welling, M., Cortes, C., Lawrence, N.D., Weinberger, K.Q. (eds.) Advances in Neural Information Processing Systems, vol. 27, pp. 3059–3067. Curran Associates, Inc. (2014). arXiv:1407.1296

38. Lin, T., Ho, N., Chen, X., Cuturi, M., Jordan, M.I.: Computational hardness and fast algorithm for fixed-support Wasserstein barycenter (2020). arXiv:2002.04783

39. Lin, T., Ho, N., Cuturi, M., Jordan, M.I.: On the Complexity of Approximating Multimarginal Optimal Transport (2019). arXiv:1910.00152

40. Lin, T., Ho, N., Jordan, M.: On efficient optimal transport: an analysis of greedy and accelerated mirror descent algorithms. In: Chaudhuri, K., Salakhutdinov, R. (eds.) Proceedings of the 36th International Conference on Machine Learning. Proceedings of Machine Learning Research, vol. 97, pp. 3982–3991. PMLR, Long Beach, California, USA (2019)

41. McCullagh, P., Nelder, J.: Generalized Linear Models. In: Monographs Statistics and Applied Probability Series, 2nd edn. Chapman & Hall (1989)

42. McLachlan, G., Krishnan, T.: The EM Algorithm and Extensions. In: Wiley Series in Probability and Statistics. Wiley (1996)

43. Nesterov, Y.: Smooth minimization of non-smooth functions. Math. Program. **103**(1), 127–152 (2005)

44. Nesterov, Y.: Efficiency of coordinate descent methods on huge-scale optimization problems. SIAM J. Optimiz. **22**(2), 341–362 (2012)

45. Nesterov, Y.: Universal gradient methods for convex optimization problems. Math. Program. **152**(1), 381–404 (2015). https://doi.org/10.1007/s10107-014-0790-0

46. Nesterov, Y., Gasnikov, A., Guminov, S., Dvurechensky, P.: Primal-dual accelerated gradient methods with small-dimensional relaxation oracle. Optim. Methods Softw. 1–28 (2020). https://doi.org/10.1080/10556788.2020.1731747, arXiv:1809.05895

47. Nesterov, Y., Stich, S.U.: Efficiency of the accelerated coordinate descent method on structured optimization problems. SIAM J. Optim. **27**(1), 110–123 (2017)

48. Ortega, J., Rheinboldt, W.: Iterative Solution of Nonlinear Equations in Several Variables. Classics in Applied Mathematics. SIAM (1970)

49. Panaretos, V.M., Zemel, Y.: Amplitude and phase variation of point processes. Ann. Statist. **44**(2), 771–812 (2016)

50. Roulet, V., d'Aspremont, A.: Sharpness, restart and acceleration. In: Guyon, I., et al. (eds.) Advances in Neural Information Processing Systems, vol. 30, pp. 1119–1129. Curran Associates, Inc. (2017). http://papers.nips.cc/paper/6712-sharpness-restart-and-acceleration.pdf

51. Rubner, Y., Tomasi, C., Guibas, L.J.: The earth mover's distance as a metric for image retrieval. Int. J. Comput. Vis. **40**(2), 99–121 (2000). https://doi.org/10.1023/A:1026543900054

52. Saha, A., Tewari, A.: On the nonasymptotic convergence of cyclic coordinate descent methods. SIAM J. Optim. **23**(1), 576–601 (2013)

53. Sandler, R., Lindenbaum, M.: Nonnegative matrix factorization with earth mover's distance metric for image analysis. IEEE TPAMI **33**(8), 1590–1602 (2011)

54. Shalev-Shwartz, S., Zhang, T.: Accelerated proximal stochastic dual coordinate ascent for regularized loss minimization. In: Xing, E.P., Jebara, T. (eds.) Proceedings of the 31st International Conference on Machine Learning. Proceedings of Machine Learning Research, vol. 32, pp. 64–72. PMLR, Bejing, China (2014). http://proceedings.mlr.press/v32/shalev-shwartz14.html, arXiv:1309.2375

55. Sinkhorn, R.: Diagonal equivalence to matrices with prescribed row and column sums. II. Proc. Am. Math. Soc. **45**, 195–198 (1974)

56. Solomon, J., Rustamov, R.M., Guibas, L., Butscher, A.: Wasserstein propagation for semi-supervised learning. In: Proceedings of the 31st International Conference on Machine Learning (ICML 2014), vol. 32, pp. I-306-I-314. PMLR (2014)

57. Stonyakin, F.S., et al.: Gradient methods for problems with inexact model of the objective. In: Khachay, M., Kochetov, Y., Pardalos, P. (eds.) Mathematical Optimization Theory and Operations Research (MOTOR 2019). Lecture Notes in Computer Science, vol. 11548, pp. 97–114. Springer, Cham (2019). https://doi.org/10.1007/978-3-030-22629-9_8, arXiv:1902.09001

58. Sun, R., Hong, M.: Improved iteration complexity bounds of cyclic block coordinate descent for convex problems. In: Proceedings of the 28th International Conference on Neural Information Processing Systems (NIPS 2015), vol. 1, pp. 1306–1314. MIT Press, Cambridge, MA, USA (2015)

59. Tupitsa, N., Dvurechensky, P., Gasnikov, A., Guminov, S.: Alternating Minimization Methods for Strongly Convex Optimization (2019). arXiv:1911.08987

60. Uribe, C.A., Dvinskikh, D., Dvurechensky, P., Gasnikov, A., Nedić, A.: Distributed computation of Wasserstein barycenters over networks. In: 2018 IEEE Conference on Decision and Control (CDC), pp. 6544–6549 (2018). arXiv:1803.02933

61. Villani, C.: Optimal Transport: Old and New, vol. 338. Springer, Heidelberg (2008). https://doi.org/10.1007/978-3-540-71050-9

62. Werman, M., Peleg, S., Rosenfeld, A.: A distance metric for multidimensional histograms. Comput. Vis. Graph. Image Process. **32**(3), 328–336 (1985)

63. Zhang, Y., Roughan, M., Lund, C., Donoho, D.L.: Estimating point-to-point and point-to-multipoint traffic matrices: an information-theoretic approach. IEEE/ACM Trans. Netw. **13**(5), 947–960 (2005)

Game Theory and Mathematical Economics

Nonlinear Models of Convergence

Konstantin Gluschenko[1,2(✉)] ⓘ

[1] Novosibirsk State University, Novosibirsk, Russia
glu@nsu.ru
[2] Institute of Economics and Industrial Engineering of the SB RAS,
Novosibirsk, Russia

Abstract. A significant issue in studies of economic development is whether economies (countries, regions of a country, etc.) converge to one another in terms of per capita income. In this paper, nonlinear asymptotically subsiding trends of the income gap in a pair of economies model the convergence process. A few specific forms of such trends are proposed: log-exponential trend, exponential trend, and fractional trend. A pair of economies is deemed converging if time series of their income gap is stationary about any of these trends. To test for stationarity, standard unit root tests are applied with non-standard test statistics that are estimated for each kind of trends.

Keywords: Income convergence · Time series econometrics · Nonlinear time-series model · Unit root

1 Introduction

A significant issue in studies of economic development is whether economies (countries, regions of a country, cities, etc.) converge to one another in terms of per capita income. There are a number of methodologies to test for the convergence hypothesis. The most widespread one in the literature is the analysis of a negative cross-section correlation between initial per capita income and its growth, the so-called beta-convergence (see, e.g., [1]). An alternative methodology is the distribution dynamics analysis that explores the evolution of cross-economy income distribution [2]. Both approaches provide only an aggregated characterization of convergence. If the whole set of economies under consideration is found to converge, it is not possible to reveal economies with a deviant behavior (e.g., diverging or randomly walking). On the other hand, if the convergence hypothesis is rejected, it is not able to detect a subset (or subsets) of converging economies.

Methodologies based on time-series analysis make it possible to overcome this problem. They consider time series of the income gap, i.e., the difference of logarithms of per capita incomes in a pair of economies r and s, $y_{rst} = y_{rt} - y_{st} = \ln(Y_{rt}/Y_{st})$, t denoting time. To discriminate between logarithmic and real (e.g., percentage) terms, $Y_{rt}/Y_{st} - 1$ is called income disparity. One element of the pair can be an aggregate, for instance, the national economy when economies under consideration are the country's regions.

Y. Kochetov et al. (Eds.): MOTOR 2020, CCIS 1275, pp. 207–215, 2020.
https://doi.org/10.1007/978-3-030-58657-7_18

Bernard and Durlauf [3] have put forward a formal definition of convergence: economies r and s converge if the long-term forecasts of per capita income (conditionally on information available by the moment of the forecast, I) for both economies are equal, that is

$$\lim_{t\to\infty} E(y_{rst}|I) = 0. \tag{1}$$

Despite this definition of convergence is general, procedures of testing for convergence applied in [3] in fact detect only a particular class of processes satisfying (1), namely, stationary processes with no trend (implying that y_{rt} and y_{st} have a common trend). Thus, such procedures are not able to classify the most interesting case of catching-up as convergence.

As a way out, [4] proposes to model the (square of) income gap by a trend $h(t)$ of a priory unknown form, approximating it by a power series of degree k. The respective econometric model looks like (ε_t denotes residuals with standard properties, α_i is a coefficient to be estimated):

$$y_{rst}^2 = h(t;k) = \alpha_0 + \alpha_1 t + \alpha_2 t^2 + \ldots + \alpha_k t^k + \varepsilon_t (t = 1, \ldots, T). \tag{2}$$

Albeit the trend may be nonlinear, Eq. (2) is linear with respect to coefficients. Convergence takes place if $dh/dt < 0$ holds for all t. This condition is supposed to be equivalent to the negativity of the time average of $dh(t)/dt$:

$$\frac{1}{T}\sum_{t=1}^T \frac{dh}{dt} = \sum_{i=1}^k \alpha_i \frac{i}{T}\sum_{t=1}^T t^{i-1} < 0. \tag{3}$$

However, the equivalence is not the fact. It is obvious, considering a continuous-time counterpart of (3):

$$\frac{1}{T}\int_1^T \frac{dh}{dt}dt = \frac{1}{T}(h(T) - h(1)) < 0.$$

Hence, the mere fact that $h(T) < h(1)$ suffices to accept the convergence hypothesis. In the general case, this does not evidence convergence. For instance, a U-shape path of the income gap may satisfy (3). Moreover, even if $dh/dt < 0$ is true for every $t = 1,\ldots T$, condition (1) knowingly does not hold, as $h(\infty; k) = \pm\infty$ for any finite k.

Thus, there is a want of developing an alternative methodology. This paper puts forward such a methodology, namely, modeling the convergence process by asymptotically subsiding trends. This leads to nonlinear econometric models that need nonstandard distributions of test statistics to test models for unit roots.

2 Modeling Convergence

Actual convergence processes are in fact a superposition of two processes that can be called long-run, or deterministic, convergence, and stochastic, or short-run, convergence. Long-run convergence is a deterministic path of the income gap y_{rst} that tends to zero over time: $y_{rst}^* = h(t), h(t) \underset{t\to\infty}{\longrightarrow} 0$. In [4], only this process is considered (albeit with no latter condition). Short-run convergence is an autocorrelated stochastic process containing no unit root (i.e., a stationary process), $v_t = \rho v_{t-1} + \varepsilon_t$, where ρ is the autocorrelation coefficient, $\rho < 1$, and $\varepsilon_t \sim N(0, \sigma^2)$ with finite σ. Intuitively, short-run convergence characterizes the behavior of transient random shocks. A unit shock deviates the income gap from its long-run path, dying out over time with half-life $\theta = \ln(0.5)/\ln(\rho)$, so that the income gap eventually returns to its long-run path. Only such processes are considered in [3] (assuming $y_{rst}^* = 0$).

The superposition of these two processes gives a process that is stationary around an asymptotically subsiding trend $h(t)$. That is, albeit random shocks force the process to deviate from the trend, it permanently tends to return to the trend, thus satisfying (1). The following econometric model of the class AR(1) describes such a process:

$$y_{rst} = h(t) + v_t(t = 0, \ldots, T - 1), \quad v_t = \rho v_{t-1} + \varepsilon_t(t = 1, \ldots, T - 1; v_0 = \varepsilon_0).$$

Applying the Cochrane-Orcutt transformation to this equation, the following model is arrived at:

$$\Delta y_{rst} = h(t) - (\lambda + 1)h(t - 1) + \lambda y_{rs,t-1} + \varepsilon_t(t = 1, \ldots, T - 1), \tag{4}$$

where $\Delta y_{rst} = y_{rst} - y_{rs,t-1}$ and $\lambda = \rho - 1$.

To make the model (4) operational, a specific function $h(t)$ has to be taken from the class of asymptotically subsiding functions. A few such functions are preferable in order to model more adequately the properties of a process under consideration. The following three functions seem convenient from the practical viewpoint: log-exponential trend $h(t) = \ln(1 + \gamma e^{\delta t})$, $\delta < 0$, exponential trend $h(t) = \gamma e^{\delta t}$, $\delta < 0$, and fractional trend $h(t) = \gamma/(1 + \delta t)$, $\delta > 0$. The respective models are nonlinear with respect to coefficients, having the forms:

$$\Delta y_{rst} = \ln\left(1 + \gamma e^{\delta t}\right) - (\lambda + 1)\ln\left(1 + \gamma e^{\delta(t-1)}\right) + \lambda y_{rs,t-1} + \varepsilon_t; \tag{4a}$$

$$\Delta y_{rst} = \gamma e^{\delta t} - (\lambda + 1)\gamma e^{\delta(t-1)} + \lambda y_{rs,t-1} + \varepsilon_t; \tag{4b}$$

$$\Delta y_{rst} = \frac{\gamma}{1 + \delta t} - (\lambda + 1)\frac{\gamma}{1 + \delta(t - 1)} + \lambda y_{rs,t-1} + \varepsilon_t. \tag{4c}$$

An advantage of the log-exponential trend is the ease of interpretation. Parameter γ is the initial (at $t = 0$) income disparity. Parameter δ characterizes the convergence rate which can be simply expressed in terms of the half-life time of the (deterministic) income disparity, i.e., the time the disparity takes to halve: $\Theta = \ln(0.5)/\delta$.

A shortcoming of this trend is that is has no symmetry properties with respect to a permutation of the economy indices. Albeit $y_{rst} = -y_{srt}$, the permutation changes absolute values of γ and δ (and may change the estimate of λ in regression (4a)).

Contrastingly, exponential and fractional trends have symmetry properties. A permutation of r and s changes only the sign of γ, leaving its absolute value and the value of δ (as well as λ in (4b), (4c)) intact. However, while the initial income disparity can be easily calculated from γ, equaling $e^{\gamma} - 1$ in both trends, the half-life of the deterministic income gap involves a mixture of γ and δ. This results in hardly interpretable expressions. For the exponential trend, $\Theta = \frac{1}{\delta} \ln\left(\frac{\ln(0.5(e^{\gamma} + 1))}{\gamma}\right)$; for the fractional trend,
$\Theta = \frac{1}{\delta}\left(\frac{\gamma}{\ln(0.5(e^{\gamma} + 1))} - 1\right)$.

Models (4a)–(4c) are also applicable to the case of deterministic divergence. It takes place if $\delta > 0$ in the log-exponential and exponential trends, or $\delta < 0$ in the fractional trend. The time the (deterministic) income disparity takes to double can characterize the divergence rate.

Model (4) encompasses two particular cases. With $h(t) = 0$, which corresponds to $\gamma = 0$ in (4a)–(4c), it degenerates to ordinary AR(1) model with no constant:

$$\Delta y_{rst} = \lambda y_{rs,t-1} + \varepsilon_t. \tag{5}$$

This implies that series y_{rt} and y_{st} are cointegrated with cointegrating vector [1, −1], i.e., they have the same trend. Intuitively, this means that convergence as such, i.e., catching-up, has completed by $t = 0$ (if it had occurred before). In the further dynamics, per capita incomes in economies r and s are equal up to random shocks (hence, only stochastic convergence takes place).

With $h(t) = $ const, which corresponds to $\delta = 0$ in (4a)–(4c), model (4) degenerates to ordinary AR(1) model with a constant:

$$\Delta y_{rst} = \alpha + y_{rs,t-1} + \varepsilon_t. \tag{6}$$

This implies that series y_{rt} and y_{st} are cointegrated with cointegrating vector [1, −γ], i.e., they have a common trend: $h_s(t) = \gamma + h_r(t)$, $\gamma = -\alpha/\lambda$. In other words, the income gap is constant (up to random shocks); y_{rt} and y_{st} move parallel to each other with the distance between their paths equaling γ. Again, only stochastic convergence takes place here. Just models (5) and (6) are considered in [3] (albeit within a more evolved framework).

Having estimated parameters of a specific model of the form (4), we need to check its adequacy. First of all, the question is whether y_{rst} is indeed stationary around the given trend (y_{rst} has no unit root). There are a number of tests for unit root (testing hypothesis $\lambda = 0$ against $\lambda < 0$, or $\lambda < 0$ against $\lambda = 0$). Most of them use t-ratio of λ, $\tau = \lambda/\sigma_{\lambda}$, as the test statistic. In the case of testing for unit roots, it has non-standard distributions, differing from the t-distribution (that is why it is designated τ, and not t). Such distributions (named the Dickey-Fuller distributions) are tabulated for AR(1) models with no constant, with a constant, and with a linear and quadratic trends, but not for models with proposed nonlinear trends. To estimate them, τ in every model with a specific trend was estimated for each of 1 million generated random walks $y_t = y_{t-1} + \varepsilon_t$. Table 1 reports some values of the τ-statistic from the obtained distributions for sample size $T = 204$ (used in the empirical analysis reported in the next section). Figure 1 plots the 10-percent tails of the distributions, comparing them with the Dickey-Fuller distributions for the cases of linear and quadratic trend from [5].

Table 1. Selected values of the τ-statistics for models with nonlinear trends, $T = 204$.

Probability	Log-exponential trend (4a)	Exponential trend (4b)	Fractional trend (4c)
1%	−3.841	−3.851	−5.152
5%	−3.220	−3.273	−3.820
10%	−2.898	−2.971	−3.297

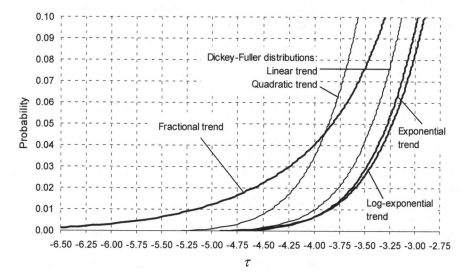

Fig. 1. Distributions of the unit root test τ-statistics for Eqs. (4a)–(4c) and selected Dickey-Fuller distributions; $T = 204$.

If the unit root test rejects the hypothesis of non-stationarity, the ordinary t-test can test parameters γ and δ for statistical significance. Given that there are three versions of the model (4), every version is estimated and tested. If they turn out to be completive, the version providing the best fit – namely, the minimal sum of squared residuals (SSR) – is accepted. Note that valid models with the "incorrect" sign of δ suggest deterministic divergence. The rejection of all versions because of the presence of unit root or insignificance of γ or δ evidences the absence of (deterministic) convergence as well. If statistical reasons for no-convergence are of interest, we can estimate and test regression (6) and then, if it is rejected, regression (5). In this case, we find whether no-convergence is due to coinciding or "parallel" dynamics of per capita incomes in a pair of economies under consideration (the same or common trend), or – if both models are rejected – it is due to a random walk.

3 Empirical Application

This section provides an illustration of the empirical application of the proposed methodology for analyzing convergence of regional incomes per capita in Russia. The time span covers January 2002 through December 2018 with a monthly frequency (204 months). The indicator under consideration is the real personal income per capita by region. The term "real" means that the income is adjusted to the respective regional price level. The cost of the fixed basket of goods and services for cross-region comparison of population's purchasing capacity serves as an indicator of the regional price level. The official statistical data on nominal incomes and the costs of the fixed basket come from [6–8].

Convergence is considered with respect to the national income per capita. Thus, index s is fixed, denoting Russia as a whole; then y_{rst} is the gap between regional and national incomes. To test models for unit roots, the Phillips-Perron test (PP test) is applied with modifications proposed in [9, 10].

Since the whole set of results is cumbersome (involving 79 regions), this section gives them only partially for illustrative purposes. It presents examples of qualitatively different cases discussed in the previous section. Table 2 reports these.

Table 2. Selected results of analyzing regional convergence in Russia.

Model	λ	PP-test p-value	γ/α in (6)	p-value of γ/α	δ	p-value of δ	SSR
Kursk Region							
(4a)	−0.484 (0.061)	0.000	−0.354 (0.018)	0.000	−0.011 (0.001)	0.000	0.550
(4b)	−0.496 (0.062)	0.000	−0.430 (0.025)	0.000	−0.013 (0.001)	0.000	0.546
(4c)	−0.361 (0.054)	0.000	−0.493 (0.066)	0.000	0.029 (0.008)	0.000	0.592
Republic of Karelia							
(4a)	−0.457 (0.059)	0.000	−0.100 (0.012)	0.000	0.005 (0.001)	0.000	0.680
(4b)	−0.462 (0.059)	0.000	−0.103 (0.013)	0.000	0.005 (0.001)	0.000	0.679
(4c)	−0.423 (0.057)	0.000	−0.122 (0.013)	0.000	−0.003 (0.000)	0.000	0.695
Saint Petersburg City							
(4a)	−0.427 (0.058)	0.000	0.236 (0.035)	0.000	−0.001 (0.001)	0.287	
(4b)	−0.427 (0.058)	0.000	0.212 (0.028)	0.000	−0.001 (0.001)	0.288	
(4c)	−0.427 (0.058)	0.000	0.212 (0.030)	0.000	0.001 (0.002)	0.365	
(6)	−0.419 (0.057)	0.000	0.078 (0.012)	0.000			
Republic of Bashkortostan							
(4a)	−0.359 (0.053)	0.000	0.018 (0.033)	0.576	0.000 (0.015)	0.976	
(4b)	−0.359 (0.053)	0.000	0.018 (0.032)	0.573	0.000 (0.015)	0.976	
(4c)	−0.359 (0.053)	0.003	0.019 (0.032)	0.564	0.000 (0.014)	0.985	
(6)	−0.359 (0.053)	0.000	0.006 (0.005)	0.249			
(5)	−0.317 (0.052)	0.000					
Moscow Region							
(4a)	−0.211 (0.043)	0.076	0.018 (0.016)	0.264	0.013 (0.005)	0.012	
(4b)	−0.209 (0.043)	0.094	0.019 (0.016)	0.246	0.013 (0.005)	0.013	
(4c)	−0.180 (0.040)	0.262	0.041 (0.019)	0.029	−0.004 (0.001)	0.000	
(6)	−0.125 (0.034)	0.357	0.010 (0.005)	0.043			
(5)	−0.091 (0.030)	0.116					

Standard errors are in parentheses.

Convergence manifests itself in the Kursk Region. All three versions of the trend model can be accepted, suggesting fast convergence. Choosing model (4b) as providing the best fit, the half-life time of the income gap equals 5.3 years (65.3 months). Figure 2(a) plots the path of the actual income gap and its estimated exponential trend. According to this trend, income per capita in the Kursk Region was below the national level by 35% at the beginning of the time span under consideration and by only 3% by its end. The log-exponential and fractional trends suggest even faster convergence with half-live times 5.1 and 3.6 years, respectively.

Divergence occurs in the Republic of Karelia. Again, all three versions of the trend model can be accepted. Model (4b) seems preferable, albeit its SSR differs from the SSR in the model (4a) only slightly. Figure 2(b) depicts the dynamical pattern. The income gap rises, doubling every 10.4 years. The income per capita in this region was 9% below the national level in January 2002 and 28% in December 2018.

The case of Saint Petersburg City (which is a separate administrative-territorial unit considered as a region) illustrates the absence of convergence that is due to the "parallel" dynamics of the national and regional incomes per capita. Figure 2(c) shows this case. Although the unit root test rejects the hypothesis of nonstationarity with confidence in all trend models, high p-values of δ suggest the absence of a trend. Model (6) proves to be valid, implying the income gap to be time-invariant. It equals 0.186 ($= -\alpha/\lambda$); in other words, real income per capita in Saint Petersburg City remains on average constant, being 20.5% above the national level.

The Republic of Bashkortostan demonstrates a similar pattern, Fig. 2(d), with the difference that there is no income gap; real income per capita here remains on average equal to the national per capita income (in fact, the regional income fluctuates around the national level). In all trend models, p-values of both γ and δ are high, thus implying rejection of these models. The constant in the model (6) has high p-value as well, which leads to the model (5). It proves to be valid; the unit root hypothesis is rejected with confidence.

At last, no one model seems to describe the behavior of the income gap in the Moscow Region, Fig. 2(e). We can reject models (4a) and (4b) because of high p-value of γ, and models (6) and (5) because of the non-rejection of a unit root. The conclusion may be that non-convergence here is due to a random walk of the income gap.

Briefly summing up the results of the full analysis of income convergence in Russia, convergence takes place in the whole of Russia, as the Gini index decreases over time. Analysis by region yields the "anatomy" of convergence. Among all 79 regions in the spatial sample, 44 regions (55.7%) are converging. In 16 regions (20.3%), non-convergence is due to common trends with the national income per capita (in three cases, regional trends coincide with the national trend). An unpleasant feature of the pattern obtained is a considerable number of diverging regions; there are 17 of them (21.5%). Besides, random walks are peculiar to two regions.

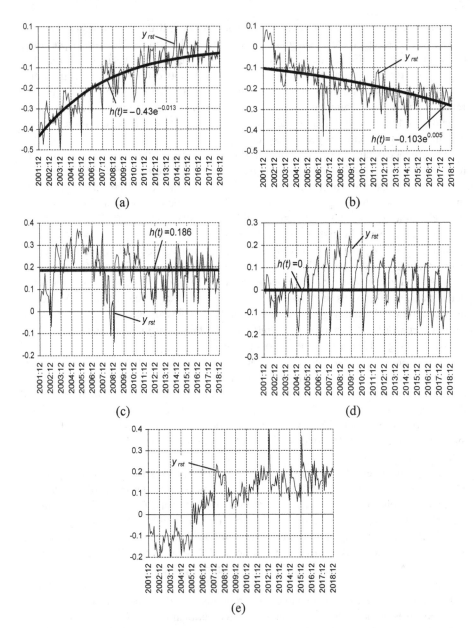

Fig. 2. Different cases of behavior of the income gap: (a) convergence (the Kursk Region); (b) divergence (Republic of Karelia); (c) a constant income gap (Saint Petersburg City); (d) no income gap (Republic of Bashkortostan); (d) random walking of income gap (the Moscow Region).

4 Conclusion

This paper develops a methodology of modeling convergence by asymptotically sub-siding trends of the income gap in a pair of economies. This way conforms to the theoretical definition of convergence. Three specific kinds of such trends are proposed, namely, log-exponential trend, exponential trend, and fractional trend. This makes it possible to select a specific model that most adequately describes properties of actual dynamics.

Transformation to testable versions generates nonlinear econometric models that represent a superposition of stochastic and deterministic convergence. Such models need additional efforts: the application of methods for estimation of nonlinear regres-sions and estimating distributions of the unit root test statistics for every specific trend. However, these efforts are repaid, providing a theoretically adequate and practically fairly flexible and helpful tool for studying processes of convergence between coun-tries, regions within a country, regions of different countries (e.g., in the European Union), etc.

The reported examples of applying the proposed methodology to the empirical analysis of convergence of real incomes per capita between Russian regions show that the results obtained look reasonable and correspond to economic intuition. As regards the whole analysis, it has yielded an interesting pattern. In spite of the fact that convergence occurs in Russia as a whole, a deviant dynamics is peculiar to a number of regions: almost a quarter of regions are found to diverge, either deterministically or stochastically.

References

1. Barro, R.J., Sala-i-Martin, X.: Convergence. J. Polit. Econ. **100**(2), 223–251 (1992)
2. Quah, D.: Galton's fallacy and tests of the convergence hypothesis. Scand. J. Econ. **95**(4), 427–443 (1993)
3. Bernard, A.B., Durlauf, S.N.: Convergence in international output. J. Appl. Economet. **10** (2), 97–108 (1995)
4. Nahar, S., Inder, B.: Testing convergence in economic growth for OECD countries. Appl. Econ. **34**(16), 2011–2022 (2002)
5. MacKinnon, J.G.: Numerical distribution functions for unit root and cointegration tests. J. Appl. Economet. **11**(6), 601–618 (1996)
6. Monthly bulletin "Socio-Economic situation of Russia", Rosstat, Moscow (2002–2007, various issues). (In Russian)
7. Rosstat Homepage. http://www.gks.ru/free_doc/new_site/population/urov/urov_11sub2008. xls. http://www.gks.ru/free_doc/new_site/population/urov/urov_11sub09-14.xls. http:// www.gks.ru/free_doc/new_site/population/urov/2013-2015kv.xls. Accessed 10 Sept 2016. (In Russian)
8. EMISS Homepage. https://fedstat.ru/indicator/57039. https://fedstat.ru/indicator/31052. Accessed 26 Jan 2020. (In Russian)
9. Perron, P., Ng, S.: Useful modifications to some unit root tests with dependent errors and their local asymptotic properties. Rev. Econ. Stud. **63**(3), 435–463 (1996)
10. Ng, S., Perron, P.: Lag length selection and the construction of unit root tests with good size and power. Econometrica **69**(6), 1519–1554 (2001)

A Game-Theoretic Approach to Team Formation in *The Voice* Show

Anna Ivashko[1,2]($\boxtimes$) (iD), Vladimir Mazalov[1,3] (iD), and Alexander Mazurov[4]

[1] Institute of Applied Mathematical Research, Karelian Research Center, Russian Academy of Sciences, ul. Pushkinskaya 11, Petrozavodsk 185910, Russia
aivashko@krc.karelia.ru, vmazalov@krc.karelia.ru
[2] Petrozavodsk State University, ul. Lenina 33, Petrozavodsk 185910, Russia
[3] Saint Petersburg State University, 7/9 Universitetskaya nab., Saint Petersburg 199034, Russia
[4] Arzamas Polytechnic Institute of R.E. Alekseev Nizhny Novgorod State Technical University, ul. Kalinina 19, Arzamas 607227, Russia
alexander.mazurov08@gmail.com

Abstract. This paper considers a game-theoretic model of a competition in which experts (players) seek to enroll two contestants into their own teams. The quality of the contestants is characterized by two random parameters, the first corresponding to the vocal talent of a contestant and the second to his appearance. The first quality parameter is known to the players, whereas the second is hidden from them. Each expert chooses an appropriate contestant based on the value of the known quality parameter only. The winner is the player whose team includes a contestant with the maximum sum of both quality parameters. This game is described by the best-choice model with incomplete information. The optimal strategies and payoffs of the players for different situations (subgames) of the game are found. The results of numerical simulation are presented.

Keywords: Best-choice game · Two-player game · Incomplete information · TV show · Threshold strategies

1 Introduction

In the study of various social and economic situations, an important issue is to analyze the behavior of participants when making some decision. The decision problem naturally arises in choice problems, e.g., when looking for a job, when buying (selling) goods or services, when choosing a mate or a business partner, and when participating in auctions or competitions. Game-theoretic best-choice problems are a suitable model for TV contests in which participants seek to choose an object or a group of objects. One of such contests is *The Voice*, a popular TV show. In this competition, a jury of several experts chooses vocalists. What is important, experts sit back to the contestants, assessing their vocal

Supported by the Russian Science Foundation (No. 17-11-01079).

Y. Kochetov et al. (Eds.): MOTOR 2020, CCIS 1275, pp. 216–230, 2020.
https://doi.org/10.1007/978-3-030-58657-7_19

talents without seeing them. During the competition, each expert can invite a fixed number of participants to his team. If the vocal talent of a next participant suits an expert, he makes an offer to the contestant to join his team. If a contestant is invited by several experts, the contestant decides himself which expert team to join. As soon as the choice procedure is complete and the teams of all experts are filled, a competition between the contestants takes place. An expert whose team member defeats the other participants becomes the winner. Note that the result of the competition depends not only on the vocal talent of the participants, but also on their appearance. A feature of the choice procedure in this competition is that the experts do not see the appearance of contestants when making their decision. Such situations are described well by best-choice models with incomplete information.

In this paper, a game-theoretic model of *The Voice* show is proposed, in which two experts (players) seek to enroll several contestants into their own teams. The players are simultaneously observing the sequence of contestants to choose two contestants into their own teams based on their qualities. The quality of each contestant is characterized by two random parameters, the first corresponding to his vocal talent and the second to his appearance. The experts observe the first quality parameter in explicit form, whereas the second quality parameter is hidden from them. The players decide to choose or reject a contestant by the known quality parameter. In this game, the winner is the player whose team includes a contestant with the maximum sum of both quality parameters.

This paper is organized as follows. In Sect. 2, the publications in this field of research are surveyed. The best-choice game with incomplete information is described in Sect. 3. The case in which only one expert is remaining in the game is discussed in Sect. 4. Different possible situations (subgames) in the two-player game are studied in Sects. 5 and 6.

2   Related Works

Best-choice games often arise in the study of behavior of participants in different auctions and competitions. An example of such a contest is *The Price is Right*. In this game, n participants spin the wheel one or two times to gain points. The goal of the participants is to collect a certain sum of points (score) not exceeding a given threshold. The game was investigated by Seregina et al. [7], Mazalov and Ivashko [8], Tenorio and Cason [9], and Bennett and Hickman [10].

The Voice show can be another attractive platform to analyze human behavior. In this show, an expert has to decide on two alternatives: choose a contestant that is performing right now, or continue the choice procedure. Therefore, an expert acts in the same way as an employee searching for a job, who has to decide whether to accept the current offer or reject it, in the hope of finding a more suitable vacant job. Similarly, when considering a series of projects, an investor has to decide whether a given project is suitable for investment or not.

The problem associated with *The Voice* show, in which each of several experts chooses only one contestant into his team, was considered by Mazalov et al. [11].

This paper presents a generalization of the problem mentioned to the case when the experts can choose no more than two contestants in their own teams.

The optimal strategies of the players in best-choice problems are often constructed using dynamic programming. With this method, a complex problem is solved by decomposing it into simpler subproblems represented by a recursive sequence. Dynamic programming has applications in various fields, such as quitting games (Solan and Vieille [1]), the house-selling problem (Sofronov [2]), the job-search problem (Immorlica et al. [3]), the mate choice problem (Alpern et al. [4]), to name a few, and has been successfully used for solving economic problems of best choice and auctions (Whitmeyer [5], Harrell et al. [6]).

This paper proposes a new game-theoretic model of competition as follows. Each of the two players seeks to choose a contestant better than the opponent. Each of the two players can accept two participants into his team and information about them is incomplete (partially available). For this problem, the optimal choice strategies of the players are found. The optimal threshold strategies and payoffs of the players are numerically simulated.

3   Two-Player Game with Incomplete Information

Consider a multistage game $\Gamma_{2,N}$ with incomplete information as follows. Two experts (players in this game) are simultaneously observing a certain sequence of contestants. Each player has to choose and invite to his team an appropriate contestant based on the latter's quality only. The quality of a contestant is characterized by two parameters x and y that reflect his vocal talent and appearance, respectively. Assume that the quality parameters of the contestants represent a sequence of independent random variables (x_i, y_i), $i = 1, \ldots, N$, with the uniform distribution on the set $[0,1] \times [0,1]$. The try-out process (called blind auditions in *The Voice*) is organized so that the experts can explicitly assess the first quality parameter, whereas the second one is hidden from them. Therefore, the players decide to accept or reject a current contestant using the known quality parameter only. The players seek to maximize the total quality of the contestant chosen (i.e., the sum of his quality parameters). The winner is the player whose team includes a contestant with the maximum sum of both quality parameters.

The choice procedure is described as follows. At stage 1, the experts are observing the quality parameter x_1 from the set (x_1, y_1) of contestant 1 and make an independent decision (accept or reject him). If contestant 1 is chosen by a single expert, he joins the latter's team. If both experts invite contestant 1, he chooses one of them equiprobably. Whenever a contestant is chosen by a player, the hidden quality parameter $(x_1 + y_1)$ becomes known to all players. Next, as soon as one of the experts chooses two contestants (e.g., at stages i and j, where $i < j$), he quits the game, and the other expert continues further choice alone. The expert remaining in the game seeks to choose a contestant l $(l = j + 1, \ldots, N)$ for his team (in fact, one or two contestants) so that his total quality is higher than the total quality of the best contestant of the expert quitted, i.e., $x_l + y_l > \max\{x_i + y_i, x_j + y_j\}$. In the case of rejecting contestant 1 by both experts, the game evolves to the next stage, and the choice procedure described is repeated again. Also, other situations are when a single expert or both experts choose a single contestant and continue their choice. All possible subgames will be listed and studied below.

In the best-choice games the player's optimal strategies always have threshold form. For this game, the optimal strategies of the players will be found in the class of threshold strategies: if the quality parameter x_i of a contestant exceeds some threshold u_i, then an expert chooses this contestant; otherwise rejects him. The optimal threshold strategies of equal-right players in the game with $N \geqslant 2$ contestants will be calculated below.

For this purpose, dynamic programming will be used. Decompose the problem under consideration into a series of simpler subproblems corresponding to different situations in the game when n contestants, $0 \leq n \leq N$, are left to perform (and hence are still available for choice).

These situations (subgames) are as follows:

$\Gamma_{2,n}^{(0,0)}$, a two-player game in which n contestants are left to perform and none of the players has chosen a contestant into his team so far;

$\Gamma_{2,n}^{(1,0)}(z)$, a two-player game in which n contestants are left to perform, player 1 has already chosen a contestant of a total quality z, whereas player 2 has chosen nobody so far;

$\Gamma_{2,n}^{(1,1)}(t,z)$, a two-player game in which n contestants are left to perform and both players have already chosen a single contestant into their teams, of a total quality t (player 1) and z (player 2);

$\Gamma_{1,n}^{(0,2)}(z)$, a one-player game (involving player 1) in which n contestants are left to perform, player 1 has chosen nobody so far, whereas player 2 has already chosen two contestants and the maximum total quality of them is z;

$\Gamma_{1,n}^{(1,2)}(t,z)$, a one-player game (involving player 1) in which n contestants are left to perform, player 1 has already chosen a contestant of a total quality t, whereas player 2 has already chosen two contestants and the maximum total quality of them is z.

In the original game $\Gamma_{2,n}^{(0,0)}$, the payoffs of the players depend on the payoffs gained in the above-mentioned subgames, $\Gamma_{2,N} = \Gamma_{2,N}^{(0,0)}$.

4   Game with Only One Player Remaining

4.1   Game $\Gamma_{1,n}^{(1,2)}(t,z)$

Consider the game $\Gamma_{1,n}^{(1,2)}(t,z)$ involving player 1 only, in which n contestants are left to perform, player 1 has already chosen a contestant of a total quality t, whereas player 2 has already chosen two contestants and the maximum total quality of them is z.

Denote by $H_{1,n}^{(1,2)}(t,z)$ the payoff of player 1 in the game $\Gamma_{1,n}^{(1,2)}(t,z)$. Construct the optimal strategies in the class of threshold strategies: if a current observation x_1 exceeds a threshold $u_{1,n}^{(1,2)} = u_{1,n}^{(1,2)}(t,z)$, then the player chooses the corresponding contestant; otherwise rejects him.

If no contestants are left to perform in the game, then

$$H_{1,0}^{(1,2)}(t,z) = 1 \cdot I_{\{t \geq z\}} + 0 \cdot I_{\{t < z\}},$$

where $I_{\{A\}} = I_{\{A\}}(\omega) = \begin{cases} 1 \text{ if } \omega \in A; \\ 0 \text{ otherwise.} \end{cases}$

Let a single contestant be available for choice. Then

$$H_{1,1}^{(1,2)}(t,z) = H_{1,0}^{(1,2)}(t,z)\mathbf{P}\{x_1 < u_{1,1}^{(1,2)}\} + 1 \cdot \mathbf{P}\{t \vee (x_1 + y_1) \geq z, x_1 \geq u_{1,1}^{(1,2)}\},$$

where $a \vee b = \max\{a, b\}$.

Obviously, $u_{1,1}^{(1,2)} = 0$, and hence

$$H_{1,1}^{(1,2)}(t,z) = \mathbf{P}\{t \vee (x_1 + y_1) \geq z, x_1 \geq 0\} = H_{1,1}(z)I_{\{t < z\}} + 1 \cdot I_{\{t \geq z\}},$$

where $H_{1,1}(z) = \mathbf{P}\{x_1 + y_1 \geq z\} = \begin{cases} 1 - \dfrac{z^2}{2}, \ z < 1; \\ \dfrac{(2-z)^2}{2}, \ z \geq 1. \end{cases}$

The same considerations for the game $\Gamma_{1,n}^{(1,2)}(t,z)$ yield

$$H_{1,n}^{(1,2)}(t,z) = H_{1,n}(z)I_{\{t < z\}} + 1 \cdot I_{\{t \geq z\}},$$

and

$$u_{1,n}^{(1,2)} = u_{1,n}I_{\{t < z\}} + 0 \cdot I_{\{t \geq z\}},$$

where $H_{1,n}(z)$ is the player's payoff in the one-player game in which n contestants are left to perform and a single contestant has to be chosen into the player's team with a total quality exceeding z. In accordance with [11], the function $H_{1,n}(z)$ has the form

$$H_{1,n}(z) = \begin{cases} H_{1,n-1}(z) \cdot u_{1,n} + \displaystyle\int_{u_{1,n}}^{z} (1 - (z - x_1))\, dx_1 + 1 - z, \ z < 1, \\ H_{1,n-1}(z) \cdot u_{1,n} + \displaystyle\int_{u_{1,n}}^{1} (1 - (z - x_1))\, dx_1, \ z \geq 1, \end{cases} \tag{1}$$

and

$$u_{1,n} = z - (1 - H_{1,n-1}(z)).$$

4.2 Game $\Gamma_{1,n}^{(0,2)}(z)$

Now, pass to the game $\Gamma_{1,n}^{(0,2)}(z)$ involving player 1 only, in which n contestants are left to perform, player 1 has chosen nobody so far, whereas player 2 has already chosen two contestants and the maximum total quality of them is z. Denote by $H_{1,n}^{(0,2)}(z)$ the player's payoff in this game.

Theorem 1. *In the game* $\Gamma_{1,n}^{(0,2)}(z)$ *the optimal strategy of player 1 has the form*

$$u_{1,n}^{(0,2)} = u_{1,n}^{(0,2)}(z) = \max\left\{0, z - \frac{1 - H_{1,n-1}^{(0,2)}(z)}{1 - H_{1,n-1}(z)}\right\},$$

and the player's payoffs are given by

$$H_{1,1}^{(0,2)}(z) = H_{1,1}(z),$$

$$H_{1,n}^{(0,2)}(z) = \begin{cases} H_{1,n-1}^{(0,2)}(z)\left(u_{1,n}^{(0,2)} + \dfrac{(z - u_{1,n}^{(0,2)})^2}{2}\right) + 1 - u_{1,n}^{(0,2)} - \dfrac{(z - u_{1,n}^{(0,2)})^2}{2}, z < 1, \\[2ex] H_{1,n-1}^{(0,2)}(z)\left(u_{1,n}^{(0,2)} + \dfrac{(1 - u_{1,n}^{(0,2)})(1 - u_{1,n}^{(0,2)} + 2z)}{2}\right) \\[2ex] + \dfrac{(1 - u_{1,n}^{(0,2)})(3 + u_{1,n}^{(0,2)} - 2z)}{2}, z \geq 1, \end{cases}$$

where $2 \leq n \leq N.$

Proof.
Assume that no contestants are left to perform in the game. Then $H_{1,0}^{(0,2)}(z) = 0.$

If a single contestant is still available for choice, then the player will choose him, gaining the payoff

$$H_{1,1}^{(0,2)}(z) = \mathbf{P}\{x_1 + y_1 \geq z\} = H_{1,1}(z).$$

Now, let $n = 2$. First of all, consider the case $z < 1$. Suppose that player 1 uses a strategy u. (Note that u cannot exceed z.) As a result,

$$H_{1,2}^{(0,2)}(z) = \int_0^u H_{1,1}^{(0,2)}(z)dx_1 + \int_u^1 H_{1,1}^{(1,2)}(x_1 + y_1, z)dx_1$$

$$= \int_0^u H_{1,1}^{(0,2)}(z)dx_1 + \int_u^z\left[\int_0^{z-x_1} H_{1,1}(z)dy_1 + \int_{z-x_1}^1 dy_1\right]dx_1 + \int_z^1 dx_1.$$

The function $H_{1,2}^{(0,2)}(z)$ is not increasing in u, since its derivative with respect to u

$$\frac{\partial H_{1,2}^{(0,2)}(z)}{\partial u} = -2H_{1,1}(z)(z - u)$$

is nonpositive. Hence, the optimal value is $u_{1,2}^{(0,2)} = u = 0.$
Next, in the case $z \geq 1,$

$$H_{1,2}^{(0,2)}(z) = \int_0^u H_{1,1}^{(0,2)}(z)dx_1 + \int_u^1 H_{1,1}^{(1,2)}(x_1 + y_1, z)dx_1$$

$$= \int_0^u H_{1,1}(z)dx_1 + \int_u^1 \left[\int_0^{z-x_1} H_{1,1}(z)dy_1 + \int_{z-x_1}^1 dy_1 \right]dx_1.$$

Calculate the derivative of $H_{1,2}^{(0,2)}(z)$ with respect to u and set it equal to 0. Consequently, the optimal value $u_{1,2}^{(0,2)} = u$ satisfies the equation

$$H_{1,1}(z) - H_{1,1}(z)(z - u) - 1 + z - u = 0.$$

Hence, it follows that $u_{1,2}^{(0,2)} = z - 1$.

Thus,

$$u_{1,2}^{(0,2)} = \begin{cases} 0, \ z < 1, \\ z - 1, \ z \geq 1, \end{cases}$$

$$H_{1,2}^{(0,2)}(z) = \begin{cases} H_{1,1}(z)\int_0^z (z - x_1)dx_1 + \int_0^z (1 - (z - x_1))dx_1 + \int_z^1 dx_1, \ z < 1, \\ H_{1,1}(z)\left(z - 1 + \int_{z-1}^1 (z - x_1)dx_1\right) + \int_{z-1}^1 (1 - (z - x_1))dx_1, \ z \geq 1, \end{cases}$$

or

$$H_{1,2}^{(0,2)}(z) = \begin{cases} H_{1,1}(z)\dfrac{z^2}{2} + 1 - \dfrac{z^2}{2}, \ z < 1, \\ H_{1,1}(z)\left(z - 1 + \dfrac{z(2 - z)}{2}\right) + \dfrac{(2 - z)^2}{2}, \ z \geq 1. \end{cases}$$

For the game $\Gamma_{1,n}^{(0,2)}(z)$,

$$H_{1,n}^{(0,2)}(z) = \int_0^u H_{1,n-1}^{(0,2)}(z)dx_1 + \int_u^1 H_{1,n-1}^{(1,2)}(x_1 + y_1, z)dx_1$$

$$= \int_0^u H_{1,n-1}^{(0,2)}(z)dx_1 + \int_u^{z\wedge 1} \left[\int_0^{z-x_1} H_{1,n-1}(z)dy_1 + \int_{z-x_1}^1 dy_1 \right]dx_1 + \int_{z\wedge 1}^1 dx_1,$$

where $a \wedge b = \min\{a, b\}$.

Calculate the derivative of $H_{1,n}^{(0,2)}(z)$ with respect to u and set it equal to 0. As a result, the optimal value $u_{1,n}^{(0,2)} = u$ satisfies the equation

$$H_{1,n-1}^{(0,2)}(z) = H_{1,n-1}(z)(z - u) + 1 - z + u.$$

Hence, it follows that

$$u_{1,n}^{(0,2)} = \max\left\{0, z - \frac{1 - H_{1,n-1}^{(0,2)}(z)}{1 - H_{1,n-1}(z)}\right\}.$$

Note that $u_{1,n}^{(0,2)} \geq z - 1$ for $z \geq 1$, because $H_{1,n-1}^{(0,2)}(z) \geq H_{1,n-1}(z)$. Consequently,

$$H_{1,n}^{(0,2)}(z) = \begin{cases} H_{1,n-1}^{(0,2)}(z)\left(u_{1,n}^{(0,2)} + \int\limits_{u_{1,n}^{(0,2)}}^{z} (z - x_1)dx_1\right) \\ + \int\limits_{u_{1,n}^{(0,2)}}^{z} (1 - z + x_1)dx_1 + \int\limits_{z}^{1} dx_1, z < 1, \\ H_{1,n-1}^{(0,2)}(z)\left(u_{1,n}^{(0,2)} + \int\limits_{u_{1,n}^{(0,2)}}^{1} (z - x_1)dx_1\right) + \int\limits_{u_{1,n}^{(0,2)}}^{1} (1 - z + x_1)dx_1, z \geq 1. \end{cases}$$

The thresholds $u_{1,n}^{(0,2)}(z)$ as functions of z for different n, $1 \leq n \leq 5$, are shown in Fig. 1.

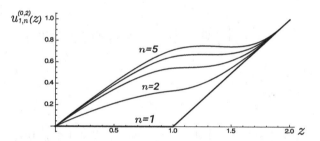

Fig. 1. Graphs of thresholds $u_{1,n}^{(0,2)}(z)$ for different n.

5 Two-Player Game

5.1 Game $\Gamma_{2,n}^{(1,1)}(t, z)$

Consider the two-player game $\Gamma_{2,n}^{(1,1)}(t, z)$, in which n contestants are left to perform and both players have already chosen a single contestant into their teams, of a total quality t (player 1) and z (player 2).

Denote by $H_{2,n}^{(1,1)}(t, z)$ the payoff of player 1 in this game. Note that $H_{2,n}^{(1,1)}(t, z) = 1 - H_{2,n}^{(1,1)}(z, t)$. Therefore, let $t < z$ without loss of generality.

If no contestants are left to perform in the game, then $H_{2,0}^{(1,1)}(t, z) = 0$.

Further, in the case $n = 1$, each of the players is interested in accepting the last contestant available. Hence, the optimal thresholds of the players coincide with each other, being equal to $u_{2,1}^{(1,1)} = 0 \cdot I_{\{z \leq 1\}} + (z - 1) \cdot I_{\{z > 1\}}$.

For $z \leq 1$, it follows that

$$H_{2,1}^{(1,1)}(t, z) = \frac{1}{2}\mathbf{P}\{x_1 + y_1 \geq z\} = \frac{1}{2}\left[\int_0^z dx_1 \int_{z-x_1}^1 dy_1 + \int_z^1 dx_1 \int_0^1 dy_1\right] = 1/2\left(1 - \frac{z^2}{2}\right);$$

for $z > 1$, $H_{2,1}^{(1,1)}(t, z) = \frac{1}{2}\left[\int_{z-1}^1 dx_1 \int_{z-x_1}^1 dy_1\right] = \frac{(z-2)^2}{4}$.

Consider the game $\Gamma_{2,n}^{(1,1)}(t, z)$. Assume that the experts have established some thresholds u and v, $u < v$, since $t < z$. Let n contestants be left to perform.

The optimal thresholds of the players can be found in the following way. Fix a threshold strategy v of player 2 and find the opponent's best response u.

The payoff of player 1 is given by

$$H_{2,n}^{(1,1)}(u, v|t, z) = \int_0^u H_{2,n-1}^{(1,1)}(t, z)dx_1 + \int_u^v dx_1 \int_0^1 (1 - H_{1,n-1}^{(1,2)}(z, t \vee (x_1 + y_1)))dy_1$$

$$+ \int_v^1 dx_1 \left[\frac{1}{2}\left(\int_0^1 (1 - H_{1,n-1}^{(1,2)}(z, t \vee (x_1 + y_1))dy_1\right)\right.$$

$$\left. + \frac{1}{2}\int_0^1 H_{1,n-1}^{(1,2)}(t, z \vee (x_1 + y_1))dy_1\right].$$

In order to find the optimal thresholds $u_{2,n}^{(1,1)}(z)$ and $v_{2,n}^{(1,1)}(z)$, calculate the derivatives of $H_{2,n}^{(1,1)}(u, v|t, z)$ with respect to u and v, setting them equal to 0.

As a matter of fact, four cases are possible, depending on the values of z.

Here is the solution of this problem for $n = 2$.

Find the optimal thresholds of the players.

a) If $z \leq u < v$, then the optimal value $u_{2,2}^{(1,1)}(z) = u$ is calculated from the equation

$$\frac{1}{2}\left(1 - \frac{z^2}{2}\right) = 1 - \left[\int_0^{1-u}\left(1 - \frac{(u+y_1)^2}{2}\right)dy_1 + \int_{1-u}^1 \frac{(u+y_1-2)^2}{2}dy_1\right]. \quad (2)$$

The optimal value $v_{2,2}^{(1,1)}(z) = v$ is calculated from the equation

$$\int_0^{1-v}\left(1 - \frac{(v+y_1)^2}{2}\right)dy_1 + \int_{1-v}^1 \frac{(v+y_1-2)^2}{2}dy_1 = 1/2, \quad (3)$$

which yields $v_{2,2}^{(1,1)}(z) = 0.5$.

b) If $u < z \leq v$, then the optimal value $u_{2,2}^{(1,1)}(z) = u$ is calculated from the equation

$$\frac{1}{2}\left(1 - \frac{z^2}{2}\right) - (1 - z + u) + \int_{z-u}^{1-u}\left(1 - \frac{(u + y_1)^2}{2}\right)dy_1 + \int_{1-u}^{1}\frac{(u + y_1 - 2)^2}{2}dy_1 = 0.$$

(4)

The optimal value $v_{2,2}^{(1,1)}(z)$ is calculated from Eq. (3), which yields $v_{2,2}^{(1,1)}(z) = 0.5$.

c) If $u < v < z \leq 1$, then the optimal value $u_{2,2}^{(1,1)}(z) = u$ is calculated from Eq. (4).

The optimal value $v_{2,2}^{(1,1)}(z) = v$ is calculated from the equation

$$(z - v)\left(1 - \frac{z^2}{2}\right) + 2\left[\int_{z-v}^{1-v}\left(1 - \frac{(v + y_1)^2}{2}\right)dy_1 + \int_{1-v}^{1}\frac{(v + y_1 - 2)^2}{2}dy_1\right] = 1 - z + v.$$

(5)

d) If $u < v < 1 < z$, then the optimal value $u_{2,1}^{(1,1)}(z) = u$ is calculated from the equation

$$\frac{1}{2}\left(1 - \frac{z^2}{2}\right) - \int_{z-u}^{1}\left(1 - \frac{(u + y_1 - 2)^2}{2}\right)dy_1 = 0.$$

(6)

The optimal value $v_{2,2}^{(1,1)}(z) = v$ is calculated from the equation

$$(z - v)\frac{(z - 2)^2}{2} + 2\int_{z-v}^{1}\frac{(v + y_1 - 2)^2}{2}dy_1 = 1 - z + v.$$

(7)

The qualitative behavior of the optimal thresholds $u_{2,2}^{(1,1)}(z)$ and $v_{2,2}^{(1,1)}(z)$ depending on z is demonstrated in Fig. 2.

Example 1. Here are the optimal thresholds for some values of z.

For $0 \leq z \leq \bar{z} \approx 0.436$, the optimal threshold $v_{2,2}^{(1,1)}(z) = 0.5$ of player 2 is given by Eq. (3), whereas the optimal threshold $u_{2,2}^{(1,1)}(z)$ of player 1 satisfies Eq. (2). For example, $u_{2,2}^{(1,1)}(0) = 0.5$, $u_{2,2}^{(1,1)}(0.4) \approx 0.45$, and $u_{2,2}^{(1,1)}(\bar{z}) = \bar{z} \approx 0.436$.

For $\bar{z} < z \leq 0.5$, the optimal threshold of player 2 is $v_{2,2}^{(1,1)}(z) = 0.5$, whereas the optimal threshold $u_{2,2}^{(1,1)}(z)$ of player 1 satisfies Eq. (4). For example, $u_{2,2}^{(1,1)}(0.45) \approx 0.434$ and $u_{2,2}^{(1,1)}(0.5) \approx 0.427$.

For $0.5 < z \leq 1$, the optimal threshold $u_{2,2}^{(1,1)}(z)$ of player 1 satisfies Eq. (4). For example, $u_{2,2}^{(1,1)}(0.75) \approx 0.392$ and $u_{2,2}^{(1,1)}(1) \approx 0.376$.

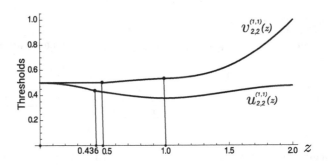

Fig. 2. Thresholds $u_{2,2}^{(1,1)}(z)$ and $v_{2,2}^{(1,1)}(z)$.

The optimal threshold $v_{2,2}^{(1,1)}(z)$ of player 2 is given by Eq. (5). For example, $v_{2,2}^{(1,1)}(0.75) \approx 0.519$ and $v_{2,12}^{(1,1)}(1) \approx 0.533$.

For $1 < z$, the optimal threshold $u_{2,2}^{(1,1)}(z)$ of player 1 satisfies Eq. (6); $u_{2,2}^{(1,1)}(1.5) \approx 0.427$ and $u_{2,2}^{(1,1)}(2) \approx 0.476$.

The optimal threshold $v_{2,2}^{(1,1)}(z)$ of player 2 is given by Eq. (7). For example, $v_{2,2}^{(1,1)}(1.5) \approx 0.634$ and $v_{2,2}^{(1,1)}(2) = 1$.

The optimal strategies of the players depending on the current maximum z in the game in which both players have already chosen a single contestant are shown in Fig. 2. The optimal strategies clearly differ, depending on which player has chosen the contestant of the maximum total quality z. The player with this contestant in his team has an increasing optimal threshold that exceeds the opponent's one. At the same time, the optimal threshold of his opponent is decreasing for $z \leq 1$ and increasing for $z > 1$.

5.2 Game $\Gamma_{2,n}^{(1,0)}(z)$

Consider the two-player game $\Gamma_{2,n}^{(1,0)}(z)$, in which n contestants are left to perform, player 1 has already chosen a contestant of a total quality z, whereas player 2 has chosen nobody so far.

Denote by $H_{2,n}^{(1,0)}(z)$ the payoff of player 1 in this game. Note that $H_{2,n}^{(1,0)}(z) = 1 - H_{2,n}^{(0,1)}(z)$. Assume that the experts have established some thresholds u and v, where $v < u$. Let n contestants be left to perform.

In the game $\Gamma_{2,1}^{(1,0)}(z)$ with no contestants available for choice, $H_{2,0}^{(1,0)}(z) = 1$. The thresholds of the players coincide with each other (the players invite any contestant) and are equal to $u_{2,1}^{(1,0)}(z) = 0 \cdot I_{\{z \leq 1\}} + (z - 1) \cdot I_{\{z > 1\}}$

and

$$H_{2,1}^{(1,0)}(z) = \frac{1}{2}\left[\int_0^1 dx_1 + \int_0^z dx_1 \int_0^{z-x_1} dy_1\right] \cdot I_{\{z<1\}}$$

$$+ \left[\int_0^{z-1} dx_1 + \frac{1}{2}\int_{z-1}^1 dx_1\left(1 + \int_0^{z-x_1} dy_1\right)\right] \cdot I_{\{z\geq1\}} = \frac{1}{2}\left(1 + \frac{z^2}{2}\right) \cdot I_{\{z<1\}} + \frac{z(4-z)}{4} \cdot I_{\{z\geq1\}}.$$

In the game $\Gamma_{2,n}^{(1,0)}(z)$ the payoff of player 1 is given by

$$H_{2,n}^{(1,0)}(u,v|z) = \int_0^v H_{2,n-1}^{(1,0)}(z)dx_1 + \int_v^u dx_1 \int_0^1 H_{2,n-1}^{(1,1)}(z,x_1+y_1)dy_1 \qquad (8)$$

$$+ \int_u^1 dx_1 \frac{1}{2}\left(\int_0^1(1 - H_{1,n-1}^{(0,2)}(x_1+y_1))dy_1\right) + \frac{1}{2}\int_0^1 H_{2,n-1}^{(1,1)}(z,x_1+y_1)dy_1.$$

In order to find the optimal thresholds $u_{2,n}^{(1,0)}(z)$ and $v_{2,n}^{(1,0)}(z)$, calculate the derivatives of $H_{2,n}^{(1,0)}(u,v|z)$ with respect to u and v, setting them equal to 0.

Consider, for example, the game $\Gamma_{2,2}^{(1,0)}(z)$. Since player 2 has chosen nobody so far, his optimal strategy is given by

$$v_{2,2}^{(1,0)}(z) = 0 \cdot I_{\{z<1\}} + (z-1) \cdot I_{\{z\geq1\}}.$$

Calculate the optimal values $u_{2,2}^{(1,0)}(z)$. For this purpose, consider the following cases:

a) If $z \leq u < 1$, then the optimal value $u_{2,2}^{(1,0)}(z) = u$ is calculated from the equation

$$\int_0^{1-u}\left(1 - \frac{(u+y_1)^2}{2}\right)dy_1 + \int_{1-u}^1 \frac{(u+y_1-2)^2}{2}dy_1 = \frac{2}{3}. \qquad (9)$$

As a result, $u_{2,2}^{(1,0)}(z) \approx 0.273$; denote it by z^*.

b) If $u < z \leq 1$, then the optimal value $u_{2,2}^{(1,0)}(z) = u$ is calculated from the equation

$$\int_0^{z-u}\frac{1}{2}\left(1 + \frac{z^2}{2}\right)dy_1 + \frac{3}{2}\int_{z-u}^{1-u}\left(1 - \frac{(u+y_1)^2}{2}\right)dy_1 + \frac{3}{2}\int_{1-u}^1 \frac{(u+y_1-2)^2}{4}dy_1$$

$$(10)$$

$$-1 + \int_0^{z-u}\left(1 - \frac{z^2}{2}\right)dy_1 = 0.$$

c) If $u < 1 < z$, then the optimal value $u_{2,2}^{(1,0)}(z) = u$ is calculated from the equation

$$\int_0^{z-u} \frac{z(4-z)}{4}dy_1 + 3 \int_{z-u}^1 \frac{(u+y_1-2)^2}{4}dy_1 - 1 + \int_0^{z-u} \frac{(z-2)^2}{2}dy_1 = 0. \quad (11)$$

The qualitative behavior of the optimal thresholds $u_{2,2}^{(0,1)}(z)$ depending on z is demonstrated in Fig. 3.

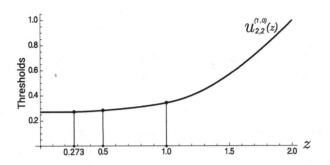

Fig. 3. Thresholds $u_{2,2}^{(1,0)}(z)$.

Example 2. Here are the optimal thresholds for some values of z.

For $0 \le z \le z^*$, the optimal threshold $u_{2,2}^{(1,0)}(z)$ is found from Eq. (9), $u_{2,2}^{(1,0)}(z) = z^* \approx 0.273$.

For $z^* < z \le 1$, the optimal threshold $u_{2,2}^{(1,0)}(z)$ is found from Eq. (10). For example, $u_{2,2}^{(1,0)}(0.5) \approx 0.284$ and $u_{2,2}^{(1,0)}(1) \approx 0.343$.

For $1 < z$, the optimal threshold $u_{2,2}^{(1,0)}(z)$ is found from Eq. (11). For example, $u_{2,2}^{(1,0)}(1.5) \approx 0.569$ and $u_{2,2}^{(1,0)}(2) = 1$.

6   Game $\Gamma_{2,n}^{(0,0)}$

Consider the game $\Gamma_{2,n}^{(0,0)}$, in which n contestants are left to perform and none of the players has chosen a contestant into his team so far.

If $n \le 2$, then the optimal thresholds of the players coincide with each other, being equal to $u_{2,n}^{(0,0)} = 0$.

Let $u < v$; in this case,

$$H_{2,n}^{(0,0)}(u,v) = \int_0^u H_{2,n-1}^{(0,0)} dx_1 + \int_u^v dx_1 \int_0^1 H_{2,n-1}^{(1,0)}(x_1 + y_1) dy_1$$

$$+ \int_v^1 dx_1 \left[\frac{1}{2} \int_0^1 H_{2,n-1}^{(1,0)}(x_1 + y_1) dy_1 + \frac{1}{2} \int_0^1 H_{2,n-1}^{(0,1)}(x_1 + y_1) dy_1 \right]$$

$$= \int_0^u \frac{1}{2} dx_1 + \int_u^v dx_1 \int_0^1 H_{2,n-1}^{(1,0)}(x_1 + y_1) dy_1 + \int_v^1 \frac{1}{2} dx_1.$$

Due to the obvious symmetry of the players, they have the same payoff $H_{2,n}^{(0,0)} = \frac{1}{2}$ and the same optimal threshold, i.e., $u_{2,n}^{(1,0)} = v_{2,n}^{(1,0)}$.
The optimal value $u_{2,n}^{(1,0)} = u$ is found from the equation

$$\int_0^1 H_{2,n-1}^{(1,0)}(u + y_1) dy_1 = \frac{1}{2}. \tag{12}$$

Calculate the thresholds of the players for $n = 3$.
Equation (12) takes the form

$$\int_0^{z^*-u} H_{2,2}^{(1,0)}(u + y_1) dy_1 + \int_{z^*-u}^{1-u} H_{2,2}^{(1,0)}(u + y_1) dy_1 + \int_{1-u}^1 H_{2,2}^{(1,0)}(u + y_1) dy_1 = \frac{1}{2}, \tag{13}$$

where z^* satisfies Eq. (9), $z^* \approx 0.273$.
The expressions for the payoffs $H_{2,2}^{(1,0)}(u + y_1)$ are derived from Eq. (8).
In accordance with the results of numerical simulation, the optimal strategies are determined by the threshold $u_{2,3}^{(0,0)} \approx 0.216$.

7   Conclusions

In this paper, a game-theoretic model of *The Voice* TV show in which two players seek to form a team of two contestants has been proposed. An important feature of this formulation is that the players have incomplete information about the quality parameters of the incoming (performing) contestants. The optimal threshold strategies and payoffs of the players in this problem have been calculated using dynamic programming.

In the future, the results can be extended to the case of several players and also to the case of several vacant places in the team of each player.

References

1. Solan, E., Vieille, N.: Quitting games. Math. Oper. Res. **26**(2), 265–285 (2001). https://doi.org/10.1287/moor.26.2.265.10549
2. Sofronov, G.: An optimal sequential procedure for a multiple selling problem with independent observations. Eur. J. Oper. Res. **225**(2), 332–336 (2013). https://doi.org/10.1016/j.ejor.2012.09.042
3. Immorlica, N., Kleinberg, R., Mahdian, M.: Secretary problems with competing employers. In: Spirakis, P., Mavronicolas, M., Kontogiannis, S. (eds.) WINE 2006. LNCS, vol. 4286, pp. 389–400. Springer, Heidelberg (2006). https://doi.org/10.1007/11944874_35
4. Alpern, S., Katrantzi, I., Ramsey, D.: Partnership formation with age-dependent preferences. Eur. J. Oper. Res. **225**(1), 91–99 (2013). https://doi.org/10.1016/j.ejor.2012.09.012
5. Whitmeyer, M.: A competitive optimal stopping game. B.E. J. Theor. Econ. **18**(1), 1–15 (2018). https://doi.org/10.1515/bejte-2016-0128
6. Harrell, G., Harrison, J., Mao, G., Wang, J.: Online auction and secretary problem. In: International Conference Scientific Computing, pp. 241–244 (2015)
7. Seregina, T., Ivashko, A., Mazalov, V.: Optimal stopping strategies in the game "The Price is Right". Proc. Steklov Inst. Math. **307**(Suppl. 1), 1–15 (2019)
8. Mazalov, V., Ivashko, A.: Equilibrium in n-person game of Showcase-Showdown. Probab. Eng. Inf. Sci. **24**, 397–403 (2010). https://doi.org/10.1017/S0269964810000045
9. Tenorio, R., Cason, T.N.: To spin or not to spin? Natural and laboratory experiments from "The Price is Right". Econ. J. **112**(476), 170–195 (2002). https://doi.org/10.1111/1468-0297.0j678
10. Bennett, R.W., Hickman, K.A.: Rationality and the "Price is Right". J. Econ. Behav. Org. **21**(1), 99–105 (1993). https://doi.org/10.1016/0167-2681(93)90042-N
11. Mazalov, V.V., Ivashko, A.A., Konovalchikova, E.N.: Optimal strategies in best-choice game with incomplete information – The Voice show. Int. Game Theor. Rev. **18**(2), 1640001 (2016). https://doi.org/10.1142/S0219198916400016

Weak Berge Equilibrium

Konstantin Kudryavtsev[1,2]($\boxtimes$) (ID), Ustav Malkov[3], and Vladislav Zhukovskiy[4]

[1] South Ural State University, Chelyabinsk, Russia
kudrkn@gmail.com
[2] Chelyabinsk State University, Chelyabinsk, Russia
[3] Central Economics and Mathematics Institute of Russian Academy of Science, Moscow, Russia
ustav-malkov@yandex.ru
[4] M.V. Lomonosov Moscow State University, Moscow, Russia
zhkvlad@yandex.ru

Abstract. Various concepts of solutions can be employed in the non-cooperative game theory. The Berge equilibrium is one of such solutions. The Berge equilibrium is an altruistic concept of equilibrium. In this concept, the players act on the principle "One for all and all for one!" The Berge equilibrium solves such well known paradoxes in the game theory as the "Prisoner's Dilemma", "Battle of the sexes" and many others. At the same time, the Berge equilibrium rarely exist in pure strategies. Moreover, in finite games, the Berge equilibrium may not exist in the class of mixed strategies. The paper proposes the concept of a weak Berge equilibrium. Unlike the Berge equilibrium, the moral basis of this equilibrium is the Hippocratic Oath "First do no harm". On the other hand, all Berge equilibria are some weak Berge equilibria. The properties of the weak Berge equilibrium have been investigated. The existence of the weak Berge equilibrium in mixed strategies has been established for finite games. A numerical weak Berge equilibrium approximate search method, based on 3LP-algorithm, is proposed. The weak Berge equilibria for finite 3-person non-cooperative games are computed.

Keywords: Three-person game · Non-cooperative game · Berge equilibrium · Weak Berge equilibrium

1 Introduction

A wide class of economic, social and political processes are well described by the methods of the game theory. Often, when decisions are made, participants in such processes can not agree among themselves that are modeled by using non-cooperative games. Certainly, the most well-known concept of a solution in the theory of non-cooperative games was proposed by John Nash in 1950 in [1]. For this work in 1994 he was awarded the Nobel Prize in Economics.

However, the application of the Nash equilibrium concept in the modelling of real socio-economic and political conflicts, in some cases, leads to paradoxical

© Springer Nature Switzerland AG 2020
Y. Kochetov et al. (Eds.): MOTOR 2020, CCIS 1275, pp. 231–243, 2020.
https://doi.org/10.1007/978-3-030-58657-7_20

results, such as the "prisoner's dilemma". One of the first who has noticed this was Claude Berge in [2]. In this book, Berge proposed a new concept of equilibrium, according to which, players are divided into coalitions, while players of one coalition can work together to maximize the payoffs of players of another coalition. Apparently, a crushing review by Martin Shubik [3] on Berge's book [2], led to the fact that Claude Berge switched his attention from the game theory to other areas of mathematics. After decades, based on Berge's ideas, V.I. Zhukovsky [4,5] and K.S. Vaisman [6,7] suggested a new altruistic concept of equilibrium which was called a *Berge equilibrium* (BE). In this concept, the players act on the principle of "One for all and all for one!" from Alexander Dumas's novel "The Three Musketeers". Another interpretation of Berge equilibrium is [8] the Golden Rule of morality: "Do things to others the way you want them did with you". The development of the Berge equilibrium concept is described in details in the review [9]. It is worth noting that the BE solves such well known paradoxes in the game theory as the "Prisoner's Dilemma", "Battle of the sexes" and many others. Also the use of BE is possible to the economics applications [10].

At the same time, the Berge equilibrium concept has some drawbacks. One of these drawbacks is that Berge equilibrium rarely exists in pure strategies. Moreover, in N- person games ($N \geq 3$) with a finite set of strategies, Berge equilibrium may not exist in the class of mixed strategies. Such example was constructed, in particular, in [11]. The lack of BE might be caused by the fact that it is often impossible to follow the Golden Rule of morality in relation to all players at the same time. For example, if the goals of two players are opposite, then the third player will not be able to apply the Golden Rule to them simultaneously. In this case, increasing the payoff of one player, simultaneously reduces the payoff of the other.

In this paper, we introduce the concept of the weak equilibrium according to Berge (Weak Berge Equilibrium or WBE), no longer based on the Golden Rule of morality, and on the Hippocratic oath "First do no harm!". Here, we will assume that, making a decision, each player adheres to the situation, one-sided deviation from which can harm although to one of the other players. Further, in Sect. 2, the concept of the weak Berge equilibrium is formalized, some of its properties are studied and sufficient conditions for the existence of such an equilibrium in N-person games are given. In Sect. 3, a numerical WBE approximate search method based on [12–14] is proposed, and numerical simulation results are given for finite games of three person.

2   The Concept of the Weak Berge Equilibrium

Let us consider a non-cooperative N-person game in normal form:

$$\Gamma = \langle \mathbf{N}, \{X_i\}_{i \in \mathbf{N}}, \{f_i(x)\}_{i \in \mathbf{N}} \rangle, \tag{1}$$

where $\mathbf{N} = \{1, 2, \ldots, N\}$ denotes the set of serial numbers of the players; the set of x_i strategies of the i-th player ($i \in \mathbf{N}$) is denoted by X_i, where $X_i \subseteq \mathbf{R}^{n_i}$.

As a result of the players choosing their strategies, the strategy profile is $x = (x_1, \ldots, x_N) \in X = X_1 \times X_2 \times \ldots \times X_N \subseteq \mathbf{R}^n$ ($n = n_1 + n_2 + \ldots + n_N$). On the set of strategy profiles X for each player i ($i \in \mathbf{N}$) the scalar payoff function $f_i(x) : X \to \mathbf{R}$ was defined. The value of $f_i(x)$ was realized on the strategy profile chosen by the players $x \in X$ was called the payoff of the i-th player.

The game Γ is played as follows. Each player i ($i \in \mathbf{N}$), without entering into a coalition with other players, chooses his strategy $x_i \in X_i$. As a result of this choice, the strategy profile is $x = (x_1, \ldots, x_N) \in X$. After that, each player i gets his payoff $f_i(x)$.

Thus, when making a decision, the player is forced to focus not only on his payoff function, but also on the possible choice of the other participants in the game.

Further, (y_i, x_{-i}) denotes the strategy profile $(x_i, \ldots, x_{i-1}, y_i, x_{i+1}, \ldots, x_N)$, which is obtained from strategy profile x by replacing the strategy of the i-th player x_i on y_i.

The most popular concept of solution in the theory of non-cooperative games is Nash equilibrium.

Definition 1. *A strategy profile* $x^e = (x_1^e, \ldots, x_N^e) \in X$ *is called a Nash equilibrium (NE) in game* (1) *if for every* $x \in X$ *the system of inequalities*

$$f_i(x^e) \geq f_i(x_i, x_{-i}^e) \quad (i \in \mathbf{N}) \tag{2}$$

is true.

The Nash equilibrium strategy profile $x^e \in X$ is stable with the respect to deviation of an individual player from his strategy which enters in x^e. Applying the concept of the Nash equilibrium, the player proceeds from his own selfish motives. He only cares about his payoff, do not take into account the interests of other players. However, this approach leads to a number of paradoxes, such as the Tucker problem in the classic game called as Prisoner's Dilemma.

Example 1. Let us consider the Prisoner's Dilemma game. Two criminals are arrested on suspicion of a crime, but the police do not have direct evidence. Therefore, the police, have isolated them from each other, and offered them the same deal: if one testifies against the other, but he keeps silence, the first one is released for helping the investigation, and the second gets 10 years - the maximum term of imprisonment. If both are silent, their deed goes through a lighter article, and each of them are sentenced to a year in prison. If both testify against each other, each receives a minimum period of 2 years. Every prisoner chooses to keep quiet or testify against another. However, none of them knows exactly what the other will do. The Nash equilibrium in this game dictates players to testify against each other, although silence will be more beneficial for them.

Thus, the players' egoism (the Nash equilibrium) in the Prisoner's Dilemma leads them to the most unprofitable solution. This is the Tucker problem.

The opposite approach to the concept of equilibrium, based on altruism, was called the Berge equilibrium.

Definition 2. *A strategy profile $x^B = (x_1^B, \ldots, x_N^B) \in X$ is called a Berge equilibrium (BE) in game (1), if for each $x \in X$ the system of inequalities*

$$f_i(x^B) \geq f_i(x_i^B, x_{-i}) \quad (i \in N) \tag{3}$$

is true.

The difference between Nash and Berge equilibria is that, in a Nash equilibrium, each player directs all efforts to increase its individual payoff as much as possible. The antipode of (2) is (3), where each player strives to maximize the payoffs of the other players, ignoring its individual interests. Such an altruistic approach is intrinsic to kindred relations and occurs in religious communities. The elements of such altruism show up in charity, sponsorship, and so on.

In Example 1, players receive the best result if they use the Berge equilibrium, thus the Berge equilibrium solves the Tucker problem in the Prisoner's Dilemma (the prisoners choose to keep quiet).

Consider a special case of game (1) with two players, i.e., the game Γ where $N = 1, 2$. Then a Berge equilibrium $x^B = (x_1^B, x_2^B)$ is defined by the equalities

$$f_1(x^B) = \max_{x_2 \in X_2} f_1(x_1^B, x_2), \quad f_2(x^B) = \max_{x_1 \in X_1} f_2(x_1, x_2^B).$$

The Nash equilibrium $x^e = (x_1^e, x_2^e)$ in this two-player game is given by the conditions

$$f_1(x^e) = \max_{x_1 \in X_1} f_1(x_1, x_2^e), \quad f_2(x^e) = \max_{x_2 \in X_2} f_2(x_1^e, x_2).$$

A direct comparison of these standalone formulas leads to the following result.

Property 1. The Berge equilibrium in game (1) with $N = \{1, 2\}$ coincides with the Nash equilibrium if both players interchange their payoff functions and then apply the concept of the Nash equilibrium to solve the game.

In view of Property 1, all results concerning the Nash equilibrium in the two-player game are automatically transferred to the Berge equilibrium (of course, with an "interchange" of the payoff functions as described by Property 1).

The differences appear when $N \geq 3$. So, the Berge equilibrium may not exist in finite 3-person games. An example of this is given in [11]. The following example is taken from [11].

Example 2. Let us consider the following 3-person game in which each of the players has two pure strategies. Pure strategies of the first, the second, and the third player are denoted A_1, A_2; B_1, B_2; C_1, C_2, respectively.

$$C_1: \begin{array}{c} \\ A_1 \\ A_2 \end{array} \begin{array}{cc} B_1 \quad\quad B_2 \\ \left(\begin{array}{cc} (2,1,0) & (1,1,1) \\ (2,0,1) & (1,0,2) \end{array} \right) \end{array} \quad\quad C_2: \begin{array}{c} \\ A_1 \\ A_2 \end{array} \begin{array}{cc} B_1 \quad\quad B_2 \\ \left(\begin{array}{cc} (1,2,0) & (0,2,1) \\ (1,1,1) & (0,1,2) \end{array} \right) \end{array}$$

The left-hand matrix refers to the pure strategy C_1 of the third player, while the right-hand matrix refers to his/her pure strategy C_2. Let us note that this game is a very special one. None of the players has any possibility to influence their own payoff, no matter if they use any of their pure or mixed strategies. On the contrary, players' payoffs depend exclusively on the choices of the remaining players.

One can easily check that the second and the third players' best support to any of the first player's (pure or mixed) strategies is a pair of pure strategies (B_1, C_1); the first and the third players' best support to any of the second player's (pure or mixed) strategies is a pair of pure strategies (A_1, C_2); and finally, the first and the second players' best support to any of (pure or mixed) strategies of the third player is a pair of pure strategies (A_2, B_2). This game has no Berge equilibria, neither in pure, nor in mixed strategies.

Then, we recall the concept of Pareto optimality, and then formalize the Weak Berge Equilibrium.

Definition 3. *The alternative x^* is a Pareto-optimal alternative in the N-criteria problem*

$$\langle X, \{f_i(x)\}_{i \in \mathbf{N}} \rangle,$$

if the system of N inequalities

$$f_i(x) \leq f_i(x^*) \quad (i \in \mathbf{N}),$$

with at least one strict inequality, is inconsistent.

The moral basis of following definition is the Hippocratic Oath "First do no harm!"

Definition 4. *Let us call the strategy profile $x^w = (x_1^w, \ldots, x_n^w)$ a weak Berge equilibrium (WBE), if for each player i $(i \in \mathbf{N})$ strategy x_i^w is Pareto-optimal alternative in the $N - 1$-criteria problem*

$$\Gamma_i = \langle X_i, \{f_j(x_i, x_{-i}^w)\}_{j \in \mathbf{N} \setminus \{i\}} \rangle.$$

Note that any BE is WBE. But the converse is not true, there are WBE that are not BE.

Let us compare the game Γ with an auxiliary game

$$\tilde{\Gamma} = \langle \mathbf{N}, \{X_i\}_{i \in \mathbf{N}}, \{g_i(x)\}_{i \in \mathbf{N}} \rangle, \tag{4}$$

where the set of players $\mathbf{N}$ and the set of strategies X_i $(i \in \mathbf{N})$ are the same as in the game (1), and the payoff functions $g_i(x)$ have the form

$$g_i(x) = \sum_{j \in \mathbf{N} \setminus \{i\}} f_j(x). \tag{5}$$

Lemma 1. *The Nash equilibrium strategy profile in the game (4) is a weak Berge equilibrium strategy profile in the game (1).*

Proof. Let x^e be a Nash equilibrium strategy profile in the game $\tilde{\Gamma}$, i.e

$$g_i(x_1^e, ..., x_{i-1}^e, x_i, x_{i+1}^e, ..., x_n^e) \leq g_i(x^e) \quad (i \in \mathbf{N}). \tag{6}$$

With regard to (5), the inequality (6) can be rewritten as

$$\sum_{j \in \mathbf{N} \setminus \{i\}} f_j(x_i, x_{-i}^e) \leq \sum_{j \in \mathbf{N} \setminus \{i\}} f_j(x^e) \quad (i \in \mathbf{N}). \tag{7}$$

Suppose x^e is not a WBE strategy profile, then there exists some number i for which the system of inequalities is consistent

$$f_j(x_i, x_{-i}^e) \geq f_j(x^e) \quad (j \in \mathbf{N} \setminus \{i\}), \tag{8}$$

of which at least one inequality is strict.

Adding inequalities (8), we obtain

$$\sum_{j \in \mathbf{N} \setminus \{i\}} f_j(x_i, x_{-i}^e) > \sum_{j \in \mathbf{N} \setminus \{i\}} f_j(x^e) \quad (i \in \mathbf{N}),$$

that contradicts (7).

Remark 1. To construct a WBE strategy profile in the game (1), we can use the following *algorithm*:

1. to compose auxiliary game $\tilde{\Gamma}$;
2. to construct a strategy profile x^e which is the Nash equilibrium strategy profile in the auxiliary game $\tilde{\Gamma}$;
3. the found strategy profile x^e will be the WBE strategy profile in the original game Γ.

As an example, let us consider the game "Snowdrift" which is proposed in [15].

Example 3. Let us consider the 3-person Snowdrift game which is shown in Table 1. The history of the game lies in the fact that A, B and C are the drivers of three cars, that stuck in a snowdrift at night, each of them has a shovel. If a solution is found for any one care, others can use it. Every driver chooses to dig or wait (in the hope that someone else will dig, or that a snowplow will come to the place of incident). Digging will cost 6 points, which are divided equally between those who perform the work; provided that there is at least one digger. If the players dug out by themselves of a snowdrift, then each player gets 4 points. Thus, if all three players dig, then everyone will get 2 points. If two players dig, they will get one point each, and the third player will earn 4 points. If one player digs, then his payoff will be negative (-2), and the payoffs of the remaining two players will be 4 points each. In the case that the players do not dig, but wait until the morning when the utilities arrive and clear the snow, their payoff will be zero.

Table 1. The 3-person Snowdrift game.

C – to wait			C – to dig	·	
A \ B	to wait	to dig	A \ B	to wait	to dig
to wait	(0, 0, 0)	(4, −2, 4)	to wait	(4, 4, −2)	(4, 1, 1)
to dig	(−2, 4, 4)	(1, 1, 4)	to dig	(1, 4, 1)	(2, 2, 2)

Here, the 3-dimensional matrices A, B, C, which determine the payoffs of the players will be

$$A: \quad A_1 = \begin{pmatrix} 0 & 4 \\ -2 & 1 \end{pmatrix}, \quad A_2 = \begin{pmatrix} 4 & 4 \\ 1 & 2 \end{pmatrix};$$

$$B: \quad B_1 = \begin{pmatrix} 0 & -2 \\ 4 & 1 \end{pmatrix}, \quad B_2 = \begin{pmatrix} 4 & 1 \\ 4 & 2 \end{pmatrix};$$

$$C: \quad C_1 = \begin{pmatrix} 0 & 4 \\ 4 & 4 \end{pmatrix}, \quad C_2 = \begin{pmatrix} -2 & 1 \\ 1 & 2 \end{pmatrix}.$$

The Nash equilibrium (NE) here will (wait, wait, wait) [15] with payoffs $(0, 0, 0)$.

We will now compile an auxiliary game, the payoff matrices in which will be: for the first player

$$A^* = B + C: \quad A_1^* = \begin{pmatrix} 0 & 2 \\ 8 & 5 \end{pmatrix}, \quad A_2^* = \begin{pmatrix} 2 & 2 \\ 5 & 4 \end{pmatrix};$$

for the second player

$$B^* = A + C: \quad B_1^* = \begin{pmatrix} 0 & 8 \\ 2 & 5 \end{pmatrix}, \quad B_2^* = \begin{pmatrix} 2 & 5 \\ 2 & 4 \end{pmatrix};$$

for the third player

$$C^* = A + B: \quad C_1^* = \begin{pmatrix} 0 & 2 \\ 2 & 2 \end{pmatrix}, \quad C_2^* = \begin{pmatrix} 8 & 5 \\ 5 & 4 \end{pmatrix}.$$

The Nash equilibrium (NE) in the auxiliary game with matrices A^*, B^*, C^* will be (dig, dig, dig), respectively, the weak Berge equilibrium (WBE) in the original game will also be (dig, dig, dig) with payoffs $(2, 2, 2)$.

Obviously, in this example, the WBE is more profitable for all players than the NE.

Remark 2. In the Snowdrift game, the Berge equilibrium (BE) [15] coincides with the WBE.

Follow to Lemma 1 and the sufficient conditions for the existence of a NE, it is easy possible to obtain sufficient conditions for the existence of a WBE under the usual restrictions for the game theory.

Theorem 1. *In a non-cooperative N-person game Γ with a finite set of strategies, a weak Berge equilibrium strategy profile in mixed strategies exists.*

Theorem 2. *If in a non-cooperative N-person game Γ, the sets of strategies X_i are convex compacts, and the payoff functions $f_i(x)$ are continuous in the aggregate of variables, then in the game Γ a weak Berge equilibrium strategy profile in mixed strategies exists.*

3   The WBE in a Finite 3-Person Game

Let us consider a non-cooperative 3-person game.

$$\Gamma_3 = \langle \{1,2,3\}, \{X_i\}_{i=1,2,3}, \{f_i(x)\}_{i=1,2,3} \rangle.$$

The strategy profile $x^w = (x_1^w, x_2^w, x_3^w)$ is the WBE strategy profile, if and only if

1) the strategy x_1^w is the Pareto-optimal alternative in the two-criterial problem

$$\langle X_1, \{f_2(x_1, x_2^w, x_3^w), f_3(x_1, x_2^w, x_3^w)\} \rangle;$$

2) the strategy x_2^w is the Pareto-optimal alternative in the two-criterial problem

$$\langle X_2, \{f_1(x_1^w, x_2, x_3^w), f_3(x_1^w, x_2, x_3^w)\} \rangle;$$

3) the strategy x_3^w is the Pareto-optimal alternative in the two-criterial problem

$$\langle X_3, \{f_1(x_1^w, x_2^w, x_3), f_2(x_1^w, x_2^w, x_3)\} \rangle.$$

Let us compose an axillary game for the game Γ_3

$$\tilde{\Gamma}_3 = \langle \{1,2,3\}, \{X_i\}_{i=1,2,3}, \{g_i(x)\}_{i=1,2,3} \rangle,$$

where, according to (5)

$$\begin{aligned}
g_1(x) &= f_2(x) + f_3(x), \\
g_2(x) &= f_1(x) + f_3(x), \\
g_3(x) &= f_1(x) + f_2(x).
\end{aligned} \tag{9}$$

The NE strategy profile in $\tilde{\Gamma}_3$ will be the WBE strategy profile in the original game Γ_3.

Below, a finite non-cooperative 3-person game Γ_3 is defined with three sets X, Y, Z of strategies of the first, second, and third player respectively, where $X = \{x = (x_1, \ldots, x_m)^T \in \mathbf{R}^m : x^T e_m = 1, x \geq 0_m\}$, $Y = \{y = (y_1, \ldots, y_n)^T \in \mathbf{R}^n : y^T e_n = 1, y \geq 0_n\}$, $Z = \{z = (z_1, \ldots, z_l)^T \in \mathbf{R}^l : z^T e_l = 1, z \geq 0_l\}$, $\omega = (x, y, z) \in \mathbf{R}^{m+n+l}$, together with their payoff functions as follows

$$f_x(\omega) = \sum_{i=1}^{m} \sum_{j=1}^{n} \sum_{k=1}^{l} a_{ijk} x_i y_j z_k,$$

$$f_y(\omega) = \sum_{i=1}^{m} \sum_{j=1}^{n} \sum_{k=1}^{l} b_{ijk} x_i y_j z_k,$$

$$f_z(\omega) = \sum_{i=1}^{m} \sum_{j=1}^{n} \sum_{k=1}^{l} c_{ijk} x_i y_j z_k.$$

Here, one has (a_{ijk}), (b_{ijk}), (c_{ijk})—the players' 3-dimensional payoff tables (without any loss of generality one can assume that all the entries of those tables are positive real numbers); the vector $\omega^T = (x^T, y^T, z^T)$, $\omega \in \Omega = X \times Y \times Z \subset$ $\subset \mathbf{R}_+^{m+n+l}$. Next, for $p = m, n, l$, we define the vectors $0_p = (0, \ldots, 0)^T \in \mathbf{R}_+^p$, $e_p = (1, \ldots, 1)^T \in \mathbf{R}^p$, as well as $\mathbf{R}_+^p$—the nonnegative orthant of the Euclidean space $\mathbf{R}^p$. The symbol ^{T} denotes the operation of transposition of a vector (matrix).

Following the algorithm in remark 1, we construct the functions (9).

$$g_x(\omega) = f_y(\omega) + f_z(\omega) = \sum_{i=1}^{m} \sum_{j=1}^{n} \sum_{k=1}^{l} (b_{ijk} + c_{ijk}) x_i y_j z_k,$$

$$g_y(\omega) = f_x(\omega) + f_z(\omega) = \sum_{i=1}^{m} \sum_{j=1}^{n} \sum_{k=1}^{l} (a_{ijk} + c_{ijk}) x_i y_j z_k,$$

$$g_z(\omega) = f_x(\omega) + f_y(\omega) = \sum_{i=1}^{m} \sum_{j=1}^{n} \sum_{k=1}^{l} (a_{ijk} + b_{ijk}) x_i y_j z_k.$$

Let us introduce the Nash function $G(\omega) = \delta_x(\omega) + \delta_y(\omega) + \delta_z(\omega)$, where

$$\delta_x(\omega) = \max_{x' \in X} g(x', y, z) - g(\omega),$$
$$\delta_y(\omega) = \max_{y' \in Y} g(x, y', z) - g(\omega),$$
$$\delta_z(\omega) = \max_{z' \in Z} g(x, y, z') - g(\omega).$$

The function $G(\omega)$ is an analogue of the Nash function defined for the bi-matrix games [16]. As the above–defined payoff functions are linear with respect to each variable x, y, z(when the other two variables are fixed), the auxiliary game $\tilde{\Gamma}_3$ is convex, hence the set of Nash points Ω^* is non-empty (but not necessarily convex).

Since $G(\omega) \geq 0$ for all $\omega \in \Omega$, and $G(\omega) = 0$ if, and only if ω is the NE of the game $\tilde{\Gamma}_3$, one can find the Nash equilibrium strategy profile of game $\tilde{\Gamma}_3$ as the global minimum (equalling zero) of the function $G(\omega)$ on Ω.

Now we turn to the approximately numerical method for the construction of WBE in the game Γ_3. In [12] this algorithm (3LP) approximately solving finite non-cooperative three-person games was proposed. The testing results illustrating the efficiency of the mentioned method's application can be found in [13,14].

The 3LP-Method for Solving the Finite 3-Persons Game

We denote $\tilde{a}_{ijk} = b_{ijk} + c_{ijk}$, $\tilde{b}_{ijk} = a_{ijk} + c_{ijk}$, $\tilde{c}_{ijk} = a_{ijk} + b_{ijk}$ and $d_{ijk} = \tilde{a}_{ijk} + \tilde{b}_{ijk} + \tilde{c}_{ijk} = 2(a_{ijk} + b_{ijk} + c_{ijk})$.

The iteration counter is set as $t = 0$. As an starting strategy, one can use any pair of the players' pure strategies (the total number of such pairs is $mn+ml+nl$); for example, fix the pair of strategies $\{y^{(0)}, z^{(0)}\}$ with the components $y_1^{(0)} = 1$, $y_j^{(0)} = 0$ $(j = 2, \ldots, n)$, $z_1^{(0)} = 1$, $z_k^{(0)} = 0$ $(k = 2, \ldots, l)$, and solve successively (for $t = 0, 1, \ldots$) the triple problem $P_x(x^{(t+1)}, y^{(t)}, z^{(t)})$, $P_y(x^{(t+1)}, y^{(t+1)}, z^{(t)})$, $P_z(x^{(t+1)}, y^{(t+1)}, z^{(t+1)})$, where

$$
P_x(x, y', z') : \quad
\begin{aligned}
&\sum_{i=1}^{m} \left(\sum_{j=1}^{n} \sum_{k=1}^{l} d_{ijk} y'_j z'_k \right) x_i - \beta - \gamma \to \max_{x, \beta, \gamma}, \\
&\sum_{i=1}^{m} \left(\sum_{k=1}^{l} \tilde{b}_{ijk} z'_k \right) x_i - \beta \leq 0, \quad j = 1, \ldots, n, \\
&\sum_{i=1}^{m} \left(\sum_{j=1}^{n} \tilde{c}_{ijk} y'_j \right) x_i - \gamma \leq 0, \quad k = 1, \ldots, l, \\
&x^T e_m = 1, \quad x \geq 0_m, \quad \beta, \gamma \in \mathbf{R}^1_+.
\end{aligned}
$$

If x^* is an optimal solution to the problem $P_x(x, y', z')$, then we set $x' := x^*$. Then we solve:

$$
P_y(x', y, z') : \quad
\begin{aligned}
&\sum_{j=1}^{n} \left(\sum_{i=1}^{m} \sum_{k=1}^{l} d_{ijk} x'_i z'_k \right) y_j - \alpha - \gamma \to \max_{y, \alpha, \gamma}, \\
&\sum_{j=1}^{n} \left(\sum_{k=1}^{l} \tilde{a}_{ijk} z'_k \right) y_j - \alpha \leq 0, \quad i = 1, \ldots, m, \\
&\sum_{j=1}^{n} \left(\sum_{i=1}^{m} \tilde{c}_{ijk} x'_i \right) y_j - \gamma \leq 0, \quad k = 1, \ldots, l, \\
&y^T e_n = 1, \quad y \geq 0_n, \quad \alpha, \gamma \in \mathbf{R}^1_+.
\end{aligned}
$$

Again, if y^* is an optimal plan for the above problem $P_y(x', y, z')$, then put $y' := y^*$, and continue solving:

$$
P_z(x', y', z) : \quad
\begin{aligned}
&\sum_{k=1}^{l} \left(\sum_{i=1}^{m} \sum_{j=1}^{n} d_{ijk} x'_i y'_j \right) z_k - \alpha - \beta \to \max_{z, \alpha, \beta}, \\
&\sum_{k=1}^{l} \left(\sum_{i=1}^{m} \tilde{b}_{ijk} x'_k \right) z_k - \alpha \leq 0, \quad j = 1, \ldots, n, \\
&\sum_{k=1}^{l} \left(\sum_{j=1}^{n} \tilde{c}_{ijk} y'_j \right) z_k - \beta \leq 0, \quad i = 1, \ldots, m, \\
&z^T e_l = 1, \quad z \geq 0_l, \quad \alpha, \beta \in \mathbf{R}^1_+.
\end{aligned}
$$

Now that z^* is an optimal solution of the problem $P_z(x', y', z)$, we denote $z' := z^*$.

The optimal objective function values $G_t = G(\omega^{(t+1)})$ are monotone non-increasing by t. The iteration process continues until the value G_t stabilizes, that is, for some t^*, the difference $G_{t^*} - G_{t^*+1}$ becomes small enough. In addition, if $G_{t^*} = 0$, it means that an (exact) Nash point has been found. If the value G_{t^*} is positive but small enough, an approximate solution of the game is reported. Otherwise, a new pair of the initial strategies is selected and the process starts again (probably, having altered the order of the solved problems P_x, P_y, P_z).

Test Results for the 3LP-Algorithms for Finding the WBE

We tested the algorithms for finding the WBE in the finite 3-person games by using the personal computer with the processor Intel(R) Core(TM) i5-3427U

(CPU @ 1.80GHz 2.300 GHz, memory 4.00 GB, 4 cores). The test codes were written in the MatLab. A series of 10 games was solved for each triple n, m, l.

We investigated 2 cases: independent matrices and mutually dependent matrices. In the first case (independent matrices) we used a pseudo-random counters to generate independently the elements of the tables a_{ijk}, b_{ijk}, c_{ijk} $(1 \leq i \leq m, 1 \leq j \leq n, 1 \leq k \leq l)$.

For the game with mutually dependent matrices, we first used pseudo-random counters to generate independently the elements of the auxiliary tables a'_{ijk}, b'_{ijk}, c'_{ijk} $(1 \leq i \leq m, 1 \leq j \leq n, 1 \leq k \leq l)$. At the second stage, we constructed the mutually dependent payoff tables by the formulas

$$a_{ijk} = a'_{ijk} - \lambda \frac{b'_{ijk} + c'_{ijk}}{2} + 1,$$
$$b_{ijk} = b'_{ijk} - \lambda \frac{a'_{ijk} + c'_{ijk}}{2} + 1,$$
$$c_{ijk} = c'_{ijk} - \lambda \frac{a'_{ijk} + b'_{ijk}}{2} + 1$$

for all $1 \leq i \leq m, 1 \leq j \leq n, 1 \leq k \leq l$, where $0 < \lambda \leq \frac{1}{2}$ is a covariance coefficient.

We solved games up to the dimension $dim = m = n = k = 100$. For comparison, using the 3LP-algorithm, we calculated the NE for the same games.

The Table 2 presents the results of the 3LP-algorithm solving the set of test games (5 series with 10 instances in each) with independent matrices. The algorithm switched to the next initial pair of strategies after having made dim iterations.

In Table 2, the following notation is used: $dim = m = n = k$ are the game's sizes (dimension); NE—the number of initial (starting) point when searching for a Nash equilibrium; WBE—the number of start points when searching for a weak Berge equilibrium; tNE—the total amount of time to search a Nash equilibrium for the series of 10 games (sec); $tWBE$—the total amount of time to search a weak Berge equilibrium for the series of 10 games (sec).

Table 2. The results of solving 5 series of games of ten problems with independent matrices

dim	NE	WBE	tNE	tWBE
20	327	85	745.85	129.88
40	230	59	539.28	99.34
60	169	40	404.43	88.1
80	129	28	373.14	92.03
100	159	41	904.85	162.62

In Table 3, for mutually dependent cases, the following notation is also used: $dim = m = n = k$ are the game's sizes (dimension); WBE—the number of start points when searching for a weak Berge equilibrium; $tWBE$—the total amount of time to search the WBE for the series of 10 games (sec); itn - the total number of steps of the 3LP algorithm. The covariance coefficient $\lambda = 0,4$ was used in the calculation of Table 3.

For mutually dependent cases, the results are given only for the WBE, so when calculating the NE for these problems take an unacceptable time or they are not solved at all.

Table 3. The results of solving 5 series of games of ten problems with mutually dependent matrices

dim	itn	WBE	tWBE
20	3173	479	1106.49
40	6532	814	2101.15
60	12826	1415	5134.66
80	10306	1017	5564.54
100	16725	1527	13049.09

It is easy to notice from the reported results (see Table 2 and Table 3), the reciprocal dependence of the payoff matrices affect much to solve a problem by the 3LP-algorithm. The reciprocal dependence sufficiently increases the complexity of problems.

It is also clear that, the search for the WBE is much faster than the search for the NE. This is most likely due to the pure weak Berge equilibrium strategy profile existing more often than the pure Nash equilibrium strategy profile.

4 Conclusion

In this paper, we formalize the conception of the WBE. The WBE follows the Hippocratic oath "First do no harm!" In contrast to the NE, the WBE always exists for every finite N-person game. As an example, we find the WBE in the finite 3-person games using the 3LP-algorithm. In the future, the authors plan to transfer the proposed numerical algorithm for finding WBE to finite games of a larger $(N = 4, 5)$ number of persons.

References

1. Nash, J.: Equilibrium points in N-person games. Proc. Nat. Acad. Sci. USA **36**, 48–49 (1950)
2. Berge, C.: Théorie générale des jeux a n personnes. Gauthier-Villar, Paris (1957)
3. Shubik, M.: Review of C. Berge, General theory of n-person games. Econometrica **29**(4), 821 (1961)
4. Zhukovskiy, V.I.: Some problems of non-antagonistic differential games. In: Mathematical Methods in Operations Research, Institute of Mathematics with Union of Bulgarian Mathematicians, Rousse, pp. 103–195 (1985)

5. Zhukovskii, V.I., Chikrii, A.A.: Linear-Quadratic Differential Games. Naukova Dumka, Kiev (1994). (in Russian)
6. Vaisman, K.S.: The Berge equilibrium for linear-quadratic differential game. In: Multiple Criteria Problems Under Uncertainty: Abstracts of the Third International Workshop, Orekhovo-Zuevo, Russia, p. 96 (1994)
7. Vaisman, K.S.: The Berge equilibrium. In: Abstract of Cand. Sci. (Phys. Math.) Dissertation St. Petersburg (1995). (in Russian)
8. Zhukovskiy, V.I., Kudryavtsev, K.N.: Mathematical foundations of the Golden Rule. I. Static case. Autom. Remote Control **78**(10), 1920–1940 (2017). https://doi.org/10.1134/S0005117917100149
9. Larbani, M., Zhukovskii, V.I.: Berge equilibrium in normal form static games: a literature review. Izv. IMI UdGU **49**, 80–110 (2017). https://doi.org/10.20537/2226-3594-2017-49-04
10. Kudryavtsev, K., Ukhobotov, V., Zhukovskiy, V.: The Berge equilibrium in Cournot oligopoly model. In: Evtushenko, Y., Jaćimović, M., Khachay, M., Kochetov, Y., Malkova, V., Posypkin, M. (eds.) OPTIMA 2018. CCIS, vol. 974, pp. 415–426. Springer, Cham (2019). https://doi.org/10.1007/978-3-030-10934-9_29
11. Pykacz, J., Bytner, P., Frackiewicz, P.: Example of a finite game with no Berge equilibria at all. Games **10**(1), 7 (2019). https://doi.org/10.3390/g10010007
12. Golshtein, E.: A numerical method for solving finite three-person games. Economica i Matematicheskie Metody **50**(1), 110–116 (2014). (in Russian)
13. Golshtein, E., Malkov, U., Sokolov, N.: Efficiency of an approximate algorithm to solve finite three-person games (a computational experience). Economica i Matematicheskie Metody **53**(1), 94–107 (2017). (in Russian)
14. Golshteyn, E., Malkov, U., Sokolov, N.: The Lemke-Howson algorithm solving finite non-cooperative three-person games in a special setting. In: 2018 IX International Conference on Optimization and Applications (OPTIMA 2018) (Supplementary Volume). DEStech Transactions on Computer Science and Engineering (2018). https://doi.org/10.12783/dtcse/optim2018/27938
15. Sugden, R.: Team reasoning and intentional cooperation for mutual benefit. J. Soc. Ontol. **1**(1), 143–166 (2015). https://doi.org/10.1515/jso-2014-0006
16. Mills, H.: Equillibrium points in finite games. J. Soc. Ind. Appl. Math. **8**(2), 397–402 (1960)

On Contractual Approach in Competitive Economies with Constrained Coalitional Structures

Valeriy Marakulin[1,2]([⊠])([iD])

[1] Sobolev Institute of Mathematics, Russian Academy of Sciences,
4 Acad. Koptyug Avenue, Novosibirsk 630090, Russia
marakulv@gmail.com
[2] Novosibirsk State University, 2 Pirogova Street, Novosibirsk 630090, Russia
http://www.math.nsc.ru/~mathecon/marakENG.html

Abstract. We establish a theorem that equilibria in an exchange economy can be described as allocations that are stable under the possibilities: (i) agents can partially and asymmetrically break current contracts, after that (ii) a new mutually beneficial contract can be concluded in a coalition of a size not more than 1 plus the maximum number of products that are not indifferent to the coalition members.

The presented result generalizes previous ones on a Pareto improvement in an exchange economy with l commodities that requires the active participation of no more than $l + 1$ traders. This concerned with Pareto optimal allocations, but we also describe equilibria. Thus according to the contractual approach to arrive at equilibrium only coalitions of constrained size can be applied that essentially raise the confidence of contractual modeling.

Keywords: Contractual economies · Coalitions of constrained size · Competitive equilibrium · Fuzzy contractual allocations

1 Introduction

I started to develop the theory of formal contractual economic interaction in the early 2000s and began to apply elaborated methods to the models of different types: Arrow–Debreu economies, incomplete markets, an economy with public goods, etc., see [1–4]. In the course of this activity, several specific characterizations of economic equilibria of different types were developed, but in all of them, the key feature was the admission of contract breakings—complete or partial. The idea of the barter exchange (contract) is by no means new in theoretical economics and seemingly goes back to classical Edgeworth results, but it usually appeared as an interpretation, in the form of net trades in a formal model. In the

The study was carried out within the framework of the state contract of the Sobolev Institute of Mathematics (project no. 0314-2019-0018).

Y. Kochetov et al. (Eds.): MOTOR 2020, CCIS 1275, pp. 244–255, 2020.
https://doi.org/10.1007/978-3-030-58657-7_21

simplest version of the pure exchange economy, a barter contract is represented as a vector of acceptable exchanges of commodities among economic agents. A partial break involves the execution of the contract in an incomplete volume. Besides, in [1] there was proposed the notion of fuzzy contractual allocation and it became clear soon that this is the most meaningful concept among other methods of the contractual interaction. Fuzzy contractual interaction means that agents are able to break contracts partially and asymmetrically, i.e., it is admitted different agents can break contracts in a different amount. There was stated that under very weak assumptions in convex economy equilibria coincide with fuzzy contractual allocations. Nevertheless, the achieved results still are not satisfactory from the modeling point of view, because they assume the existence of agreements in many unrealistic coalitions between agents living at great distances, etc. This paper aims to fill this gap.

In this paper we consider a possibility to restrict the number of participants in the exchange transactions. We show that certain constrains of this type can be used without prejudice to its equilibrium properties of the final allocation. The idea goes back to [5–8], where it was found that Pareto optimal allocation can be achieved via mutually beneficial exchanges carried out in coalitions limited by the dimension of the commodity space, see also [9,10]. In these works, the contractual approach itself was not developed and the possibilities of individuals to break contracts were not considered. As a result, the obtained characterization does not appeal to Walrasian equilibria. Doing the admission of partially breaking of current contracts, we also take into account the fact that an agent may not be interested in absolutely all existing products. We show that the analysis can be reduced to an effective products' area of lower dimensionality—by eliminating products that are not of interest to the contracting parties. As a result, a coalition has a specific product space which dimension can be applied to restrict the size of coalitions. We will see that such restrictions on the size of coalitions do not prevent so-called fuzzy contractual allocations to be Walrasian equilibria.

The paper is organized as follows. In the second section, I present a contractual economic model and formulate some preliminary results that are the basis for the subsequent considerations. In the third one, I present the main result: new theorems on characterization of equilibria and other contractual allocations implemented via contracts of limited number of participants.

2   Contractual Exchange Economy

We consider a typical exchange economy in which $L = \mathbb{R}^l$ denotes the (finite dimensional) *space of commodities* (l is a number of commodities). Let $\mathcal{I} = \{1, \ldots, n\}$ be a set of agents (traders or consumers). A consumer $i \in \mathcal{I}$ is characterized by a consumption set $X_i \subset L$, an initial endowment $\mathbf{e}_i \in L$, and a preference relation described by a point-to-set mapping $\mathcal{P}_i : X_i \Rightarrow X_i$ where $\mathcal{P}_i(x_i)$ denotes the set of all consumption bundles strictly preferred by the i-th agent to the bundle x_i. The notation $y_i \succ_i x_i$ is equivalent to $y_i \in \mathcal{P}_i(x_i)$.

So, the pure exchange model may be represented as a triplet

$$\mathcal{E} = \langle \mathcal{I}, L, (X_i, \mathcal{P}_i, \mathbf{e}_i)_{i \in \mathcal{I}} \rangle.$$

Let us denote by $\mathbf{e} = (\mathbf{e}_i)_{i \in \mathcal{I}}$ the vector of initial endowments of all traders of the economy. Denote $X = \prod_{i \in \mathcal{I}} X_i$ and let

$$\mathcal{A}(X) = \{ x \in X \mid \sum_{i \in \mathcal{I}} x_i = \sum_{i \in \mathcal{I}} \mathbf{e}_i \}$$

be the set of all *feasible allocations*. Everywhere below we assume that the model under study satisfies the following assumption.

(**A**) *For each $i \in \mathcal{I}$, X_i is a convex solid[1] closed set, $\mathbf{e}_i \in X_i$ and for every $x_i \in X_i$ there exists an open convex $G_i \subset L$ such that $\mathcal{P}_i(x_i) = G_i \cap X_i$ and if $\mathcal{P}_i(x_i) \neq \emptyset$ then $x_i \in \overline{\mathcal{P}}_i(x_i) \setminus \mathcal{P}_i(x_i)$.*[2]

Notice that due to (**A**) preferences may be satiated, i.e., $\mathcal{P}_i(x_i) = \emptyset$ is possible for some agent i and $x_i \in X_i$. However if $\mathcal{P}_i(x_i) \neq \emptyset$, then preference is *locally non-satiated* at the point x_i and this implies $\lambda(\mathcal{P}_i(x_i) - x_i) \subseteq \mathcal{P}_i(x_i) - x_i$ $\forall \lambda \in (0, 1]$. Next I recall some standard definitions and notions.

A pair (x, p) is said to be a *quasi-equilibrium* of $\mathcal{E}$ if $x \in \mathcal{A}(X)$ and there exists a linear functional $p \neq 0$ onto L such that

$$\langle p, \mathcal{P}_i(x_i) \rangle \geq px_i = p\mathbf{e}_i, \quad \forall i \in \mathcal{I}.$$

A quasi-equilibrium such that $x_i' \in \mathcal{P}_i(x_i)$ actually implies $px_i' > px_i$ is a *Walrasian or competitive equilibrium*.

An allocation $x \in \mathcal{A}(X)$ is said to be dominated (blocked) by a nonempty coalition $S \subseteq \mathcal{I}$ if there exists $y^S \in \prod_{i \in S} X_i$ such that $\sum_{i \in S} y_i^S = \sum_{i \in S} \mathbf{e}_i$ and $y_i^S \in \mathcal{P}_i(x_i)$ $\forall i \in S$.

The *core* of $\mathcal{E}$, denoted by $\mathcal{C}(\mathcal{E})$, is the set of all $x \in \mathcal{A}(X)$ that are blocked by no (nonempty) coalition.

Weak Pareto boundary for $\mathcal{E}$, denoted by $\mathcal{PB}^w(\mathcal{E})$, is the set of all $x \in \mathcal{A}(X)$ that cannot be dominated by the coalition $\mathcal{I}$ of all agents.

An allocation $x \in \mathcal{A}(X)$ is called *individual rational* if it cannot be dominated by singleton coalitions. $\mathcal{IR}(\mathcal{E})$ denotes the set of all these allocations.

Let $\mathfrak{L} = L^{\mathcal{I}}$ denote the space of all allocations of the economy $\mathcal{E}$. In the framework of model $\mathcal{E}$, we are going to introduce and study a formal mechanism of contractual interaction. This mechanism reflects the idea that any group of agents can find and realize some (permissible) within-the-group exchanges of commodities, referred to as contracts. The mechanism defines rules of contracting.

---

[1] Here "solid" is equivalent to "having nonempty interior."

[2] The symbol $\overline{A}$ denotes the closure of A and $\setminus$ is set for the set-theoretical difference.

By the formal definition, any reallocation of commodities $v = (v_i)_{i \in \mathcal{I}} \in \mathfrak{L}$, i.e., any vector $v \in \mathfrak{L}$ satisfying $\sum_{i \in \mathcal{I}} v_i = 0$, is called a *contract*.

Not every kind of possible reallocation may be realized in the economy; there are some institutional, physical, and behavioral restrictions in the economic models of different types. This is why we equip the abstract contractual economy model with a new element, the set of *permissible* contracts $\mathcal{W} \subset \mathfrak{L}$. Thus, the contractual (exchange) economy under study may be shortly represented by the 4-tuple

$$\mathcal{E}^c = \langle \mathcal{I}, L, \mathcal{W}, (X_i, \mathcal{P}_i, \mathbf{e}_i)_{i \in \mathcal{I}} \rangle.$$

For a contractual economy we study the sets of contracts which represent *feasible* allocations and introduce the operation of breaking a part of a given set of contracts. This motivates the following definition.

A finite collection V of permissible contracts is called *a web of contracts* iff

$$x_{\mathbf{e}}(U) = \mathbf{e} + \sum_{v \in U} v \in X, \quad \forall U \subseteq V.$$

So V being a web means that $\forall U \subseteq V$ its generated allocation $x_{\mathbf{e}}(U)$ is feasible one. Clearly, this notion can be considered with respect to any another allocation $y \in \mathcal{A}(X)$ chosen instead of $\mathbf{e}$. Note that $V = \emptyset$ is a web relative to every $y \in \mathcal{A}(X)$ (by convention $\sum_{v \in \emptyset} v = 0$).

Now we introduce the breaking operation of existing contracts and the signing of new ones. For any contract $v \in V$, let us set

$$S(v) = \operatorname{supp}(v) = \{i \in \mathcal{I} \mid v_i \neq 0\},$$

the support of the contract v. It is assumed that any contract $v \in V$ may be *broken* by any trader in $S(v)$, since he/she may not keep his/her contractual obligations. Also a non-empty group (coalition) of consumers can *sign* any number of new contracts. Being applied jointly, i.e., as a simultaneous procedure, these operations allow coalition $T \subseteq \mathcal{I}$ to yield new webs of contracts. The set of all such webs is denoted by $F(V,T)$.

Notice also that due to the definition of a web of contracts, a coalition can break any subset of contracts of a given web.[3]

Further, for the webs of contracts the notion of domination via a coalition is introduced that allows to consider different forms of web stabilities. This property, being written as $U \underset{T}{\succ} V$ (U dominates V via coalition T), means that

(i) $U \in F(V,T)$,
(ii) $x_i(U) \underset{i}{\succ} x_i(V)$ for all $i \in T$.

Definition 1. *A web of contracts V is called stable if there is no web U and no coalition $T \subseteq \mathcal{I}$, $T \neq \emptyset$ such that $U \underset{T}{\succ} V$.*

*An allocation x is called **contractual** if $x = x(V)$ for a stable web V.*

---

[3] Otherwise, it would occur that an allocation realized via breaking contracts is not feasible.

The property that a web of contracts is stable may be relaxed as well as strengthened. The most important possibilities are described below.

Definition 2. *A web of contracts V is called:*

(i) **lower** *stable if there is no web U and no coalition $T \subseteq \mathcal{I}$, $T \neq \emptyset$ such that* $U \underset{T}{\succ} V$ *and* $U \subset V$;

(ii) **upper** *stable if there is no web U and no coalition $T \subseteq \mathcal{I}$, $T \neq \emptyset$ such that* $U \underset{T}{\succ} V$ *and* $V \subset U$.

(iii) *An upper and lower stable web of contracts V is called* **weakly** *stable.*

An allocation x is called lower, upper, or weakly contractual if $x = x(V)$ for some lower, upper, or weakly stable web V, respectively.

The next possibility to strengthen contractual stability is to allow agents to break contracts partially. Partial breaking of the contract $v = (v_i)_{i \in \mathcal{I}}$ in the amount of $\alpha \in [0, 1]$ means that contract v is replaced by the contract $(1 - \alpha)v$. System (web) of contracts is called *proper* if no one is interested in the partial break off contracts: for each agent partial break (potentially different for different contracts) does not lead to the increase of utility. Only the proper web of contracts can be long-lived. Clearly, to admit agents apply partial breaking we have to assume the set $\mathcal{W}$ is a *star-shaped* at zero in $\mathfrak{L}$, i.e.,

$$v \in \mathcal{W} \;\Rightarrow\; \lambda v \in \mathcal{W}, \quad \forall 0 \leq \lambda \leq 1.$$

Allocation $x(V) = \mathbf{e} + \sum_{v \in V} v$, implemented by the web of contracts V is called *properly contractual* if the partial breaking of contracts is allowed to dominate and V is proper one.

One more notion is quite important in our analysis, it is the concept of *fuzzy contractual* allocation. To present it in a simplest way let us assume that the web consists on the only contract, i.e., $V = \{v\}$. So one has a feasible allocation to which the gross contract $x - \mathbf{e} = v = (v_i)_{i \in \mathcal{I}}$ (net trade) corresponds. It is assumed that the agents of the economy can (fuzzy and asymmetrically) break contract $v = (v_i)_{i \in \mathcal{I}}$, decreasing the individual consumption (fragment) from this contract in shares $(1 - t_i)_{i \in \mathcal{I}}$, $t_i \in [0, 1]$ forming a tuple[4]

$$v^t = (t_1 v_1, t_2 v_2, \ldots, t_n v_n)$$

of commodity bundles, which can be used in subsequent exchange transactions together with the initial endowments. After the conclusion of a new contract $w^S = (w_i)_{\mathcal{I}} \in L^{\mathcal{I}}$, $\sum_{\mathcal{I}} w_i = 0$ by a coalition $S \subseteq \mathcal{I}$ ($i \notin S \Rightarrow w_i = 0$) they yield (possibly unfeasible!) "allocation"

$$\xi(t, v, w) = w + v^t + \mathbf{e} = (w_1 + t_1 v_1^t + \mathbf{e}_1, \ldots, w_n + t_n v_n^t + \mathbf{e}_n).$$

---

[4] This is not a contract, because its key property $\sum_{\mathcal{I}} t_i v_i = 0$ is violated.

Definition 3. *An allocation $x \in \mathcal{A}(X)$ is called **fuzzy contractual** if for every $t = (t_i)_{i \in \mathcal{I}}$, $0 \le t_i \le 1$, $\forall i \in \mathcal{I}$ and for $x - \mathbf{e} = v$ there is no barter contract $w = (w_1, \ldots, w_n) \in L^{\mathcal{I}}$, $\sum_{\mathcal{I}} w_i = 0$, such that*

$$\xi_i = \xi_i(t, v, w) = w_i + t_i v_i + \mathbf{e}_i, \quad i \in \mathcal{I} \tag{1}$$

$$\xi_i \succ_i x_i \quad \forall i : \ \xi_i \ne x_i. \tag{2}$$

Note that by virtue of (2) $w = 0$ is permissible, i.e. only partial breaking of contracts is possible. Denying the possibility of such domination means that the web of contracts is *proper* and the allocation is *stable* with respect to asymmetric *partial break* of contracts.

Depending on the structure of permissible contracts, specified as a new element $\mathcal{W} \subset L^{\mathcal{I}}$ of the model, one can describe well known economic theoretical notions in terms of a stable web of contracts. In a standard exchange model (every contract is permissible) they are the core (contractual allocations, only full break off contracts), competitive equilibria (admission of partial break), the Pareto frontier (upper contractual allocations), etc. The most interesting among others is the presentation of competitive equilibrium as a fuzzy contractual allocation, described in the following technical lemma and proposition.

The following characteristic lemma can be directly produced from Definition 3.

Lemma 1. *Suppose $\mathcal{W} = L^{\mathcal{I}}$. Then an allocation $x \in \mathcal{A}(X)$ is fuzzy contractual if and only if[5]*

$$\mathcal{P}_i(x_i) \cap [x_i, \mathbf{e}_i] = \emptyset \quad \forall i \in \mathcal{I} \tag{3}$$

and

$$\prod_{\mathcal{I}} [(\mathcal{P}_i(x_i) + [0, \mathbf{e}_i - x_i]) \cup \{\mathbf{e}_i\}] \bigcap \{(z_i)_{\mathcal{I}} \in L^{\mathcal{I}} \mid \sum_{i \in \mathcal{I}} z_i = \sum_{i \in \mathcal{I}} \mathbf{e}_i\} = \{\mathbf{e}\}. \tag{4}$$

Here condition (3) indicates that a partial break off contracts without signing of a new one cannot be beneficial. The requirement (4) denies the existence of a dominating coalition after the partial asymmetric break of the contract $v = (x - \mathbf{e})$. Now applying separation theorem one can easily state (see [1] for details) the following

Proposition 1. *Every equilibrium is a fuzzy contractual allocation and vice versa: any non-satiated fuzzy contractual allocation is a nontrivial quasi-equilibrium.*

So, if the model is such that every nontrivial quasi-equilibrium is an equilibrium[6] then the notion of competitive equilibrium and fuzzy contractual allocation is equivalent. This and similar statements from [1–4] allow us to state that our

---

[5] *A linear segment with ends $a, b \in L$ is the set $[a, b] = \text{conv}\{a, b\} = \{\lambda a + (1 - \lambda)b \mid 0 \le \lambda \le 1\}$.*

[6] Conditions, providing this fact are well known in the literature, e.g. it can be *irreducibility*.

contractual approach presents a model of perfect competition (simplest among others).

The sketch of the proof of Proposition 1. Separating sets in (4) by a (non-zero) linear functional $\pi = (p_1, \ldots, p_n) \in L^{\mathcal{I}}$ one can conclude:

(i) $p_i = p_j = p \neq 0$ for each $i, j \in \mathcal{I}$; this is so because π is bounded on

$$\mathcal{A}(L^{\mathcal{I}}) = \{(z_1, \ldots, z_n) \in L^{\mathcal{I}} \mid \sum_{i \in \mathcal{I}} z_i = \sum_{i \in \mathcal{I}} \mathbf{e}_i\}.$$

So, one can take p as a price vector.

(ii) Due to construction and in view of preferences are locally non-satiated at the point $x \in \mathcal{A}(X)$ the points x_i and $\mathbf{e}_i$ belong to the closure of

$$\mathcal{P}_i(x) + \text{conv}\{0, \mathbf{e}_i - x_i\}.$$

Therefore via separating property we have

$$\sum_{j \neq i} p\mathbf{e}_j + px_i \geq \sum_{\mathcal{I}} p\mathbf{e}_j \quad \Rightarrow \quad px_i \geq p\mathbf{e}_i \quad \forall i \in \mathcal{I},$$

that is possible only if $px_i = p\mathbf{e}_i \ \forall i \in \mathcal{I}$. So, we obtain budget constrains for consumption bundles.

(iii) By separation property for each i we also have

$$\langle p, \mathcal{P}_i(x) + \text{conv}\{0, \mathbf{e}_i - x_i\}\rangle \geq p\mathbf{e}_i,$$

that by (ii) implies $\langle p, \mathcal{P}_i(x)\rangle \geq px_i = p\mathbf{e}_i$. So we proved that p is quasi-equilibrium prices for allocation $x = (x_i)_{i \in \mathcal{I}}$.

As a result one can see that if an economic model is such that every quasi-equilibrium is equilibrium, then fuzzy contractual allocation is an equilibrium one. Conditions delivering this fact are well known in literature; for example, it is the case when an economy is irreducible. ∎

3   Result

In a real economy, consumers may not be interested in all existing products, i.e., individuals may be indifferent to some products[7]. Excluding them from consideration, one can reduce the dimension of the actual product space for each agent. The exact definition is given below.

Definition 4. *A commodity j is **indifferent** for $i \in \mathcal{I}$ if $\forall x \in \mathcal{A}(X)$[8]*

$$\forall y_i = ((y_i)_{-j}, y_i^j) \in \mathcal{P}_i(x_i) \iff ((y_i)_{-j}, \mathbf{e}_i^j) \in \mathcal{P}_i((x_i)_{-j}, \mathbf{e}_i^j) = \mathcal{P}_i(x_i).$$

---

[7] For example, an ordinary consumer on the market is not interested in all kinds of spare parts, parts and structural elements (bolts, nuts, gears, transistors ...).

[8] Here we indirectly assume that all bundles we need belong to consumption set, i.e., $((y_i)_{-j}, \mathbf{e}_i^j), ((x_i)_{-j}, \mathbf{e}_i^j) \in X_i$; it is a specific constraint for $X_i, i \in \mathcal{I}$.

Here $y_i = (y_i^j)_{j=1,\ldots,l} \in \mathbb{R}^l$ and $(y_i)_{-j} = (y_i^k)_{k \neq j, k=1,\ldots,l}$ is a vector consisting of all components of y_i excluding y_i^j.

Two properties are postulated in this definition: a product j is indifferent to a given individual i, if in any consumption bundle $y_i = ((y_i)_{-j}, y_i^j) \in X_i$ his/her consumption can be replaced by the initial one (to nullify?), i.e., one goes to a bundle $((y_i)_{-j}, \mathbf{e}_i^j)$ such that $((y_i)_{-j}, \mathbf{e}_i^j) \in X_i$ and this does not lead to the change of consumption properties of $y_i \in L$. Clearly, for preferences specified via utility functions for indifferent commodity j we have $\forall y_i \in X_i \; u_i((y_i)_{-j}, y_i^j) = u_i((y_i)_{-j}, \mathbf{e}_i^j)$, $i \in \mathcal{I}$.

Let $L_i \subseteq L$ be the space of *non-indifferent* commodities (interesting) for individual i and let $L_S \subseteq L$ be a subspace of commodities that are interesting for the members of coalition $S \subseteq \mathcal{I}$:

$$L_S = \sum_{i \in S} L_i.$$

In this section, the notation z^S means the projection of the vector $z \in L$ onto the subspace $L_S \subseteq L$. Recall that for contracts $v \in \mathcal{W}$ there is defined $S(v) = \text{supp}(v)$, this is the support of the contract. Given the possible indifference to some products, as a product space for a coalition $S(v)$, one can specify

$$L_{S(v)} = \sum_{i \in S(v)} L_i.$$

Now let us consider the following restriction for the set of all permissible contracts.

$$v = (v_i)_{i \in \mathcal{I}} \in \mathcal{W} \iff v_i \in L_{S(v)}, \; i \in S(v), \; |S(v)| \leq \dim(L_{S(v)}) + 1. \quad (5)$$

This specification restricts the size of permissible contracting coalitions.

Remark 1. In the process of manufacturing high-tech products, a huge number of elements are used, the range of which can be counted in millions—for example, in modern aircraft construction. However, the final user needs the resulting product (the plane!), and not some of its components, bolts, nuts, ailerons, and other structural elements, the existence of which he may not know at all. However, this is important for service companies, etc. Production unions enter into contracts for the supply of the element base of the final product can be very large, but consumer unions can be much smaller—this fact can be concluded from Theorem 1 and Corollary 1 below. Formal examples also can be easily constructed. Indeed one can consider several exchange economies $\mathcal{E}_1, \ldots, \mathcal{E}_k$ having product ranges $S_1, \ldots, S_k \subseteq \{1, 2, \ldots, l\}$. Assume that utilities of individuals from $\mathcal{E}_\xi$ may depend of only commodities from S_ξ. One can consider extended commodity space $L = \mathbb{R}^l$ and formally extend these utilities to this space, supposing that they do not depend of new variables. Now we consider the united economy $\mathcal{E} = \bigcup_{\xi=1}^{k} \mathcal{E}_\xi$. The first result below describes Pareto frontier and says us that if coalition contains only individuals of one economy $\mathcal{E}_\xi$, then number of coalition members may be restricted by $card(S_\xi)$; for coalition of two individual kinds from $\mathcal{E}_{\xi_1}, \mathcal{E}_{\xi_2}$ the number of its members may be not more of $card(S_{\xi_1} \cup S_{\xi_2})$ and so on. Similar conclusion is done for equilibria.   ∎

Further we first discuss the concept of upper stable web and upper contractual allocation, see Definition 2. Now let us consider a slightly modified classical concept of Pareto optimality[9]. We call an allocation $x \in \mathcal{A}(X)$ *strictly* Pareto optimal iff

$$\nexists S \subseteq \mathcal{I} \ \& \ y^S \in \prod_{i \in S} X_i \ \text{ such that } \ \sum_{i \in S} y_i^S = \sum_{i \in S} x_i \ \& \ y_i^S \in \mathcal{P}_i(x_i) \ \forall i \in S.$$

It is easy to see, that according to the definitions if there are no permissibility constrains for contracts, the notions of upper contractual and *strictly* Pareto optimal allocation are equivalent.

It is said that a vector (consumption bundle) $\kappa \in L$ is *extremely desirable* if for each $x_i \in X_i$ one has

$$x_i + \kappa \succ_i x_i, \ \ i \in \mathcal{I}.$$

In the literature, it is standardly assumed that cumulative initial endowments $\sum_{i \in \mathcal{I}} \mathbf{e}_i = \bar{\mathbf{e}}$ presents an extremely desirable bundle.

Recall that binary relation $\succ$ is transitive iff

$$\forall x, y, z \in \mathrm{Dom}(\succ) \ \ x \succ y \succ z \ \Rightarrow \ x \succ z.$$

Theorem 1. *If $\mathcal{W}$ obeys (5) then every upper contractual allocation is strictly Pareto optimal. Moreover, if preferences of $\mathcal{E}$ are transitive and there is an extremely desirable bundle $\kappa \in L$, then (5) can be weakened and one can require*

$$v \in \mathcal{W} \ \ \Longleftrightarrow \ \ v_i \in L_{S(v)}, \ i \in S(v) \ \& \ |S(v)| \le \dim(L_{S(v)}).$$

So, the Theorem states that the economic system can arrive at Pareto optimal commodity allocation via a contractual process with coalitions size constrained by (5). In further analysis, we apply the following

Theorem 2 (Carathéodory, 1907). *Let $A \subset L$ be a subset of a vector space L. If $\dim \mathrm{aff}(A) = d < \infty$, then any element $x \in \mathrm{conv} A$ can be presented as a convex hull of not more than $d + 1$ elements of A.*

Proof of Theorem 1. Suppose that an upper contractual allocation $x \in \mathcal{A}(X)$ is not strictly Pareto optimal. Therefore, there exists a coalition $S \subset \mathcal{I}$ and contract $v = (v_i)_{i \in \mathcal{I}} \in \mathcal{L} = L^{\mathcal{I}}$, $\mathrm{supp}(v) = S$ such that

$$\forall i \in S \ \ x_i + v_i \in \mathcal{P}_i(x). \tag{6}$$

---

[9] Under classical assumptions they are equivalent, but it is not so in general case.

Here by Definition 4 for each member of the coalition S, the components of v_i corresponding to indifferent products can be considered as zero, i.e., $v_i \in L_{S(v)}$ $\forall i \in \mathcal{I}$. Since x is upper contractual, then $v \notin \mathcal{W}$ (here $v_i = 0$, $i \in \mathcal{I} \setminus S$) and, therefore, $|S(v)| > \dim(L_{S(v)}) + 1$. Now we can assume that S is a coalition of minimal size among those having this property. We have $\frac{1}{|S|} \sum_{i \in S} v_i = 0$, $S = S(v)$. Using the Caratheodory theorem, one can find a coalition $T \subset S$ such that

$$\forall i \in T \ \exists \alpha_i \in (0,1]: \quad \sum_{i \in T} \alpha_i = 1, \quad \sum_{i \in T} \alpha_i v_i = 0 \ \& \ |T| \leq \dim(L_S) + 1.$$

Define $w_i = \alpha_i v_i \neq 0$, and think without loss of generality that $w_i \in L_T$, $i \in T$ (if necessary, one replaces some components with zeros). Now due to the main assumption $(\mathbf{A})$ one has $\lambda(\mathcal{P}_i(x_i) - x_i) \subseteq \mathcal{P}_i(x_i) - x_i \ \forall \lambda \in (0,1]$, that implies $x_i + w_i \in \mathcal{P}_i(x)$, $i \in T$. Since $\sum_{i \in T} w_i = 0$ and $|T| < |S|$, we come to a contradiction with the choice of S as a coalition of minimal size. Therefore, there are no such coalitions at all and x is a strictly Pareto optimal allocation.

In the second part of the statement of the Theorem, we again argue from the contrary and find a coalition $S \subset \mathcal{I}$ of minimal size and a contract $v \in \mathfrak{L}$, $\operatorname{supp}(v) = S$, $v_i \in L_S$, $i \in \mathcal{I}$ satisfying (6) and such that $|S| > \dim L_S$. Let us specify

$$\Gamma = \operatorname{conv}\{v_i \in L_S \mid i \in S\}.$$

By construction one has $\frac{1}{|S|} \sum_{i \in S} v_i = 0 \in \Gamma$. Next, we take an extremely desirable $\kappa \in L$, consider its projection κ^S onto L_S and find a real $\lambda \geq 0$ such that $-\lambda \kappa^S$ belongs to the face of (bounded) polyhedron Γ. This can be done from the condition

$$\lambda = \max\{\lambda' \mid -\lambda' \kappa^S \in \Gamma\}.$$

Since the dimension of any proper face is at most $\dim L_S - 1$, there is a coalition $T \subset S$ such that $|T| \leq \dim L_S$ and

$$\forall i \in T \ \exists \alpha_i \in (0,1]: \quad \sum_{i \in T} \alpha_i = 1, \quad \sum_{i \in T} \alpha_i v_i = -\lambda \kappa^S.$$

Next one defines $w_i = \alpha_i(v_i + \lambda \kappa^S)$, $i \in T$ and $w_i = 0$, $i \in \mathcal{I} \setminus T$. As a result one has:

$$x_i \prec_i x_i + \alpha_i v_i \prec_i x_i + \alpha_i v_i + \alpha_i \lambda \kappa^S = x_i + w_i, \quad i \in T,$$

$$\sum_{i \in T} w_i^S = \sum_{i \in T} \alpha_i v_i^S + (\sum_{i \in T} \alpha_i) \lambda \kappa^S = 0.$$

These relations indicate that $w = (w_i)_{i \in \mathcal{I}}$ is a mutually beneficial contract, the support of which is the coalition T, no larger than $\dim(L_S)$. Thus, we again have found the coalition that dominates the current allocation, and its size is strictly less than $|S|$, which is impossible. $\blacksquare$

Let us turn now to the characterization of fuzzy contractual allocation, which represents the main result of the section.

Lemma 2. *Let x be a fuzzy contractual allocation and $\mathcal{W}$ obey (5). Then (4) is true:*

$$\prod_{i \in \mathcal{I}} [(\mathcal{P}_i(x_i) + \mathrm{co}\{0, \mathbf{e}_i - x_i\}) \cup \{\mathbf{e}_i\}] \bigcap \mathcal{A}(L^{\mathcal{I}}) = \{\mathbf{e}\}.$$

Now by virtue of the characterization presented in Proposition 1 we directly conclude

Corollary 1. *Let $\mathcal{W}$ obey (5). Then every non-satiated fuzzy contractual allocation is a quasi-equilibrium one.*

So, these Lemma and Corollary state that applying partial break and contracts specified in (5), a contractual process can arrive the economy to Walrasian equilibrium.

Proof of Lemma 2. Let x be a fuzzy contractual allocation, $\mathcal{W}$ obey (5) and conclusion of the Lemma be false. Let us consider the left part of intersection (4). Now we first show that there is no $y = (y_i)_{\mathcal{I}} \neq \mathbf{e}$ such that the coalition

$$T(y) = \{i \in \mathcal{I} \mid y_i \neq \mathbf{e}_i\} \neq \emptyset \tag{7}$$

satisfies $|T(y)| \leq \dim(L_T) + 1$. Indeed, otherwise according to the construction one can find $z_i \in \mathcal{P}_i(x_i)$, $\alpha_i \in [0,1]$, $i \in T$ such that

$$y_i = z_i + \alpha_i(\mathbf{e}_i - x_i) \neq \mathbf{e}_i, \quad i \in T, \quad \sum_T y_i = \sum_T \mathbf{e}_i.$$

Applying now Definition 4, we may think that $z_i, x_i \in X_i \cap (L_T + \mathbf{e}_i)$ (for i and the bundle x_i one has to change indifferent components with his/her initial endowments and do not change all other). Now, specifying $v_i = (y_i - \mathbf{e}_i) \in L_T$, $i \in \mathcal{I}$, via construction and Definition 4 we obtain

$$z_i = v_i + \alpha_i(x_i - \mathbf{e}_i) + \mathbf{e}_i \succ_i x_i, \quad i \in T, \quad \sum_{i \in \mathcal{I}} v_i = 0 \ \& \ \mathrm{supp}\,(v) = T,$$

that contradicts Definition 3 and condition (5).

Thus, if the conclusion of the Lemma is false, then $|T(y)| > \dim(L_T) + 1$ for each coalition specified by (7). But in the (finite) set of all such coalitions there is a coalition of minimal size, which we denote $S \subset \mathcal{I}$. Again, one can think $x_i \in X_i \cap (L_S + \mathbf{e}_i)$, $i \in S$. By construction there are $z_i \in \mathcal{P}_i(x_i) \cap (L_S + \mathbf{e}_i)$, $\alpha_i \in [0,1]$, $i \in S$ such that

$$y_i = z_i + \alpha_i(\mathbf{e}_i - x_i) \neq \mathbf{e}_i, \quad i \in S, \quad \sum_S y_i = \sum_S \mathbf{e}_i.$$

We have $\frac{1}{|S|} \sum_S (y_i - \mathbf{e}_i) = 0$. Since by assumption $|S(y)| > \dim(L_S) + 1$, then using Caratheodory theorem one concludes there exists $R \subset S$ and $\beta_i \in (0,1]$, $i \in R$ such that $|R| < |S|$ and

$$\sum_{i \in R} \beta_i(y_i - \mathbf{e}_i) = 0, \quad \sum_{i \in R} \beta_i = 1 \ \Rightarrow \ \sum_{i \in R}(\beta_i z_i + \beta_i \alpha_i(\mathbf{e}_i - x_i) - \beta_i \mathbf{e}_i) = 0.$$

Since $\lambda(\mathcal{P}_i(x_i) - x_i) \subseteq \mathcal{P}_i(x_i) - x_i \ \forall \lambda \in (0,1]$, the terms on the left-hand side of the latter equality can be rewritten in the form

$$\beta_i(z_i - x_i) + \beta_i \alpha_i(\mathbf{e}_i - x_i) - \beta_i(\mathbf{e}_i - x_i) = \xi_i - x_i - \beta_i(1 - \alpha_i)(\mathbf{e}_i - x_i) = v_i$$

for some $\xi_i \succ_i x_i$, $i \in R$. By construction $\sum_{i \in R} v_i = 0$ and defining $y_i' = v_i + \mathbf{e}_i$, $i \in R$ and $y_i' = \mathbf{e}_i$ for $i \in \mathcal{I} \setminus R$ one obtains $\sum_{i \in \mathcal{I}} y_i' = \sum_{i \in \mathcal{I}} \mathbf{e}_i$ and

$$y_i' = \xi_i + (1 - \beta_i(1 - \alpha_i))(\mathbf{e}_i - x_i) \in \mathcal{P}_i(x_i) + \mathrm{co}\{0, \mathbf{e}_i - x_i\}, \quad i \in R.$$

Thus, we found y' such that under condition (7) we have $T(y') = R$, where $|R| < |S|$, which contradicts the minimality of $S \subset \mathcal{I}$. This contradiction completes the proof. ∎

References

1. Marakulin, V.M.: Contracts and domination in competitive economies. J. New Econ. Assoc. **9**, 10–32 (2011). (in Russian)
2. Marakulin, V.M.: On the Edgeworth conjecture for production economies with public goods: a contract-based approach. J. Math. Econ. **49**(3), 189–200 (2013)
3. Marakulin, V.M.: On contractual approach for Arrow-Debreu-McKenzie economies. Econ. Math. Methods **50**(1), 61–79 (2014). (in Russian)
4. Marakulin, V.M.: Contracts and domination in incomplete markets: what is a true core? Econ. Theory Bull. **5**(1), 81–108 (2016). https://doi.org/10.1007/s40505-016-0105-0
5. Green, J.R.: The stability of Edgeworth's recontracting process. Econometrica **42**, 21–34 (1974)
6. Graham, D.A., Weintraub, E.R.: On convergence to Pareto allocations. Rev. Econ. Stud. **42**, 469–472 (1975)
7. Madden, P.J.: Efficient sequences of non-monetary exchange. Rev. Econ. Stud. **42**(4), 581–595 (1975)
8. Graham, D.A., Jennergren, L.P., Peterson, D.P., Weintraub, E.R.: Trader-commodity parity theorems. J. Econ. Theory **12**, 443–454 (1976)
9. Polterovich, V.M.: Mathematical models of resources allocations. Central Economic and Mathematical Institute, Moscow, 107 p. (1970). (in Russian)
10. Goldman, S.M., Starr, R.M.: Pairwise, t-wise, and Pareto optimalities. Econometrica **50**(3), 593–606 (1982)

Pontryagin's Maximum Principle for Non-cooperative Differential Games with Continuous Updating

Ovanes Petrosian[1,2]([✉]) [ID], Anna Tur[1] [ID], and Jiangjing Zhou[2]

[1] St.Petersburg State University, 7/9, Universitetskaya nab., Saint-Petersburg 199034, Russia
petrosian.ovanes@yandex.ru, a.tur@spbu.ru
[2] School of Mathematics and Statistics, Qingdao University, 308 Ningxia Road, Qingdao 266071, People's Republic of China
260777549@qq.com

Abstract. The paper is devoted to the optimality conditions as determined by Pontryagin's maximum principle for a non-cooperative differential game with continuous updating. Here it is assumed that at each time instant players have or use information about the game structure defined for the closed time interval with a fixed duration. The major difficulty in such a setting is how to define players' behavior as the time evolves. Current time continuously evolves with an updating interval. As a solution for a non-cooperative game model, we adopt an open-loop Nash equilibrium within a setting of continuous updating. Theoretical results are demonstrated on an advertising game model, both initial and continuous updating versions are considered. A comparison of non-cooperative strategies and trajectories for both cases are presented.

Keywords: Differential games with continuous updating ·
Pontryagin's maximum principle · Open-loop Nash equilibrium ·
Hamiltonian

1 Introduction

Most conflict-driven processes in real life evolve continuously in time, and their participants continuously receive updated information and adapt accordingly. The principal models considered in classical differential game theory are associated with problems defined for a fixed time interval (players have all the information for a closed time interval) [10], problems defined for an infinite time interval with discounting (players have all information specified for an infinite time interval) [1], problems defined for a random time interval (players have information for a given time interval, but the duration of this interval is a random variable) [27]. One of the first works in the theory of differential games

Research of the first author was supported by a grant from the Russian Science Foundation (Project No 18-71-00081).

Y. Kochetov et al. (Eds.): MOTOR 2020, CCIS 1275, pp. 256–270, 2020.
https://doi.org/10.1007/978-3-030-58657-7_22

was devoted to a differential pursuit game (a player's payoff depends on when the opponent gets captured) [23]. In all the above models and approaches it is assumed that at the onset players process all information about the game dynamics (equations of motion) and about players' preferences (cost functions). However, these approaches do not take into account the fact that many real-life conflict-controlled processes are characterized by the fact that players at the initial time instant do not have all the information about the game. Therefore such classical approaches for defining optimal strategies as the Nash equilibrium, the Hamilton-Jacobi-Bellman equation [2], or the Pontryagin maximum principle [24], for example, cannot be directly used to construct a large range of real game-theoretic models. Another interesting application of dynamic and differential games is for networks, [5].

Most real conflict-driven processes continuously evolve over time, and their participants constantly adapt. This paper presents the approach of constructing a Nash equilibrium for game models with continuous updating using a modernized version of Pontryagin's maximum principle. In game models with continuous updating, it is assumed that

1. at each current time $t \in [t_0, +\infty)$, players only have or use information on the interval $[t, t + \overline{T}]$, where $0 < \overline{T} < \infty$ is the length of the information horizon,
2. as time $t \in [t_0, +\infty)$ goes by, information related to the game continues to update and players can receive this updated information.

In the framework of the dynamic updating approach, the following papers were published [17], [18], [20],[21], [22], [29]. Their authors set the foundations for further study of a class of games with dynamic updating. It is assumed that information about motion equations and payoff functions is updated in discrete time instants and the interval for which players know information is defined by the value of the information horizon. A non-cooperative setting with dynamic updating was examined along with the concept of the Nash equilibrium with dynamic updating. Also in the papers above cooperative cases of game models with dynamic updating were considered and the Shapely value for this setting was constructed. However, the class of games with continuous updating provides new theoretical results. The class of differential games with continuous updating was considered in the papers [11], [19], here it is supposed that the updating process evolves continuously in time. In the paper [19], the system of Hamilton-Jacobi-Bellman equations are derived for the Nash equilibrium in a game with continuous updating. In the paper [11] the class of linear-quadratic differential games with continuous updating is considered and the explicit form of the Nash equilibrium is obtained.

The approach of continuous updating has some similarities with Model Predictive Control (MPC) theory which is worked out within the framework of numerical optimal control [6], [14], [26], [28], and which has also been used as a human behavior model in [25]. In the MPC approach, the current control action is achieved by solving a finite-horizon open-loop optimal control problem at each sampling instant. For linear systems there exists a solution in explicit form [3], [7]. However, in general, the MPC approach demands the solution of several optimization problems. Another related series of papers corresponds to the class of stabilizing control [12], [13], [16], here similar approaches were considered for the class of linear quadratic optimal control problems. But in the current paper and in papers about the continuous updating approach, the main goal is different: to model players' behavior when information about the course of the game updates continuously in time.

In this paper the optimality conditions for the Nash equilibrium in the form of Pontryagin's maximum principle are derived for a class of non-cooperative game models with continuous updating. In the previous papers on this topic, [19], [11] the optimality conditions were formulated in the form of the Hamilton-Jacobi-Bellman equation and for the special case of a linear quadratic model. From the authors' point of view, formulating Pontryagin's maximum principle for the continuous updating case is the final step for the Nash equilibrium's range of optimality conditions under continuous updating. In future the authors will focus on convex differential games with continuous updating and on the uniqueness of the Nash equilibrium with continuous updating. The concept of the Nash equilibrium for the class of games with continuous updating is defined in the paper [19], and constructed here using open-loop controls and the Pontryagin maximum principle with continuous updating. The corresponding trajectory is also derived. The approach here presented is tested with the advertising game model consisting of two firms. It is interesting to note that in this particular game model the equilibrium strategies are constant functions of time t, unlike the equilibrium strategies in the initial game model.

The paper is organized as follows. Section 2 starts by describing the initial differential game model. Section 3 demonstrates the game model with continuous updating and also defines a strategy for it. In Sect. 4, the classical optimality principle Nash equilibrium is adapted for the class of games with continuous updating. In Sect. 5, a new type of Pontryagin's maximum principle for a class of games with continuous updating is presented. Section 6 presents results of the proposed modeling approach based on continuous updating, such as a logarithmic advertising game model. Finally, we draw conclusions in Sect. 7.

2   Initial Game Model

Consider differential n-player game with prescribed duration $\Gamma(x_0, T - t_0)$ defined on the interval $[t_0, T]$.

The state variable evolves according to the dynamics:

$$\dot{x}(t) = f(t, x, u), \quad x(t_0) = x_0, \tag{1}$$

where $x \in \mathbb{R}^l$ denotes the state variables of the game, $u = (u_1, \ldots, u_n)$, $u_i = u_i(t, x_0) \in U_i \subset \text{comp}\mathbb{R}^k$, $t \in [t_0, T]$, is the control of player i.

The payoff of player i is then defined as

$$K_i(x_0, T - t_0; u) = \int_{t_0}^{T} g^i[t, x(t), u(t, x_0)]dt, \ i \in N, \tag{2}$$

where $g^i[t, x, u]$, $f(t, x, u)$ are the integrable functions, $x(t)$ is the solution of Cauchy problem (1) with fixed $u(t, x_0) = (u_1(t, x_0), \dots, u_n(t, x_0))$. The strategy profile $u(t, x_0) = (u_1(t, x_0), \dots, u_n(t, x_0))$ is called admissible if the problem (1) has a unique and continuable solution. The existence and global asymptotic stability of the open-loop equilibrium for a game with strictly convex adjustment costs was dealt with by Fershtman and Muller [4].

Using the initial differential game with prescribed duration of T, we construct the corresponding differential game with continuous updating.

3   Differential Game Model with Continuous Updating

In differential games with continuous updating players do not have information about the motion equations and payoff functions for the whole period of the game. Instead at each moment t players get information at the interval $[t, t + \overline{T}]$, where $0 < \overline{T} < +\infty$. When choosing a strategy at moment t, this is the only information they can use. Therefore, we consider subgames $\Gamma(x, t, t + \overline{T})$ in which players find themselves at each moment t.

Let us start with the subgame $\Gamma(x_0, t_0, t_0 + \overline{T})$ defined on the interval $[t_0, t_0 + \overline{T}]$. The initial conditions in this subgame coincide with the starting point of the initial game.

Furthermore, assume that the evolution of the state can be described by the ordinary differential equation:

$$\dot{x}^{t_0}(s) = f(s, x^{t_0}, u^{t_0}), \quad x^{t_0}(t_0) = x_0, \tag{3}$$

where $x^{t_0} \in \mathbb{R}^l$ denotes the state variables of the game that starts from the initial time t_0, $u^{t_0} = (u_1^{t_0}, \dots, u_n^{t_0})$, $u_i^{t_0} = u_i^{t_0}(s, x_0) \in U_i \subset \text{comp}\mathbb{R}^k$ is the vector of actions chosen by the player i at the instant time s.

The payoff function of player i is defined in the following way:

$$K_i^{t_0}(x_0, t_0, \overline{T}; u^{t_0}) = \int\limits_{t_0}^{t_0 + \overline{T}} g^i[s, x^{t_0}(s), u^{t_0}(s, x_0)]ds, \ i \in N, \tag{4}$$

where $x^{t_0}(s)$, $u^{t_0}(s, x_0)$ are trajectory and strategies in the game $\Gamma(x_0, t_0, t_0 + \overline{T})$, $\dot{x}^{t_0}(s)$ is the derivative of s.

Now let us give a description of subgame $\Gamma(x, t, t + \overline{T})$ starting at an arbitrary time $t > t_0$ from the situation x.

The motion equation for the subgame $\Gamma(x, t, t + \overline{T})$ has the form:

$$\dot{x}^t(s) = f(s, x^t, u^t), \quad x^t(t) = x, \tag{5}$$

where $\dot{x}^t(s)$ is the derivative of s, $x^t \in \mathbb{R}^l$ is the state variables of the subgame that starts from time t, $u^t = (u_1^t, \dots, u_n^t)$, $u_i^t = u_i^t(s, x) \in U_i \subset \text{comp}\mathbb{R}^k$, $s \in [t, t + \overline{T}]$, denotes the control vector of the subgame that starts from time t at the current time s.

The payoff function of player i for the subgame $\Gamma(x, t, t + \overline{T})$ has the form:

$$K_i^t(x, t, \overline{T}; u^t) = \int\limits_t^{t+\overline{T}} g^i[s, x^t(s), u^t(s, x)]ds, \ i \in N, \tag{6}$$

where $x^t(s)$, $u^t(s, x)$ are the trajectories and strategies in the game $\Gamma(x, t, t+\overline{T})$.

A differential game with continuous updating is developed according to the following rule:

Current time $t \in [t_0, +\infty)$ evolves continuously and as a result players continuously obtain new information about motion equations and payoff functions in the game $\Gamma(x, t, t + \overline{T})$.

The strategy profile $u(t, x)$ in a differential game with continuous updating has the form:

$$u(t, x) = u^t(s, x)|_{s=t}, \ t \in [t_0, +\infty), \tag{7}$$

where $u^t(s, x)$, $s \in [t, t + \overline{T}]$ are strategies in the subgame $\Gamma(x, t, t+\overline{T})$.

The trajectory $x(t)$ in a differential game with continuous updating is determined in accordance with

$$\begin{aligned} \dot{x}(t) &= f(t, x, u), \\ x(t_0) &= x_0, \\ x &\in \mathbb{R}^l, \end{aligned} \tag{8}$$

where $u = u(t, x)$ are strategies in the game with continuous updating (7) and $\dot{x}(t)$ is the derivative of t. We suppose that the strategy with continuous updating obtained using (7) is admissible, or that the problem (8) has a unique and continuable solution. The conditions of existence, uniqueness and continuability of open-loop Nash equilibrium for differential games with continuous updating are presented as follows, for every $t \in [t_0, +\infty)$

1. right-hand side of motion equations $f(s, x^t, u^t)$ (5) is continuous on the set $[t, t + \overline{T}] \times X^t \times U_1^t \times \cdots \times U_n^t$
2. right-hand side of motion equations $f(s, x^t, u^t)$ satisfies the Lipschitz conditions for x^t with the constant $k_1^t > 0$ uniformly regarding to u^t:

$$||f(s, (x^t)', u^t) - f(s, (x^t)'', u^t)|| \le k_1^t ||(x^t)' - (x^t)''||, \ \forall \ s \in [t, t + \overline{T}],$$
$$(x^t)', (x^t)'' \in X^t, u^t \in U^t$$

3. exists such a constant k_2^t that function $f(s, x^t, u^t)$ satisfies the condition:

$$||f(s, x^t, u^t)|| \le k_2^t(1 + ||x||), \ \forall \ s \in [t, t + \overline{T}], \ x^t \in X^t, u^t \in U^t$$

4. for any $s \in [t, t + \overline{T}]$ and $x_t \in X_t$ set

$$G(x^t) = \{f(s, x^t, u^t)|u^t \in U^t\}$$

is a convex compact from R^l.

The essential difference between the game model with continuous updating and a classic differential game with prescribed duration $\Gamma(x_0, T - t_0)$ is that players in the initial game are guided by the payoffs that they will eventually obtain on the interval $[t_0, T]$, but in the case of a game with continuous updating, at the time instant t they orient themselves on the expected payoffs (6), which are calculated based on the information defined for interval $[t, t + \overline{T}]$ or the information that they have at the instant t.

4 Nash Equilibrium in a Game with Continuous Updating

In the framework of continuously updated information, it is important to model players' behavior. To do this, we use the Nash equilibrium concept in open-loop strategies. However, for the class of differential games with continuous updating, modeling will take the following form:

For any fixed $t \in [t_0, +\infty)$, $u^{NE}(t, x) = (u_1^{NE}(t, x), ..., u_n^{NE}(t, x))$ coincides with the Nash equilibrium in game (5), (6) defined for the interval $[t, t + \overline{T}]$ at instant t.

However, direct application of classical approaches for the definition of the Nash equilibrium in open-loop strategies is not possible, consider two intervals $[t, t + \overline{T}]$, $[t + \epsilon, t + \overline{T} + \epsilon]$, $\epsilon << \overline{T}$. Then according to the problem statement:

$-u^{NE}(t)$ at instant t coincides with the open-loop Nash equilibrium in the game defined for interval $[t, t + \overline{T}]$,

$-u^{NE}(t + \epsilon)$ at instant $t + \epsilon$ coincides with the open-loop Nash equilibrium in the game defined for interval $[t + \epsilon, t + \overline{T} + \epsilon]$.

In order to construct such strategies, we consider the concept of generalized Nash equilibrium in open-loop strategies as the principle of optimality

$$\widetilde{u}^{NE}(t, s, x) = (\widetilde{u}_1^{NE}(t, s, x), ..., \widetilde{u}_n^{NE}(t, s, x)), t \in [t_0, +\infty), s \in [t, t + \overline{T}], \quad (9)$$

which we are going to use further for construction of strategies $u^{NE}(t, x)$.

Definition 1. *Strategy profile* $\widetilde{u}^{NE}(t, s, x) = (\widetilde{u}_1^{NE}(t, s, x), ..., \widetilde{u}_n^{NE}(t, s, x))$ *is a generalized Nash equilibrium in the game with continuous updating, if for any fixed* $t \in [t_0, +\infty)$, *strategy profile* $\widetilde{u}^{NE}(t, s, x)$ *is the open-loop Nash equilibrium in game* $\Gamma(x, t, t + \overline{T})$.

Using a generalized open-loop Nash equilibrium, it is possible to define a solution concept for a game model with continuous updating.

Definition 2. *Strategy profile* $u^{NE}(t,x) = (u_1^{NE}(t,x), ..., u_n^{NE}(t,x))$ *is called an open-loop-based Nash equilibrium with continuous updating if it is defined in the following way:*

$$u^{NE}(t,x) = \widetilde{u}^{NE}(t,s,x)|_{s=t}$$
$$= (\widetilde{u}_1^{NE}(t,s,x)|_{s=t}, ..., \widetilde{u}_n^{NE}(t,s,x)|_{s=t}), \quad t \in [t_0, +\infty), \tag{10}$$

where $\widetilde{u}^{NE}(t,s,x)$ *is the generalized open-loop Nash equilibrium defined in Definition 1.*

Strategy profile $u^{NE}(t,x)$ will be used as a solution concept in a game with continuous updating.

5   Pontryagin's Maximum Principle with Continuous Updating

In order to define strategy profile $u^{NE}(t,x)$, it is necessary to determine the generalized Nash equilibrium in open-loop strategies $\widetilde{u}^{NE}(t,s,x)$ of a game with continuous updating. To do this, we will use a modernized version of Pontryagin's maximum principle. Let us start by defining a real-valued function H_i^t by

$$H_i^t(\tau, x^t, u^t, \lambda^t) = \overline{T}g^i(\overline{T}\tau + t, x^t, u^t) + \lambda_i^t \overline{T} f(\overline{T}\tau + t, x^t, u^t). \tag{11}$$

The function $H_i^t, i \in N$ is called the (current-value) Hamiltonian function and plays a prominent role in Pontryagin's Maximum Principle. The variable λ_i^t is called the (current-value) costate variable associated with the state variable x^t, or the (current-value) adjoint variable.

The following theorem is applied:

Theorem 1. *Let* $f(s, \cdot, u^t)$ *be continuously differentiable on* R^l, $\forall s \in [t, t+\overline{T}]$ *and* $g^i(s, \cdot, u^t)$ *be continuously differentiable on* R^l, $\forall s \in [t, t+\overline{T}]$, $i \in N$. *Then, if* $\widetilde{u}^{NE}(t,s,x)$ *provides generalized open-loop Nash equilibrium in a differential game with continuous updating, and for all* $t \in [t_0, +\infty)$ $\widetilde{x}^t(s)$, *with* $s \in [t, t+\overline{T}]$, *is the corresponding state trajectory in the game* $\Gamma(x,t,t+\overline{T})$, *then for all* $t \in [t_0, +\infty)$ *exist* n *costate functions* $\lambda_i^t(\tau,x)$, *where* $\tau \in [0,1]$, $i \in N$, *such that the following relations are satisfied:*

1. for all $\tau \in [0,1]$

$$H_i^t(\tau, \widetilde{x}^t, \widetilde{u}^{NE}(t,\tau,x), \lambda^t) = \max_{\phi_i}\{H_i^t(\tau, \widetilde{x}^t, \widetilde{u}_{-i}^{NE}(t,\tau,x), \lambda^t)\}, i \in N, \tag{12}$$

where $\widetilde{u}_{-i}^{NE} = (\widetilde{u}_1^{NE}, ..., \phi_i, ..., \widetilde{u}_n^{NE})$,

2. $\lambda_i^t(\tau, x)$ *is a decision of the system of adjoint equations*

$$\frac{d\lambda_i^t(\tau, x)}{d\tau} = -\frac{\partial H_i^t(\tau, \widetilde{x}^t(\tau), \widetilde{u}^{NE}(t, \tau, x), \lambda^t)}{\partial x^t} =$$

$$= -\overline{T}\frac{\partial g^i(\overline{T}\tau + t, \widetilde{x}^t, \widetilde{u}^{NE})}{\partial x^t} - \lambda_i^t(\tau, x)\overline{T}\frac{\partial f(\overline{T}\tau + t, \widetilde{x}^t, \widetilde{u}^{NE})}{\partial x_t}, \ i \in N, (13)$$

where the transversality conditions are

$$\lambda_i^t(1, x) = 0, \ i \in N \tag{14}$$

3. *for all* $t \in [t_0, +\infty)$

$$\dot{\widetilde{x}}^t(\tau) = \overline{T}f(\overline{T}\tau + t, \widetilde{x}^t, \widetilde{u}^{NE}), \quad \widetilde{x}^t(0) = x, \ \tau \in [0, 1]. \tag{15}$$

Proof: Let fix $t \geq t_0$ and consider game $\Gamma(x, t, t + \overline{T})$.

Using following substitution

$$\tau = \frac{s - t}{\overline{T}}, \tag{16}$$

we get the motion equation (5) in the form:

$$\dot{x}^t(\tau) = \overline{T}f(\overline{T}\tau + t, x^t, u^t), \quad x^t(0) = x, \quad \tau \in [0, 1]. \tag{17}$$

And payoff function of player $i \in N$ has the form

$$K_i^t(x, t, \overline{T}; u^t) = \int_0^1 \overline{T}g^i[\overline{T}\tau + t, x^t(\tau), u^t(\tau, x)]d\tau, \ i \in N. \tag{18}$$

For the optimization problem (17)–(18) Hamiltonian has the form

$$H_i^t(\tau, x^t, u^t, \lambda^t) = \overline{T}g^i(\overline{T}\tau + t, x^t(\tau), u^t(\tau, x)) + \lambda_i^t(\tau, x)\overline{T}f(\overline{T}\tau + t, x^t(\tau), u^t(\tau, x)). \tag{19}$$

If $\widetilde{u}^{NE}(t, \tau, x)$ – generalized open-loop Nash equilibrium in the differential game with continuous updating, then, according to Definition 1, for every fixed $t \geq t_0$, $\widetilde{u}^{NE}(t, \tau, x)$ is an open-loop Nash equilibrium in the game $\Gamma(x, t, t+\overline{T})$. Therefore for any fixed $t \geq t_0$ conditions 1–3 of the theorem are satisfied as necessary conditions for Nash equilibrium in open-loop strategies (see [1]). The Theorem is proved.

It can been mentioned also that if for every $t \geq t_0$ functions H_i^t are concave in (x_t, u_t) for all $i \in N$, then the conditions of the theorem are sufficient for a Nash open-loop solution [15].

6   Differential Game of Logarithmic Advertising Game Model with Continuous Updating

As an illustrative example, we consider a logarithmic excess-advertising model of a duopoly proposed by Jørgensen in [8]. There are two firms operating in a market. It is assumed that market potential is constant over time. The only marketing instrument used by the firms is advertising. Advertising has diminishing returns since it suffers from increasing marginal costs. Nash optimal open-loop advertising strategies are determined in [8]. Here we obtain open-loop Nash equilibrium with continuous updating by means of Theorem 1.

6.1   Initial Game Model

Consider the model investigated in [8]. Let $x_i(t)$ denote the rate of sales of firm i at the instant time t, $(i = 1, 2)$ and assume that $x_1 + x_2 = M$, implying that the market potential is fully exhausted at each instant of time. The game is played on interval $[0, T]$, where T is an arbitrary but fixed positive number. Because of the assumption $x_1 + x_2 = M$, so $\dot{x}_2 = -\dot{x}_1$. The state equation is

$$\dot{x}_1 = k \log \frac{u_1}{u_2} = k(\log u_1 - \log u_2),$$
$$\dot{x}_2 = -\dot{x}_1 = k(\log u_2 - \log u_1), \tag{20}$$
$$x_1(0) = x_1^0, x_2(0) = x_2^0,$$

where k is a positive constant, $x_i(0)$ is a given initial rate of sales of firm i. The state equation (20) model describes a market where buyers are perfectly mobile and switch instantaneously to the firm which has the largest rate of advertising expenditure, that is, advertises in excess of the other. In the model, market share increases linearly according to the amount of excess advertising. Performance indices are given by

$$K_i = \int_0^T (\varphi_i x_i - u_i) \exp\{-r_i t\} dt, \quad i = 1, 2, \tag{21}$$

where $x_2 = M - x_1$. Assume that $r_i > 0$, $i = 1, 2$. The open-loop Nash equilibrium in its explicit form was constructed in [8]:

$$u_i^{initial,NE} = \frac{k\varphi_i}{r_i}[1 - exp\{-r_i(T - t)\}]. \tag{22}$$

For the case $r_1 = r_2$, the optimal trajectories are given by

$$x_1(t) = (k \log \frac{\varphi_1}{\varphi_2})t + x_1(0),$$
$$x_2(t) = M - (k \log \frac{\varphi_1}{\varphi_2})t - x_1(0). \tag{23}$$

If $r_1 \neq r_2$, then trajectory x_1 is the solution of

$$\dot{x}_1 = k \log \frac{\varphi_1 r_2 [1 - exp\{-r_1(T-t)\}]}{\varphi_2 r_1 [1 - exp\{-r_2(T-t)\}]}. \tag{24}$$

The solution of Eq. (24) is given by

$$x_1(t) = x_1(0) + k \log \frac{\varphi_1 r_2}{\varphi_2 r_1} t + k \int_0^t \log \frac{1 - exp\{-r_1(T-s)\}}{1 - exp\{-r_2(T-s)\}} ds.$$

6.2   Game Model with Continuous Updating

Now consider this model as a game with continuous updating. It is assumed that information about motion equations and payoff functions is updated continuously in time. At every instant $t \in [0, +\infty)$, players have information only at interval $[t, t + \overline{T}]$.

Therefore, for every time instant t, we can get the payoff function of player i for the interval $[t, t + \overline{T}]$. The payoff functions are given as follows:

$$K_i^t = \int_t^{t+\overline{T}} (\varphi_i x_i^t - u_i^t) \exp(-r_i s) ds, \quad i = 1, 2.$$

In order to simplify the problem that we desire to solve, we can do a transfer $\tau = \frac{s-t}{\overline{T}}$. Furthermore, restate the problem to be solved:

$$\begin{aligned}
&\dot{x}_1^t(\tau) = \overline{T} k \log \frac{u_1^t(\tau, x)}{u_2^t(\tau, x)} = \overline{T} k (\log u_1^t(\tau) - \log u_2^t(\tau)), \quad \tau \in [0, 1], \\
&\dot{x}_2^t(\tau) = -\dot{x}_1^t(\tau), \\
&x_1^t(0) = x_1, \quad x_2^t(0) = x_2, \\
&K_i^t = \int_0^1 \overline{T}(\varphi_i x_i^t(\tau) - u_i^t(\tau, x)) \exp\{-r_i(\overline{T}\tau + t)\} d\tau, \quad i = 1, 2.
\end{aligned} \tag{25}$$

The Hamiltonian functions are given by

$$H_1^t(t, \tau, x, u^t, \lambda^t) = (\varphi_1 x_1^t - u_1^t)\overline{T} + \lambda_1^t(\tau, x)\overline{T} k(\log u_1^t - \log u_2^t), \tag{26}$$

$$H_2^t(t, \tau, x, u^t, \lambda^t) = (\varphi_2 x_2^t - u_2^t)\overline{T} - \lambda_2^t(\tau, x)\overline{T} k(\log u_1^t - \log u_2^t). \tag{27}$$

Note that the current-value Hamiltonian is simply $exp(r_i(\overline{T}\tau + t))$ times the conventional Hamiltonian. Necessary conditions for the maximization of H_i^t, for $u_i^t \in (0, +\infty)$ are given by

$$\frac{\partial H_1^t}{\partial u_1^t} = -\overline{T} + \lambda_1^t(\tau, x)\overline{T} k \frac{1}{u_1^t} = 0,$$

$$\frac{\partial H_2^t}{\partial u_2^t} = -\overline{T} + \lambda_2^t(\tau, x)\overline{T}k\frac{1}{u_2^t} = 0.$$

Therefore, the optimal control u_i^t is given by

$$u_1^t(\tau, x) = \lambda_1^t(\tau, x)k, \quad u_2^t(\tau, x) = \lambda_2^t(\tau, x)k. \tag{28}$$

The adjoint variables $\lambda_i^t(\tau)$ should satisfy the following equations

$$\dot{\lambda}_1^t(\tau, x) = -\frac{\partial H_1^t}{\partial x_1^t} + \overline{T}r_1\lambda_1^t(\tau, x) = -\varphi_1\overline{T} + \overline{T}r_1\lambda_1^t(\tau, x),$$
$$\dot{\lambda}_2^t(\tau, x) = -\frac{\partial H_2^t}{\partial x_2^t} + \overline{T}r_2\lambda_2^t(\tau, x) = -\varphi_2\overline{T} + \overline{T}r_2\lambda_2^t(\tau, x). \tag{29}$$

Note that these equations are uncoupled. The transversality conditions are

$$\lambda_i^t(1, x) = 0, \quad i = 1, 2.$$

By solving the above differential equations about the adjoint variables, the solutions are given by

$$\lambda_1^t(\tau, x) = \frac{\varphi_1}{r_1}[1 - exp\{\overline{T}r_1(\tau - 1)\}],$$
$$\lambda_2^t(\tau, x) = \frac{\varphi_2}{r_2}[1 - exp\{\overline{T}r_2(\tau - 1)\}]. \tag{30}$$

Substituting (30) into (28) yields

$$u_i^{tNE}(\tau, x) = \frac{k\varphi_i}{r_i}[1 - exp\{\overline{T}r_i(\tau - 1)]\}. \tag{31}$$

Note that x is the initial state in the subgame $\Gamma(x, t, t + \overline{T})$. The open-loop strategies $u_i^{tNE}(\tau, x)$ in our example in fact do not depend on initial state x.

Let us show that the solution obtained satisfies sufficiency conditions. Since $\frac{\partial^2 H_i^t}{\partial x^t \partial x^t} = 0$, $\frac{\partial^2 H_i^t}{\partial x^t \partial u_i^t} = 0$, $\frac{\partial^2 H_i^t}{\partial u_i^t \partial u_i^t} = -\lambda_i^t(\tau)\overline{T}k\frac{1}{(u_i^t)^2} \le 0$, then, according to [9], $u^{tNE}(\tau, x)$ is indeed a Nash equilibrium in the subgame $\Gamma(x, t, t + \overline{T})$.

Finally, we convert τ to t, s. Then the generalized open-loop Nash equilibrium strategies have the following form:

$$\tilde{u}_1^{NE}(t, s, x) = \frac{k\varphi_1}{r_1}[1 - exp\{r_1(s - t - \overline{T})\}],$$
$$\tilde{u}_2^{NE}(t, s, x) = \frac{k\varphi_2}{r_2}[1 - exp\{r_2(s - t - \overline{T})\}]. \tag{32}$$

According to Definition 2, we construct an open-loop-based Nash equilibrium with continuous updating :

$$u_i^{NE}(t, x) = \tilde{u}_i^{NE}(t, s, x)|_{s=t} = \frac{k\varphi_i}{r_i}[1 - exp\{-r_i\overline{T}\}] \quad i = 1, 2. \qquad (33)$$

Note that in the example under consideration, strategies u_i^{tNE} are independent of the initial values of the state variables of subgame $\Gamma(x, t, t + \overline{T})$, so strategies $u_i^{NE}(t, x)$ in fact do not depend on x.

Consider the difference between optimal strategies in the initial game and in a game with continuous updating:

$$u_i^{initial, NE} - u_i^{NE} = \frac{k\varphi_i}{r_i} exp\{-r_i\overline{T}\}[1 - exp\{-r_i(T - t - \overline{T})\}]$$

We can see that the amounts of players' advertising expenditure is less in a game with continuous updating for $t < T - \overline{T}$.

The optimal trajectories $x_1^{NE}(t)$, $x_2^{NE}(t)$ in a game with continuous updating are the solutions of

$$\dot{x}_1(t) = k\log(\frac{\varphi_1 r_2}{r_1\varphi_2}\frac{[1 - exp\{-r_1\overline{T}\}]}{[1 - exp\{-r_2\overline{T}\}]}),$$
$$\dot{x}_2(t) = -\dot{x}_1(t),$$
$$x_1(0) = x_1^0, \qquad (34)$$
$$x_2(0) = x_2^0,$$

where $r_i > 0, i = 1, 2$. Therefore, the state dynamics of the system are given as follows:

$$x_1^{NE}(t) = x_1^0 + k\log(\frac{\varphi_1 r_2[1 - exp\{-r_1\overline{T}\}]}{r_1\varphi_2[1 - exp\{-r_2\overline{T}\}]})t,$$
$$x_2^{NE}(t) = M - x_1^0 - k\log(\frac{\varphi_2 r_1[1 - exp\{-r_2\overline{T}\}]}{r_2\varphi_1[1 - exp\{-r_1\overline{T}\}]})t. \qquad (35)$$

It can be noted, that if $r_1 = r_2$, then optimal trajectories in initial model and in the game with continuous updating are the same.

Figures 1, 2 represent a comparison of results obtained in the initial model and in the model with continuous updating for the following parameters: $\frac{\varphi_1}{r_1} = 0.1$, $\frac{\varphi_2}{r_2} = 0.5$, $k = 1$, $T = 10$, $\overline{T} = 0.2$, $r_1 = 5$, $r_2 = 3$, $x_1^0 = 8$, $x_2^0 = 10$.

We see that the rate of sales for player 1 in the game with continuous updating is less than in the initial model.

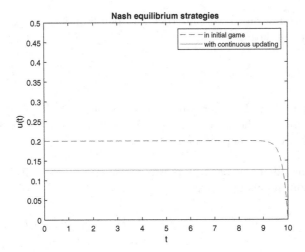

Fig. 1. Comparison of Nash equilibrium strategies in the initial model and in the game with continuous updating

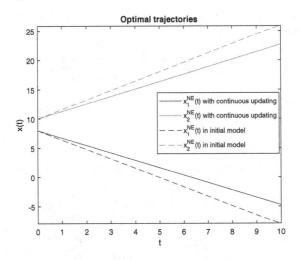

Fig. 2. Comparison of optimal trajectories in the initial model and in the game with continuous updating

7 Conclusion

A differential game model with continuous updating is presented and described. The definition of the Nash equilibrium concept for a class of games with continuous updating is given. Optimality conditions on the form of Pontryagin's maximum principle for the class of games with continuous updating are presented for the first time and the technique for finding the Nash equilibrium is described.

The theory of differential games with continuous updating is demonstrated by means of an advertising model with a logarithmic state dynamic. Ultimately, we present a comparison of the Nash equilibrium and the corresponding trajectory in both the initial game model as well as in the game model with continuous updating and conclusions are drawn.

References

1. Başar, T., Olsder, G.J.: Dynamic Non Cooperative Game Theory. Classics in Applied Mathematics, 2nd edn. SIAM, Philadelphia (1999)
2. Bellman, R.: Dynamic Programming. Princeton University Press, Princeton (1957)
3. Bemporad, A., Morari, M., Dua, V., Pistikopoulos, E.: The explicit linear quadratic regulator for constrained systems. Automatica **38**(1), 3–20 (2002)
4. Fershtman, C., Muller, E.: Turnpike properties of capital accumulation games. J. Econ. Theor. **38**(1), 167–177 (1986)
5. Gao, H., Petrosyan, L., Qiao, H., Sedakov, A.: Cooperation in two-stage games on undirected networks. J. Syst. Sci. Complex. **30**(3), 680–693 (2017). https://doi.org/10.1007/s11424-016-5164-7
6. Goodwin, G., Seron, M., Dona, J.: Constrained Control and Estimation: An Optimisation Approach. Springer, New York (2005). https://doi.org/10.1007/b138145
7. Hempel, A., Goulart, P., Lygeros, J.: Inverse parametric optimization with an application to hybrid system control. IEEE Trans. Autom. Control **60**(4), 1064–1069 (2015)
8. Jørgensen, S.: A differential games solution to a logarithmic advertising model. J. Oper. Res. Soc. **33**, 425–432 (1982)
9. Jørgensen, S.: Sufficiency and game structure in Nash open-loop differential games. J. Optim. Theor. Appl. **1**(50), 189–193 (1986)
10. Kleimenov, A.: Non-antagonistic positional differential games. Science, Ekaterinburg (1993)
11. Kuchkarov, I., Petrosian, O.: On class of linear quadratic non-cooperative differential games with continuous updating. In: Khachay, M., Kochetov, Y., Pardalos, P. (eds.) MOTOR 2019. LNCS, vol. 11548, pp. 635–650. Springer, Cham (2019). https://doi.org/10.1007/978-3-030-22629-9_45
12. Kwon, W., Bruckstein, A., Kailath, T.: Stabilizing state-feedback design via the moving horizon method. In: 21st IEEE Conference on Decision and Control (1982). https://doi.org/10.1109/CDC.1982.268433
13. Kwon, W., Pearson, A.: A modified quadratic cost problem and feedback stabilization of a linear system. IEEE Trans. Autom. Control **22**(5), 838–842 (1977). https://doi.org/10.1109/TAC.1977.1101619
14. Kwon, W., Han, S.: Receding Horizon Control: Model Predictive Control for State Models. Springer, New York (2005). https://doi.org/10.1007/b136204
15. Leitmann, G., Schmitendorf, W.: Some sufficiency conditions for pareto optimal control. ASME J. Dyn. Syst. Meas. Control **95**, 356–361 (1973)
16. Mayne, D., Michalska, H.: Receding horizon control of nonlinear systems. IEEE Trans. Autom. Control **35**(7), 814–824 (1990). https://doi.org/10.1109/9.57020

17. Petrosian, O., Kuchkarov, I.: About the looking forward approach in cooperative differential games with transferable utility. In: Frontiers of Dynamic Games: Game Theory and Management, St. Petersburg, pp. 175–208 (2019). https://doi.org/10.1007/978-3-030-23699-1_10

18. Petrosian, O., Shi, L., Li, Y., Gao, H.: Moving information horizon approach for dynamic game models. Mathematics **7**(12), 1239 (2019). https://doi.org/10.3390/math7121239

19. Petrosian, O., Tur, A.: Hamilton-Jacobi-Bellman equations for non-cooperative differential games with continuous updating. In: Mathematical Optimization Theory and Operations Research, pp. 178–191 (2019). https://doi.org/10.1007/978-3-030-33394-2_14

20. Petrosian, O.: Looking forward approach in cooperative differential games. Int. Game Theor. Rev. **18**, 1–14 (2016)

21. Petrosian, O., Barabanov, A.: Looking forward approach in cooperative differential games with uncertain-stochastic dynamics. J. Optim. Theor. Appl. **172**, 328–347 (2017)

22. Petrosian, O., Nastych, M., Volf, D.: Non-cooperative differential game model of oil market with looking forward approach. In: Petrosyan, L.A., Mazalov, V.V., Zenkevich, N., Birkhäuser (eds.) Frontiers of Dynamic Games, Game Theory and Management, St. Petersburg, Basel (2018)

23. Petrosyan, L., Murzov, N.: Game-theoretic problems in mechanics. Lith. Math. Collect. **3**, 423–433 (1966)

24. Pontryagin, L.S.: On the theory of differential games. Russ. Math. Surv. **21**(4), 193 (1966)

25. Ramadan, A., Choi, J., Radcliffe, C.J.: Inferring human subject motor control intent using inverse MPC. In: 2016 American Control Conference (ACC), pp. 5791–5796, July 2016

26. Rawlings, J., Mayne, D.: Model Predictive Control: Theory and Design. Nob Hill Publishing, Madison (2009)

27. Shevkoplyas, E.: Optimal solutions in differential games with random duration. J. Math. Sci. **199**(6), 715–722 (2014)

28. Wang, L.: Model Predictive Control System Design and Implementation using MATLAB. Springer, New York (2005). https://doi.org/10.1007/978-1-84882-331-0

29. Yeung, D., Petrosian, O.: Cooperative stochastic differential games with information adaptation. In: International Conference on Communication and Electronic Information Engineering (2017)

Feasible Set Properties of an Optimal Exploitation Problem for a Binary Nonlinear Ecosystem Model with Reducible Step Operator

Alexander I. Smirnov$^{(\boxtimes)}$ (iD) and Vladimir D. Mazurov

Krasovskii Institute of Mathematics and Mechanics UB RAS, Ekaterinburg, Russia
asmi@imm.uran.ru, vldmazurov@gmail.com

Abstract. Previously, the authors proposed a formalization of renewable resources rational use problem based on the representation of controlled system as a discrete dynamical system. In the particular case of structured ecosystem described by Leslie's binary model, despite its nonlinearity, it turned out that all optimal controls preserving this system belong to the certain hyperplane. This paper explores the conditions under which the positive boundary of a feasible set of problem with so-called quasi-preserving controls also contain a part of some hyperplane. In the process, we used a generalization of classical concept of map irreducibility—the concept of local irreducibility. Statements about the influence of the irreducibility property of discrete dynamical system step operator on the properties of an feasible set positive boundary are proved.

Keywords: Rational exploitation of ecosystems · Binary Leslie's model · Concave programming · Irreducible map

1 Introduction

The renewable resources rational use problem is currently extremely acute. Current estimates are that overfishing has impacted over 85% of the world's fish resources and that most fisheries are fished far beyond their sustainable capacity. The depletion of forest resources is increasing; the net loss of the global forest area (deforestation plus reforestation) in the last decade of the 20th century was about 94 million hectares, the equivalent of 2.4% of total world forests [10].

We study here some formalization for a problem of sustainable exploitation of renewable resources. The problem statement is inspired by problems of sustainable management of fisheries, agriculture, forestry and other renewable resources, including the problems of non-destructive exploitation of ecological populations (through partial removal of biomass).

The first studies in this direction considered only the total population biomass. Later, the need to take into account a structure of exploited populations led to the use of matrix models. The dynamic aspect of the problem was

© Springer Nature Switzerland AG 2020
Y. Kochetov et al. (Eds.): MOTOR 2020, CCIS 1275, pp. 271–286, 2020.
https://doi.org/10.1007/978-3-030-58657-7_23

not taken into account; in the vast majority of studies, the problem of exploiting a population in a stationary state was considered. A comprehensive review of these studies is given in monograph [4].

The first results for non-linear models of exploited populations were obtained in pioneering papers [1,9], where a density-dependent model of the structured population was studied.

As a basic model, in studies taking into account the structure of exploited populations, as a rule, various generalizations of so-called Leslie's model of the population age structure were used (biological aspects of this model, as well as a comprehensive description of its properties, are given in [5]). These papers became the basis for numerous further publications (see review in [2]).

In the vast majority of studies, even if additive control is used in the initial formulation, as a rule, the transition to the proportional removal, that is, to multiplicative control, is subsequently carried out, which simplifies the search and analysis of optimal strategies. A typical approach is described in [4]. Although the iterative process with additive control is considered first, then the fractions to be withdrawn of the structural units are determined. Thus, there is a return to the multiplicative control (typically, in an equilibrium state of the model used). Usually, a step operator of dynamical system is multiplied by a diagonal matrix with these fractions on the main diagonal. In this paper, we use a more natural, in our opinion, the additive setting the problem under consideration.

This study is a continuation of our series of publications [7,11,12] on the ecosystems exploitation problem. The results obtained in them will be presented in the following sections as necessary; in this section, we note only a few of them.

Let us note that for a more complete acquaintance with the history and the current state on modeling the sustainable exploitation of ecosystems, one can also use the reviews available in our publications mentioned above.

Although the exact mathematical formulation of the problem of ecosystem exploitation varied among different authors, there is a common characteristic property of optimal solutions, consisting in the number of age classes to be exploited. It was established that there is a bimodal optimal control, which allows the exploitation (withdrawal, partially or completely) of no more than two age (stage) classes: the partial withdrawal of one age class and the complete withdrawal of another (older) one.

All of the above studies examined populations with a one-dimensional (age or stage) structure. In the series of papers, we consider the problem of exploitation for a population with a binary structure, when there is an additional criterion for structuring the population, different from the age or stage of development. We characterized the properties of a feasible set of the population exploitation problem for this generalization of Leslie's model. In particular, we obtained a generalization of the bimodality property for the binary population structure [12].

Our model of the population belongs to the class of general models that some authors call compact or global. As emphasized in [3], such models are necessary at the initial stages of ecosystem modeling, because they must be basic for more detailed models of specific natural systems. A comprehensive study of the properties of global models is necessary since the basic properties of detailed concrete models follow from the corresponding properties of the underlying global models.

In previous papers, we dealt with the generally accepted so-called "regular" case when the dynamical system step operator was irreducible (the definition and properties of the local map irreducibility can be found in [6]; the classical irreducibility [8] here will be called as global irreducibility).

The goal of this paper is to study the features of the feasible set of the optimal exploitation problem for the binary Leslie's model in the absence of assumptions about the irreducibility of the dynamical system step operator.

2   Some Definitions, Notation and Preliminary Results

Used notation: $\mathbb{R}_+^q$—the nonnegative orthant of $\mathbb{R}^q$; $\overline{M}$—the closure of a set M; $\text{co}\,(M)$—the convex hull of M; $|M|$—the number of elements of a finite set M; $\overline{m,n} = \{i \in \mathbb{Z} \mid m \leq i \leq n\}$; $\mathbb{Z}$—the set of integers; $I^+(x) = \{i \in \overline{1,q} \mid x_i > 0\}$. Sometimes we briefly write $x = (x_i)$ instead of $x = (x_1, x_2, \ldots, x_q)$.

We say that x, y are in *strict dominance relation* or in *partial dominance relation* if $x < y$ or $x \lneqq y$, $x \not< y$, respectively; $x \lneqq y$ means $x \leq y$, $x \neq y$.

The ecosystem that is being exploited is modeled by the iterative process

$$x_{t+1} = F_u(x_t), \qquad t = 0, 1, 2, \ldots, \tag{1}$$

where $F_u(x) = F(x) - u$, $x_t \geq 0$—the population state at time $t = 0, 1, 2, \ldots$. The components of x_t are the biomass values of ecosystem structural units; the components of u determine the volumes of withdrawn biomass.

It is assumed that the original system (in the absence of control), along with the trivial equilibrium ($F(0) = 0$), also has a nontrivial equilibrium (i.e., its step operator F also has a nonzero fixed point $\bar{x}_F$).

In [7] we posed the problem of maximization of ecosystem exploitation effect $c(u)$ on the feasible set $\overline{U}$ that is the closure of preserving controls set U. The control u is *preserving* if, for at least one initial state x_0, the trajectory of process (1) is separated from zero, so all structural units of the system stably exist indefinitely.

The map F was also assumed to be concave on $\mathbb{R}_+^q$ and irreducible at zero. Under these conditions, was proved [7] the equivalence of posed optimal exploitation problem to the mathematical programming problem

$$\max\{c(u) \mid x = F(x) - u, \ x \geq 0, \ u \geq 0\}, \tag{2}$$

where $c(u)$—nonnegative monotone increasing function.

We denote by N_u and N_u^+ the sets of nonzero and positive fixed points of F_u, respectively. A nonempty set N_u contains the largest element $\bar{x}_u$, and the map $\bar{x}(u) \colon u \to \bar{x}_u$ is *concave* and *monotone (strictly) decreasing* on $\overline{U}$ [7]:

$$N_u \neq \varnothing, \ 0 \leq v \leq u \Rightarrow N_v \neq \varnothing, \ \bar{x}_v \geq \bar{x}_u; \quad 0 \leq v < u \Rightarrow \bar{x}_v > \bar{x}_u. \tag{3}$$

The feasible control u is preserving if $\bar{x}(u) > 0$. If $\bar{x}(u) \gneq 0$, but $\bar{x}(u) \not> 0$, then the control u is called *quasi-preserving*. The feasible set $\overline{U}$ of the problem (2) is the closure of the preserving controls set U. These sets are representable [7] as

$$U = \{u \in \mathbb{R}^q_+ \mid N^+_u \neq \varnothing\}, \quad \overline{U} = \{u \in \mathbb{R}^q_+ \mid N_u \neq \varnothing\}. \tag{4}$$

Since $c(u)$ is monotone increasing, optimal controls of (2) belong to the set

$$D = \{u \in \mathbb{R}^q_+ \mid N_u \neq \varnothing, \ N_v = \varnothing \ (\forall v > u)\}. \tag{5}$$

This set is a part of the boundary of U; some authors call it as a *positive boundary* of U. Clear that $D = D' \cup D''$, where $D' = D \cap U$, $D'' = D \setminus U$, so that

$$D' = \{u \mid N^+_u \neq \varnothing, \ N_v = \varnothing \ (\forall v > u)\}, \ D'' = \{u \in \mathbb{R}^q_+ \mid N_u \neq \varnothing, \ N^+_u = \varnothing\}. \tag{6}$$

Thus, all optimal preserving controls belong to D'; accordingly, all optimal quasi-preserving controls belong to D''.

In the sequel, we consider only those preserving and quasi-preserving controls that belong to the positive boundary $D = D' \cup D''$ of the feasible set (19).

3   The Optimal Exploitation Problem for an Ecosystem Modeled by Nonlinear Binary Leslie Model

Let us describe our generalization of the Leslie model. The population consists of m structural subdivisions, each of which, in turn, contains individuals of n ages (stages). If we denote by $x^{(t)}_{i,j}$ the number of individuals of the structural subdivision $i \in \overline{1,m}$ of age (stage) $j \in \overline{1,n}$ at time $t = 0, 1, 2 \ldots$, then the relations of this model will take the following form:

$$x^{(t+1)}_{i,1} = f_i(a_t), \quad x^{(t+1)}_{i,j+1} = \alpha_{i,j} x^{(t)}_{i,j} \quad (i \in \overline{1,m}, j \in \overline{1,n-1}). \tag{7}$$

Here $\alpha_{i,j} > 0$ and $\beta_{i,j} \geq 0$ are the survival and fertility rates in the relevant subdivisions, $a_t = \sum_{i=1}^m \sum_{j=1}^n \beta_{i,j} x^{(t)}_{i,j}$—the number of newborns at time t.

The population state vector has a block form $x = (x^{(1)}; x^{(2)}; \ldots; x^{(m)})$ with the blocs $x^{(i)} = (x_{i,1}, \ldots, x_{i,n})$ $(i \in \overline{1,m})$. The step operator $F(x) = (f_{i,j}(x))$ of the iterative process (7) has the following components:

$$f_{i,1}(x) = f_i(a(x)), \quad f_{i,j+1}(x) = \alpha_{i,j} x_{i,j} \ (i \in \overline{1,m}, \ j \in \overline{1,n-1}). \tag{8}$$

We assume that among the functions $f_i(a)$ there are no identically equal to zero and they satisfy the following assumptions:

$$f_i(0) = 0, \quad f_i(a) \ \text{are concave on} \ \mathbb{R}_+ \ (\forall i \in \overline{1,m}). \tag{9}$$

So, these functions are monotone increasing and positive for $a > 0$. Denote

$$\sigma(a) = \sum_{i=1}^m \sigma^{(i)} f_i(a), \sigma^{(i)} = \sigma^{(i)}_n, \sigma^{(i)}_j = \sum_{k=1}^j \beta_{i,k} \prod_{\ell=1}^{k-1} \alpha_{i,\ell} \ (i \in \overline{1,m}, j \in \overline{0,n}). \tag{10}$$

The condition of positive equilibrium existence is the following [11]:

$$\sigma'(+\infty) < 1 < \sigma'(+0), \tag{11}$$

The irreducibility at zero of the map (8) is equivalent to the following condition:

$$\beta_{i,n} > 0 \quad (\forall i \in \overline{1,m}). \tag{12}$$

The global irreducibility of the map (8) is equivalent to irreducibility only at zero under the following requirement:

$$f_i(a) \text{ are strictly increasing on } [0, +\infty) \ (\forall i \in \overline{1,m}). \tag{13}$$

We introduce the following notation (here everywhere $i \in \overline{1,m}, \ j, k \in \overline{1,n}$):

$$\pi^{(i)} = \pi_n^{(i)}, \quad \pi_j^{(i)} = p_{1,j}^{(i)}, \quad p_j^{(i)} = p_{j,n}^{(i)}, \quad p_{j,k}^{(i)} = \prod_{\ell=j}^{k-1} \alpha_{i,\ell}, \tag{14}$$

$$p^{(i)}(u) = p_n^{(i)}(u), \quad p_j^{(i)}(u) = \sum_{k=1}^{j} p_{k,j}^{(i)} u_{i,k}, \tag{15}$$

$$q(u) = \sum_{i=1}^{m} \sum_{j=1}^{n} q_j^{(i)} u_{i,j}, \quad q_j^{(i)} = q_{j,n}^{(i)}, \quad q_{j,k}^{(i)} = \sum_{s=j}^{k} \beta_{i,s} \prod_{t=j}^{s-1} \alpha_{i,t}, \tag{16}$$

$$\mu(a) = \sigma(a) - a, \quad \mu^* = \max_{a \geq 0} \mu(a), \quad \lambda_i(a) = \pi^{(i)} f_i(a). \tag{17}$$

Note that $q(u) = \langle q, u \rangle$, where the symbol $\langle \cdot, \cdot \rangle$ means the scalar product,

$$q = (q^{(1)}; q^{(2)}; \ldots; q^{(m)}), \quad q^{(i)} = (q_1^{(i)}, q_2^{(i)}, \ldots, q_n^{(i)}). \tag{18}$$

The feasible set $\overline{U}$ of (2) for the model (7) is given by the restrictions

$$x_{i,1} = f_i(a(x)) - u_{i,1}, \quad x_{i,j+1} = \alpha_{i,j} x_{i,j} - u_{i,j+1} \quad (i \in \overline{1,m}, \ j \in \overline{1,n-1}), \tag{19}$$

where x, u are nonnegative, $a(x) = \sum_{i=1}^{m} \sum_{j=1}^{n} \beta_{i,j} x_{i,j}$.

It is easy to get explicit expressions for coordinates of the feasible vector x:

$$x_{i,j} = \pi_j^{(i)} f_i(a) - p_j^{(i)}(u) \ (j \in \overline{1,n}), \ x_{i,n} = \lambda_i(a) - p^{(i)}(u) \ (i \in \overline{1,m}), \tag{20}$$

where $a = a(x)$. We write x with coordinates (19) as $x = x(a, u)$. The following properties are a consequence of the Eqs. (20):

$$q(u) = \mu(a) \ (a = a(x)), \quad q(u) \leq \mu^* \ (\forall u \in \overline{U}). \tag{21}$$

For $u \in \overline{U}, \ \bar{x}_u = (\bar{x}_{i,j}(u))$ we introduce the following indices sets:

$$I_0(u) = \{k \in \overline{1,m} \mid \bar{x}_{k,n}(u) = 0\}, \ I_1(u) = \{k \in \overline{1,m} \mid \exists j \in \overline{1,n}: u_{k,j} > 0\}. \tag{22}$$

The set $I_0(u)$ shows blocks $x^{(k)}$ whose last age (stage) groups are completely eliminated. The fact $I_1(u) = \varnothing$ means that $u = 0$; otherwise, the set $I_1(u)$ shows blocks $u^{(k)}$ containing positive coordinates. Clear that $I_0(u) \subseteq I_1(u)$.

From (20), (21) we obtain the following representations (see [11] for details):

$$D' = \{u \mid p^{(i)}(u) < \lambda_i^* \ (i \in \overline{1,m}), \ q(u) = \mu^*, \ u \geq 0\}, \tag{23}$$

$$D'' = \{u \mid I_0(u) \neq \varnothing, \ p^{(i)}(u) \begin{cases} = \lambda_i(a), \ i \in I_0(u), \\ < \lambda_i(a), \ i \notin I_0(u), \end{cases} q(u) = \mu(a), \ u \geq 0\}. \tag{24}$$

A remarkable property of D' was proved in [11]: it turns out that the set of potentially optimal preserving controls D' is not empty and lies entirely on the hyperplane $\Gamma = \{u \mid q(u) = \mu^*\}$. Moreover,

$$D' = \Gamma \cap U, \quad D'' \cap \Gamma \subset \overline{D'}, \quad \overline{D'} = \Gamma \cap D. \tag{25}$$

This property allows even before solving the problem (2) to determine whether optimal preserving controls exist, and if they exist, simplifies their finding.

Now we give (without proof) the auxiliary proposition about the properties of the function $\bar{a}(u) = a(\bar{x}(u))$. This proposition allows us, in particular, to determine the change boundaries of the parameter a in the representation (24).

Lemma 1. *Let the assumptions* (9) *and* (11) *hold. Then the following statements are true:*

(i) *The function $\bar{a}(u)$ is nonnegative, monotone decreasing, and concave on $\overline{U}$.*
(ii) *The inequality $\bar{a}(u) \geq a^*$ $(\forall u \in \overline{U})$ holds; more specifically,*

$$\bar{a}(u) = a^* \ (\forall u \in D'), \quad \bar{a}(u) \begin{cases} = a^*, \ u \in \Gamma, \\ > a^*, \ u \notin \Gamma \end{cases} (\forall u \in D''). \tag{26}$$

As mentioned, the set of preserving controls D' for the problem (2) in the case of the generalization of Leslie's model is always nonempty. To characterize the conditions when $D'' \neq \varnothing$, we need the following notation.

We will consider ordered subsets of $\overline{1,m}$, $\overline{0,n}$, for the designation of which it is reasonable to use variable-length row vectors: the notation $I = (i_1, \ldots, i_\ell)$ means that there is the set $I = \{i_1, \ldots, i_\ell\}$ with a fixed order of elements $i_1, \ldots, i_\ell$.

Let $I = (i_1, \ldots, i_\ell) \subseteq \overline{1,m}$, $J = (j_1, \ldots, j_\ell) \subseteq \overline{0,n}$, and the second of these sets allow repetition of elements, in contrast to the first: $|I| = \ell$, $|J| \in \overline{1,\ell}$. We introduce the following notation:

$$\bar{S}_J^I(a) = \bar{S}_{j_1,\ldots,j_\ell}^{i_1,\ldots,i_\ell}(a) = S_J^I(a) + \sum_{i \notin I} \sigma^{(i)} f_i(a), \tag{27}$$

where $S_J^I(a) = S_{j_1,\ldots,j_\ell}^{i_1,\ldots,i_\ell}(a) = \sum_{k=1}^{\ell} \sigma_{j_k}^{(i_k)} f_{i_k}(a)$. In particular,

$$\bar{S}_j^{(i)}(a) = \sigma_j^{(i)} f_i(a) + \sum_{k \neq i} \sigma^k f_k(a). \tag{28}$$

The set $D'' \neq \varnothing$ if and only if $J^* \neq \varnothing$ [12, Theorem 1], where

$$J^* = \{k \in \overline{1,m} \mid J_k^* \neq \varnothing\}, \quad J_k^* = \{j \in \overline{1,n} \mid \bar{S}_{j-1}^{(k)}(a^*) \geq a^*\} \quad (k \in \overline{1,m}). \quad (29)$$

We denote by $a_j^{(i)}$ the solution of the equation

$$\bar{S}_{j-1}^{(i)}(a) = a \quad (i \in \overline{1,m}, \ j \in \overline{1,n}). \quad (30)$$

This equation is solvable if and only if $j \in J_i^*$ [12, Lemma 1].

We characterize now the structure of D''. For this we consider the polyhedron

$$\overline{U}(a) = \{u \mid p^{(i)}(u) \leq \lambda_i(a) \ (\forall i \in \overline{1,m}), \ q(u) = \mu(a), \ u \geq 0\}. \quad (31)$$

Clear that $\overline{U} = \cup\{\overline{U}(a) \mid a \in [0, \bar{a}_F]\}$. We introduce the sets

$$V_{D'} = V(a^*) \cap D', \quad V_{D''}(a) = V(a) \cap D'',$$

where $V(a)$ is the set of vertex of $\overline{U}(a)$, $a \in [a^*, \bar{a}_{D''}]$. For $i \in \overline{1,m}$, let us denote

$$L_i(a) = \{u \geq 0 \mid p^{(j)}(u) \begin{cases} = \lambda_j(a), \ j = i, \\ \leq \lambda_j(a), \ j \neq i, \end{cases} q(u) = \mu(a)\}, \ L_i = \cup_a L_i(a). \quad (32)$$

Obviously, $L_i(a)$ is the nonnegative part of $(mn-2)$-dimensional affine variety. If $D(a) = \overline{U}(a) \cap D''$, then using the parameterization (31), we obtain:

$$D'' = \bigcup_{a \in [a^*, \bar{a}_{D''}]} D(a), \quad D(a) = \overset{m}{\underset{i=1}{\cup}} L_i(a). \quad (33)$$

For $V \subseteq \overline{U}$ we denote $\sup\{\bar{a}(u) \mid u \in V\} = \bar{a}_V$. Note that $\bar{a}_{\overline{U}} = \bar{a}_F$, $\bar{a}_{D'} = a^*$; by Lemma 1, $\bar{a}_D = \bar{a}_{D''}$ (if $D'' \neq \varnothing$). It is easy to derive the following equalities:

$$L_i \neq \varnothing \Rightarrow \bar{a}_i = \bar{a}_{L_i} = a_n^{(i)} \ (\forall i \in \overline{1,m}), \quad \bar{a}_{D''} = \max_{i \in J^*} \bar{a}_i, \quad (34)$$

where $a_n^{(i)}$ is the solution of the Eq. (30) with $j = n$.

We use also the following notation (see (27); here $i \in \overline{1,m}$, $j,k \in \overline{0,n}$):

$$\bar{\Delta}_{j,k}^i(I,J,a) = (\bar{S}_{j_1-1,\ldots,j_\ell-1,j}^{i_1,\ldots,i_\ell,i}(a), \bar{S}_{j_1-1,\ldots,j_\ell-1,k}^{i_1,\ldots i_\ell,i}(a)]. \quad (35)$$

For a given population state vector $x = (x^{(1)}, x^{(2)}, \ldots, x^{(m)})$, we collect the positive coordinates indices of its blocs $x^{(k)} = (x_{k,1}, \ldots, x_{k,n})$ in the sets $I_k^+(x)$:

$$I_k^+(x) = \{j \in \overline{1,n} \mid x_{k,j} > 0\} \ (k \in \overline{1,m}).$$

In [12], the elements of $V_{D''}(a)$ were found explicitly. If $I_0(u) \subseteq I \cup \{i\}$ and $I_i^+(u) \subseteq \{j,k\}$, where $I = (i_1, i_2, \ldots, i_\ell) \subseteq \overline{1,m} \setminus \{i\}$, $J = (j_1, j_2, \ldots, j_\ell) \subseteq \overline{1,n}$, $j \leq k$, then positive coordinates of $u \in V_{D''}(a)$ are determined as follows:

$$u_{i_r,j_r} = \pi_{j_r}^{(i_r)} f_{i_r}(a) \ (r \in \overline{1,\ell}), \quad u_{i,j} = (q_{j,n}^{(i)})^{-1}(\bar{S}_{j_1-1,\ldots,j_\ell-1}^{i_1,\ldots i_\ell}(a) - a). \quad (36)$$

in the case $\sum\limits_{s=j}^{n} \beta_{i,s} > 0$, $a \in \bar{\Delta}^i_{j-1,n}(I, J, a)$, $I_0(u) = I$, $I_i^+(u) = \{j\}$, and

$$\left.\begin{aligned}
u_{i_r,j_r} &= \pi^{(i_r)}_{j_r} f_{i_r}(a) \ (r \in \overline{1, \ell}), \\
u_{i,j} &= (q^{(i)}_{j,k-1})^{-1} (\bar{S}^{i_1,\dots i_\ell,i}_{j_1-1,\dots,j_\ell-1,k-1}(a) - a), \\
u_{i,k} &= p^{(i)}_{j,k}(q^{(i)}_{j,k-1})^{-1}(a - \bar{S}^{i_1,\dots i_\ell,i}_{j_1-1,\dots,j_\ell-1,j-1}(a)).
\end{aligned}\right\} \tag{37}$$

in the case $\sum\limits_{s=j}^{k-1} \beta_{i,s} > 0$, $a \in \bar{\Delta}^i_{j-1,k-1}(I, J, a)$, $I_0(u) = I_1(u)$.

We denote by $u^i_j(I, J, a)$ and $u^i_{j,k}(I, J, a)$ the controls with coordinates (36) and (37), respectively.

4   The Case of Irreducibility Assumption Absence

We consider here some "degenerate situations" for the feasible set (19) the problem under consideration and show that they lead to the reducibility of the maps used. The first situation is related to the existence on the positive boundary D of the feasible set of controls that are in partial dominance relation; the second— with the existence of linear sections on its nonlinear part D''.

Recall that the strict dominance relation was used in the definition (5) of D. The following statement—shows the adequacy of this approach—it turns out that the set D may contain the elements that are in partial dominance relation.

Theorem 1. *Let the assumptions* (9) *and* (11) *hold. Then* D' *(resp.,* D'') *contains* u, v *with* $u \ngeq v$ *if and only if the condition* (38) *is satisfied (resp., at least one of the conditions* (38), (39) *is satisfied*) :

$$\exists i \in \overline{1, m}\colon \beta_{i,n} = 0, \tag{38}$$

$$m \geq 2; \quad \exists i \in J^*\colon f_i(a) = const \ (\forall a \in [a_f, +\infty), \ a_f \in [a^*, \min_{j \in J_i^*} a_j^{(i)})). \tag{39}$$

Proof. Necessity. If $u, v \in D'$ then, by (23), $q(u) = \mu^* = q(v)$. Since linearity of $q(\cdot)$ (see (16)), then $q(w) = 0$ for $w = u - v$. If $u \ngeq v$ then $w \ngeq 0$, so $I_{i_0}^+(w) \neq \varnothing$ for some $i_0 \in \overline{1, m}$. If $j_0 \in I_{i_0}^+(w)$, then from $q(w) = 0$ we get, by (16), $q^{(i_0)}_{j_0,n} w_{i_0,j_0} = 0$, so $\beta_{i_0,j} = 0$ ($\forall j \in \overline{j_0, n}$). Thus, (38) is satisfied.

Now let $u, v \in D''$. It follows from $D'' \neq \varnothing$ that $J^* \neq \varnothing$ [12, Theorem 1]. If $u \ngeq v$ and $a_1 = \bar{a}(u)$, $a_2 = \bar{a}(v)$, then $a_1 \leq a_2$ by Lemma 1. If $i \in I_0(v)$ then we see from (3) and (22), that $\bar{x}_{i,n}(u) \leq \bar{x}_{i,n}(v) = 0$, so $I_0(v) \subseteq I_0(u)$.

We introduce, as above, $w = u - v$, and show that $I = I_0(v) \cap I_1(w) = \varnothing$, where the sets $I_0(v)$, $I_1(w)$ are defined by equalities (22).

Indeed, otherwise the equalities $\bar{x}_{i,n}(u) = 0$, $\bar{x}_{i,n}(v) = 0$ imply, by (20), the equalities $\lambda_i(a_1) = p^{(i)}(u)$, $\lambda_i(a_2) = p^{(i)}(v)$ ($\forall i \in I$). Since positivity of all $p^{(i)}_{j,n}$ ($j \in \overline{1, n}$) in definition (15) of $p^{(i)}(u)$, we get $\lambda_i(a_1) = p^{(i)}(u) > p^{(i)}(v) = \lambda_i(a_2)$, so $\lambda_i(a_1) > \lambda_i(a_2)$ ($\forall i \in I$). But this inequality contradicts the condition $a_1 \leq a_2$ due to the monotonicity of $\lambda_i(a)$ (see (17)). Therefore, $I = I_0(v) \cap I_1(w) = \varnothing$.

For $m = 1$, this implies that either $I_0(v) = \varnothing$ or $I_1(w) = \varnothing$. But both of these equalities contradict our assumptions: in the first case we get $v \notin D''$, in the second—$w = 0$, so $u = v$. Therefore, the inequality $u \gneq v$ for $u, v \in D''$ is possible only in the case of $m \geq 2$.

If $a_1 = a_2 = a$, then, as above, from the equality $q(u) = \mu(a) = q(v)$ we get $\beta_{i,n} = 0$ for some $i \in \overline{1,m}$, so the condition (38) is satisfied.

Suppose now that $a_1 < a_2$. We show that then $I_0(v) \subseteq J^*$ (see notation (29)).

Let $i \in I_0(v)$. Then $i \in I_0(u)$ and it follows from $I = \varnothing$ that $i \notin I_1(w)$, so that $u^{(i)} = v^{(i)}$. Since $u \neq v$ this imply $I_0(v) \neq \overline{1,m}$. Next, $\lambda_i(a_1) = p^{(i)}(u) = p^{(i)}(v) = \lambda_i(a_2)$, i.e. $\lambda_i(a_1) = \lambda_i(a_2)$ ($\forall i \in I_0(u)$). Then, by (17), $f_i(a_1) = f_i(a_2)$. For a concave monotone increasing function, this means constancy over the entire interval $[a_1, +\infty)$. In this case, since $a^* \leq a_1$ (see (26)), we get:

$$(a^*)^{-1} f_i(a^*) \geq (a_1)^{-1} f_i(a_1) > (a_2)^{-1} f_i(a_2). \tag{40}$$

We see from (28) that $\bar{S}_{n-1}^{(i)}(a) = \sigma_{n-1}^{(i)} f_i(a) + \sum_{k \neq i} \sigma^{(k)} f_k(a)$. Summing up the equalities (20) with the coefficients $\beta_{i,j}$ (taking into account $x_{i,n} = 0$ for $i \in I_0(u)$), we derive the inequality $\bar{S}_{n-1}^{(i)}(a) \geq a$, where $a = \bar{a}(u)$, $u \in D''$. Using this inequality with $a = a_2$, thanks to (40) we obtain: $1 \leq (a_2)^{-1} \bar{S}_{n-1}^{(i)}(a_2) < (a_1)^{-1} \bar{S}_{n-1}^{(i)}(a_1) \leq (a^*)^{-1} \bar{S}_{n-1}^{(i)}(a^*)$, so $\bar{S}_{n-1}^{(i)}(a^*) \geq a^*$ and, by (29), $i \in J^*$.

Thus, the inclusion $I_0(v) \subseteq J^*$ is proved. Therefore, the Eq. (30) is solvable for some $j \in J_i^*$ [12, Lemma 1]. If $a_j^{(i)}$ is its solution, then, as proved above, $(a_j^{(i)})^{-1} \bar{S}_{n-1}^{(i)}(a_j^{(i)}) = 1 < (a_1)^{-1} \bar{S}_{n-1}^{(i)}(a_1)$, so that $a_1 < a_j^{(i)}$. Since a_1 belongs to the interval of constancy of $f_i(a)$, we have $a_f < \min\{a_j^{(i)} \mid i \in J^*, j \in J_i^*\}$. Thus, for $a_1 < a_2$ the condition (39) is satisfied.

Sufficiency. Suppose first that $v \in D'$ and the assumption $\beta_{i_0,n} = 0$ is satisfied for some $i_0 \in \overline{1,m}$. Then, by (23), $q(v) = \mu^*$, $p^{(i)}(v) < \lambda_i^*$ ($\forall i \in \overline{1,m}$). By (16), we have $q_{n,n}^{(i_0)} = 0$, i.e. the coefficient of $v_{i_0,n}$ on the L.H.S. of the equality $q(v) = \mu^*$ is zero. Take v^0 having a single nonzero coordinate $v_{i_0,n}^0$, then for $u(\alpha) = v + \alpha v_0$, by (15), we have: $p^{(i)}(u(\alpha)) = p^{(i)}(v) < \lambda_i^*$ ($\forall i \in \overline{1,m} \setminus \{i_0\}$). Further, $p^{(i_0)}(u(\alpha)) = p^{(i_0)}(v) + \alpha p^{(i_0)}(v_0) = p^{(i_0)}(v) + \alpha v_{i_0,n}^0$. It follows that for $\alpha \in (0, \alpha_0)$, where $\alpha_0 = (v_{i_0,n}^0)^{-1}(\lambda_{i_0}^* - p^{(i_0)}(v))$, the condition $p^{(i_0)}(u(\alpha)) < \lambda_{i_0}^*$ also holds. Next, by (16) we have: $q(u(\alpha)) = q(v) + \alpha q(v_0) = q(v) + \alpha q_{n,n}^{(i_0)} v_{i_0,n}^0 = q(v) = \mu^*$, so that all the conditions (23) guaranteeing $u(\alpha) \in D'$ are met for $\alpha \in (0, \alpha_0)$. Therefore, as the vector u, we can take any vector $u(\alpha)$ with $\alpha \in (0, \alpha_0)$. Thus, $u, v \in D'$ are found that satisfy the condition $u \gneq v$.

Proof of the sufficiency of (38) for the existence of u, v from D'' with $u \gneq v$, is completely analogous to the corresponding proof for the case of the set D'; the control v in this case must be selected in accordance with the representation (33) from the set $\cup_{i \in \overline{1,m}} L_i(a) \setminus \cap_{i \in \overline{1,m}} L_i(a)$ ($a \in [a^*, \bar{a}_D)$).

Now, suppose that the assumption (39) is satisfied for some $i = i_0$. Then it follows from (29), due to the monotonicity of $S_j^{(i)}(a)$ with respect to subscript, that $i_0 \in J^*$, $n \in J^*_{i_0}$, and the Eq. (30) for $i = i_0$, $j = n$ has the solution $a_n^{(i_0)}$. The sufficiency in the case of (38) is proved, therefore we can assume that $\beta_{i,n} > 0$ ($\forall i \in \overline{1,m}$). We show that for $m \geq 2$ there exists $u \in D''$ from $L_{i_0}(a) \backslash \cup_{i \neq i_0} L_i(a)$ ($a = \bar{a}(u) \in [a_f, \bar{a}_D)$) that has a positive coordinate u_{i_1,j_1} in some other block $u^{(i_1)}$ of u ($i_1 \in \overline{1,m}$, $j_1 \in \overline{1,n}$, $i_1 \neq i_0$).

Since the function $a^{-1} \bar{S}_{n-1}^{(i_0)}(a)$ is strictly decreasing for $a \in [a^*, a_n^{(i_0)})$, it follows from $(a_n^{(i_0)})^{-1} \bar{S}_{n-1}^{(i_0)}(a_n^{(i_0)}) = 1$ the inequality $a^{-1} \bar{S}_{n-1}^{(i_0)}(a) > 1$, or, equivalently, $\bar{S}_{n-1}^{(i_0)}(a) > a$. For $u = u_1^{i_1}(I, J, a)$ with coordinates (36), where $I = \{i_0\}$, $J = \{n\}$, we have: $u_{i_1,1} = (\sigma^{(i_1)})^{-1}(\bar{S}_{n-1}^{(i_0)}(a) - a)$, $x_{i_1,1} = (\sigma^{(i_1)})^{-1}(a - \bar{S}_{n-1,0}^{i_0,i_1}(a))$, $u_{i_0,n} = \lambda_{i_0}(a)$. If $a \in \bar{\Delta}_{0,n}^{i_1}(I, J, a) = (\bar{S}_{n-1,0}^{i_0,i_1}(a), \bar{S}_{n-1,n}^{i_0,i_1}(a)] = (\bar{S}_{n-1,0}^{i_0,i_1}(a), \bar{S}_{n-1}^{(i_0)}(a)]$ (see (35)) then, by (36), $u_{i_1,1} > 0$, $x_{i_1,1} > 0$. We saw above that $a < \bar{S}_{n-1}^{(i_0)}(a)$ for $a \in [a^*, a_n^{(i_0)})$. Let us show that there are $a \in [a^*, a_n^{(i_0)})$ such that $a > \bar{S}_{n-1,0}^{i_0,i_1}(a)$.

Indeed, since $\beta_{i_1,n} > 0$, the length d of the interval $\bar{\Delta}_{0,n}^{i_1}(I, J, a)$ is positive: $d = \bar{S}_{n-1}^{(i_0)}(a) - \bar{S}_{n-1,0}^{i_0,i_1}(a) = \sigma^{(i_1)} f_{i_1}(a) > 0$, we get $\bar{\Delta}_{0,n}^{i_1}(I, J, a) \neq \varnothing$. By Lemma 1, $a = \bar{a}(u) \geq a^*$, so $f_{i_1}(a) \geq f_{i_1}(a^*)$ and $d \geq \sigma^{(i_1)} f_{i_1}(a^*)$. This means that for all $a \in (a_0, a_n^{(i_0)})$, where $a_0 \geq a_n^{(i_0)} - \sigma^{(i_1)} f_{i_1}(a^*)$, we have $a > \bar{S}_{n-1,0}^{i_0,i_1}(a)$, hence $a \in \bar{\Delta}_{0,n}^{i_1}(I, J, a)$. In this case $x_{i_1,n}(a, u) > 0$, therefore, $u = u_1^{(i_1)}(I, J, a)$ is the sought-for vector. It remains to find $v \in D''$ with $u \not\geq v$.

Note that since $a_f < a_n^{(i_0)}$, we can assume that the function $f_{i_0}(a)$ is constant over the entire interval $(a_0, a_n^{(i_0)})$, i.e. $\bar{a}(u) > a_f$.

Only one coordinate $\bar{x}_{i_0,n}(u)$ of all the last coordinates $\bar{x}_{i,n}(u)$ ($i \in \overline{1,m}$) of the blocks $\bar{x}^{(i)}(u)$ is zero: $\bar{x}_{i_0,n}(u) = 0$, $\bar{x}_{i,n}(u) > 0$ ($\forall i \in \overline{1,m} \backslash \{i_0\}$), so $u \notin \cup_{i \neq i_0} L_i(a)$. Denote by $u_0 = (u_{i,j}^0)$ a vector having a single nonzero coordinate $u_{i_1,1}^0 = 1$. Let $v(\alpha) = u - \alpha u_0$. We show that $v(\alpha) \in D''$ for sufficiently small $\alpha > 0$. To do this, it is enough to verify that the constraints (24) are met.

Indeed, $v(\alpha)$ is nonnegative for $0 < \alpha < \alpha_1 = u_{i_1,1}^0$. Next, using (16), we obtain: $q(v(\alpha)) = q(u) - \alpha q(u_0) = \mu(\bar{a}(u)) - \alpha q_{1,n}^{(i_1)} u_{i_1,j_1}^0 = \mu(\bar{a}(u)) - \alpha \sigma^{(i_1)}$. Since $\sigma^{(i_1)} > 0$, for $\mu = \mu(\bar{a}(u)) - \alpha \sigma^{(i_1)}$ we have: $0 < \mu < \mu(\bar{a}(u)) \leq \mu^*$ for all $\alpha \in (0, \alpha_2)$, where $\alpha_2 = (\sigma^{(i_1)})^{-1} \mu(\bar{a}(u))$.

Therefore, the equation $q(v(\alpha)) = \mu(a)$ has a solution [11, Lemma 1] for $0 < \alpha < \alpha_1$; we denote it by $a = \bar{a}$. The function $\mu(a)$ is monotone decreasing for $a \geq a^*$, therefore, $\bar{a} \geq \bar{a}(u) > a_f$. It follows that $f_{i_0}(\bar{a}) = f_{i_0}(\bar{a}(u))$ and that $p^{(i_0)}(v(\alpha)) = p^{(i_0)}(u) = \lambda_{i_0}(\bar{a}(u)) = \lambda_{i_0}(\bar{a})$. Next, we obtain from (15): $p^{(i_1)}(v(\alpha)) = p^{(i_1)}(u) - \alpha p^{(i_1)}(u_0) = \lambda_{i_1}(\bar{a}(u)) - \alpha \pi^{(i_1)}$. Thus, the equalities $0 < p^{(i_1)}(v(\alpha)) < \lambda_{i_1}(\bar{a})$ are fulfilled for $0 < \alpha < \alpha_3 = (\pi^{(i_1)})^{-1} \lambda_{i_1}(\bar{a}(u))$.

Finally, $p^{(i)}(v(\alpha)) = p^{(i)}(u) = \lambda_i(\bar{a}(u)) \leq \lambda_i(\bar{a})$ ($\forall i \neq i_0, i_1$), because of $\bar{a}(u) \leq \bar{a}$. So, all conditions for $v(\alpha)$ to belong to D'' are satisfied for $\alpha \in (0, \alpha_0)$, where $\alpha_0 = \min\{\alpha_1, \alpha_2, \alpha_3\}$. Therefore, as the vector v, we can take any vector $u(\alpha)$ with $\alpha \in (0, \alpha_0)$. Thus, in this case too, vectors u, v were found that satisfy the condition $u \not\geq v$. All statements of Theorem are proved.

Remembering that $\bar{a}_i = \sup\{\bar{a}(u) \mid u \in L_i\} = a_n^{(i)}$ and given that $L_i(a_n^{(i)})$ contains a single element, we get the following criterion for $|L_i(a)| \geq 1$:

$$L_i(a) \neq \varnothing \Leftrightarrow a^* \leq a \leq \bar{a}_i, \quad |L_i(a)| > 1 \Leftrightarrow a^* \leq a < \bar{a}_i \ (i \in \overline{1,m}). \quad (41)$$

As noted above, the set of preserving potentially optimal controls D' is entirely contained in the hyperplane Γ. Let us characterize situations when the set of quasi-preserving controls D'' contains a part of some hyperplane (when we say that a certain set contains a part of some hyperplane, it is understood that this set contains the convex hull of some linearly independent vectors belonging to this hyperplane, the number of which is equal to the dimension of space).

Theorem 2. *Let the assumptions (9) and (11) hold. Then the set D'' contains a part of some hyperplane if and only if at least one of the conditions is satisfied:*

(i) *Some function $f_i(a)$ $(i \in \overline{1,m})$ is constant on $[a_1, +\infty)$, where $a_1 \in [a^*, \bar{a}_i)$.*
(ii) *All functions $f_i(a)$ $(\forall i \in \overline{1,m})$ are affine on some interval $[a_1, a_2] \subseteq [a^*, \bar{a}_D]$.*

Proof. Necessity. Let the set D'' contain a part of hyperplane Π. We show first that there exist $u_1, u_2 \in \Pi \cap D''$ with $a_1 \neq a_2$, where $\bar{a}(u_1) = a_1$, $\bar{a}(u_2) = a_2$. Suppose to the contrary that $\bar{a}(u) = \bar{a}$ for all $u \in \Pi \cap D''$. Then, denoting $\bar{\mu} = \mu(\bar{a})$, in view of (24) we obtain $q(u) = \bar{\mu}$. This equality is the equation of the hyperplane Π that (by (24)) is parallel to the hyperplane Γ. Note that since the inclusion $D'' \cap \Gamma \subset \overline{D'}\backslash D'$ (see (25)), the hyperplane Π cannot coincide with Γ (whose equation is $q(u) = \mu^*$), because the set $D'' \cap \Pi$ contains (relatively) internal points (unlike the set $\overline{D'}\backslash D'$). This means, in view of (21), that $\bar{\mu} < \mu^*$.

Take any vector v_1 from the (relative) interior of $D'' \cap \Pi$ and an arbitrary vector $v_2 \in D'$. Due to the convexity of $\overline{U}$ this set contains $v = (1 - \alpha)v_1 + \alpha v_2$ $(\forall \alpha \in (0,1))$ together with v_1, v_2. Since $v_2 \in D'$, by (23), $\bar{x}_{i,n}(v_2) > 0$ $(\forall i \in \overline{1,m})$. From the concavity of $\bar{x}_{i,n}(v)$ we get: $\bar{x}_{i,n}(v) \geq (1 - \alpha)\bar{x}_{i,n}(v_1) + \alpha\bar{x}_{i,n}(v_2) > 0$ $(\forall i \in \overline{1,m})$, so, by (19), $\bar{x}(v) > 0$ and, by (4), $v \in U$.

According to our way of choosing v_1, the projection u of v onto the hyperplane Π for a sufficiently small $\alpha > 0$ belongs to $D'' \cap \Pi$. Since q from (18) is the normal vector to the hyperplane Π, we have $v = u + \beta q$, where $\beta > 0$. Therefore, $v \geq u$ and, by (3), $\bar{x}(u) \geq \bar{x}(v) > 0$, so $\bar{x}(u) > 0$. It follows from (4) that $u \in U$. But this contradicts, by virtue of (6), the condition $u \in D''$.

Thus, the presence of $u_1, u_2 \in \Pi \cap D''$ is proved with $\bar{a}(u_1) \neq \bar{a}(u_2)$. Denote

$$u(\alpha) = (1 - \alpha)u_1 + \alpha u_2, \quad a(\alpha) = (1 - \alpha)a_1 + \alpha a_2, \quad \bar{a}(\alpha) = \bar{a}(u(\alpha)).$$

By Lemma 1, the function $\bar{a}(u)$ is also concave, therefore, $\bar{a}(\alpha) = \bar{a}(u(\alpha)) \geq (1 - \alpha)\bar{a}(u_1) + \alpha\bar{a}(u_2) = (1 - \alpha)a_1 + \alpha a_2 = a(\alpha)$, so $\bar{a}(\alpha) \geq a(\alpha)$.

By assumption that $u_1, u_2 \in \Pi \cap D''$, we have $u(\alpha) \in D''$ for all $\alpha \in [0,1]$. Let us show that u_1, u_2 cannot lie in different sets L_i (recall the notation (32)).

Indeed, otherwise, the condition $I_0 = I_0(u_1) \cap I_0(u_2) = \varnothing$ would be satisfied (see (22)). But then $\bar{x}_{i,n}(u_1)$ and $\bar{x}_{i,n}(u_2)$ could not be zero at the same time $(\forall i \in \overline{1,m})$. Then, as above, Then, as above, we could get the inequality

$$\bar{x}_{i,n}(u(\alpha)) \geq (1 - \alpha)\bar{x}_{i,n}(u_1) + \alpha\bar{x}_{i,n}(u_2) \quad (\forall i \in \overline{1,m}),$$

so that $\bar{x}_{i,n}(u(\alpha)) > 0$ $(\forall i \in \overline{1,m})$. As we saw above, the last inequality contradicts the condition $u(\alpha) \in D''$.

Thus, the set $\Pi \cap D''$ is entirely contained in each of the sets L_i, $i \in I_0$. In addition, according to the last inequality above, the condition $\bar{x}_{i,n}(u(\alpha)) = 0$ implies the equalities $\bar{x}_{i,n}(u_1) = \bar{x}_{i,n}(u_2) = 0$, so $I_0(u(\alpha)) \subseteq I_0(u_1) \cap I_0(u_2)$. Let $i_0 \in I_0(u(\alpha))$, then $i_0 \in I_0(u_1)$, $i_0 \in I_0(u_2)$, and, by (20), the equalities $p^{(i_0)}(u_1) = \lambda_{i_0}(a_1)$, $p^{(i_0)}(u_2) = \lambda_{i_0}(a_2)$, $p^{(i_0)}(u(\alpha)) = \lambda_{i_0}(\bar{a}(\alpha))$ hold.

Hence, by (15), $\lambda_{i_0}(\bar{a}(\alpha)) = p^{(i_0)}(u(\alpha)) = (1 - \alpha)p^{(i_0)}(u_1) + \alpha p^{(i_0)}(u_2) = (1 - \alpha)\lambda_{i_0}(a_1) + \alpha\lambda_{i_0}(a_2)$, so $\lambda_{i_0}(\bar{a}(\alpha)) = (1 - \alpha)\lambda_{i_0}(a_1) + \alpha\lambda_{i_0}(a_2)$. In view the notation (17), we obtain the equivalence of this inequality to the following: $f_{i_0}(\bar{a}(\alpha)) = (1 - \alpha)f_{i_0}(a_1) + \alpha f_{i_0}(a_2)$. Due to the concavity of $f_{i_0}(a)$, we get: $f_{i_0}(\bar{a}(\alpha)) = (1 - \alpha)f_{i_0}(a_1) + \alpha f_{i_0}(a_2) \leq f_{i_0}(a(\alpha))$, so that $f_{i_0}(\bar{a}(\alpha)) \leq f_{i_0}(a(\alpha))$.

On the other hand, it follows from the relation $\bar{a}(\alpha) \geq a(\alpha)$ proved above the opposite inequality $f_{i_0}(\bar{a}(\alpha)) \geq f_{i_0}(a(\alpha))$. Therefore, $f_{i_0}(a(\alpha)) = f_{i_0}(\bar{a}(\alpha)) = (1 - \alpha)f_{i_0}(a_1) + \alpha f_{i_0}(a_2)$. Thus, the following equality holds:

$$f_{i_0}((1 - \alpha)a_1 + \alpha a_2) = (1 - \alpha)f_{i_0}(a_1) + \alpha f_{i_0}(a_2).$$

Further, it follows from equalities $q(u_1) = \mu(a_1)$, $q(u_2) = \mu(a_2)$ since linearity of $q(u)$ that $q(u(\alpha)) = (1 - \alpha)q(u_1) + \alpha q(u_2) = (1 - \alpha)\mu(a_1) + \alpha\mu(a_2)$. Because of $q(u(\alpha)) = \mu(\bar{a}(\alpha))$, we get $\mu(\bar{a}(\alpha)) = (1 - \alpha)\mu(a_1) + \alpha\mu(a_2)$.

If $\bar{a}(\alpha) = a(\alpha)$ then, using the equality $\sigma(a) = \mu(a) + a$ (see (17)), we find: $\sigma(a(\alpha)) = \sigma(\bar{a}(\alpha)) = \mu(\bar{a}(\alpha)) + \bar{a}(\alpha) = (1 - \alpha)\mu(a_1) + \alpha\mu(a_2) + a(\alpha) = (1 - \alpha)\sigma(a_1) + \alpha\sigma(a_2)$, so that $\sigma(a(\alpha)) = (1 - \alpha)\sigma(a_1) + \alpha\sigma(a_2)$. This equality, due to the concavity of $\sigma(a)$ (see (10)), is possible only under the condition

$$f_i((1 - \alpha)a_1 + \alpha a_2) = (1 - \alpha)f_i(a_1) + \alpha f_i(a_2) \ (\forall i \in \overline{1,m}).$$

Thus, we proved the affinity of all functions $f_i(a)$ $(\forall i \in \overline{1,m})$ on $[a_1, a_2]$ in the case $\bar{a}(\alpha) = a(\alpha)$.

If $\bar{a}(\alpha) \neq a(\alpha)$, then $\bar{a}(\alpha) > a(\alpha)$, and $f_{i_0}(\bar{a}(\alpha)) = f_{i_0}(a(\alpha))$ means, thanks to the concavity and monotonicity of $f_{i_0}(a)$, its constancy on $[a(\alpha), +\infty)$.

For definiteness, let $a_1 < a_2$; then $a_1 < a(\alpha) < a_2$ $(\forall \alpha \in (0,1))$ and $f_{i_0}(a(\alpha)) = f_{i_0}(a_2)$. Since $i_0 \in I_0(u(\alpha))$ we get from (20): $0 = \bar{x}_{i_0,n}(u(\alpha)) = \lambda_{i_0}(a(\alpha)) - p^{(i_0)}(u(\alpha)) = \pi^{(i_0)}f_{i_0}(a(\alpha)) - p^{(i_0)}(u(\alpha)) = \pi^{(i_0)}f_{i_0}(a_2) - p^{(i_0)}(u(\alpha)) = p^{(i_0)}(u_2) - [(1 - \alpha)p^{(i_0)}(u_1) + \alpha p^{(i_0)}(u_2)] = (1 - \alpha)[p^{(i_0)}(u_2) - p^{(i_0)}(u_1)] = (1 - \alpha)[\lambda_{i_0}(a_2) - \lambda_{i_0}(a_1)]$. So, $\lambda_{i_0}(a_1) = \lambda_{i_0}(a_2)$, and, by (17), $f_{i_0}(a_1) = f_{i_0}(a_2)$. But then $f_{i_0}(a)$ is constant on $[a_1, +\infty)$. Since $u_1, u_2 \in L_{i_0}$, by (41), $a_1 \in [a^*, \bar{a}_{i_0})$.

Sufficiency. (i) If $f_i(a)$ is constant on $[a_1, +\infty)$ for some i then $\lambda_i(a) = \pi^{(i)}f_i(a)$ is constant on $[a_1, +\infty)$ and all elements of $L_i^0 = \{u \in L_i(a) \mid a \in [a_1, \bar{a}_i]\}$ satisfy the constraint $p^{(i)}(u) = \lambda_i(a) = const$ from (24) that defines the hyperplane.

(ii) Now let all the functions $f_i(a)$ $(\forall i \in \overline{1,m})$ be affine on $[a_1, a_2]$. Fix the index $i_0 \in \overline{1,m}$. Let $S = \{u_1, u_2, \ldots, u_{mn-1}\}$ be an arbitrary linearly independent system of vectors from $L_{i_0}(a_1)$ and let v be an arbitrary vector from $L_{i_0}(a_2)$. We show that the system $S' = S \cup \{v\}$ is linearly independent.

Indeed, otherwise $v = \sum_{j=1}^{mn-1} \alpha_j u_j$ for some α_j, where $\sum_{j=1}^{mn-1} |\alpha_j| \neq 0$. Denoting $\alpha = \sum_{j=1}^{mn-1} \alpha_j$, since $u_j \in L_{i_0}(a_1)$ we get from (32): $\lambda_{i_0}(a_2) = p^{(i_0)}(v) = \sum_{j=1}^{mn-1} \alpha_j p^{(i_0)}(u_j) = \lambda_{i_0}(a_1) \sum_{j=1}^{mn-1} \alpha_j = \alpha \lambda_{i_0}(a_1)$, i.e. $\lambda_{i_0}(a_2) = \alpha \lambda_{i_0}(a_1)$. Hence, due to the monotonicity of $\lambda_{i_0}(a)$ (see (17)) we get $\alpha \geq 1$.

Similarly, $\mu(a_2) = q(v) = \sum_{j=1}^{mn-1} \alpha_j q(u_j) = q(u_j) \sum_{j=1}^{mn-1} \alpha_j = \alpha \mu(a_1)$, so $\mu(a_2) = \alpha \mu(a_1)$. Since $\mu(a)$ is strictly decreasing on $[a^*, a_{D''}]$, it follows, on the contrary, that $\alpha < 1$.

This contradiction shows that the system S' is linearly independent; therefore, it defines a hyperplane in R^{mn}. We denote it by Π.

Let $u(\alpha) = (1-\alpha)u + \alpha v$ ($\alpha \in (0,1)$), where $u \in \text{co } S$. If we show that $u(\alpha) \in D''$, this will mean that D'' contains the part co S' of the hyperplane Π.

Since the set $\overline{U}$ is convex, $u(\alpha) \in \overline{U}$; therefore, there exists a solution $x(\alpha)$ of the system (20) corresponding to $u = u(\alpha)$. We show that this solution is $x(\alpha) = (1-\alpha)\bar{x}(u) + \alpha\bar{x}(v)$ with $a(x(\alpha)) = a(\alpha) = (1-\alpha)a_1 + \alpha a_2$.

Indeed, we have: $x_{i,1}(\alpha) = (1-\alpha)\bar{x}_{i,1}(u) + \alpha\bar{x}_{i,1}(v) = (1-\alpha)(f_i(a_1) - u_{i,1}) + \alpha(f_i(a_2) - v_{i,1}) = [(1-\alpha)f_i(a_1) + \alpha f_i(a_1)] - [(1-\alpha)u_{i,1} + \alpha v_{i,1}] = f_i(a(\alpha)) - u_{i,1}(\alpha)$. Similarly, $x_{i,j+1}(\alpha) = \alpha_{i,j}x_{i,j}(\alpha) - u_{i,j}(\alpha)$ for $j \in \overline{1,n-1}$ ($i \in \overline{1,m}$). Thus, $x(\alpha)$, $u(\alpha)$, $a(\alpha)$ satisfy the constraints (20).

Hence, in particular $x_{i_0,n}(\alpha) = (1-\alpha)\bar{x}_{i_0,n}(u) + \alpha\bar{x}_{i_0,n}(v) = 0$, so that $u(\alpha) \in L_{i_0}(a(\alpha))$ and $u(\alpha) \in D''$. This means that $u(\alpha)$ belongs to the hyperplane Π; therefore, its part co S' is contained in D''. The proof is complete.

We give now an illustrative example.

Example 1. Consider a constraint system of the form (19) with $m = n = 2$, so $x = (x_{1,1}, x_{1,2}; x_{2,1}, x_{2,2})$, $u = (u_{1,1}, u_{1,2}; u_{2,1}, u_{2,2})$, $\alpha_{1,1} = \alpha_{2,1} = 1/2$, $\beta_{i,j} = 1$ ($i, j = 1, 2$). As functions $f_1(a)$, $f_2(a)$ we take the following:

$$f_1(a) = \frac{1}{4}a, \quad f_2(a) = \begin{cases} \sqrt{a}, & 0 \leq a \leq a_f, \\ \sqrt{a_f}, & a > a_f. \end{cases}$$

Here a_f is a parameter, varying which, one can obtain various situations considered in the previous theorems. The functions $f_1(a)$, $f_2(a)$ are concave, all the coefficients $\beta_{i,j}$ are positive, $\sigma'(0) = +\infty$, $\sigma'(+\infty) = 3/8 < 1$, so that all the assumptions (9), (11), (12) are satisfied.

First we consider the case when $f_2(a)$ has no intervals of constancy: $a_f = +\infty$, so $f_2(a) = \sqrt{a}$ on R_+. Using (10), (14)–(17), we obtain: $\sigma_0^{(1)} = \sigma_0^{(2)} = 0$, $\sigma_1^{(1)} = \sigma_1^{(2)} = 1$, $\sigma_2^{(1)} = \sigma_2^{(2)} = \sigma^{(1)} = \sigma^{(1)} = 3/2$, $\sigma(a) = 3a/8 + 3\sqrt{a}/2$, $\mu(a) = 3\sqrt{a}/2 - 5a/8$, $a^* = 36/25$, $\mu^* = 9/10$, $f_1(a^*) = 9/25$, $f_2(a^*) = 6/5$, $\lambda_1(a) = a/8$, $\lambda_2(a) = \sqrt{a}/2$, $p_1^{(1)} = p_2^{(1)} = 1/2$, $p_1^{(2)} = p_2^{(2)} = 1$, $p^{(1)}(u) = \frac{1}{2}u_{1,1} + u_{1,2}$, $p^{(2)}(u) = \frac{1}{2}u_{2,1} + u_{2,2}$, $q(u) = \frac{3}{2}u_{1,1} + u_{1,2} + \frac{3}{2}u_{2,1} + u_{2,2}$.

Next, by (28), $\bar{S}_0^{(1)}(a) = \sigma^{(2)}f_2(a) = 3\sqrt{a}/2$, $\bar{S}_0^{(2)}(a) = \sigma^{(1)}f_1(a) = 3a/8$. Since $\bar{S}_0^{(1)}(a^*) = 9/5 > 36/25$ and $\bar{S}_0^{(2)}(a^*) = 27/50 < 36/25$, we see from (29) that $J_1^* = \{1,2\}$, $J_2^* = \{2\}$, $J^* = \{1,2\}$. Therefore [12, Theorem 1], $D'' \neq \emptyset$.

Solving the Eqs. (30) for $i \in J^*$, $j \in J_i^*$, we find: $a_1^{(1)} = 9/4$, $a_2^{(1)} = 4$, $a_2^{(2)} = 64/25$, therefore, by (34), $\bar{a}_D = \bar{a}_{D''} = a_2^{(1)} = 4$.

Now we consider the case $a_f = 25/9$. Then for $a \geq a_f$ we have $f_2(a) = 5/3$, $\lambda_2(a) = 5/6$. Compared with the previous case, only the values $S_0^{(1)}(a) = 5/2$, $S_1^{(1)}(a) = a/4 + 5/2$, $S_1^{(2)}(a) = 3a/8 + 5/3$, $a_2^{(1)} = 10/3$ are changed; so now $\bar{a}_D = a_2^{(1)} = 10/3$. Since $a_f = 25/9 > a_2^{(2)} = 64/25$, the condition (39) is not satisfied, and there are no vectors with partial dominance in D''. This is because $L_2(a) = \varnothing$ for $a \in [25/9, \infty)$, since $a_2^{(2)} = 64/25 < a_f = 25/9$ (see (41)).

Thus, this example shows the essentiality of the condition $a_f < \min_{j \in J_i^*} a_j^{(i)}$ for the validity of the conclusion of Theorem 2. Note that the condition (39) violates the global irreducibility condition (13) of the map (8).

Now let $a_f = 49/25$. Then for $a \geq a_f$ we have $f_2(a) = 7/5$, $\lambda_2(a) = 7/10$. Compared with the previous case, only the following values are changed: $S_0^{(1)}(a) = 21/10$, $S_1^{(1)}(a) = a/4 + 21/10$, $S_1^{(2)}(a) = 3a/8 + 7/5$, $a_1^{(1)} = 21/10$, $a_2^{(1)} = 14/5$, $a_2^{(2)} = 56/25$. Therefore, in this case $\bar{a}_D = \bar{a}_{D''} = a_2^{(1)} = 14/5$.

Now from (36), (37) we find all controls $u \in V_{D''}$, using process of exhaustion of all possible indices i_1, j_1, i, j, k their non-zero coordinates:

$$u_1(a) = \left(\tfrac{21}{10} - \tfrac{3}{4}a, \tfrac{1}{2}a - \tfrac{21}{20}; 0, 0\right), \quad u_2(a) = \left(\tfrac{1}{4}a, 0; \tfrac{7}{5} - \tfrac{2}{3}a, 0\right), \quad u_3(a) = \left(\tfrac{1}{4}a, 0; 0, \tfrac{21}{10} - a\right),$$

$$u_4(a) = \left(0, \tfrac{1}{8}a; 0, \tfrac{7}{5} - \tfrac{1}{2}a, 0\right), \quad u_5(a) = \left(0, \tfrac{1}{8}a; 0, \tfrac{21}{10} - \tfrac{3}{4}a\right), \quad u_6(a) = \left(0, 0; \tfrac{7}{5} - \tfrac{5}{8}a, \tfrac{5}{16}a\right),$$

$$u_7(a) = \left(\tfrac{14}{15} - \tfrac{5}{12}a, 0; 0, \tfrac{7}{10}\right), \quad u_8(a) = \left(0, \tfrac{7}{5} - \tfrac{5}{8}a; 0, \tfrac{7}{10}\right).$$

The controls $u_1(a) - u_5(a)$ belong to L_1, the controls $u_6(a) - u_8(a)$ belong to L_2. Note that $L_1(a) \cap L_2(a) = \varnothing$ for all $a \in [a_f, \bar{a}_D]$ in this case.

We turn now to Theorem 1. Since $a_f = 49/25 < a_2^{(2)} = 56/25$, the condition (39) is satisfied and, therefore, there are vectors with partial dominance. From the proof of this theorem it is clear, that such vectors must belong to L_2 since the function $f_2(a)$ has a constant part. Indeed, we see that $u_7(a)$ has three of four identical coordinates, and therefore, the controls $u_7(a_1)$, $u_7(a_2)$ with two different admissible values a_1, a_2 of parameter a are in partial dominance relation: for example, $u_7(2) = (1/10, 0; 0, 7/10) \not\geq u_7(11/5) = (1/60, 0; 0, 7/10)$.

This also applies, for example, to the controls $u_8(a)$ with $a_1 = 49/25$ and $a_2 = 56/25$: $u_8(49/25) = (1/10, 0; 0, 7/10) \not\geq u_8(56/25) = (1/60, 0; 0, 7/10)$. Thus, the surface D'' contains segments of lines parallel to axes u_1, u_2. The reason for this fact will become clearer when we proceed to illustrate Theorem 2.

Further, consider the following controls from L_2:

$$u_6(2) = \left(0, 0; \tfrac{3}{20}, \tfrac{5}{8}\right), \quad u_7(2) = \left(\tfrac{1}{10}, 0; 0, \tfrac{7}{10}\right), \quad u_8(2) = \left(0, \tfrac{3}{20}; 0, \tfrac{7}{10}\right), \quad u_8\left(\tfrac{11}{5}\right) = \left(0, 0; \tfrac{1}{40}; 0, \tfrac{7}{10}\right).$$

These vectors form a linearly independent system and, therefore, define some hyperplane. It is easy to verify that whose equation $5u_{2,1} + 10u_{2,2} = 7$ coincides for $a \in [49/25, 56/25]$ with the equation $p^{(2)}(u) = \lambda_2(a)$ from (24) defining the set $L_2(a)$. Indeed, due to the constancy of $f_2(a)$ on $[49/25, 56/25]$, the R.H.S. $\lambda_2(a)$ of this equality is independent of a.

The functions $f_1(a)$, $f_2(a)$ are affine on $[49/25, 56/25]$, therefore, as seen from the proof of Theorem 2, there is a hyperplane, a part of which is contained in L_1. Consider, for example, the following controls:

$u_1(\frac{14}{5}) = (0, \frac{7}{20}; 0, 0)$, $u_2(2) = (\frac{1}{2}, 0; \frac{1}{15}, 0)$, $u_3(2) = (\frac{1}{2}, 0; 0, \frac{1}{10})$, $u_4(2) = (0, \frac{1}{4}; \frac{2}{5}, 0)$.

These vectors define the hyperplane $8u_{1,1} + 12u_{1,2} + 3u_{2,1} + 2u_{2,2} = 21/5$. It is easy to verify that the controls $u_1(a) - u_5(a)$ for $a \in [49/25, 56/25]$ belong to this hyperplane. Thus, the convex hull of $u_1(a)-u_5(a)$ is entirely contained in L_1.

Note that in the case $a_f = 49/25$ the positive boundary of the feasible set contains a nonlinear section corresponding to a change of a in $[a^*, 49/25]$. This nonlinear section disappears if we take $a_f = a^*$. In this case, the feasible set is a polyhedron, despite the presence of a nonlinear constraint. Indeed, the set D', as already mentioned, is part of the hyperplane Γ (see (25)), and for $a_f = a^*$ the set D'' consist of the union of hyperplanes portions corresponding to L_1, L_2.

5   Conclusion

Thus, in this article, we studied the features of the feasible set for the ecological population optimal exploitation problem in the "degenerate" case when the dynamical system step operator is reducible, or (in some region) piecewise linear. It turns out that in the first case, according to Theorem 1, the positive boundary D of the feasible set contains controls that are in the partial dominance relation, and, in the second case, by Theorem 2, the part of the positive boundary D'' consisting of quasi-preserving controls contains a part of some hyperplane. Along the way, we revealed a class of nonlinear models of ecological populations, for which the feasible set of the optimal exploitation problem is a polyhedron.

Next, the results on the properties of the feasible set for the considered optimal exploitation problem obtained here allow us to proceed to the development of an algorithm for its solution, taking into account the found features of this problem. In addition, the fact that the set of preserving controls is contained in the hyperplane Γ (see (25)) makes it possible to obtain a criterion for the existence of an optimal preserving control for a given objective function.

Finally, the results obtained are of methodological significance - thanks to Theorem 1, it becomes more clear why strict dominance relation is used in definition (5) of the positive boundary D.

References

1. Allen, R., Basasibwaki, P.: Properties of age structure models for fish populations. J. Fish. Res. Board Can. **31**, 1119–1125 (1974)
2. Boucekkine, R., Hritonenko, N., Yatsenko, Y.: Age-structured modeling: past, present, and new perspectives. In: Optimal Control of Age structured Populations in Economy. Demography, and the Environment, pp. 1–19. Routledge, London (2011)
3. De Lara, M., Doyen, L.: Sustainable Management of Natural Resources: Mathematical Models and Methods. ESE. Springer, Heidelberg (2008). https://doi.org/10.1007/978-3-540-79074-7
4. Getz, W.M., Haight, R.G.: Population Harvesting: Demographic Models of Fish, Forest, and Animal Resources. Princeton University Press, Princeton (1989)

5. Logofet, D.: Matrices and Graphs. Stability Problems in Mathematical Ecology. CRC Press, Boca Raton (2018)
6. Mazurov, V.D., Smirnov, A.I.: The conditions of irreducibility and primitivity monotone subhomogeneous mappings. Trudy Instituta Matematiki i Mekhaniki UrO RAN **22**(3), 169–177 (2016). (in Russian)
7. Mazurov, V., Smirnov, A.: On the reduction of the optimal non-destructive system exploitation problem to the mathematical programming problem. In: Evtushenko, Y.G., Khachay, M.Y., Khamisov, O.V., Malkova, V., Posypkin, M. (eds.) OPTIMA-2017. CEUR Workshop Proceedings, vol. 1987, pp. 392–398 (2017)
8. Nikaido, H.: Convex Structures and Economic Theory. Academic Press, NY (1968)
9. Reed, W.J.: Optimal harvesting policy for an age specific population. Biometrics **36**(4), 579–593 (1980)
10. Rowntree, L., Lewis, M., Price, M., Wyckoff, W.: Globalization and Diversity: Geography of a Changing World, 4th edn. Pearson, New York (2015)
11. Smirnov, A., Mazurov, V.: On existence of optimal non-destructive controls for ecosystem exploitation problem applied to a generalization of leslie model. In: Evtushenko, Y., Jaćimović, M., Khachay, M., Kochetov, Y., Malkova, V., Posypkin, M. (eds.) OPTIMA-2018. DEStech Trans. Comput. Sci. Eng. pp. 199–213 (2019). https://doi.org/10.12783/dtcse/optim2018/27933
12. Smirnov, A., Mazurov, V.: Generalization of controls bimodality property in the optimal exploitation problem for ecological population with binary structure. In: Jaćimović, M., Khachay, M., Malkova, V., Posypkin (eds.) OPTIMA-2019. CCIS, vol. 1145, pp. 206–221 (2020). https://doi.org/10.1007/978-3-030-38603-0_16

Monopolistic Competition Model with Retailing

Olga Tilzo[1] and Igor Bykadorov[1,2,3]

[1] Novosibirsk State University, Novosibirsk, Russia
kidanovaola@gmail.com, bykadorov.igor@mail.ru
[2] Sobolev Institute of Mathematics SB RAS, Novosibirsk, Russia
[3] Novosibirsk State University of Economics and Management, Novosibirsk, Russia

Abstract. We study the two-level interaction "producer - retailer - consumer" in the monopolistic competition frame. The industry is organized by Dixit-Stiglitz type, the retailer is the only monopolist. The utility function is quadratic. The case of the retailer's leadership is considered. Two types of retailer behavior are studied, namely: with/without free entry conditions. It turned out that the government needs to stimulate the retailer by paying him subsidies to increase social welfare. A similar result was obtained with respect to consumer surpluses.

Keywords: Monopolistic competition · Retailing · Equilibrium · Taxation · Social welfare · Consumer surplus

1 Introduction

Now, there are many works on the vertical interaction between the producer and the retailer, which describe the impact of such interaction on the economy.

In Spengler's early work [1], the simplest case of Stackelberg game with two players is studied, one of the players is a leader while the other one is a follower. In the first step, the leader sets his price. Then the follower, having analyzed the actions of the leader, makes his move. As a result, the price increases twice by each monopolist, respectively, which results in a decrease in social welfare.

Further, two classes of models can be singled out: spatial Hotelling models [2] and models of Dixit-Stiglits [3] type.

A striking example of the first class of models is the Salop model [4] (a circular city model) with one manufacturer and several retailers, located along the circle (street) equidistant from each other. As a result, retailers and consumers interact, so that each consumer is served, and for retailers there is no need to unite with the manufacturer. This model was modified by Dixit [5], by introducing two-level production (up stream and down stream industries), i.e., now the retailer also has the means of production. In other words, the monopolist-producer sells the intermediate goods to retailers, who then process and sell them in the form of finished goods. In this case, the integration is justified, because it increases social welfare by optimizing the number of retailers.

© Springer Nature Switzerland AG 2020
Y. Kochetov et al. (Eds.): MOTOR 2020, CCIS 1275, pp. 287–301, 2020.
https://doi.org/10.1007/978-3-030-58657-7_24

The second class of models is based on the idea of a representative consumer in the style of Dixit-Stiglits. In particular, Parry and Groff [6] use the *CES* function and show that the integration leads to a deterioration of social welfare. The work of Chen [7] can be considered as the next step in this direction. In his model, the monopolist-producer is considered, who first chooses the number of manufactured products (and, accordingly, the number of retailers). Further, a supply contract is concluded with each retailer separately, namely, wholesale prices per unit of goods and an initial one-time payment are established. Finally, the retailer sets the retail price. The main result is that the number of product names is less than "socially optimal".

The combination of models of the type of Hotelling and Dixit-Stiglits gives Hamilton and Richards model [8]. It considers two types of commodity varieties: competing supermarkets and competing products within each supermarket. It turned out that the increase in product differentiation does not necessarily lead to an increase in the equilibrium length of product lines.

In our paper, the industry of producers is organized according to Dixit-Stiglits type with quadratic utility [9], and the monopolist is the only retailer. The relevance of the chosen approach is determined by modern realities. The majority of sales today fall on well-developed supermarket chains. Selling their value to suppliers, retailers begin to dictate tough conditions to them. In order to sell products, manufacturers have to agree these conditions. We explored two types of behavior of the retailer, namely, with and without the conditions of free entry (zero-profit).

The paper is organized as follows.

In Sect. 2 we define the main assumptions of monopolistic competition, formulate the model, find the condition when the free entry happens and when not, see Proposition 1; describe the equilibrium, see Proposition 2.

In Sect. 3 we introduce the taxation of Pigouvian type and recalculate the equilibrium, see Proposition 3.

In Sect. 4 we get the equilibrium social welfare (see Sect. 4.1, Proposition 4) and consumer surplus (see Sect. 4.2, Proposition 5). Moreover, here we formulate the main result of the paper: the optimal taxation is negative, i.e., the best is to subsidize the retailer (see Sect. 4.3, Proposition 6).

We omit the proofs of Propositions 1–5 which are rather technical. Instead, the main Proposition 6 we prove carefully, see Appendix A.

Section 6 concludes.

Note that the paper continues the authors' research [10–12].

2   Model

Consider the monopolistic competition model with two-level interaction "manufacturer - retailer - consumer". As it is usual on monopolistic competition, we assume the assumptions (cf. [3])

- firms produce the goods of the same nature ("product variety"), but not absolute substitutes;

- each firm produces only one type of product variety and chooses its price;
- the number (mass) of firms is quite large;
- the free entry (zero-profit) condition is fulfilled.

In addition to product diversity, there are other products on the market, "numéraire". Moreover, there are L identical consumers, each of them delivers one unit of labor to the market.

In this paper, we consider the situation when manufacturers sell products through a monopolist-retailer.

Consider the utility function in the case of linear demand, so-called OTT-function, proposed by Ottaviano, Tabucci, and Tisse [9]:

$$U(\mathbf{q}, N, A) = \alpha \int_0^N q(i)di - \frac{\beta - \gamma}{2} \int_0^N (q(i))^2\, di - \frac{\gamma}{2} \left(\int_0^N q(i)di \right)^2 + A, \quad (1)$$

where

- $\alpha > 0, \beta > \gamma > 0$ are some parameters[1];
- N is a continuum of firms or the length of a product line, reflecting the range (interval) of diversity;
- $q(i) \geq 0$ is a demand function, i.e., the consumption of i-th variety, $i \in [0, N]$;
- $\mathbf{q} = (q(i))_{i \in [0,N]}$ is infinite-dimensional vector;
- $A \geq 0$ is the consumption of other, aggregated products (numéraire).

Now we formulate the budget constraint. Let

- $p(i)$ be the wholesale price of i-th product variety, i.e., the price in the case without retailer;
- $r(i)$ be the retailer's premium on the i-th product variety, so $p(i) + r(i)$ is the price of the i-th variety for the consumer;
- $w \equiv 1$ be the wage rate in the economy normalized to 1;
- P_A be the price numéraire.

Then the budget constraint is

$$\int_0^N (p(i) + r(i))q(i)di + P_A A \leq wL + \int_0^N \pi_{\mathcal{M}}(i)di + \pi_{\mathcal{R}}, \quad (2)$$

---

[1] Due to [9] (see p. 413), "... α expresses the intensity of preferences for the differentiated product, whereas $\beta > \gamma$ means that consumers are biased toward a dispersed consumption of varieties. ... the quadratic utility function exhibits love of variety as long as $\beta > \gamma$. ... for a given value of β, the parameter γ expresses the substitutability between varieties: The higher γ, the closer substitutes the varieties. When $\beta = \gamma$, substitutability is perfect." Two quadratic terms ensure strict concavity in two dimensions: definite consumer's choice among commodities and between the two sectors. The main feature achieved by this constructions is that this utility generates the system of linear demands for each variety and linear demand for the whole differentiated sector.

where $\pi_{\mathcal{M}}(i)$ is the profit of firm $i \in [0, N]$ while $\pi_{\mathcal{R}}$ is retailer profit. The right side of (2) is Gross Domestic Product (GDP) by income, while the left side is costs.

This way, the problem of the representative consumer is

$$U(\mathbf{q}, N, A) \to \max_{\mathbf{q}, A}$$

subject to (2), where $U(\mathbf{q}, N, A)$ is defined in (1).

Solving this problem, we can find the demand function for each $i \in [0, N]$:

$$q(i) = a - (b + cN)(p(i) + r(i)) + cP, \qquad (3)$$

where coefficients a, b, c are defined as

$$a = \frac{\alpha}{\beta + (N-1)\gamma}, \quad b = \frac{1}{\beta + (N-1)\gamma}, \quad c = \frac{\gamma}{(\beta - \gamma)(\beta + (N-1)\gamma)},$$

and P is a price index

$$P = \int_0^N (p(j) + r(j))dj.$$

Let

- d be marginal costs, i.e., the number of units of labor required to each firm to produce a unit of differentiated product;
- F be fixed costs, i.e., the number of units of labor required by each firm to produce differentiated product.

Then the problem of maximizing the profit of firm $i \in [0, N]$ is

$$\pi_{\mathcal{M}}(i) = (p(i) - d)q(i) - F \to \max_{p(i)}, \qquad (4)$$

where $q(i)$ is (3).

Note that problem (4) is quadratic on $p(i)$.

Now let us formulate the problem of the retailer. Similar to the firm's problem (4), let

- $d_{\mathcal{R}}$ be marginal costs, i.e., the number of units of labor required to retailer to sale a unit of differentiated product of each firm;
- $F_{\mathcal{R}}$ be fixed costs, i.e., the number of units of labor required to retailer to sale the differentiated product of each firm.

This way the problem of maximizing the profit of the retailer is

$$\pi_{\mathcal{R}} = \int_0^N (r(j) - d_{\mathcal{R}})\, q(j)dj - \int_0^N F_{\mathcal{R}} dj \to \max_{\mathbf{r}, N}, \qquad (5)$$

$$\pi_{\mathcal{M}}(i) \geq 0, \quad i \in [0, N], \qquad (6)$$

where $\mathbf{r} = (r(i))_{i \in [0, N]}$.

Since the model is homogeneous (the firms are identical), it is possible to show that only two cases can happen, namely,

- in the solution of problem (5), (6), the profit of each firm $i \in [0, N]$ is positive: $\pi_{\mathcal{M}}(i) > 0$ (ignoring the free entry conditions[2]);
- in the solution of problem (5), (6), the profit of each firm $i \in [0, N]$ is zero: $\pi_{\mathcal{M}}(i) = 0$ (taking into account the free entry conditions).

In this paper, we study the Stackelberg equilibrium under the leadership of a retailer, i.e., the retailer maximizes its profit under the best response of firms.

We call the case of the retailer's leadership with ignoring the free entry conditions as RL, while we call the case of the retailer's leadership with taking into account the free entry conditions as $RL(I)$.

Let us describe these cases explicitly.

Case RL. In this case the retailer at the same time chooses trade markup $\mathbf{r} = (r(i))_{i \in [0,N]}$ and scale of product diversity N, correctly predicting a subsequent response from manufacturers. There are stages of solving the problem:

1. Solving the problem of a consumer, we get $q = q(i, p(i), r(i), N)$.
2. Solving the problem of the manufacturer

$$\pi_{\mathcal{M}}(i) = (p(i) - d)q(i, p(i), r(i), N) - F \rightarrow \max_{p(i)},$$

we get $p = p(i, r(i), N)$ and $q = q(i, r(i), N)$.
3. Solve the problem of the retailer

$$\pi_{\mathcal{R}} = \int_0^N (r(i) - d_{\mathcal{R}}) q(\mathbf{r}, N) di - \int_0^N F_{\mathcal{R}} di \rightarrow \max_{\mathbf{r}, N},$$

$$\pi_{\mathcal{M}}(\mathbf{r}, N) \geq 0.$$

Case RL(I). In this case the retailer first uses the free entry condition to calculate $N = N(\mathbf{r})$, given the subsequent manufacturers response and then maximizes their profits by $\mathbf{r}$:

1. Solving the problem of a consumer, we get $q = q(i, p(i), r(i), N)$.
2. Solving the problem

$$\pi_{\mathcal{M}}(i) = (p(i) - d)q(i, p(i), r(i), N) - F \rightarrow \max_{p(i)},$$

we get $p = p(i, r(i), N)$, $q = q(i, r(i), N)$.
3. The free entry condition $\pi_{\mathcal{M}}(i, r(i), N) = 0$ gives $N = N(\mathbf{r})$.
4. Solve the problem of the retailer

$$\pi_{\mathcal{R}} = \int_0^N (r(i) - d_{\mathcal{R}}) q(\mathbf{r}) di - \int_0^N F_{\mathcal{R}} di \rightarrow \max_{\mathbf{r}}.$$

---

[2] Ignoring the free entry conditions means that we are somewhat expanding the traditional concept of monopolistic competition.

The question arises: when is case RL realized, and when is case $RL(I)$ realized? It turns out the answer is uniquely determined by the value

$$\mathcal{F} = \frac{F_{\mathcal{R}}}{2F}. \tag{7}$$

Proposition 1. *1. The case RL happens if and only if $\mathcal{F} > 1$.*
2. The case $RL(I)$ happens if and only if $\mathcal{F} \leq 1$.

Now we can describe the Stackelberg equilibrium when retailer is leader. Let

$$\Delta = \sqrt{\frac{F}{\beta - \gamma}} > 0, \qquad \varepsilon = \frac{\beta - \gamma}{\gamma} > 0, \tag{8}$$

$$f = \sqrt{F \cdot (\beta - \gamma)} > 0, \qquad D = \frac{\alpha - d - d_{\mathcal{R}}}{\sqrt{F \cdot (\beta - \gamma)}}. \tag{9}$$

Proposition 2. *In cases RL and $RL(I)$, the equilibrium demand q, price p, markup r, mass of firms N and profit of retailer $\pi_{\mathcal{R}}$ are as in Table 1, where $\mathcal{F}, \Delta, \varepsilon, f, D$ are defined in (7)–(9).*

Table 1. The equilibrium

	q	p	r	N	$\pi_{\mathcal{R}}$
RL	$\Delta\sqrt{\mathcal{F}}$	$d + f\sqrt{\mathcal{F}}$	$d_{\mathcal{R}} + \dfrac{fD}{2}$	$\dfrac{\varepsilon}{2} \cdot \left(\dfrac{D}{\sqrt{\mathcal{F}}} - 4\right)$	$\dfrac{f^2}{4\gamma} \cdot \left(D - 4\sqrt{\mathcal{F}}\right)^2$
RL(I)	Δ	$d + f$	$d_{\mathcal{R}} + f \cdot \left(\dfrac{D}{2} + \mathcal{F} - 1\right)$	$\dfrac{(D - 2\mathcal{F} - 2)\varepsilon}{2}$	$\dfrac{f^2}{4\gamma} \cdot (D - 2\mathcal{F} - 2)^2$

3   Taxation

Let the government stimulate producers in the following way: let the retailer pay the tax τ from each unit of the sold product. Then profit of the retailer is modified as

$$\pi_{\mathcal{R}} = \int_0^N (r(i) - (d_{\mathcal{R}} + \tau))q(i)di - \int_0^N F_{\mathcal{R}}di.$$

The taxes collected are distributed among the producers by a one-time payment method (of Pigouvian type). The case of negative τ means that the retailer needs subsidies, paid from consumer taxes in the amount of $\tau \int_0^N q(i)di$.

Proposition 3. *With taxation τ, the equilibrium demand q, price p, markup r, and mass of firms N are as in Table 2, where*

$$S = -\frac{\tau}{2f} + \sqrt{\left(\frac{\tau}{2f}\right)^2 + 1} > 0 \tag{10}$$

while $\mathcal{F}, \Delta, \varepsilon, f, D$ are defined in (7)–(9).

Table 2. Equilibrium with taxation

	q	p	r		N
RL	$\Delta\sqrt{\mathcal{F}}$	$d + f\sqrt{\mathcal{F}}$	$d_{\mathcal{R}} + \dfrac{fD}{2} + \dfrac{\tau}{2}$		$\dfrac{\varepsilon}{2\sqrt{\mathcal{F}}} \cdot \left(D - \dfrac{\tau}{f} - 4\sqrt{\mathcal{F}} \right)$
RL(I)	ΔS	$d + fS$	$d_{\mathcal{R}} + \dfrac{f}{2} \cdot \left(D + \dfrac{2\mathcal{F}+1}{S} - 3S \right)$		$\dfrac{\varepsilon}{2S} \left(D - \dfrac{2\mathcal{F}+1}{S} - S \right)$

4   Social Welfare and Consumer Surplus. Optimal Taxation

In this section, we consider social welfare, consumer surplus and we calculate the equilibrium social welfare and equilibrium consumer surplus in two cases: with and without taxation. Besides, we compare two kinds of optimal taxation.

4.1   Social Welfare

Consider the function of social welfare W, a measure of the well-being of society.

$$
W = \alpha \int_0^N q(i)di - \frac{\beta - \gamma}{2} \cdot \int_0^N (q(i))^2 \, di - \frac{\gamma}{2} \cdot \left(\int_0^N q(i)di \right)^2
$$
$$
- \int_0^N (d + d_{\mathcal{R}})q(i)di - \int_0^N (F + F_{\mathcal{R}})di.
$$

In the symmetric case, it has the form:

$$
W = (\alpha - d - d_{\mathcal{R}})Nq - \frac{\beta - \gamma}{2} \cdot q^2 N - \frac{\gamma}{2} \cdot N^2 q^2 - (F + F_{\mathcal{R}})N.
$$

Substituting the equilibrium from Proposition 2 and Proposition 3, we get

Proposition 4. *The equilibrium welfare is as in Table 3 and Table 4, where*

$$
H = \frac{F \cdot (\beta - \gamma)}{2\gamma} > 0, \tag{11}
$$

while $\mathcal{F}, f, D, S$ are defined in (7), (9)–(10).

4.2   Consumer Surplus

Consumer surplus (CS) is a measure of well-being that people derive from the consumption of goods and services. It is the difference between the maximum price a consumer is willing to pay and the market price:

$$
CS = \alpha \int_0^N q(i)di - \frac{\beta - \gamma}{2} \cdot \int_0^N (q(i))^2 \, di - \frac{\gamma}{2} \cdot \left(\int_0^N q(i)di \right)^2 - \int_0^N (p(i) + r(i))q(i)di.
$$

Table 3. Social welfare without taxation

	W
RL	$\left(D - 4\sqrt{\mathcal{F}}\right) \cdot \left(\dfrac{3}{4} \cdot (D - 2\sqrt{\mathcal{F}}) - \dfrac{1}{\sqrt{\mathcal{F}}}\right) \cdot H$
RL(I)	$(D - 2\mathcal{F} - 2) \cdot \left(\dfrac{3}{4} \cdot (D - 2\mathcal{F}) - 1\right) \cdot H$

Table 4. Social welfare with taxation

	W
RL	$\left(D - \dfrac{\tau}{f} - 4\sqrt{\mathcal{F}}\right) \cdot \left(3D + \dfrac{\tau}{f} - 6\sqrt{\mathcal{F}} - \dfrac{4}{\sqrt{\mathcal{F}}}\right) \cdot \dfrac{H}{4}$
RL(I)	$\left(D - \dfrac{2\mathcal{F}+1}{S} - S\right) \cdot \left(3D - 3 \cdot \dfrac{2\mathcal{F}+1}{S} - S\right) \cdot \dfrac{H}{4}$

In symmetric case, it is

$$CS = qN \cdot \left(\alpha - \frac{\beta - \gamma}{2} \cdot q - \frac{\gamma}{2} \cdot Nq - (p+r)\right).$$

Proposition 5. *The equilibrium consumer surplus is as in Table 5, where* $\mathcal{F}, f, D, S, H$ *are defined in (7), (9)–(11).*

Table 5. Consumer surplus

	CS (without taxation)	CS (with taxation)
RL	$\dfrac{f^2}{8\gamma} \cdot \left(D - 4\sqrt{\mathcal{F}}\right) \cdot \left(D - 2\sqrt{\mathcal{F}}\right)$	$\left(D - \dfrac{\tau}{f} - 4\sqrt{\mathcal{F}}\right) \cdot \left(D - 2\sqrt{\mathcal{F}} - \dfrac{\tau}{f}\right) \cdot \dfrac{H}{4}$
RL(I)	$\dfrac{f^2}{8\gamma} \cdot (D - 2\mathcal{F} - 2) \cdot (D - 2\mathcal{F})$	$\left(D - \dfrac{2\mathcal{F}+1}{S} - S\right) \cdot \left(D - \dfrac{2\mathcal{F}+1}{S} + S\right) \cdot \dfrac{H}{4}$

4.3   Optimal Taxation

We consider two concepts of optimal taxation. Namely,

- maximization of welfare W with respect to τ which leads to optimal τ_W, it allows the government to determine optimal fiscal policy;
- maximization of consumer surplus CS with respect to τ, which leads to optimal τ_{CS}.

It seems nature to assume that "market exists at $\tau = 0$", i.e., the condition

$$N|_{\tau=0} \geq 0 \tag{12}$$

holds.

It turns out that in case RL, an explicit formula for τ_W can be found. Moreover, in both cases, RL and $RL(I)$, it is possible to determine the sign of τ_W. As to τ_{CS}, we were not able to find τ_{CS}, but in case $RL(I)$ we can to determine its sign. Finally, in case $RL(I)$ we can compare τ_W and τ_{CS}.

Let us summarize these findings in

Proposition 6. *Let condition (12) holds.*

1. In case RL, the optimal tax from social welfare point of view is

$$\tau_W = \left(\frac{2 + \mathcal{F}}{\sqrt{\mathcal{F}}} - D \right) f < 0, \tag{13}$$

while the optimal tax from consumer surplus point of view is

$$\tau_{CS} = -\infty. \tag{14}$$

2. In case RL(I), the optimal tax, from social welfare point of view, is negative,

$$\tau_W < 0, \tag{15}$$

while the optimal tax from consumer surplus point of view is

$$\tau_{CS} < 0. \tag{16}$$

3. The optimal tax from social welfare point of view is the optimal tax from consumer surplus point of view is less than the optimal taxation from social welfare point of view, i.e.,

$$\tau_{CS} < \tau_W < 0. \tag{17}$$

5 Numerical Example

The example below illustrates Proposition 6. Parameters:

$$\alpha = 12, \beta = 2, \gamma = 1, F = 1, d = 1, d_{\mathcal{R}} = 2, f = 1, D = 9, \varepsilon = 1, \Delta = 1, H = \frac{1}{2}.$$

For the case RL: $F_{\mathcal{R}} = 7$, $\mathcal{F} = \dfrac{7}{2}$.

For the case RL(I): $F_{\mathcal{R}} = 1$, $\mathcal{F} = \dfrac{1}{2}$.

Figure 1 shows that (13)–(17) holds.

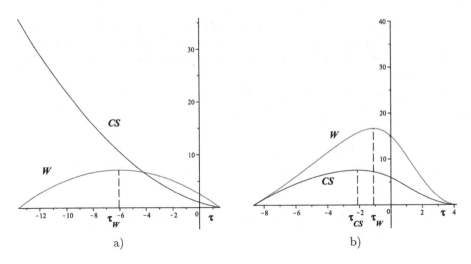

Fig. 1. Example: a) RL with taxation, b) RL(I) with taxation.

6 Conclusion

Our research is based on a theoretical model, which supports several conclusions.

1. The concentration of retailing per se (or, more precisely, the increase in bargaining power of the retailer against numerous manufacturers) generally *enhances* total welfare and, under realistic cost function, consumer surplus as well. The consumer gains operate either through lower retail prices, or through increasing variety, or both. From this viewpoint, the undertaken restriction of market shares looks unnecessary and even harmful.
2. The redistribution of profits to consumers in the form of Pigovian taxation levied on the retailer (with taxes proportional to the volume of sales) has been found *inefficient*. For the sake of total public welfare, the government should rather subsidize the volume of retailing, a policy which appears politically infeasible.

The economic forces leading to such (surprising for the Russian legislature) conclusions are more or less familiar to economists. Any consumer-goods industry produces many varieties of food, clothes, or other goods. Therefore it is organized as a "monopolistic competition" industry: each producer behaves as a monopolist for her brand, but entry is open. Each shop behaves monopolistically for similar reasons. Then, monopolistic behavior in production combined with a monopolistic retailer imply two-tier monopoly and a harmful "double marginalization" effect. When an increase in bargaining power on the retailer side occurs (the other side is too dispersed), it brings the bargaining relations closer to a vertically integrated industry, thus reducing the deadweight loss.

Of course, our theoretical conclusions need more thorough study from different viewpoints of reality, in particular from re-distributional considerations:

which groups of consumers may be affected positively or negatively by market concentration or its regulation (we did not take into account groups so far). Nevertheless, we provide several important hypotheses to be discussed in relation to current legislation. Our results also indicate the need for further research before imposing new market regulations[3].

Further increase in the retailer's bargaining power, in the sense of possible entrance fees levied on manufacturers, also *enhances* profit, consumer surplus and therefore total social welfare. Moreover, if the retailers were forced to pay from the government for the right to use entrance fees. By this or another similar redistribution tool, the additional retailer's profit could be transferred to consumers (through increasing public goods availability or decreasing taxes). Then such a welfare-improving pricing strategy as entrance fees could become more politically feasible. As to possible direct governmental regulation of retailing via capping the markup (not actually practiced so far), while this measure generally *enhances* total welfare and consumer surplus when the cap is tailored optimally (in the absence of entrance fees), it does *not* do so *as much as entrance fees combined with transfers to consumers.*

It seems interesting to extend this approach to the case of additive separable utility, as well as nonlinear cost functions, investments in R&D [13–16].

Acknowledgments. The authors are very grateful to the anonymous reviewers for very useful suggestions for improving the text. The study was carried out within the framework of the state contract of the Sobolev Institute of Mathematics (project no. 0314-2019-0018) and within the framework of Laboratory of Empirical Industrial Organization, Economic Department of Novosibirsk State University. The work was supported in part by the Russian Foundation for Basic Research, projects 18-010-00728 and 19-010-00910 and by the Russian Ministry of Science and Education under the 5–100 Excellence Programme.

A   Appendix. Proof of Proposition 6

A.1   Optimal Taxation: The Case RL

We first consider the case RL, i.e.,

$$\mathcal{F} \geq 1. \tag{18}$$

---

[3] It seems that further increase in the retailer's bargaining power, in the sense of possible entrance fees levied on manufacturers, also *enhances* profit, consumer surplus and therefore total social welfare. Moreover, if the retailers were forced to pay from the government for the right to use entrance fees. By this or another similar redistribution tool, the additional retailer's profit could be transferred to consumers (through increasing public goods availability or decreasing taxes). Then such a welfare-improving pricing strategy as entrance fees could become more politically feasible. As to possible direct governmental regulation of retailing via capping the markup (not actually practiced so far), while this measure generally *enhances* total welfare and consumer surplus when the cap is tailored optimally (in the absence of entrance fees), it does *not* do so *as much as entrance fees combined with transfers to consumers.*

Let us note that in this case (12) means, see Table 1,

$$D \geq 4\sqrt{\mathcal{F}}. \tag{19}$$

Recall that, see Tables 2, 4 and 5,

$$N = -\frac{(\tau - \tau_1)\,\varepsilon}{2f\sqrt{\mathcal{F}}} \geq 0 \iff \tau \leq \tau_1, \tag{20}$$

$$W = -\frac{(\tau - \tau_1)\,(\tau - \tau_2)\,H}{4f^2}, \tag{21}$$

$$CS = \frac{(\tau - \tau_1)\,(\tau - \tau_3)\,H}{4f^2}, \tag{22}$$

where

$$\tau_1 = \left(D - 4\sqrt{\mathcal{F}}\right)f, \qquad \tau_2 = \left(\frac{6\mathcal{F} + 4}{\sqrt{\mathcal{F}}} - 3D\right)f, \qquad \tau_3 = \left(D - 2\sqrt{\mathcal{F}}\right)f.$$

Note that, due to (18) and (19),

$$\tau_2 < 0 \leq \tau_1 < \tau_3. \tag{23}$$

Function (21) is strictly concave. Therefore, due to (20) and (23), we get

$$\tau_W = \frac{\tau_1 + \tau_2}{2} = \left(\frac{\mathcal{F} + 2}{\sqrt{\mathcal{F}}} - D\right)f.$$

Moreover, due to (19) and (18),

$$\tau_W \leq \left(\frac{\mathcal{F} + 2}{\sqrt{\mathcal{F}}} - 4\sqrt{\mathcal{F}}\right)f = \frac{(2 - 3\mathcal{F})\,f}{\sqrt{\mathcal{F}}} \leq -\frac{f}{\sqrt{\mathcal{F}}} < 0.$$

Thus, we get (13).

Besides, function (22) is strictly convex. Therefore, due to (20) and (23), we get (14).

A.2   Optimal Taxation: The Case $RL(I)$

Consider the case $RL(I)$, i.e.,

$$0 \leq \mathcal{F} < 1. \tag{24}$$

Let us note that in this case (12) means

$$D \geq 2 \cdot (\mathcal{F} + 1). \tag{25}$$

Recall that, see Tables 2, 4 and 5,

$$N = -\frac{(S - S_1)\,(S - S_2)\,\varepsilon}{2S^2} \geq 0 \iff S_1 \leq S \leq S_2, \tag{26}$$

$$W = \frac{(S - S_1)(S - S_2)(S - S_3)(S - S_4) H}{4S^2}, \tag{27}$$

$$CS = -\frac{(S - S_1)(S - S_2)(S - S_5)(S - S_6) H}{4S^2}, \tag{28}$$

where $S_1, S_2, S_3, S_4, S_5, S_6$ are (real due to (25))

$$S_1 = \frac{D - \sqrt{D^2 - 4 \cdot (2\mathcal{F} + 1)}}{2}, \qquad S_2 = \frac{D + \sqrt{D^2 - 4 \cdot (2\mathcal{F} + 1)}}{2},$$

$$S_3 = \frac{3D - \sqrt{9D^2 - 12 \cdot (2\mathcal{F} + 1)}}{2}, \qquad S_4 = \frac{3D + \sqrt{9D^2 - 12 \cdot (2\mathcal{F} + 1)}}{2},$$

$$S_5 = \frac{-D + \sqrt{D^2 + 4 \cdot (2\mathcal{F} + 1)}}{2}, \qquad S_6 = \frac{-D - \sqrt{D^2 + 4 \cdot (2\mathcal{F} + 1)}}{2},$$

and $S > 0$ is defined in (10). Since $\dfrac{\partial S}{\partial \tau} < 0$, S is monotone with respect to τ. Therefore, it is enough to examine the behavior of the functions (27) and (28) with respect to S. Note that, due to (24) and (25),

$$S_6 < 0, \quad 0 < S_5 < S_3 < S_1 \le 1 < S_2 < S_4. \tag{29}$$

One has

$$\frac{\partial W}{\partial S} = \frac{H}{2S^3} \cdot \left(S^4 - 2D \cdot S^3 + 3D \cdot (2\mathcal{F} + 1) \cdot S - 3 \cdot (2\mathcal{F} + 1)^2 \right), \tag{30}$$

$$\frac{\partial}{\partial S}(CS) = -\frac{H}{2S^3} \cdot \left(S^4 - (2\mathcal{F} + 1) D \cdot S + (2\mathcal{F} + 1)^2 \right). \tag{31}$$

Due to (25), (29), and Descartes' theorem, the number of positive roots of the equation $\dfrac{\partial W}{\partial S} = 0$ is either three or one, while the number of positive roots of the equation $\dfrac{\partial}{\partial S}(CS) = 0$ is either two or zero. Due to Rolle's theorem, on each of the intervals $[S_3, S_1], [S_1, S_2], [S_2, S_4]$ there is a point at which $\dfrac{\partial W}{\partial S} = 0$, while on each of the intervals $[S_5, S_1], [S_1, S_2]$ there is a point at which $\dfrac{\partial}{\partial S}(CS) = 0$.

Therefore, $\dfrac{\partial W}{\partial S}$ has three positive roots, two of which do not lie in $[S_1, S_2]$, while $\dfrac{\partial}{\partial S}(CS)$ has two positive roots, one of which does not lie in $[S_1, S_2]$. Therefore, W has a single maximum S_W on $[S_1, S_2]$, while CS has a single maximum S_{CS} on $[S_1, S_2]$. Moreover, see (26),

$$\frac{\partial W}{\partial S} \begin{cases} > 0, & S \in (S_1, S_W), \\ < 0, & S \in (S_W, S_2), \end{cases} \qquad \frac{\partial}{\partial S}(CS) \begin{cases} > 0, & S \in (S_1, S_{CS}), \\ < 0, & S \in (S_{CS}, S_2). \end{cases}$$

Due to (25),

$$\frac{\partial W}{\partial S}\Big|_{S=1} = \frac{H}{2} \cdot \left((6\mathcal{F}+1)\,D - 2\cdot\left(6\mathcal{F}^2 + 6\mathcal{F}+1\right)\right)$$

$$\geq \frac{H}{2} \cdot \left(2\cdot(\mathcal{F}+1)\,(6\mathcal{F}+1) - 2\cdot\left(6\mathcal{F}^2 + 6\mathcal{F}+1\right)\right) = H\mathcal{F} > 0$$

and

$$\frac{\partial}{\partial S}\,(CS)\Big|_{S=1} = -\frac{H}{2}\cdot\left(1 - (2\mathcal{F}+1)\,D + (2\mathcal{F}+1)^2\right)$$

$$\geq -\frac{H}{2}\cdot\left(1 - 2\,(2\mathcal{F}+1)\,(\mathcal{F}+1) + (2\mathcal{F}+1)^2\right) = H\mathcal{F} > 0.$$

Therefore, due to (29), $S_W > 1$ and $S_{CS} > 1$. Hence, see (10),

$$\tau_W = f\cdot\frac{1 - (S_W)^2}{S_W} < 0, \qquad \tau_{CS} = f\cdot\frac{1 - (S_{CS})^2}{S_{CS}} < 0.$$

Thus we get (15) and for the case (16).

A.3   Optimal Taxation: Comparison of τ_W and τ_{CS}

Now we get (17). For the case RL, (17) follows from (13) and (14).

Consider the case $RL(I)$. As we have got in Appendix A.2, on $[S_1, S_2]$, function W has a single maximum S_W, while function CS has a single maximum S_{CS}. To get (17), it is sufficient to show that $S_{CS} > S_W$, i.e., that

$$\frac{\partial W(S_{CS})}{\partial S} < 0. \tag{32}$$

From (31), we get

$$D = \frac{(S_{CS})^4 + (2\mathcal{F}+1)^2}{(2\mathcal{F}+1)S_{CS}}. \tag{33}$$

Substituting (33) in (30), we get

$$\frac{\partial W\,(S_{CS})}{\partial S} = \frac{H}{2}\cdot\left(S_{CS} + \frac{3\cdot(2\mathcal{F}+1) - 2\,(S_{CS})^2}{(S_{CS})^2}\cdot D - 3\cdot\frac{(2\mathcal{F}+1)^2}{(S_{CS})^3}\right)$$

$$= -H\cdot\frac{\left(2\mathcal{F}+1 - (S_{CS})^2\right)^2}{(2\mathcal{F}+1)\,(S_{CS})^2} < 0,$$

i.e., (32) holds. Thus, we get (17).

References

1. Spengler, J.: Vertical integration and antitrust policy. J. Polit. Econ. **58**(4), 347–352 (1950)
2. Hotelling, H.: Stability in competition. Econ. J. **39**(153), 41–57 (1929)
3. Dixit, A., Stiglitz, J.: Monopolistic competition and optimum product diversity. Am. Econ. Rev. **67**(3), 297–308 (1977)
4. Salop, S.: Monopolistic competition with outside goods. Bell J. Econ. **10**(1), 141–156 (1979)
5. Dixit, A.: Vertical integration in a monopolistically competitive industry. Int. J. Ind. Organ. **1**(1), 63–78 (1983)
6. Perry, M., Groff, R.: Resale price maintenance and forward integration into a monopolistically competitive industry. Quart. J. Econ. **100**(4), 1293–1311 (1985)
7. Chen, Z.: Monopoly and production diversity: the role of retailer countervailing power. Discussion Paper of Carleton University, 8 November (2004)
8. Hamilton, S., Richards, T.: Comparative statics for supermarket oligopoly with applications to sales taxes and slotting allowances. Selected Paper Prepared for Presentation at the American Agricultural Economics Association Annual Meeting, Portland, OR, 29 July–1 August 2017, 31 May 2007, 22 p.
9. Ottaviano, G.I.P., Tabuchi, T., Thisse, J.-F.: Agglomeration and trade revised. Int. Econ. Rev. **43**(2), 409–435 (2002)
10. Bykadorov, I., Ellero, A., Funari, S., Kokovin, S., Pudova, M.: Chain store against manufacturers: regulation can mitigate market distortion. In: Kochetov, Y., Khachay, M., Beresnev, V., Nurminski, E., Pardalos, P. (eds.) DOOR 2016. LNCS, vol. 9869, pp. 480–493. Springer, Cham (2016). https://doi.org/10.1007/978-3-319-44914-2_38
11. Bykadorov, I.A., Kokovin, S.G., Zhelobod'ko, E.V.: Product diversity in a vertical distribution channel under monopolistic competition. Autom. Remote Control **75**(8), 1503–1524 (2014). https://doi.org/10.1134/S0005117914080141
12. Tilzo, O., Bykadorov, I.: Retailing under monopolistic competition: a comparative analysis, pp. 156–161. IEEE Xplore (2019). https://doi.org/10.1109/OPCS.2019.8880233
13. Antoshchenkova, I.V., Bykadorov, I.A.: Monopolistic competition model: the impact of technological innovation on equilibrium and social optimality. Autom. Remote Control **78**(3), 537–556 (2017). https://doi.org/10.1134/S0005117917030134
14. Belyaev, I., Bykadorov, I.: International trade models in monopolistic competition: the case of non-linear costs, pp. 12–16. IEEE Xplore (2019). https://doi.org/10.1109/OPCS.2019.8880237
15. Bykadorov, I.: Monopolistic competition with investments in productivity. Optim. Lett. **13**(8), 1803–1817 (2018). https://doi.org/10.1007/s11590-018-1336-9
16. Bykadorov, I., Kokovin, S.: Can a larger market foster R&D under monopolistic competition with variable mark-ups? Res. Econ. **71**(4), 663–674 (2017)

Heuristics and Metaheuristics

Mutation Rate Control in the $(1 + \lambda)$ Evolutionary Algorithm with a Self-adjusting Lower Bound

Kirill Antonov[1], Arina Buzdalova[1(✉)], and Carola Doerr[2]

[1] ITMO University, Saint Petersburg, Russia
abuzdalova@gmail.com
[2] Sorbonne Université, CNRS, LIP6, Paris, France

Abstract. We consider the 2-rate $(1 + \lambda)$ Evolutionary Algorithm, a heuristic that evaluates λ search points per each iteration and keeps in the memory only a best-so-far solution. The algorithm uses a dynamic probability distribution from which the radius at which the λ "offspring" are sampled. It has previously been observed that the performance of the 2-rate $(1 + \lambda)$ Evolutionary Algorithm crucially depends on the threshold at which the mutation rate is capped to prevent it from converging to zero. This effect is an issue already when focusing on the simple-structured OneMax problem, the problem of minimizing the Hamming distance to an unknown bit string. Here, a small lower bound is preferable when λ is small, whereas a larger lower bound is better for large λ.

We introduce a secondary parameter control scheme, which adjusts the lower bound during the run. We demonstrate, by extensive experimental means, that our algorithm performs decently on all OneMax problems, independently of the offspring population size. It therefore appropriately removes the dependency on the lower bound. We also evaluate our algorithm on several other benchmark problems, and show that it works fine provided the number of offspring, λ, is not too large.

Keywords: Parameter setting · Evolutionary computation · Metaheuristic · Algorithm configuration · Mutation rate

1 Introduction

Evolutionary algorithms (EAs) are a class of iterative optimization heuristics that are aimed to produce high-quality solutions for complex problems in a reasonably short amount of time [5,9,16]. They are applied to solve optimization problems in various areas, such as industrial design and scheduling [2], search-based software engineering [12], bioinformatical problems (for example, protein folding or drug design) [11], and many more.

The reported study was funded by RFBR and CNRS, project number 20-51-15009.

Y. Kochetov et al. (Eds.): MOTOR 2020, CCIS 1275, pp. 305–319, 2020.
https://doi.org/10.1007/978-3-030-58657-7_25

EAs perform black-box optimization, i.e. they learn the information about the problem instance by querying the *fitness function* value of a solution candidate (also referred to as *individual* within the evolutionary computation community). The structure of an EA is inspired by the ideas of natural evolution: *mutation* operators make changes to individuals, *crossover* operators create offspring from parts of the existing individuals (*parents*), while a *selection* operator decides which individuals to keep in the memory (*population*) for the next iteration (*generation*).

The performance of an EA depends on the setting of its parameters. Existing works on choosing these parameters can be classified into two main categories: **Parameter tuning** aims at fitting the parameter values to the concrete problem at hand, typically through an iterated process to solve this meta-optimization problem. While being very successful in practice [13], parameter tuning does not address an important aspect of the parameter setting problem: for many optimization problems, the optimal parameter settings are not static but change during the optimization stages [3,10,14]. **Parameter control** addresses this observation by entailing methods that adjust the parameter values during the optimization process, so that they not only aim at *identifying* well-performing settings, but to also *track* these while they change during the run [6].

Our work falls in the latter category. More precisely, in this work we address a previously observed shortcoming of an otherwise successful parameter control mechanism, and suggest ways to overcome this shortcoming. We then perform a thorough experimental analysis in which we compare our algorithm with previously studied ones.

We continue previous work summarized in [18], where we have studied the so-called 2-rate $(1 + \lambda)$ $\mathrm{EA}_{r/2,2r}$ algorithm suggested in [4]. The 2-rate $(1 + \lambda)$ $\mathrm{EA}_{r/2,2r}$ is an EA of $(1 + \lambda)$ EA type with self-adjusting mutation rates. Intuitively, the mutation rate is the expected value of the search radius at which the next solution candidates are sampled. More precisely, this radius is sampled from a binomial distribution $\mathrm{Bin}(n, p)$, where n is the problem dimension and p the mutation rate. After sampling the search radius from this distribution, the "mutation" operator samples the next solution candidate uniformly at random among all points at this distance around the selected parent (which, in the context of our work, is always a best-so-far solution).

It has been proven in [4] that the 2-rate $(1 + \lambda)$ $\mathrm{EA}_{r/2,2r}$ achieves the asymptotically best possible expected running time on the ONEMAX problem, the problem of minimizing the Hamming distance to an unknown bit string $z \in \{0,1\}^n$. In evolutionary computation, as in other black-box settings, the running time is measured by number of function evaluations performed before evaluating for the first time an optimal solution.

The main idea of the 2-rate $(1 + \lambda)$ $\mathrm{EA}_{r/2,2r}$ algorithm is to divide the offspring population in two equal subgroups and to create the offspring in each subgroup with a group-specific mutation rate. The offspring in the first group are generated by mutating each bit of the parent individual with probability $r/2$, whereas a mutation rate of $2r$ is applied to the second group. At the end of each iteration, the mutation rate r is updated as follows: with probability $1/2$, it is updated to the rate used in the subpopulation that contains the best among all offspring (ties broken uniformly at random), and with probability $1/2$ the mutation rate is updated to $r/2$ or $2r$ uniformly at random.

When running the 2-rate $(1 + \lambda)$ $\text{EA}_{r/2,2r}$, it is convenient to define a lower bound lb for the mutation rate, to prevent it from going to zero, in which case the algorithm would get stuck. The original algorithm in [4] uses as lower bound lb $= 1/n$. In our previous work [18], however, we observed that a more generous lower bound of $1/n^2$ can be advantageous. In particular, we observed the following effects on ONEMAX, for problem dimension n up to 10^5 and offspring population size λ up to $32 \cdot 10^3$:

- it is more preferable to use $1/n^2$ as a lower bound for the mutation rate when the population size is small $(5 \leq \lambda \leq \lambda_1 \in (50, 100))$, while for large population sizes $(\lambda_1 \leq \lambda \leq \lambda_2 \in (800, 3200))$ a lower bound of $1/n$ seems to work better;
- these results hold for all considered problem dimensions.

We consider the first property as a disadvantage of the 2-rate $(1+\lambda)$ $\text{EA}_{r/2,2r}$ because its efficiency depends on the lower bound and for different values of λ different lower bounds should be used. The aim of the current paper is to propose and to test an algorithm based on the 2-rate $(1 + \lambda)$ $\text{EA}_{r/2,2r}$, which manages to adapt the lower bound lb automatically.

The main idea of the 2-rate $(1 + \lambda)$ $\text{EA}_{r/2,2r}$ improvement which we propose in this paper could be briefly described in the following way. First we start with the higher lower bound of $1/n$. Then in each population, we count the amount of individuals that were better than the parent separately for the cases of a higher mutation rate and for a lower one, which we call as the number of *votes* for a certain mutation rate. If a certain number of total votes is reached, we check if there are enough votes for the lower mutation rate among them, and decrease the value of the lower bound lb in this case. Hence, the lower bound tend to become more generous closer to the end of optimization, as it never increases and has a chance to decrease.

The proposed algorithm is shown to be efficient on ONEMAX for all considered population sizes λ, which we vary from 5 to 3200. Our approach therefore solves the disadvantageous behavior of the 2-rate $(1 + \lambda)$ $\text{EA}_{r/2,2r}$ described above. We also tested our modification on other benchmark problems and observed good efficiency at least on sufficiently small population sizes, i.e. on $5 \leq \lambda \leq 20$ for LEADINGONES and on $5 \leq \lambda \leq 16$ for W-MODEL transformations of ONEMAX.

2 Description of the Proposed Algorithm

2.1 Preliminaries

Throughout the paper we consider the maximization of a problem that is expressed as a "fitness function" $f : \{0,1\}^n \to \mathbb{R}$. That is, we study single-objective optimization of problems with n binary decision variables. Particularly, we study maximization of the benchmark functions ONEMAX, LEADING-ONES, and some problems obtained through so-called W-MODEL transformations applied to ONEMAX. These functions will be formally introduced in Sects. 3.1, 3.2, and 3.3, respectively.

All the considered algorithms are based on the $(1 + \lambda)$ $EA_{0 \to 1}$. Just like the 2-rate $(1 + \lambda)$ $EA_{r/2,2r}$, the algorithm keeps in its memory only one previously evaluated solution, the most recently evaluated one with best-so-far "fitness" (ties among the λ points evaluated in each iteration are broken uniformly at random). Each of the λ "offspring" is created by the *shift mutation* operator discussed in [17], which – instead of sampling the search radius from the plain binomial distribution $\text{Bin}(n,p)$ – shifts the probability mass from 0 to 1. That is, the search radius is sampled from the distribution $\text{Bin}_{0 \to 1}(n, p)$, which assigns to each integer $0 \leq k \leq n$ the value $\text{Bin}_{0 \to 1}(n, p)(k) = \text{Bin}(n, p)(k)$ for $1 < k \leq n$, but sets $\text{Bin}_{0 \to 1}(n, p)(0) = 0$ and $\text{Bin}_{0 \to 1}(n, p)(1) = \text{Bin}(n, p)(1) + \text{Bin}(n, p)(0)$. Given a search radius k, the offspring is then sampled uniformly at random among all the points at Hamming distance k from the parent (i.e., the point in the memory). For each offspring, the search radius k is sampled from $\text{Bin}_{0 \to 1}(n, p)$ independently of all other decisions that have been made so far. Note that shifting the probability mass from 0 to 1 can only improve $(1+\lambda)$-type algorithms, as evaluating the same solution candidate is pointless in our static and non-noisy optimization setting.

As mentioned previously, our main performance criterion is optimization time, i.e., the number of evaluations needed before the algorithm evaluates an optimal solution. In all our experiments, however, the value of λ remains fixed, so that – for a better readability of the plots – we report *parallel optimization times* instead, i.e., the number of iterations (generations) until the algorithm finds an optimal solution. Of course, the parallel optimization time is just the (classical, i.e., sequential) optimization time divided by λ.

2.2   General Description

Using the mentioned observations from the paper [18], we developed the 2-rate $(1 + \lambda)$ $EA_{r/2,2r}$ with voting that is aimed to solve the issues of the conventional 2-rate $(1 + \lambda)$ $EA_{r/2,2r}$. The pseudocode of the proposed algorithm is shown in Algorithm 1. Let us first introduce the notation and then explain the algorithm following its pseudocode:

- Voting - individuals from a population *vote* for the mutation rates. Only individuals which are better than the parent vote. Each voting individual votes for the mutation rate with which it was obtained;
- v - the number of individuals voted for decreasing of the mutation rate used on the current optimization stage;
- cnt - the total number of voted individuals;
- quorum - a constant value which is calculated as described in Sect. 2.3. It actually depends on n and λ but those are fixed during the optimization stage;

Algorithm 1: 2-rate $(1 + \lambda)$ $\text{EA}_{r/2,2r}$ with voting

---

1 **Initialization:** Sample $x \in \{0,1\}^n$ uniformly at random and evaluate $f(x)$;
2 $r \leftarrow 2/n$; $v \leftarrow 0$; $\text{cnt} \leftarrow 0$; $\text{lb} \leftarrow 1/n$;
3 **Optimization: for** $t = 1, 2, 3, \ldots$ **do**
4      $\text{voices}[r/2] \leftarrow 0$; $\text{voices}[2r] \leftarrow 0$;
5   **for** $i = 1, \ldots, \lfloor \lambda/2 \rfloor$ **do**
6   Sample $\ell^{(i)} \sim \text{Bin}_{0 \to 1}(n, r/2)$, create $y^{(i)} \leftarrow \text{flip}_{\ell^{(i)}}(x)$, and evaluate $f(y^{(i)})$;
7   **if** $f(y^{(i)}) > f(x)$ **then** $\text{voices}[r/2] \leftarrow \text{voices}[r/2] + 1$;
8   **for** $i = \lfloor \lambda/2 \rfloor + 1, \ldots, \lambda$ **do**
9   Sample $\ell^{(i)} \sim \text{Bin}_{0 \to 1}(n, 2r)$, create $y^{(i)} \leftarrow \text{flip}_{\ell^{(i)}}(x)$, and evaluate $f(y^{(i)})$;
10   **if** $f(y^{(i)}) > f(x)$ **then** $\text{voices}[2r] \leftarrow \text{voices}[2r] + 1$;
11      $v \leftarrow v + \text{voices}[r/2]$;
12      $\text{cnt} \leftarrow \text{cnt} + \text{voices}[r/2] + \text{voices}[2r]$;
13      $x^* \leftarrow \arg\max\{f(y^{(1)}), \ldots, f(y^{(\lambda)})\}$ (ties broken u.a.r.);
14   **if** $f(x^*) \geq f(x)$ **then** $x \leftarrow x^*$;
15   Perform one of the following two actions equiprobably;
16   ▶ replace r with the mutation rate that x^* has been created with;
17   ▶ replace r with either $2r$ or $r/2$ equiprobably.;
18   **if** $r < \text{lb}$ **then**
19   **if** $\text{cnt} \geq \text{quorum}$ **then**
20   **if** $v \geq d \cdot \text{quorum}$ **then**
21                  $\text{lb} \leftarrow \max(k \cdot \text{lb}, \text{LB})$;
22              $\text{cnt} \leftarrow 0$; $v \leftarrow 0$;
23          $r \leftarrow \text{lb}$;
24   **if** $r > \text{UB}$ **then**
25          $r \leftarrow \text{UB}$;

- lb - the current lower bound of the mutation rate. In our algorithm, we interpret this bound as a parameter and adjust it during the optimization. Offspring votes are used as feedback for the adjustment;
- LB - the all-time lower bound of the mutation rate. It stays constant during all the execution time;
- UB - the all-time upper bound of the mutation rate. Analogically with LB it stays constant during all the execution time;
- voices - the map where we store voices;
- $0 < d < 1$ - the portion of votes which is needed to decrease the current lower bound lb;
- $0 < k < 1$ - the multiplier which is used to decrease lb.

Let us explain the algorithm following the pseudocode. At the initialization stage a current solution x is generated as a random bit string of length n and the mutation rate r is initialized with an initial value $2/n$ (we use this value following [4]). Also we initialize the variables that are used at the optimization stage.

The optimization stage works until we end up with the optimal solution or until a certain budget is exhausted. In the first cycle in line 6 we perform mutation of the first half of the current population using the mutation rate $r/2$. Also we calculate the number of individuals which are better than their parents (line 7). The next cycle in line 9 does the same for the second half of the population using the mutation rate $2r$.

After all the mutations are done we update the number of individuals v voted for decreasing the mutation rate (line 11) and the total amount cnt of offspring who appeared to be better than the parent (line 12). Then we pick the individual with the best fitness function among the generated offspring as a candidate to become a new parent (line 13). In line 14 we check if the candidate is eligible for that by comparing its fitness $f(x^*)$ with the fitness of the current parent $f(x)$. After that in lines 15–17 we perform the usual for 2-rate $(1 + \lambda)$ EA$_{r/2,2r}$ stage which adapts the value of the mutation rate.

Then in line 18 we check whether the adapted mutation rate r is less than the current lower bound lb. In this case we update r in line 23 after adapting the lower bound lb in lines 19–22.

The adaptation of the lower bound is the key contribution of our approach. It is adapted on the basis of offspring votes. We decrease lb in line 21 if there are enough voted individuals (cnt $\geq$ quorum) and a sufficient portion of them voted for decreasing of the mutation rate ($v \geq d \cdot$ quorum). Note that at the same time lb may not become lower than the all-time lower bound LB. If the required number of voted individuals (cnt $\geq$ quorum) is reached, regardless whether lb was changed or not, we also always start the calculation of cnt and v votes from scratch in line 22. Finally, in lines 24, 25 we make sure that the mutation rate does not exceed the all-time upper bound UB.

Let us notice that the described approach is not restricted to be applied only in the 2-rate $(1 + \lambda)$ EA$_{r/2,2r}$ algorithm, but in principle may also be applied in other parameter control algorithms with simple update rules, such as one-fifth rule, to control lower bounds of the adapted parameters.

2.3   Selection of quorum

The quorum value used in Algorithm 1 is calculated as quorum$(n, \lambda) = A(n, \lambda) \cdot B(\lambda)$, where $A(n, \lambda) = (8n/9000 + 10/9)\lambda$ and $B(\lambda) = (1 + (-0.5)/(1 + (\lambda/100)^2)^2)$. Below we describe how this dependency was figured out.

In general the formula for quorum consists of two parts: $A(n, \lambda)$ and $B(\lambda)$ which were obtained experimentally. Part $A(n, \lambda)$ is linear in n and λ, while part $B(\lambda)$ depends on $1/\lambda^4$.

In early experiments on ONEMAX only the $A(n, \lambda)$ part was used. Let us describe how we obtained it. We observed $n = 100$, $n = 1000$, $n = 10000$ and run experiments on ONEMAX with various values of λ and different values of quorum. For every tested pair (n, λ) we gained the quorum that gave the best

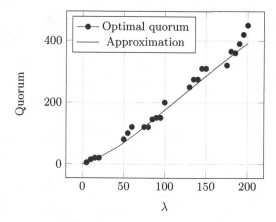

Fig. 1. Dependency of the optimal quorum on λ for $n = 1000$ (Color figure online)

runtime and plotted the dependency quorum(λ) for a fixed n. We noticed that this dependency was close to linear. One of such dependencies is shown in Fig. 1 with blue points (which were acquired during experiments). We also plotted the dependency quorum(n) for constant λ and noticed that it was close to linear as well. That gave us the target formula for $A(n, \lambda)$.

However, it turned out that quorum(n, λ) $= A(n, \lambda)$ is not very efficient on the LEADINGONES problem. Multiplying $A(n, \lambda)$ by $B(\lambda)$ allows the formula to work on small λ on both LEADINGONES and ONEMAX problems. At the same time, on large values of λ, $B(\lambda)$ part does not affect $A(n, \lambda)$ because $B(\lambda)$ approaches its limit which is 1. Both of these parts were chosen during the experiments but $B(\lambda)$ was chosen later on to make our method work on LEADINGONES at least for small values of λ and to not mess anything up on ONEMAX. The final approximation is shown with the black line in Fig. 1.

3 Empirical Analysis

In this section we present the results of empirical analysis of the proposed algorithm on ONEMAX, LEADINGONES, and some W-MODEL benchmark problems.

In all the experiments, the same parameter setting is used. The all-time bounds are LB $= 1/n^2$ and UB $= 1/2$. The parameters used for the lower bound adaptation are $d = k = 0.7$, they were determined in a preliminary experiment. We use quorum $= (8n/9000 + 10/9)\lambda \cdot (1 + (-0.5)/(1 + (\lambda/100)^2)^2)$, as described in Sect. 2.3. All the reported results are averaged over 100 runs.

3.1 Benchmarking on ONEMAX

The classical ONEMAX problem is that of counting the number of ones in the string, i.e., OM(x) $= \sum_{i=1}^{n} x[i]$. The algorithms studied in our work are "unbiased" in the sense introduced in [15], i.e., their performance is invariant with

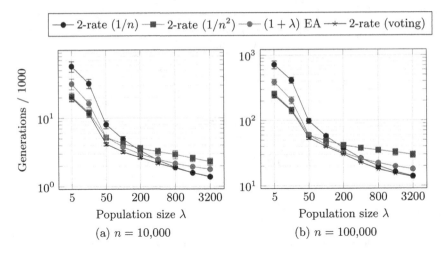

Fig. 2. Average parallel optimization time and its standard deviation for the different $(1 + \lambda)$ EA variants to find the optimal solution on ONEMAX problem. The averages are for 100 independent runs each.

respect to all Hamming automorphisms of the hypercube $\{0, 1\}^n$. From the point of view of the algorithms, the ONEMAX problem is therefore identical to that of minimizing the Hamming distance $H(\cdot, z) : \{0, 1\}^n \rightarrow \mathbb{R}, x \mapsto |\{i \mid x[i] = z[i]\}|$ – we only need to express the latter as the maximization problem $n - H(\cdot, z)$. Since the algorithms treat all these problems indistinguishably, we will in the following focus on OM only. All results, however, hold for any ONEMAX instance $n - H(\cdot, z)$.

Comparison with the Existing Methods. We compare the proposed 2-rate $(1 + \lambda)$ $\text{EA}_{r/2,2r}$ with voting, 2-rate $(1 + \lambda)$ $\text{EA}_{r/2,2r}$ with two different fixed lower bounds, and the conventional $(1 + \lambda)$ $\text{EA}_{0 \rightarrow 1}$ with no mutation rate adaptation. All the algorithms run on ONEMAX using the same population sizes as in [18]. The corresponding results are presented in Fig. 2 for $n = 10,000$ (left) and 100,000 (right) problem sizes. The plots show average number of generations, or parallel optimization time, needed to find the optimum for each population size λ. An algorithm is more efficient if it has a lower parallel optimization time and hence if its plot is lower.

Let us first compare the previously known algorithms: the $(1+\lambda)$ $\text{EA}_{0 \rightarrow 1}$ with a fixed mutation rate $1/n$, the 2-rate $(1 + \lambda)$ $\text{EA}_{r/2,2r}$ which adjusts mutation rate with respect to the $1/n$ lower bound and its 2-rate $(1 + \lambda)$ $\text{EA}_{r/2,2r}(1/n^2)$ version which uses a more generous $1/n^2$ lower bound. For both considered problem sizes, a similar pattern is observed:

- for $5 \leq \lambda \leq 50$ the 2-rate $(1 + \lambda)$ $\text{EA}_{r/2,2r}$ performed the best;
- then for $50 < \lambda \leq 400$ the $(1 + \lambda)$ $\text{EA}_{0 \rightarrow 1}$ became the best performing algorithm;

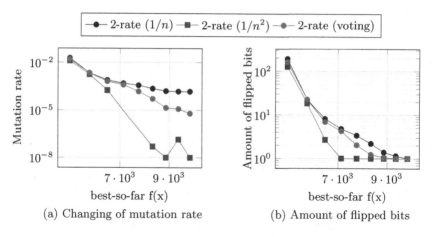

(a) Changing of mutation rate   (b) Amount of flipped bits

Fig. 3. Changing of mutation rate and number of flipped bits for $\lambda = 50$ and $n = 10,000$

– for $400 < \lambda \leq 3200$ the previous winner gave way to the 2-rate $(1 + \lambda)$ EA$_{r/2,2r}(1/n^2)$.

These observations confirm the results from [18], i.e. efficiency of the considered methods are strongly dependent on λ.

Let us now consider the results obtained with the proposed 2-rate $(1 + \lambda)$ EA$_{r/2,2r}$ with voting. *For both problem sizes and each considered value of λ the proposed algorithm is at least as good as the previous leader observed in the corresponding area of λ values.* Hence, it is quite efficient regardless the value of λ. Particularly, for $\lambda = 50, 100, 200$ and $n = 10,000$ (Fig. 2, left) the proposed algorithm is substantially better than the $(1 + \lambda)$ EA$_{0 \to 1}$, which was the previous best performing algorithm in this area. And the proposed algorithm is never worse than both 2-rate $(1 + \lambda)$ EA$_{r/2,2r}$ and 2-rate $(1 + \lambda)$ EA$_{r/2,2r}(1/n^2)$ in all the considered cases.

According to these observations, the proposed algorithm seem to be quite promising. In the next sections, we analyze its behavior in more deep on ONE-MAX and then test it on other benchmark problems.

Analysis of Bound Switching. Let us analyze the mutation rate dependency on the current best offspring to get deeper understanding of how the switching between the lower bounds really happens. Let us observe Fig. 3, where the average mutation rates chosen during the optimization process (a) and the corresponding average number of flipped bits (b) are shown.

In the beginning of optimization, the mutation rate obtained using the 2-rate $(1 + \lambda)$ EA$_{r/2,2r}$ with voting is the same as in the 2-rate $(1 + \lambda)$ EA$_{r/2,2r}$ (see Fig. 3(a)) and so the amount of flipped bits is the same approximately until the point $7 \cdot 10^3$ on the horizontal axis (Fig. 3(b)). When more and more offspring generated with rate $r/2$ are better than the parent, algorithm detects this and relaxes the lower bound to allow lower mutation rates. The switching process

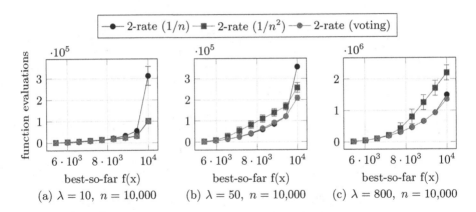

Fig. 4. Any-time performance on ONEMAX

is shown between points $7 \cdot 10^3$ and $9 \cdot 10^3$ on the horizontal axes. You can see how mutation rate is changing during this in Fig. 3(a). This leads to decrease of number of flipped bits (which is shown in Fig. 3(b)) and consequently switching to the 2-rate $(1 + \lambda)$ $EA_{r/2,2r}(1/n^2)$. Voting algorithm finally starts to flip the same amount of bits as the 2-rate $(1 + \lambda)$ $EA_{r/2,2r}(1/n^2)$ after the point $9 \cdot 10^3$.

Note that while mutation rates obtained with 2-rate $(1 + \lambda)$ $EA_{r/2,2r}$ with voting and 2-rate $(1 + \lambda)$ $EA_{r/2,2r}(1/n^2)$ in the end of optimization process are still different, the number of flipped bits tends to be the same for both algorithms, namely, one bit. This is explained by the fact that after reaching a sufficiently small mutation rate, both algorithms enter the regime of flipping exactly one randomly chosen bit according to the shift mutation operator described in Sect. 2.1. In this regime actual mutation rate values do not influence the resulting performance, so 2-rate $(1 + \lambda)$ $EA_{r/2,2r}(1/n^2)$ does not get sufficient information to control the rate. Thus the rates are chosen randomly and the deviation of the mutation rate increases. This may explain the excess of the 2-rate $(1 + \lambda)$ $EA_{r/2,2r}(1/n^2)$ plot in Fig. 3(a) at the end of optimization process.

Any-Time Performance Analysis. To further investigate how the optimization process goes, let us observe the number of fitness function evaluations which an algorithm needs to perform in order to reach a particular fitness value. We refer to this point of view as *any-time performance*, as shown in Fig. 4 for the 2-rate $(1+\lambda)$ $EA_{r/2,2r}$, 2-rate $(1+\lambda)$ $EA_{r/2,2r}(1/n^2)$ and 2-rate $(1+\lambda)$ $EA_{r/2,2r}$ with voting.

Let us first observe the Fig. 4(b) which corresponds to a medium population size $\lambda = 50$. One can see that starting at some point the 2-rate $(1 + \lambda)$ $EA_{r/2,2r}(1/n^2)$ works better than the 2-rate $(1 + \lambda)$ $EA_{r/2,2r}$, and the proposed 2-rate $(1 + \lambda)$ $EA_{r/2,2r}$ with voting detects this and tries to act like the best algorithm on each segment.

The any-time performance on a small population size $\lambda = 10$ is shown in Fig. 4(a). One can see that the segment of the 2-rate $(1 + \lambda)$ $EA_{r/2,2r}$ leadership

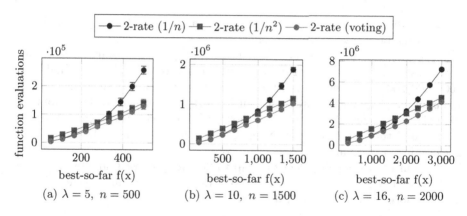

Fig. 5. Any-time performance on LEADINGONES

is quite small, hence most of the time 2-rate $(1 + \lambda)$ $EA_{r/2,2r}$ with voting keeps 2-rate $(1 + \lambda)$ $EA_{r/2,2r}(1/n^2)$ and so the final runtime of the proposed adaptive algorithm is close to the 2-rate $(1 + \lambda)$ $EA_{r/2,2r}(1/n^2)$ here.

Finally, we consider the performance on large population sizes on the example of $\lambda = 800$ in Fig. 4(c). The larger λ is the shorter becomes the segment of 2-rate $(1 + \lambda)$ $EA_{r/2,2r}(1/n^2)$ efficiency. Hence 2-rate $(1 + \lambda)$ $EA_{r/2,2r}(1/n^2)$ does not converge to the optimum faster than the 2-rate $(1 + \lambda)$ $EA_{r/2,2r}$ at any segment except probably the final mutation stage. The 2-rate $(1 + \lambda)$ $EA_{r/2,2r}$ with voting detects this and never switches to the 2-rate $(1+\lambda)$ $EA_{r/2,2r}(1/n^2)$ so the runtime is close to the 2-rate $(1 + \lambda)$ $EA_{r/2,2r}$. To conclude, the observations of this section illustrate how 2-rate $(1 + \lambda)$ $EA_{r/2,2r}$ with voting turns out to be never worse than 2-rate $(1 + \lambda)$ $EA_{r/2,2r}$ and 2-rate $(1 + \lambda)$ $EA_{r/2,2r}(1/n^2)$ for different population sizes, as was previously seen in Fig. 2.

3.2   Benchmarking on LEADINGONES

LEADINGONES is another set of benchmark functions that is often used in theoretical analysis of evolutionary algorithms [5]. The classical LEADINGONES function assigns to each bit string $x \in \{0,1\}^n$ the function value $\mathrm{LO}(x) := \max \{i \mid \forall j \leq i : x[j] = 1\}$. As mentioned in Sect. 3.1, the algorithms studied in our work are invariant with respect to Hamming automorphisms. Their performance is hence identical on all functions $\mathrm{LO}_{z,\sigma} : \{0,1\}^n \to \mathbb{R}, x \mapsto \max \{i \mid \forall j \leq i : x[\sigma(j)] = z[\sigma(j)]\}$. This problem has also been called "the hidden permutation problem" in [1].

Unlike ONEMAX, the LEADINGONES problem is not separable; the decision variables essentially have to be optimized one by one. Most standard EAs have a quadratic running time on this problem, and the best possible query complexity is $\Theta(n \log \log n)$ [1].

We have analyzed the adaptation process with parameters $\lambda = 10$ and $n = 1500$. The results of any-time performance are shown in Fig. 5. We see that the adaptation works nice here as well and 2-rate $(1 + \lambda)$ $\text{EA}_{r/2,2r}$ with voting achieves the best any-time performance.

In order to make sure that such positive results are invariant to λ and n we tested 2-rate $(1 + \lambda)$ $\text{EA}_{r/2,2r}$ with voting on $\lambda = 5, 10, 20$ and $n = 500, 1500, 2000$. On every pair of these parameters we gained the similar results. Example for corner cases are shown in Fig. 5(b). Nevertheless, the adaptation with the same quorum formula does not work for larger values of λ. Actually, it appeared that quorum grows much slower when optimizing LEADINGONES because in this case much less offspring vote.

To sum up, for small values of λ up to 20, the proposed adaptation works well on both ONEMAX and LEADINGONES. However, for greater values of λ the quorum formula has to be tuned to work well on LEADINGONES.

3.3   Benchmarking on W-MODEL Problems

Problem Description. W-MODEL problems were proposed in [19] as a possible step towards a benchmark set which is both accessible for theoretical analysis and also captures features of real-world problems. These features are:

- **Neutrality.** This feature means that different offspring may have the same fitness and be unrecognizable for EA. In our experiments, the corresponding function is implemented by excluding from the fitness calculation 10% of randomly chosen bits.
- **Epistasis.** In the corresponding problems the fitness contribution of a gene depends on other genes. In the implementation of the fitness function the offspring string is perturbed. The string is divided in the blocks of subsequent bits and special function is applied to each block to perform this perturbations. A more detailed description is given in [8].
- **Ruggedness and deceptiveness.** The problem is rugged when small changes of offspring lead to large changes of fitness. Deceptiveness means that gradient information might not show the right direction to the optimum. We used ruggedness function r_2 from [8], which maps the values of the initial fitness function $f(x)$ to $r_2(f(x)) := f(x) + 1$ if $f(x) \equiv n \mod 2$ and $i < n$, $r_2(f(x)) := \max\{f(x) - 1, 0\}$ for $f(x) \equiv n + 1 \mod 2$ and $i < n$, and $r_2(n) := n$.

These properties are supported in the IOHprofiler [7] and we benchmarked our algorithms with help of this tool. Details about each function implementation are described in [19]. We used IOHprofiler for our experiments, and all the parameter values are the same as described in [8]. The $F5$, $F7$ and $F9$ functions from the IOHprofiler were used, which means that the described W-MODEL transformations were applied to the ONEMAX function.

Results. We have analyzed problem dimensions $n = 500, .., 2000$ with step 500 and population sizes $\lambda = 5, .., 20$. In Figs. 6, 7 we show our results on the example of $\lambda = 10$, similar results hold for all $5 \leq \lambda \leq 16$. For $\lambda > 16$ the results of the

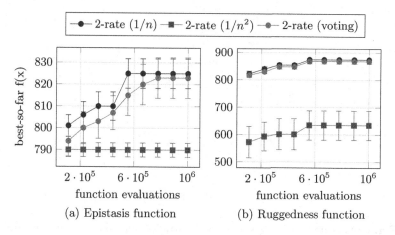

Fig. 6. Fixed budget results for the Epistasis (a) and the Ruggedness (b) functions, $\lambda = 10$, $n = 1000$

proposed algorithm get worse, we probably again need to additionally tune the quorum formula, as it was observed for LeadingOnes.

Due to the high complexity of Epistasis and Ruggedness, 2-rate $(1 + \lambda)$ EA$_{r/2,2r}$ takes too long to find the optimal solution, so we analyzed its performance from the fixed budget perspective. In the fixed budget approach, we do not pursue the goal to find the optimal solution, but we only limit the amount of fitness function evaluations and observe the best fitness values which can be reached within the specified budget. In the current work, we considered budget equal to the squared dimension size.

According to the fixed budget perspective, we had to use transposed axes compared to the previous plots in the paper, as shown for the Epistasis and Ruggedness functions in Fig. 6(a), (b) correspondingly. In this case, a higher plot corresponds to a better performing algorithm. The considered functions appeared to be too hard for 2-rate $(1 + \lambda)$ EA$_{r/2,2r}(1/n^2)$, so we did not find any segment, where 2-rate $(1 + \lambda)$ EA$_{r/2,2r}(1/n^2)$ appeared to be better than 2-rate $(1 + \lambda)$ EA$_{r/2,2r}$. Hence, for this situation results are positive when 2-rate $(1 + \lambda)$ EA$_{r/2,2r}$ with voting does not make 2-rate $(1 + \lambda)$ EA$_{r/2,2r}$ worse. And this is true for our case.

Finally, let us look at the Neutrality problem (see Fig. 7). This problem was easier to solve in a reasonable time, so we were able to use the usual stopping criterion of reaching the optimal solution, and the order of axes here is the same as in the rest of the paper. The Neutrality problem appeared to be not so hard for the 2-rate $(1 + \lambda)$ EA$_{r/2,2r}(1/n^2)$, so the 2-rate $(1 + \lambda)$ EA$_{r/2,2r}$ with voting used its advantages and worked well. Unlike for the two above problems, the promising behavior of 2-rate $(1 + \lambda)$ EA$_{r/2,2r}$ with voting was observed for $\lambda > 16$ as well, which may be explained by the fact that the considered Neutrality transformation does not change the complexity of the OneMax function drastically.

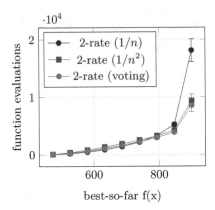

Fig. 7. Any-time performance on the Neutrality function, $\lambda = 10$, $n = 1000$

4 Conclusion

We proposed a new mutation rate control algorithm based on the 2-rate $(1 + \lambda)$ $\text{EA}_{r/2,2r}$. It automatically adapts the lower bound of the mutation rate and demonstrates an efficient behavior on the ONEMAX problem for all considered population sizes $5 \leq \lambda \leq 3200$, while the efficiency of the initial 2-rate $(1 + \lambda)$ $\text{EA}_{r/2,2r}$ algorithm with a fixed lower bound strongly depends on λ.

We have also applied the proposed algorithm to the LEADINGONES problem and noticed that adaptation of the lower bound worked there as well for $5 \leq \lambda \leq 20$. Finally, we tested the proposed algorithm on the W-MODEL problems with different fitness landscape features and observed that it behaves similarly to the best algorithm mostly on sufficiently small values of the population size $5 \leq \lambda \leq 16$.

Our next important goal is to further improve the efficiency of our algorithm for large population sizes λ. In the long term, we are particularly interested in efficient techniques for controlling two and more algorithm parameters. Despite some progress in recent years, the majority of works still focus on controlling a single parameter [3,14].

References

1. Afshani, P., Agrawal, M., Doerr, B., Doerr, C., Larsen, K.G., Mehlhorn, K.: The query complexity of a permutation-based variant of Mastermind. Discrete Appl. Math. **260**, 28–50 (2019)
2. Bäck, T., Emmerich, M., Shir, O.M.: Evolutionary algorithms for real world applications [application notes]. IEEE Comput. Intell. Mag. **3**(1), 64–67 (2008)
3. Doerr, B., Doerr, C.: Theory of parameter control for discrete black-box optimization: provable performance gains through dynamic parameter choices. In: Doerr, B., Neumann, F. (eds.) Theory of Evolutionary Computation. NCS, pp. 271–321. Springer, Cham (2020). https://doi.org/10.1007/978-3-030-29414-4_6. https://arxiv.org/abs/1804.05650

4. Doerr, B., Giessen, C., Witt, C., Yang, J.: The $(1+\lambda)$ evolutionary algorithm with self-adjusting mutation rate. In: Proceedings of the Genetic and Evolutionary Computation Conference, pp. 1351–1358. ACM, New York (2017)
5. Doerr, B., Neumann, F. (eds.): Theory of Evolutionary Computation: Recent Developments in Discrete Optimization. NCS. Springer, Cham (2020). https://doi.org/10.1007/978-3-030-29414-4
6. Doerr, C.: Non-static parameter choices in evolutionary computation. In: Proceedings of Genetic and Evolutionary Computation Conference Companion, pp. 736–761. ACM, New York (2017)
7. Doerr, C., Wang, H., Ye, F., van Rijn, S., Bäck, T.: IOHprofiler: a benchmarking and profiling tool for iterative optimization heuristics (2018). https://arxiv.org/abs/1810.05281. IOHprofiler. https://github.com/IOHprofiler
8. Doerr, C., Ye, F., Horesh, N., Wang, H., Shir, O.M., Bäck, T.: Benchmarking discrete optimization heuristics with IOHprofiler. In: Proceedings of the Genetic and Evolutionary Computation Conference Companion, pp. 1798–1806. ACM, New York (2019)
9. Eiben, A.E., Smith, J.E.: Introduction to Evolutionary Computing. NCS. Springer, Heidelberg (2015). https://doi.org/10.1007/978-3-662-44874-8
10. Eiben, Á.E., Hinterding, R., Michalewicz, Z.: Parameter control in evolutionary algorithms. IEEE Trans. Evol. Comput. **3**, 124–141 (1999)
11. Fogel, G., Corne, D. (eds.): Evolutionary Computation in Bioinformatics. Morgan Kaufmann, San Francisco (2003)
12. Harman, M., Mansouri, A., Zhang, Y.: Search-based software engineering: trends, techniques and applications. ACM Comput. Surv. **45**, 1–78 (2012)
13. Hutter, F., Kotthoff, L., Vanschoren, J. (eds.): Automated Machine Learning: Methods, Systems, Challenges. SSCML. Springer, Heidelberg (2019). https://doi.org/10.1007/978-3-030-05318-5
14. Karafotias, G., Hoogendoorn, M., Eiben, A.: Parameter control in evolutionary algorithms: trends and challenges. IEEE Trans. Evol. Comput. **19**, 167–187 (2015)
15. Lehre, P.K., Witt, C.: Black-box search by unbiased variation. Algorithmica **64**(4), 623–642 (2012). https://doi.org/10.1007/s00453-012-9616-8
16. Mitchell, M.: An Introduction to Genetic Algorithms. MIT Press, Cambridge (1996)
17. Pinto, E.C., Doerr, C.: Towards a more practice-aware runtime analysis of evolutionary algorithms (2018). https://arxiv.org/abs/1812.00493
18. Rodionova, A., Antonov, K., Buzdalova, A., Doerr, C.: Offspring population size matters when comparing evolutionary algorithms with self-adjusting mutation rates. In: Proceedings of the Genetic and Evolutionary Computation Conference, pp. 855–863. ACM, New York (2019)
19. Weise, T., Wu, Z.: Difficult features of combinatorial optimization problems and the tunable w-model benchmark problem for simulating them. In: Proceedings of the Genetic and Evolutionary Computation Conference Companion, pp. 1769–1776. ACM, New York (2018)

An Experimental Study of Operator Choices in the $(1 + (\lambda, \lambda))$ Genetic Algorithm

Anton Bassin and Maxim Buzdalov[✉]

ITMO University, Saint Petersburg, Russia
anton.bassin@gmail.com, mbuzdalov@gmail.com

Abstract. We study the influence of the particular choice of operators on the running time of the recently proposed $(1 + (\lambda, \lambda))$ genetic algorithm. In particular, we consider three choices for the mutation operator, six choices of the crossover operator, three strategies for sampling population sizes based on non-integer parameter values, and four choices of what to do when the best mutant is better than the parent.

We test all these 216 configurations on four optimization problems and in three adjustment flavours: the fixed λ, the unlimited self-adjustment of λ and the logarithmically capped one. For each of these configurations, we consider both the default values of the hyperparameters and the ones produced by the `irace` parameter tuning tool.

The result of our experimental analysis showed that there exists a configuration that is robust on linear functions and is roughly two times faster compared to the initially proposed one and 12% faster on the ONEMAX problem compared to one of the similar previous studies. An even more robust configuration exists, which has a slightly worse performance on ONEMAX but is better on satisfiability problems. Such configurations can be the default choices for the practical evaluation of the $(1 + (\lambda, \lambda))$ GA.

Keywords: $(1 + (\lambda, \lambda))$ genetic algorithm · Algorithm configuration · Parameter tuning

1 Introduction

The $(1+(\lambda, \lambda))$ genetic algorithm (GA), proposed in [6], is a crossover-based evolutionary algorithm with several remarkable properties. First of all, it achieves the expected running time of $o(n \log n)$ on a simple toy problem called ONEMAX (a family of optimization problems of the form $f_z(x) = |\{i \mid x_i = z_i\}|$), which is provably impossible for unbiased mutation-based algorithms whatever mutation operators they use [7]. Second, it solves this problem in time $\omega(n)$ for any parameter choice that is fixed for the entire run (but may depend on the problem

Supported by the Russian Scientific Foundation, agreement No. 17-71-20178.

size n), however, it takes only $O(n)$ time to do it with a simple self-adjustment of the parameter λ using the 1/5-th rule [5,6].

This algorithm seems to be quite efficient in settings closer to practice. For instance, it was found to be competitive in solving maximum satisfiability problems [9], taking the second place after the algorithm proposed by the authors of [9]. This efficiency was also supported theoretically in [2], however, for an easier version of the maximum satisfiability problem. However, in both these papers a few changes were introduced to the algorithm. In [9], for instance, the authors, in certain contexts, never sample individuals that are equal to their parents, which is a good concern in noise-free settings, and also restart the algorithm when it appears to be stuck in a local optimum. On the other hand, in [2] an additional threshold for the adjusted parameter λ was introduced, which prevents the algorithm from the performance degradation. A similar degradation observed in different contexts motivated further research in that direction [1].

The $(1 + (\lambda, \lambda))$ GA apparently features a single parameter λ, however, there are more quantities inside the algorithm which are tied to that parameter. For instance, the intermediate population sizes are typically equal to λ, the bit flip probability during the mutation is taken as λ/n, and the crossover exchanges the bits with the probability of $1/\lambda$. It was proven in [5] that any super-constant deviation from these values is harmful already for ONEMAX. However, the technique employed to prove this cannot be used to give statements about the optimal constant factors, which can be treated as hyperparameters. For this reason, a separate research was done in [4] to investigate how the performance depends on these constant factors. They have shown that a speed-up of roughly 15% is possible due to the appropriate choice of the constant factors. The tuning was done in the five-dimensional hyperparameter space using the `irace` tool [11].

However, the improvements can come not only from the tuning of hyperparameters, but also from the appropriate choices of distributions that define the variation operators (namely, mutation and crossover). The paper [3] advocates to use, whenever appropriate, mutation operators that never sample offspring identical to its parent, such as the so-called *shift mutation* that forces the smallest change to be made in any case, and the *resampling mutation* that repeats the sampling procedure whenever a copy of the parent is to be sampled. We shall note that the paper [4] also deals not with the original version of the $(1 + (\lambda, \lambda))$ GA, but with the one which uses the resampling mutation, and which also ignores evaluation of crossover offspring that are identical to parents. This is also done in some of the earlier-mentioned papers [1,9], however, a systematical study of the influence of these choices on the running time of the $(1 + (\lambda, \lambda))$ GA, including a study of the robustness of such a choice regarding different problems to be solved, was so far missing.

In this paper, we make a first contribution of this sort. We consider three choices for the mutation operator and six choices of the crossover operator, but also three strategies for sampling population sizes based on non-integer parameter values, and four choices of what to do when the best mutant is better than the parent. This amounts to 216 different choices, which we further augment with three different self-adjustment strategies for the parameter λ and to four different pseudo-Boolean problems. Furthermore, following the proposal of [4], we treat the constant factors as hyperparameters and consider their default values as well as the values produced by `irace`.

2   The $(1 + (\lambda, \lambda))$ GA and Its Modifications

The $(1+(\lambda,\lambda))$ GA is outlined in Algorithm 1. An iteration of the $(1+(\lambda,\lambda))$ GA consists of two phases, namely, the mutation phase and the crossover phase. On the *mutation phase*, a number of mutants is sampled with the higher-than-usual mutation rate, and, additionally, all the mutants have the same Hamming distance to their parent. The best of the mutants, together with the parent, enter the *crossover phase*, when they also produce a number of crossover offspring. The crossover is asymmetrical, however, as the bits are taken from the best mutant with a much smaller probability. The best of the crossover offspring competes directly with the parent, which completes the iteration.

The logic above is controlled by the parameter λ as follows: the mutation rate is proportional to λ/n, the population sizes for both phases are proportional to λ, and the probability to take a mutant's bit in the crossover, called the *crossover bias* in many papers, is proportional to $1/\lambda$. Following the suggestions from [4], we use the following notation for the proportionality quotients:

- α: the mutation probability quotient; the mutation probability is $\alpha\lambda/n$;
- β: the crossover population size quotient: the population size in the crossover phase is derived from $\beta\lambda$;
- γ: the crossover probability quotient: the crossover bias is γ/λ.

Note that the population size in the mutation phase is still bound to λ directly without quotients, as in [4]. This makes sense, otherwise all the quotients could have been multiplied by the same factor while λ is divided by the same factor. The default values of these hyperparameters are $\alpha = \beta = \gamma = 1$. The paper [4] also introduces two parameters, A and b, that control the self-adjustment of λ when the latter takes place. More precisely, when the best individual is updated to a strictly better one, $\lambda \leftarrow b\lambda$, otherwise $\lambda \leftarrow A\lambda$. The default values from the one-fifth rule are $b = 1/C$ and $A = C^{1/4}$ for some $C \in (1; 2)$.

Additional to these five hyperparameters, we suggest to consider a few more options. First of all, there can be different distributions that control the Hamming distance ℓ during the mutation phase. We consider three of them, two of which have been introduced in [3]:

- "standard": the binomial distribution $\mathcal{B}(n, p)$, where p is the mutation probability as above;
- "shift": $\max\{1, \mathcal{B}(n, p)\}$, that is, whenever the binomial distribution samples zero, the value of one is forced;
- "resampling": this can be written as $\ell \sim \mathcal{B}(n, p) \mid \ell > 0$, meaning that whenever the binomial distribution samples zero, sampling is repeated.

Algorithm 1. The $(1 + (\lambda, \lambda))$ GA with modifications

---

Hyperparameters:
- $\mathcal{D}_M(n, p)$: distribution of the number of flipped bits in mutation
- $\mathcal{D}_C(n, m, p)$: distribution of the number of flipped bits in crossover
- $\rho(n)$: the rounding distribution to determine population size
- α: the mutation probability quotient
- β: the crossover population size quotient
- γ: the crossover probability quotient
- A, b: the self-adjustment parameters (if applicable)
- $\xi \in \{\mathtt{I}, \mathtt{S}, \mathtt{C}, \mathtt{M}\}$: the strategy to handle mutants that are better than parents

Algorithm:
$n \leftarrow$ the problem size, $f \leftarrow$ the function on $\{0, 1\}^n$ to maximize
$\lambda \leftarrow \lambda_0$ (in self-adjusting versions, $\lambda_0 = 1$, otherwise an algorithm-wise constant)
$x \leftarrow$ sample a bit string uniformly; evaluate $f(x)$
while true do
    $s_M \leftarrow \rho(\lambda)$ ▷ The mutation population size
    $\ell \sim \mathcal{D}_M(n, \alpha\lambda/n)$ ▷ The number of bits to flip in mutation
    for $i \in 1, 2, \ldots, s_M$ **do**   ▷ Mutation phase
        $x^{(i)} \leftarrow x$ with ℓ random different bits flipped   ▷ Sample an offspring
        Evaluate $f(x^{(i)})$
    end for
    $x' \leftarrow$ best from $\{x^{(1)}, x^{(2)}, \ldots, x^{(s_M)}\}$ chosen uniformly
    if $\xi = \mathtt{S}$ **and** $f(x') > f(x)$ **then**
        $x \leftarrow x'$, adjust λ ▷ If best mutant is better than parent...
        continue   ▷ ...and ξ is "skip crossover", skip it
    end if
    $s_C \leftarrow \rho(\beta\lambda)$ ▷ The crossover population size
    for $i \in 1, 2, \ldots, s_C$ **do**   ▷ Crossover phase
        $\delta \leftarrow \mathcal{D}_C(\lambda, \ell, \gamma/\lambda)$ ▷ Number of bit exchanges in crossover
        if $\delta = \ell$ **then**   ▷ x' is to be sampled again
            switch ξ:
                case $\mathtt{M}$: $i \leftarrow i - 1$ ▷ ξ is "do not sample identical to mutant", rollback
                case $\mathtt{C}$: do nothing ▷ ξ is "do not count identical to mutant", ignore
                case $\mathtt{I}, \mathtt{S}$: evaluate $f(x')$ again
            end switch
        else
            $y^{(i)} \leftarrow$ crossover of x and x' with δ differing bits taken from x'
            Evaluate $f(y^{(i)})$
        end if
    end for
    $Y = \{y^{(1)}, y^{(2)}, \ldots, y^{(s_C)}\}$
    if $\xi = \mathtt{C}$ **or** $\xi = \mathtt{M}$ **then**
        $Y \leftarrow Y \cup \{x'\}$ ▷ Take the best mutant to the comparison
    end if
    $x \leftarrow$ best from x and Y, adjust λ
end while

Second, the same ideas can be applied to the crossover. More precisely, the corresponding distribution would tell how many bits that are different in the parent and in the best mutant would be taken from the mutant. It is clear that this number cannot be greater than ℓ, so the first argument in the binomial distribution will be ℓ. However, the second argument (the success probability) can be derived from either ℓ or from λ. Although, when introduced as such, the first choice is more natural, only the second one was actually used in all the previous research. For this reason, we consider six versions of the crossover distribution:

$$\{\text{standard}, \text{shift}, \text{resampling}\} \times \{\text{from } \ell, \text{from } \lambda\}.$$

Third, the population sizes are derived from, generally, real values, and choices are possible. The original paper [6] suggested rounding towards the larger value, while some other papers used a different way. We use three options:

- always round down;
- always round up;
- round probabilistically as follows: if the real value x is not integral, we sample $\lfloor x \rfloor$ with probability $(\lceil x \rceil - x)$ and we sample $\lceil x \rceil$ with probability $(x - \lfloor x \rfloor)$. This way, the smaller x, the larger the probability to round it down.

Finally, the algorithm may encounter that the best mutant is already better than the parent. This may happen with a decent probability when λ is small and in easier parts of the fitness landscape. The original algorithm does not treat this case in a special way, however, in [9] the author decided to skip the crossover phase altogether once this happens. We introduce a few other options in the hope that crossing over such a mutant with the parent may potentially produce an even better individual. In total, we have the following options:

- "ignore" (letter I): similar to the original algorithm, do nothing special;
- "skip crossover" (letter S): similar to [9], skip the crossover once the best mutant is better than the parent;
- "do not count identical to mutant" (letter C): add the best mutant to the comparison with the parent, which follows the crossover phase, and skip the crossover offspring which are sampled identically to the best mutant;
- "do not sample identical to mutant" (letter M): same as above, but also do not increment the number of already sampled crossover offspring, so that there will always be the required number of crossover offspring different from the best mutant (and from the parent if required).

The latter option is, in fact, quite tricky. There can be the cases when the crossover distribution will always sample the maximum possible difference, of which the most obvious case is when $\ell = 1$ and the distribution cannot sample zeros because it is either "shift" or "resampling". In these cases, which can fortunately be detected without querying individuals, the entire crossover phase is skipped as well, and the parent is updated only if the best mutant is not worse than the parent. This case is not illustrated in Algorithm 1 for brevity.

With all these options, most of the flavours of the $(1 + (\lambda, \lambda))$ GA from the literature can be expressed as Algorithm 1 with certain values of the hyperparameters. In the following, we evaluate all the 216 combinations of the qualitative hyperparameters on several benchmark problems. The five quantitative parameters, on the other hand, come in two variants: the default ones and the ones produced by the tool called `irace` [11].

3   Experiments

We perform experiments with the following benchmark problems: ONEMAX, linear pseudo-Boolean functions with random weights from $[1; w_{max}]$, where $w_{max} = 2$ and $w_{max} = 5$, and the easy maximum satisfiability problems. For each of the problem, the details will be given in the corresponding subsection.

The complete output of experiments is available on Zenodo[1].

3.1   General Methodology

For each problem, we choose two reasonably large problem sizes, which are noticeably different and which still allow the intended volume of experimentation. For example, these two sizes are $n = 2^{13} = 8192$ and $n = 2^{16} = 65536$ for ONEMAX. For each problem size, we consider three adjustment strategies for λ:

- a constant $\lambda = 8$;
- a self-adjusting λ with the lower limit of 1 and the upper limit of n, controlled by hyperparameters A and b;
- the same with the upper limit of $2\ln(n + 1)$.

Next, for each of the 216 configurations we perform 100 independent runs with the default real-valued hyperparameters (namely, $\alpha = \beta = \gamma = 1$, $b = 2/3$ and $A = (3/2)^{1/4}$) and collect the statistics. We report medians and quartiles instead of means and standard deviations for the following reasons:

- the running times of evolutionary algorithms are typically quite far away from being normally distributed;
- the running times of the $(1 + (\lambda, \lambda))$ GA, at least on simple problems, are very well concentrated, and as a result means and medians nearly coincide;
- order statistics are more robust to outliers.

These quantities are presented visually as specified on Fig. 1. This figure also introduces the short five-letter notation that we use to describe the hyperparameter choices: for instance, the originally proposed version of the $(1 + (\lambda, \lambda))$ GA is denoted as `SSLIU` for standard binomial distributions for crossover and mutation, the base for the crossover probably being λ as opposite to ℓ, the ignorant good mutant strategy and population sizes rounded up.

---

[1] Location: https://zenodo.org/record/3871043, DOI: 10.5281/zenodo.3871043.

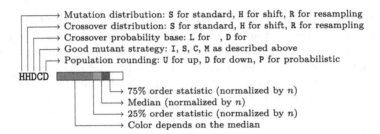

Fig. 1. The explanation of the column elements in the results of experiments

We also run `irace` to determine the near-best values for the real-valued hyperparameters. The hyperparameter ranges are $[0; 3]$ for α, β and γ, while $b \in [0; 1]$ and $A \in [1; 3]$. We use a budget of 1000 tested hyperparameter vectors for each configuration, and average the performance of the algorithm over 25 independent runs before reporting it to `irace`. The training instances feature a smaller problem size, and we also limit the number of evaluations (the particular values will be given in problem descriptions). Once `irace` presents the results, we take the best one and run it in the same way as the configuration with the default parameters.

The results are presented, side by side, in large one-page pictures, where each column corresponds to a particular combination of the problem size, the self-adjustment strategy and the property of being tuned by `irace`. The elements of each column are sorted from the best to the worst result. The matching elements in otherwise identical columns with different sizes, as well as the default and tuned maximum-sized columns, are connected by gray lines. This is a visual indicator of the stability of the order: few intersections mean a better stability, while a lot of intersections means that the order changes significantly. One can also spot the groups of configurations which are similar inside the group but are different from all other configurations. Note that because of the huge quantity of configurations to be tuned, the quality of each tuning produced by `irace` may be inferior to what could be possible with only a few configurations and a larger budget. We leave further investigations in this direction to the future work.

3.2   Experimental Results: OneMax

This section describes the results of the experiments on ONEMAX, which is a family of pseudo-Boolean functions defined as follows:

$$\text{ONEMAX}_f : \{0,1\}^n \to \mathbf{R}; \quad x \mapsto \sum_{i=1}^{n}[x_i = f_i],$$

where f is a hidden bit string representing the optimum, and $[.]$ is the Iverson bracket which gives 1 for logical truth and 0 otherwise. We use $f = 1^n$ without loss of generality. The experiment parameters are as follows:

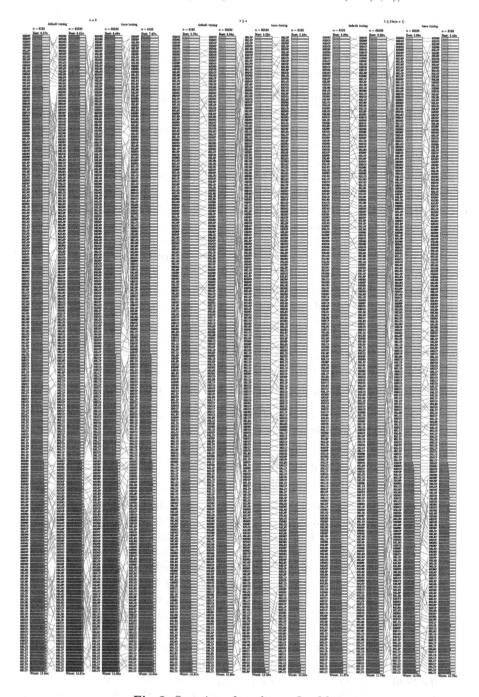

Fig. 2. Overview of results on OneMax

- small problem size: $2^{13} = 8192$;
- large problem size: $2^{16} = 65536$;
- training problem size for `irace`: 1000;
- evaluation limit for `irace`: 20000.

The results are presented in Fig. 2 with the stripe span of 15.

From these data we can do the first few observations. First of all, the configuration corresponding to the original version of the algorithm, `SSLIU`, is one of the worst: in 7 out of 12 columns it is the worst, in two more it is the second worst, and it is in the 10 worst configurations on all other cases. Second, we see that `irace` generally helps, except for a few worst configurations where its tuning is worse than the default one. Furthermore, `irace` heavily influences the order of the configurations, typically much more than the problem size. Regarding the problem size, the self-adjusting version of $\lambda \leq n$ is very stable, $\lambda \leq \log n$ comes second, and $\lambda = 8$ is much less stable, although the sets of top 18 default configurations and roughly top 30 `irace` configurations are both similar in performance and do not change much.

We also see that the most significant impact comes from the crossover distribution, where the use of the standard choice introduces a visible penalty in $\lambda = 8$ and in the irace version of $\lambda \leq \log n$. Finally, in $\lambda \leq n$, the best irace configuration (with runtime of roughly $5.22n$ on $n = 2^{16}$) is twice as fast as the worst default configuration, which is the above mentioned `SSLIU`.

3.3   Experimental Results: Random Linear Functions

This section describes the results of the experiments on linear pseudo-Boolean functions with random weights defined as follows:

$$\text{LIN}_{f,w} : \{0,1\}^n \to \mathbf{R}; \quad x \mapsto \sum_{i=1}^{n} w_i \cdot [x_i = f_i],$$

where f is a hidden bit string representing the optimum, and w is the weight vector. Similar to ONEMAX, we use $f = 1^n$ without loss of generality. The elements of w are sampled uniformly from $[1; w_{max}]$, where w_{max} is the problem's parameter. We consider $w_{max} = 2$ and $w_{max} = 5$. Note that [1] used random integer weights, which generally results in harder problems for the same weight limits, and [6] used the same technique as we do here for $w_{max} = 2$ only.

We used experiment parameters similar to ONEMAX, except that the training problem size was set to 500, and evaluation limits were 20000 for $w_{max} = 2$ and 40000 for $w_{max} = 5$. Furthermore, we do not present the results for $\lambda \leq n$ and large problem size $n = 2^{16}$, because this self-adjustment technique diverges for the same reasons as explained in [1], which leads to much worse running times (the quotient typically exceeds 30 for $w_{max} = 2, n = 2^{16}$), and to even worse wall-clock running times. The results are presented in Fig. 3 for $w_{max} = 2$ and in Fig. 4 for $w_{max} = 5$ with stripe spans of 25 and 50 correspondingly.

In these data, we also observe a few trends which were already noticed for ONEMAX. First of all, the standard crossover distribution (the second letter

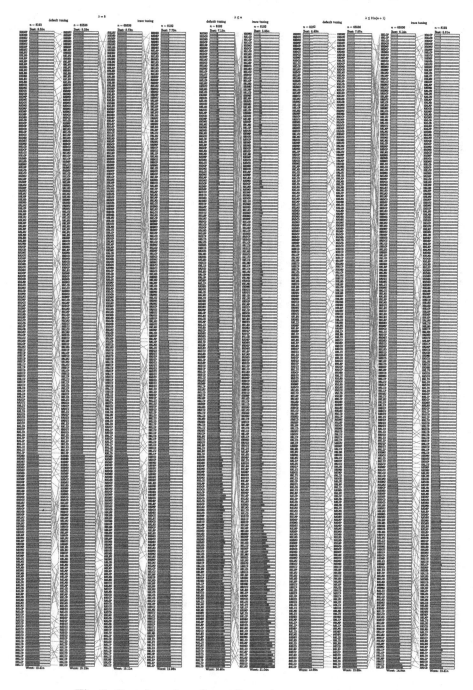

Fig. 3. Overview of results on linear functions with $w_{\max} = 2$

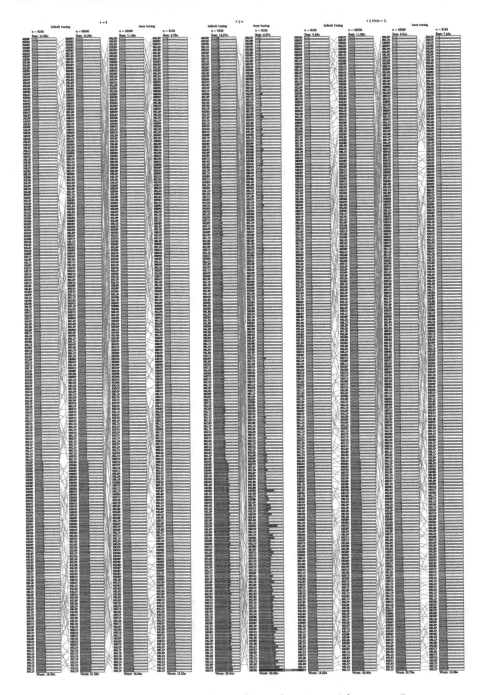

Fig. 4. Overview of results on linear functions with $w_{max} = 5$

S in the configuration descriptions) is the worst choice in nearly all columns: this trend is very strict for $\lambda = 8$ and, additionally, in the non-tuned $\lambda \leq n$, and only slightly less so in all other columns. Second, the configuration SSLIU that describes the original version of the algorithm is again one of the worst configurations. Third, the best irace configuration, compared to the worst non-tuned configuration, is more than twice as fast, especially in the more promising $\lambda \leq \log n$ self-adjustment mode.

One of the things different to ONEMAX is that the interquartile ranges for $\lambda \leq n$ are now much wider. This can be explained by the fact that the divergence of the self-adjusting strategy happens at random times and for poorly concentrated periods, which introduces a lot of noise to the running time. This seems to be detrimental to irace, which fails to find a good configuration even compared to the default one for a slightly larger portion of runs.

3.4   Experimental Results: Easy Maximum Satisfiability Problems

This section describes the results of the experiments on easy maximum satisfiability problems. These problems were investigated in the context of evolutionary computation in [2,8,12] theoretically and also experimentally in a number of papers [9]. A satisfiability problem in a conjunctive normal form is written as follows for a vector x of n Boolean variables:

$$\mathrm{CNF}(x) = \wedge_{i=1}^{m}(v_{i_1} \vee v_{i_2} \vee \ldots \vee v_{i_{c_i}}),$$

where v_j is either x_j or its negation. This problem is NP-complete, and even the problem where all $c_i = 3$, the so-called 3CNF problem, is also NP-complete. The maximization version of this problem, which is often called MAX-SAT and is of course NP-hard, aims at maximizing the number of clauses:

$$\mathrm{MAX\text{-}SAT}(x) = \sum_{i=1}^{m}[v_{i_1} \vee v_{i_2} \vee \ldots \vee v_{i_{c_i}}].$$

Following [2,8,12], we consider the easy instances of the MAX-SAT problem with all $c_i = 3$. First, the clauses are sampled uniformly at random from the clauses on three variables that satisfy some pre-defined *planted assignment* (which is all ones without loss of generality). Second, the number of clauses, m, is chosen to be $4n \ln n$, which amounts to the logarithmic density of clauses, as opposed to the well-known hard random instances with the density of 4.27 [10]. These choices make the problem quite similar to ONEMAX, although not exactly the same, which poses a moderate difficulty to the $(1 + (\lambda, \lambda))$ GA when it attempts applying high mutation rates.

We use the following experiment parameters:

- small problem size: $2^{10} = 1024$;
- large problem size: $2^{13} = 8192$;
- training problem size for irace: 128;

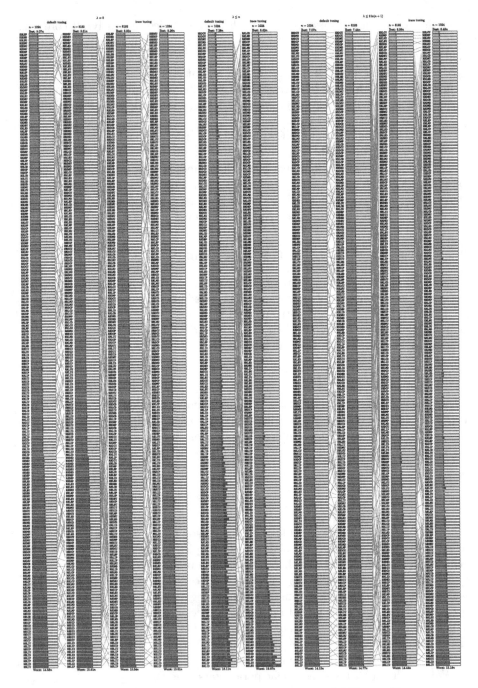

Fig. 5. Overview of results on easy maximum satisfiability problems

– evaluation limit for `irace`: 5000.

The chosen problem sizes are much smaller than for linear functions, because incremental fitness evaluation, which we employ to speed-up our experiments, introduces an additional factor of $\Theta(\log n)$ to the wall-clock running times. The results are presented in Fig. 5 with the stripe span of 25. The overall trends seem to be roughly the same as in the case of linear functions, with the exception that the performance of the self-adjusting configurations tuned by `irace` appears to be even less stable regarding the problem size. Furthermore, the set of the best `irace` configurations for $\lambda \leq \log n$ seems to be noticeably different from the similar set produced for linear functions.

Table 1. Summary of analysis of robustness

Non-tuned				irace			
Rank		Ratio to best		Rank		Ratio to best	
Avg	Max	Avg	max	Avg	Max	Avg	Max
HHDCU 0.20	HHDCU 1	HHDCU 1.000	HHDCU 1.002	HHDMD 19.00	HHDMD 35	HHLCU 1.046	HHDMD 1.075
HHDMU 1.20	HHDMU 2	HHDMU 1.007	HHDMU 1.019	HHDCP 23.00	RHDMU 43	HHDMD 1.053	RHLCP 1.079
HHDCP 3.00	HHDCP 5	HHDCP 1.020	HHDCP 1.038	RHLMU 24.80	RHLMU 48	RHLCP 1.058	HHLCU 1.089
HHDMD 4.20	HHDMD 6	HHDMD 1.023	HHDMD 1.043	RHLCP 24.80	HRLMD 53	HHDCP 1.061	RRLMP 1.118
RHDCU 5.40	RHDMU 7	RHDCU 1.026	RHDMU 1.045	HHLCU 26.40	HHDCP 56	RHLMU 1.062	RHLMU 1.123

Table 2. Best self-adjusting configurations

Config	Median on ONEMAX	α	β	γ	A	b	Comment
HHDMD	$5.945n$	0.3577	0.5273	0.3991	1.2220	0.5297	Robust on all tested functions
HHLCP	$5.242n$	1.1083	1.1601	0.0290	1.2803	0.5605	Robust on tested linear functions

3.5 Analysis of Robustness

To understand which configuration is generally better, we conducted a simple analysis of robustness. We chose the following pairs of a problem and a self-adjusting technique, which are representatives of approaches that work well: $\lambda \leq n$ on ONEMAX and $\lambda \leq \log n$ on all the problems (again including ONEMAX). For each such pair we take the maximum available problem size and evaluate the configurations, taking non-tuned and `irace` configurations separately. For each configuration we evaluate its rank, as well as the ratio of the median running time to the minimum observed median. Then we compute the average and the maximum values of these quantities across all problem-technique pairs. Five best configurations for each quantity are presented in Table 1.

The table reveals that for the non-tuned configurations the ultimate choice is the configuration called HHDCU, which wins in all nominations. For the irace configurations, there is more diversity, but the configuration HHDMD can also be seen as a winner, arriving second only for the average median ratio. The latter, however, does not yield the best running times on ONEMAX, finishing only 26th and running in approximately $5.95n$ median time. When considering only linear functions, however, another configuration HHLCP is stable enough in producing good results and finishes only second on ONEMAX with median time of $5.24n$. Table 2 presents their irace tunings for ONEMAX.

4   Conclusion

We presented an experimental study of 216 different configurations of the recently proposed $(1 + (\lambda, \lambda))$ GA, where each configuration represents a choice of a particular set of distributions that tell the mutation and crossover operators what to do, as well as a strategy of coping with a best mutant that can be better than its parent and a strategy of rounding the population sizes. We also compared the default hyperparameter choices and the choices tuned by the irace tool. Our findings can be shortly summarized as follows:

1) The choice of the crossover distribution has the largest impact on the performance. While the "shift" and "resampling" strategies perform almost the same, the default choice is definitely worse. This means that the original $(1 + (\lambda, \lambda))$ GA spends too much time in creating crossover offspring that are identical to the parent, and the best single improving move is to avoid this.

2) The default configuration, which corresponds to the original version of the algorithm as in [6], is always among the worst ones, and the best configurations are typically at least twice as fast.

3) There exists a configuration which is 12% more efficient than the tuning found in [4]. We also presented an even more robust configuration which performs well both on linear functions and on easy satisfiability problems.

4) The irace tool is noticeably good at finding good tunings for good configurations, however, for the few worst configurations it fails to find even the tuning which is as good as the default one, which we attribute to budget limitations.

References

1. Bassin, A., Buzdalov, M.: The 1/5-th rule with rollbacks: on self-adjustment of the population size in the $(1+(\lambda,\lambda))$ GA. In: Proceedings of Genetic and Evolutionary Computation Conference Companion, pp. 277–278 (2019). https://arxiv.org/abs/1904.07284

2. Buzdalov, M., Doerr, B.: Runtime analysis of the $(1 + (\lambda, \lambda))$ genetic algorithm on random satisfiable 3-CNF formulas. In: Proceedings of Genetic and Evolutionary Computation Conference, pp. 1343–1350 (2017)
3. Carvalho Pinto, E., Doerr, C.: Towards a more practice-aware runtime analysis of evolutionary algorithms (2018). https://arxiv.org/abs/1812.00493
4. Dang, N., Doerr, C.: Hyper-parameter tuning for the $(1 + (\lambda, \lambda))$ GA. In: Proceedings of Genetic and Evolutionary Computation Conference, pp. 889–897 (2019)
5. Doerr, B., Doerr, C.: Optimal static and self-adjusting parameter choices for the $(1 + (\lambda, \lambda))$ genetic algorithm. Algorithmica **80**(5), 1658–1709 (2018). https://doi.org/10.1007/s00453-017-0354-9
6. Doerr, B., Doerr, C., Ebel, F.: From black-box complexity to designing new genetic algorithms. Theoret. Comput. Sci. **567**, 87–104 (2015)
7. Doerr, B., Johannsen, D., Kötzing, T., Lehre, P.K., Wagner, M., Winzen, C.: Faster black-box algorithms through higher arity operators. In: Proceedings of Foundations of Genetic Algorithms, pp. 163–172 (2011)
8. Doerr, B., Neumann, F., Sutton, A.M.: Improved runtime bounds for the (1+1) EA on random 3-CNF formulas based on fitness-distance correlation. In: Proceedings of Genetic and Evolutionary Computation Conference, pp. 1415–1422 (2015)
9. Goldman, B.W., Punch, W.F.: Parameter-less population pyramid. In: Proceedings of Genetic and Evolutionary Computation Conference, pp. 785–792 (2014)
10. Kirkpatrick, S., Selman, B.: Critical behavior in the satisfiability of random Boolean expressions. Science **264**(5163), 1297–1301 (1994)
11. López-Ibáñez, M., Dubois-Lacoste, J., Cáceres, L.P., Stützle, T., Birattari, M.: The irace package: iterated racing for automatic algorithm configuration. Oper. Res. Perspect. **3**, 43–58 (2016)
12. Sutton, A.M., Neumann, F.: Runtime analysis of evolutionary algorithms on randomly constructed high-density satisfiable 3-CNF formulas. In: Bartz-Beielstein, T., Branke, J., Filipič, B., Smith, J. (eds.) PPSN 2014. LNCS, vol. 8672, pp. 942–951. Springer, Cham (2014). https://doi.org/10.1007/978-3-319-10762-2_93

Optimal Investment in the Development of Oil and Gas Field

Adil Erzin[1,2]([⊠]) [ID], Roman Plotnikov[1] [ID], Alexei Korobkin[3], Gregory Melidi[2],
and Stepan Nazarenko[2]

[1] Sobolev Institute of Mathematics, SB RAS, Novosibirsk 630090, Russia
adilerzin@math.nsc.ru
[2] Novosibirsk State University, Novosibirsk 630090, Russia
[3] Gazpromneft, St. Petersburg, Russia

Abstract. Let an oil and gas field consists of clusters in each of which an investor can launch at most one project. During the implementation of a particular project, all characteristics are known, including annual production volumes, necessary investment volumes, and profit. The total amount of investments that the investor spends on developing the field during the entire planning period we know. It is required to determine which projects to implement in each cluster so that, within the total amount of investments, the profit for the entire planning period is maximum.

The problem under consideration is NP-hard. However, it is solved by dynamic programming with pseudopolynomial time complexity. Nevertheless, in practice, there are additional constraints that do not allow solving the problem with acceptable accuracy at a reasonable time. Such restrictions, in particular, are annual production volumes. In this paper, we considered only the upper constraints that are dictated by the pipeline capacity. For the investment optimization problem with such additional restrictions, we obtain qualitative results, propose an approximate algorithm, and investigate its properties. Based on the results of a numerical experiment, we conclude that the developed algorithm builds a solution close (in terms of the objective function) to the optimal one.

Keywords: Investment portfolio optimization · Production limits

1 Introduction

The founder of the mathematical theory of portfolio optimization is G. Markowitz, who, in 1952, published an article [16] with the basic definitions and approaches for evaluating investment activity. He developed a methodology for the formation of an investment portfolio, aimed at the optimal choice of assets, based on a given ratio of profitability/risk. The ideas formulated by him form the basis of modern portfolio theory [8,16,17].

The research is carried out within the framework of the state contract of the Sobolev Institute of Mathematics (project 0314–2019–0014).

Y. Kochetov et al. (Eds.): MOTOR 2020, CCIS 1275, pp. 336–349, 2020.
https://doi.org/10.1007/978-3-030-58657-7_27

The author of [21] gave a review of portfolio selection methods and described the prospects of some open areas. At first, the author described the classical Markowitz model. Then comes the "intertemporal portfolio choice" developed by Merton [18,19], the fundamental concept of dynamic hedging and martingale methods. Pliska [21], Karatzas [11], as well as Cox and Huang [6] made the main contribution to the development of this direction. The authors of [7] and [20] proposed the formulas for the optimal portfolio for some private productions. These formulas have the form of conditional expectation from random variables.

In most well-known studies, the problem of optimal investment is solved numerically [3,4], which does not allow us to identify the contribution of portfolio components to the optimal solution. In [2], a new approach is proposed for dynamic portfolio selection, which is not more complicated than the Markowitz model. The idea is to expand the asset space by including simple, manageable portfolios and calculate the optimal static portfolio in this extended space. It is intuitively assumed that a static choice among managed portfolios is equivalent to a dynamic strategy.

If we consider investing in specific production projects, then each of them is either implemented or not. In contrast to the classical Markowitz's problem, a discrete statement, arises, and the mathematical apparatus developed for the continuous case is not applicable. In [15], the authors examined a two-criterion problem of maximizing profit and minimizing risk. The characteristics of each project, mutual influence, and the capital available to the investor are known. For the Boolean formulation of the problem, the authors proved NP-hardness and found special cases when the problem is solved with pseudopolynomial time complexity.

The portfolio optimization problems described above relate to the stock market. For companies operating in the oil and gas sector, optimization problems are relevant. In these problems it is necessary to maximize total profit and minimize risks for a given period, taking into account additional restrictions, for example, on production volume, as well as problems in which it is necessary to maximize production (or profit) for a given amount of funding. In [1], the author presented an approach aimed at improving the efficiency of the management of the oil and gas production association. Two control loops are distinguished: macroeconomic, which is responsible for optimizing policies at the aggregated level (industry and regional), and microeconomic, which is responsible for optimizing the organizational and functional structure of the company. The first circuit implemented using the author developed computable models of general economic equilibrium and integrated matrices of financial flows. The second circuit performed using an approach based on simulation of the business processes of an enterprise.

The author of [9] considers the problem of forming a portfolio of investment projects, which required to obtain maximum income under given assumptions regarding risks. A method is proposed based on a comprehensive multidimensional analysis of an investment project. The authors of [10,12] consider a problem of minimizing the deposit costs with restrictions on the volume of the production. They propose an algorithm for building an approximate solution by dynamic programming. The accuracy of the algorithm depends on the discretization step of the investment volume. The authors of [10] formulate the problem of minimizing various costs associated with servicing wells, with limitations related to the amount of oil produced, as a linear programming problem, and find the optimal solution using the simplex method.

For the decision-maker, the main concern is how to allocate limited resources to the most profitable projects. Recently, a new management philosophy, Beyond NPV (Net Present Value), has attracted more and more international attention. Improved portfolio optimization model presented in [23]. It is an original method, in addition to NPV, for budgeting investments. In the proposed model, oil company executives can compromise between profitability and risk concerning their acceptable level of risk. They can also use the "operating bonus" to distinguish their ability to improve the performance of major projects. To compare optimized utility with non-optimized utility, the article conducted a simulation study based on 19 foreign upstream assets owned by a large oil company in China. The simulation results showed that the optimization model, including the "operating bonus", is more in line with the rational demand of investors.

The purpose of the paper [5] is to offer a tool that might support the strategic decision-making process for companies operating in the oil industry. Their model uses Markowitz's portfolio selection theory to construct an efficient frontier for currently producing fields and a set of investment projects. These relate to oil and gas exploration projects and projects aimed at enhancing current production. The net present value obtained for each project under a set of user-supplied scenarios. For the base-case scenario, the authors also model oil prices through Monte Carlo simulation. They run the model for a combination of portfolio items, which include both currently producing assets and new exploration projects, using data characteristics of a mature region with a high number of low-production fields. The objective is to find the vector of weights (equity stake in each project), which minimizes portfolio risk, given a set of expected portfolio returns.

Due to the suddenness, uncertainty, and colossal loss of political risks in overseas projects, the paper [14] considers the time dimension and the success rate of project exploitation for the goal of optimizing the allocation of multiple objectives, such as output, investment, efficiency, and risk. A linear portfolio risk decision model proposed for multiple indicators, such as the uncertainty of project survey results, the inconsistency of project investment time, and the number of projects in unstable political regions. Numerical examples and the results test the model and show that the model can effectively maximize the portfolio income within the risk tolerance range under the premise of ensuring the rational allocation of resources.

This paper discusses the problem of optimal investment of oil and gas field development consisting of subfields – *clusters*. For each cluster, there are several possibilities for its development, which we call the *projects*. Each project characterized by cost, lead time, resource intensity, annual production volumes, and profit from its implementation. Also, there are restrictions on the annual production volumes of the entire field. This requirement leads to the need for a later launch of some projects so that the annual production volume does not exceed the allowable volumes. Assuming that a project launched later is another project, we proposed a statement of the problem in the form of a Boolean linear programming (BLP) problem. We estimated the maximum dimension at which CPLEX solves the BLP in a reasonable time. For a large-dimensional problem, we developed a method that constructs an approximate solution in two stages. At the first stage, the problem is solved without limitation on the volume of annual production. This problem remains NP-hard, but it is solvable by a pseudopoly-nomial dynamic programming algorithm. As a result, one project is selected for each cluster. The project is characterized, in particular, by the year of launch and production volumes in each subsequent year. If we start the project later, the annual production volumes shift. In the second stage, the problem of deter-mining the start moments of the projects selected at the first stage is solved, taking into account the restrictions on annual production volumes, and the profit is maximal. We developed a local search algorithm for partial enumeration of permutations of the order in which projects are launched. At the same time, for each permutation, the algorithm of tight packing of production profiles devel-oped by us (we call it a greedy algorithm), which builds a feasible solution, is applied. A numerical experiment compared our method and CPLEX.

The rest of the paper has the following organization. In Sect. 2, we state the problem as a BLP. In Sect. 3, the problem without restrictions on the vol-ume of production reduced to a nonlinear distribution problem, which is solved by dynamic programming. As a result, a "best" project found for each clus-ter. Section 4 describes the method for constructing an approximate solution by searching for the start times for the "best" projects. The next section presents the results of a numerical experiment. We identify the maximum dimension of the problem, which is solved by the CPLEX package in a reasonable time, and compare the accuracy of the developed approximate algorithms. Section 6 con-tains the main conclusions and describes the directions for further research.

2   Formulation of the Problem

For the mathematical formulation of the problem, we introduce the following notation for the parameters:

- $[1, T]$ is the planning period;
- C is the total amount of investment;
- K is the set of clusters ($|K| = n$);
- P_k is the set of projects for the development of the cluster $k \in K$ ($\max_k |P_k| = p$) taking into account the shift at the beginning of each project;
- $d_k^i(t)$ is the volume of production in the cluster $k \in K$ per year $t = 1, \ldots, T$, if the project $i \in P_k$ is implemented there;
- q_k^i is the profit for the entire planning period from the implementation of project i in cluster k;

– c_k^i is the cost of implementing project i in cluster k;
– $D(t)$ is the maximum allowable production per year t;

and for the variables:

$$x_k^i = \begin{cases} 1, \text{ if project } i \text{ is selected for cluster } k; \\ 0, \text{ else.} \end{cases}$$

Then the problem under consideration can be written as follows.

$$\sum_{k \in K} \sum_{i \in P_k} q_k^i x_k^i \to \max_{x_k^i \in \{0,1\}}; \tag{1}$$

$$\sum_{k \in K} \sum_{i \in P_k} c_k^i x_k^i \leq C; \tag{2}$$

$$\sum_{i \in P_k} x_k^i \leq 1, \ k \in K; \tag{3}$$

$$\sum_{k \in K} \sum_{i \in P_k} d_k^i(t) x_k^i \leq D(t), \ t \in [1, T]. \tag{4}$$

Remark 1. Each project has various parameters, among which the annual production volumes. If we start the project later, then the graphic of annual production will shift entirely. Suppose $d_k^i(t)$ is the volume production per year t if the project i is implemented in the cluster k. If this project is launched τ years later, the annual production during year t will be $d_k^i(t - \tau)$. So, each project in the set P_k is characterized, in particular, by its beginning.

However, not all characteristics retain their values at a later launch of the project. Profit from project implementation depends on the year of its launch, as money depreciates over the years. One way to account for depreciation is to use a discount factor. The value of money decreases with each year by multiplying by a discount factor that is less than 1. In this regard, at the stage of preliminary calculations, we recount values associated with investment and profit.

As a result, the set P_k consists of the initial projects, and the shifted projects for different years as well. So, having solved the problem (1)–(4), we will choose for each cluster not only the best project but also the time of its start.

Problem (1)–(4) is an NP-hard BLP. For the dimension which we define in Sect. 5, a software package, for example, CPLEX, can be used to solve it. In order to solve the problem of a large dimension, it is advisable to develop an approximate algorithm. To do this, in the next section we consider the problem (1)–(3).

3   The Problem Without Restrictions on Production Volumes

If there are no restrictions on production volumes, then instead of the variables x_k^i, we can use the variables c_k, which are equal to the amount of the investment allocated for the development of the cluster k. To do this, for each cluster k, we introduce a new profit function $q_k(c_k)$, which does not depends on the selected project but depends on the amount of investment. For each k, the function $q_k(c_k)$ is obviously non-decreasing piecewise constant. Moreover, if we know the value c_k, then one project is uniquely will be used to develop the cluster k, and all its characteristics will be known. Indeed, the more money is required to implement the project, the more efficient it is (more profitable). If this is not a case, then a less effective but more expensive project can be excluded. Obviously, the values of all functions $q_k(c_k)$, $c_k \in [0, C]$, $k \in K$, are not difficult to calculate in advance. The complexity of this procedure does not exceed $O(KpC)$.

Given the previous, we state the problem of maximizing profit without restrictions on production volumes, assuming that all projects start without delay, in the following form.

$$\sum_{k \in K} q_k(c_k) \to \max_{c_k \in [0,C]};$$ (5)

$$\sum_{k \in K} c_k \leq C.$$ (6)

Although problem (5)–(6) become easier than the problem (1)–(4), it remains NP-hard. However, it is a distribution problem, for the solution of which we apply the dynamic programming method, the complexity of which is $O(n(C/\delta)^2)$, where δ is the step of changing the variable c_k. Solving the problem (5)–(6), we choose the "best" project for each cluster. If it turns out that at the same time, all the restrictions (4) fulfilled, then this solution is *optimal* for the original problem (1)–(4). If at least one inequality (4) violated, then we will construct a feasible solution in the manner described in the next section.

4   Consideration of Restrictions on Production Volumes

We will not change the projects selected for each cluster as a result of solving the problem (5)–(6). We will try to determine the moments of launching these projects so that inequalities (4) fulfill, and profit takes maximal value. The project selected for the cluster k is characterized by the production volumes $d_k(t)$ in each year $t \in [1, T]$. It is necessary to shift the beginning of some projects to a later time so that in each year $t \in [1, T]$ the total production is at most $D(t)$:

$$\sum_{k \in K} d_k(t) \leq D(t).$$

Assume that the cluster k development project, whose beginning is shifted by $i \in [0, t_k]$ years, is another project. Then for each cluster, there is a set of projects, which we denote as before by P_k ($|P_k| = t_k + 1$). As a result, to determine the shift in the start of the project launch for each cluster, it is enough to solve the problem (1)–(4) without restriction (2), in which the Boolean variable $x_k^i = 1$ if and only if the start of the cluster project k is shifted by i years. Then the solution to the small-dimension problem can be found using a CPLEX. However, for a large-sized problem, it is necessary to use an approximate algorithm.

4.1   Greedy Algorithm

Suppose we order the projects according to the years of their launch. A shift in the start of projects changes this order. The order in which projects start uniquely determined by the permutation π of the cluster numbers $\{1, 2, \ldots, n\}$. For a given permutation, we describe informally a greedy algorithm that constructs a feasible solution to the problem.

Denote by $P(\pi)$ the list of ordered projects. The first project starts without delay (with zero shift). We exclude it from the set $P(\pi)$. For the first project of the updated set $P(\pi)$, we determine its *earliest* start time, which is no less than the start time of a previous project, to comply with the production order and restrictions (4) in each year and exclude this project from the set $P(\pi)$. We continue the process until the start year of the last project, $\pi(n)$ is found.

The greedy algorithm will construct a feasible solution for the given permutation π, if it exists, with the time complexity of $O(nT)$. In the oil and gas industry, profiles (graphs) of annual production volumes have a log-normal distribution [22], which is characterized by a rapid increase, and then a slight decrease. This observation and the following lemma, to some extent, justify why we use the greedy algorithm.

Proposition 1. *If the order of launching the projects is known, the annual production schedules for all projects are not-increasing, and $D(t) = D = const$, $t \in [1, T]$, then the greedy algorithm determines the optimal start years for all projects.*

Proof. In the problem under consideration, time is discrete (measured in years). Therefore, the value of production in each cluster is a certain real number that does not change for one year. A greedy algorithm for a given order of projects determines the earliest start time for each project, which is not less than the start time of the previous project. Suppose that in all optimal solutions, there is at least one project that begins later than the year determined by the greedy algorithm. Consider some optimal solution and let k be the first project that we can start earlier (Fig. 1a). Since the project k can start earlier, then move it as much as possible to the left to maintain validity (Fig. 1b). Notice that it is enough to check the value of production d_k^1 only in the first year of the project k because it is not less than production in subsequent years ($d_k^t \leq d_k^1$, $t > 1$). The solution obtained after shifting the project k to the left is no worse (and taking into account the discount coefficient, even better), but the project k starts earlier, which contradicts the assumption. The proof is over.

So, with a particular order of projects, a greedy algorithm builds a solution close to optimal with $O(nT)$ time complexity. A complete enumeration of permutations requires $O(n!)$ operations. However, it is reasonable to develop a local search algorithm in which, at each step, the best permutation is searched in the vicinity of the current permutation. In order to obtain a solution for a given permutation, a greedy algorithm is used. The higher the profit in a particular order of projects, the better the permutation. In the next subsection, we develop a local search algorithm.

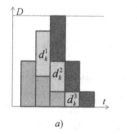
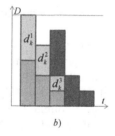

a)   b)

Fig. 1. Illustration to the Lemma 1 proof (Bars of the same color belong to the same project, and the height of the bar is the volume of production in the corresponding year). *a*) Project k (yellow) may start earlier; *b*) After shifting the project k to the left. (Color figure online)

4.2 Local Search

Using the greedy algorithm described in the previous subsection, one can construct a solution for each permutation of the cluster numbers. Therefore, it is essential to find a permutation where the solution constructed in such a way is near-optimal. For this reason, we suggest a local search procedure for permutations starting from some promising one.

In order to obtain the first permutation for the local search procedure we perform the following greedy algorithm. At the first step we choose such cluster that yields maximum value of income if its development is started at the first year. The number of this cluster becomes the first value of the permutation. Then, at each step of the algorithm, we choose among the unprocessed clusters such cluster, that if its development is started at the earliest year (taking into account the per-year production bound and already chosen clusters), then the total income increment will be maximum. After such cluster is found, we assign the corresponding shift for it and set its number to the next permutation value. The permutation obtained by the described greedy procedure becomes the first permutation of the local search algorithm.

For each permutation in the local search procedure, we construct a solution using the greedy algorithm, that is described in previous subsection, with time complexity $O(nT)$. As a movement operation of the local search procedure, we perform the best possible exchange of two different elements of a permutation. Then the cardinality of the neighborhood of the current permutation is $O(n^2)$, and the time complexity of the searching the best solution is this neighborhood is $O(n^3 T)$.

5 Simulation

The proposed algorithms have been implemented in the C++ programming language and launched on the randomly generated test instances. We also used the IBM ILOG CPLEX package (version 12.10) in order to obtain optimal or near-optimal solutions together with guaranteed upper bounds for the comparison.

The numerical experiment performed on an Intel Core i5-4460 (3.2GHz) 8Gb machine.

For the generation of the test data, we supposed that the distribution function of production volumes by the planned period is log-normal [22]. The parameters of the distribution density μ and σ were chosen randomly with uniform distribution on the intervals $[1, 2]$ and $[1, 1.4]$ correspondingly. We defined the maximum value of the production volume for each project at random with uniform distribution on the interval $[30, 200]$ (in thousands of tons) and then multiplied each per-year volume by the corresponding scaling factor.

We assumed that the profit is proportional to the production volume. For each project, in order to generate the profit per each year, we, at first, took the random coefficient (the cost of one ton in millions of rubles) uniformly distributed on the interval $[4, 6]$. This value may vary depending on the differences in the condition of the production, overhead costs, remoteness of the cluster. After that, we multiplied the per-year production volumes by this coefficient. We also added the noise to the generated values multiplying them by the random values uniformly distributed on the interval $[0.95, 1.05]$.

Like the investments, we generated random values uniformly distributed on the interval $[250, 1500]$ (in millions of rubles). We assumed that the obtained amount of money spent in the first year of the project exploitation. In about 10% of the cases, other investments made in the second year of the project exploitation. The second investment is taken as a random part of the first investment from 10 to 50%. As the upper bounds of investments and per-year production volumes, we took the one-third part of the sum by clusters of the maximum values per project and per year. That is,

$$C = 1/3 \sum_{k \in K} \max_{i \in P_k} c_k^i,$$

and

$$D = 1/3 \sum_{k \in K} \max_{i \in P_k, t \in [1,T]} d_k^i(t), D(t) = D \ \forall t \in [1, T].$$

For solving the problem (5)–(6) we set δ equal ten thousands of rubles, because, according to our preliminary experiments, further decrease of δ does not improve the solution significantly.

We generated instances for four different variants of the number of clusters: $n = 10, 25, 50$, and 100. For each value of n, we generated four instances with different maximum and a minimum number of projects per cluster: 1) from 1 to 10; 2) from 10 to 25; 3) from 25 to 50, and 4) from 50 to 100. We launched our algorithm and CPLEX on each instance. The results presented in Table 1. In this table, CPLEX stands for the results obtained by CPLEX launched on the problem (1)–(4). $CPLEX_{fp}$ stands for the results obtained by CPLEX launched on the restricted problem with the fixed project per each cluster found by the dynamic programming method described in Sect. 3. Notations *obj*, *ub*, and *gap* stand for, correspondingly, the objective function of the incumbent, the upper bound of the value of objective function, and the relative difference between *obj*

Table 1. Comparison of the proposed algorithm A with CPLEX

n	p_{min}	p_{max}	CPLEX			$CPLEX_{fp}$				A			
			Obj	Ub	Gap	Obj	Ub	Gap	Decline (%)	Obj	r_1	r_2	Time (sec.)
10	1	10	12851.93	12851.93	0	12072	12072	0	6.07	12072	0.94	1	0.006
	10	25	16460.95	16460.95	0	15502.23	15502.23	0	5.82	14710.6	0.89	0.95	0.005
	25	50	16988.53	16988.53	0	14862.28	14862.28	0	12.51	14764.3	0.87	0.99	0.004
	50	100	17140.36	17140.36	0	15607.92	15607.92	0	8.94	15374.1	0.9	0.99	0.003
25	1	10	30465.26	30465.26	0	29849.9	29849.9	0	2.02	29571.9	0.97	0.99	0.082
	10	25	42501.57	42501.57	0	39728.94	39728.94	0	6.52	38930.6	0.92	0.98	0.077
	25	50	46508.15	46849.94	0.007	43646.48	43646.48	0	6.15	42407.4	0.9	0.97	0.056
	50	100	47432.09	47906.72	0.01	44307.47	44307.47	0	6.59	43324	0.9	0.98	0.023
50	1	10	70568.99	70568.99	0	69609.18	69609.18	0	1.36	66328.2	0.94	0.95	0.752
	10	25	86529.19	86659.92	0.002	80616.31	80778.51	0.002	6.83	76845.5	0.89	0.95	0.627
	25	50	93928.28	94290.32	0.003	88415.05	88415.05	0	5.87	86861.8	0.92	0.98	0.612
	50	100	95201.48	95621.93	0.004	88532.86	88661.56	0.001	7	86380.7	0.9	0.97	0.39
100	1	10	139928.34	140023.11	0.0007	136420.24	136586.12	0.001	2.51	128679	0.92	0.94	5.22
	10	25	173898.61	174065.77	0.001	163833.54	163932.06	0.0006	5.79	154452	0.89	0.94	5.35
	25	50	189722.3	190000.79	0.001	177064.12	177138.3	0.0004	6.67	171431	0.9	0.97	5.84
	50	100	195223.21	195686.39	0.0023	180375.66	180476.58	0.0005	7.61	176830	0.9	0.98	9.44
250	250	500	—	—	—	515525.8	516003.7	0.0009	—	495366	—	0.96	366.4

and *ub. decline* stands for the decline (in percents) of the objective function value of the incumbent of the problem with the fixed set of projects concerning the objective function of the incumbent of the entire problem. The last four columns represent the results obtained by our algorithm, which is named A in the table. r_1 denotes the ratio $obj(A)/ub(CPLEX)$, and r_2 denotes the ratio $obj(A)/ub(CPLEX_{fp})$. The last column stands for the total running time of our algorithm. The running time of CPLEX was limited by 60 s for all the cases except the last one of the largest size,—in the last case CPLEX was given for 1 hour. It also should be noted that CPLEX was parallelized on four threads.

As it follows from the table, in the cases of small and moderate size CPLEX solves the problem rather precisely within 60 s. In these cases it always constructs a solution on which the value of the objective function differs from the optimal one by at most 1%. Algorithm A constructs a less accurate solution. As it is seen at the column r_1, in the worst case, the objective value of the obtained solution differs from the optimal by 13%, in the best case — by 3%, and on average, this difference does not exceed 9%. As one can see at the column *decline*, the choice of the projects obtained by solution of the problem without restriction on the production volumes deteriorates the solution of the entire problem by up to 12.5%. On average, this decline is about 6%. The quality of our local search procedure applied to the solution obtained by the greedy heuristic is estimated in the column r_2. On average, the ratio does not exceed 3%. In a case of large size, when the number of clusters is 250 and the number of projects in each cluster varies from 250 to 500, CPLEX failed to construct any feasible solution within 1 hour, but the algorithm A constructed an approximate solution within about 6 min. When we set the projects found by algorithm A to CPLEX for this instance, it successfully found the solution with rather small value of gap (less than 0.1%) within 1 hour. In this case, the local search procedure found a solution that differs from the optimal one by not more than 4%.

6   Conclusion

In this paper, we studied the NP-hard problem of maximizing profit by choosing long-term cluster development projects within the oil and gas field, with restrictions on the total investment and maximum annual production. We proposed a statement of the problem in the form of Boolean linear programming and set ourselves three goals. First, to investigate the effectiveness of application software packages, such as CPLEX, for solving the BLP problem. Secondly, develop a fast approximate algorithm. Thirdly, compare the effectiveness of the CPLEX package and the approximate algorithm.

The approximate algorithm consists of two stages, which are partially dictated by the specifics of the problem. At the first stage, profit is maximized by selecting one project for each cluster without taking into account the restrictions on the volume of annual production. The distribution problem arising, in this case, is solved by the dynamic programming algorithm with acceptable running time. Projects selected at the first stage can be launched later (with a delay).

Therefore, at the second stage, the moments of the start of the selected projects are determined in such a way that the annual production volumes do not exceed the set values, and the profit is maximum. The problem of the second stage also formulated in the form of the BLP. It makes sense without the first stage because, in practice, development projects for each cluster often known, and it is only necessary to determine the moments of their launch.

Production profiles have a characteristic shape, which is determined by the log-normal distribution law and has the form of a graph that first overgrows, reaches its maximum value, and then slowly decreases [22]. With a certain degree of assumption, we assumed that the profiles are non-increasing. We proved that in the case of non-increasing production profiles and for a given order of project start (which is determined by the permutation of cluster numbers), the greedy algorithm constructs the optimal solution. The algorithm of permutations sorting is justified, and the greedy algorithm used for each permutation. Iterating over all permutations is time-consuming, and for large dimensions, it is just not applicable, so we used a relatively simple local search algorithm.

The results of the numerical experiment on randomly generated examples surprised us (see Table 1). For $25 \leq n \leq 100$, the CPLEX package was not able to build an optimal solution, but it turned out that CPLEX within one minute builds a feasible solution quite close to the optimal one. The approximate algorithm that we developed also builds a solution close to optimal, but CPLEX turned out to be more efficient for such dimension. Thus, we conclude that for the considered problem when $n \leq 100$, it is advisable to use a package of application programs CPLEX instead of our algorithm. In a case of large size, for example when $n \geq 250$, CPLEX failed to construct any feasible solution within 1 hour, but the algorithm A constructed an approximate solution within 6 min. When we set the projects found by solving the problem (5)–(6) with $n = 250$ to CPLEX, it successfully found the solution with gap less than 0.1% within 1 hour.

Perhaps the situation will change if we consider some additional restrictions. In practice, it is necessary to produce annually at least a given volume and no more than a predetermined quantity. Moreover, there are restrictions on the size of annual investments. Furthermore, annual production volumes are random variables, so the need to take into account the probabilistic nature of the source data can ruin the problem so that the use of CPLEX will become inappropriate.

In future research, we plan to take into account the additional restrictions and specifics, as well as to develop a more efficient approximate algorithm based on a genetic algorithm in which an effective local search, for example, VNS [13], will be used at the mutation stage.

References

1. Akopov, A.S.: Metodi povisheniya effektivnosti upravleniya neftegazodobi-vaiushimi ob'edinenuyami. Ekonomicheskaya nauka sovremennoy Rossii. **4**, 88–99 (2004). (in Russian)
2. Brandt, M.W., Santa-Clara, P.: Dynamic portfolio selection by augmenting the asset space. NBER Working Paper. 10372, JEL No. G0, G1 (2004)
3. Brandt, M.W., Goyal, A., Santa-Clara, P., Stroud, J.R.: A simulation approach to dynamic portfolio choice with an application to learning about return predictability. Rev. Financ. Stud. **18**, 831–873 (2005)
4. Brennan, M., Schwartz, E., Lagnado, R.: Strategic asset allocation. J. Econ. Dyn. Control. **21**, 1377–1403 (1997)
5. Bulai, V.C., Horobet, A.: A portfolio optimization model for a large number of hydrocarbon exploration projects. In: Proceedings of the 12th International Conference on Business Excellence, vol. 12(1), pp. 171–181 (2018). https://doi.org/10.2478/picbe-2018-0017
6. Cox, J.C., Huang, C.-F.: Optimal consumption and portfolio policies when asset prices follow a diffusion process. J. Econ. Theory **49**, 33–83 (1989)
7. Detemple, J.B., Garcia, R., Rindisbacher, M.: A Monte-Carlo method for optimal portfolios. J. Financ. **58**, 401–446 (2003)
8. Detemple, J.: Portfolio selection: a review. J. Optim. Theory Appl. **161**(1), 1–21 (2012). https://doi.org/10.1007/s10957-012-0208-1
9. Dominikov, A., Khomenko, P., Chebotareva, G., Khodorovsky, M.: Risk and profitability optimization of investments in the oil and gas industry. Int. J. Energy Prod. Manage. **2**(3), 263–276 (2017)
10. Goncharenko, S.N., Safronova, Z.A.: Modeli i metody optimizacii plana dobychi i pervichnoy pererabotki nifti. Gorniy informacionno-analiticheskiy biulleten **10**, 221–229 (2008). (in Russian)
11. Karatzas, I., Lehoczky, J.P., Shreve, S.E.: Optimal portfolio and consumption decisions for a "mall investor" on a finite horizon. SIAM J. Control Optim. **25**, 1557–1586 (1987)
12. Konovalov, E.N., Oficerov, V.P., Smirnov, S.V.: Povishenie effectivnosti investiciy v neftedobyche na osnove modelirovaniya. Problemi upravleniya i modelirovaniya v slognih sistemsh: Trudi V negdunarodnoy konferencii. Samara, pp. 381–385 (2006). (in Russian)
13. Hansen, P., Mladenovic, N.: Variable neighborhood search: principles and applications. Eur. J. Oper. Res. **130**, 449–467 (2001)
14. Huang, S.: An improved portfolio optimization model for oil and gas investment selection considering investment period. Open J. Soc. Sci. **7**, 121–129 (2019)
15. Malah, S.A., Servah, V.V.: O slognosti zadachi vybora investicionnih proektov. Vestnik Omskogo universiteta. **3**, 10–15 (2016). (in Russian)
16. Markowitz, H.M.: Portfolio selection. J. Financ. **7**(1), 71–91 (1952)
17. Markowitz, H.M.: Portfolio Selection: Efficient Diversification of Investment. Wiley, New York (1959)
18. Merton, R.C.: Lifetime portfolio selection under uncertainty: the continuous time case. Rev. Econ. Stat. **51**, 247–257 (1969)
19. Merton, R.C.: Optimum consumption and portfolio rules in a continuous-time model. J. Econ. Theory **3**, 273–413 (1971)
20. Ocone, D., Karatzas, I.: A generalized Clark representation formula, with application to optimal portfolios. Stoch. Stoch. Rep. **34**, 187–220 (1991)

21. Pliska, S.: A stochastic calculus model of continuous trading: optimal portfolios. Math. Oper. Res. **11**, 371–382 (1986)
22. Power, M.: Lognormality in the observed size distribution of oil and gas pools as a consequence of sampling bias. Math. Geol. **24**, 929–945 (1992). https://doi.org/10.1007/BF00894659
23. Xue, Q., Wang, Z., Liu, S., Zhao, D.: An improved portfolio optimization model for oil and gas investment selection. Pet. Sci. **11**(1), 181–188 (2014). https://doi.org/10.1007/s12182-014-0331-8

Genetic Algorithms with the Crossover-Like Mutation Operator for the k-Means Problem

Lev Kazakovtsev[1,2]($\boxtimes$) (ID), Guzel Shkaberina[1], Ivan Rozhnov[1,2] (ID), Rui Li[1] (ID), and Vladimir Kazakovtsev[3]

[1] Reshetnev Siberian State University of Science and Technology, prosp. Krasnoyarskiy Rabochiy 31, Krasnoyarsk 660031, Russia
levk@bk.ru
[2] Siberian Federal University, prosp. Svobodny 79, Krasnoyarsk 660041, Russia
[3] ITMO University, Kronverksky pr. 49, St. Petersburg 197101, Russia

Abstract. Progress in the development of automatic grouping (clustering) methods, based on solving the p-median and similar problems, is mainly aimed at increasing the computational efficiency of the algorithms, their applicability to larger problems, accuracy, and stability of their results. The researchers' efforts are focused on the development of compromise heuristic algorithms that provide a fairly quick solution with minimal error. The Genetic Algorithms (GAs) with greedy agglomerative crossover procedure and other special GAs for the considered problems demonstrate the best values of the objective function (sum of squared distances) for many practically important problems. Usually, such algorithms do not use any mutation operator, which is common for other GAs.

We propose new GAs for the k-means problem, which use the same procedures as both the crossover and mutation operators. We compared a simple GA for the k-means problem with one-point crossover and its modifications with the uniform random mutation and our new crossover-like mutation. In addition, we compared the GAs with greedy heuristic crossover procedures to their modifications which include the crossover-like mutation. The comparison results show that the idea of our new mutation operator is able to improve significantly the results of the simplest GA as well as the genetic algorithms with greedy agglomerative crossover operator.

Keywords: Clustering · k-Means · Genetic algorithm · Greedy agglomerative procedure

1 Introduction and Problem Statement

The k-means problem [1] can be described as finding a set of k cluster centroids $X_1, ... X_k$ in a d-dimensional space with the minimal sum of squared distances

Supported by the Ministry of Science and Higher Education of the Russian Federation (Project FEFE-2020-0013).

Y. Kochetov et al. (Eds.): MOTOR 2020, CCIS 1275, pp. 350–362, 2020.
https://doi.org/10.1007/978-3-030-58657-7_28

from them to the given N points (vectors) A_i (SSE, sum of squared errors):

$$\arg\min_{X_1,...,X_k \in \mathbb{R}^d} F(X_1,...,X_k) = \sum_{i=1}^{N} \min_{j \in \{\overline{1,k}\}} \|X_j - A_i\|^2. \tag{1}$$

In the continuous p-median problem, a sum of distances (instead of squared distances) is calculated, and the searched points are called centers or medians. If a sum of Manhattan (L_1, rectilinear) distances is used as the minimized function, the problem is referred to as the k-means problem [2].

An algorithm of the same name (Algorithm 1) [3,4], also known known as Lloyd algorithm, sequentially improves a known solution, looking for a local minimum of (1). This local search algorithm (LSA) is simple, fast, and applicable to the widest class of problems. The algorithm has a limitation: the number of groups (clusters) k must be known. The result is highly dependent on the initial solution chosen at random.

Algorithm 1. k-Means (Lloyd, ALA: Alternating Location-Allocation)

Require: data vectors $A_1...A_N$, k initial cluster centers (centroids) $X_1,...,X_k$.
   repeat
      Step 1: For each of centers X_i, compose clusters C_i of data vectors so that each of the data vectors is assigned to the nearest center.
      Step 2: Calculate new center X_i for each of the clusters.
   until Steps 1,2 result in no modifications.

Both Steps 1 and 2 improve the objective function (1) value.

In this research, we try to improve the accuracy of the k-means problem result (1) and its stability within a fixed, limited run time. By the accuracy of the algorithm, we mean the achieved value of (1). We do not consider other important issues in the fields of cluster analysis such as adequacy of the model (1) and correspondence of the algorithm result to the actual partition [5].

The idea of Genetic Algorithms (GAs) is based on a recombination of elements of some candidate solutions set called "population". Each candidate solution is called an "individual" encoded by a "chromosome" represented by a vector of bits, integers or real numbers depending on the algorithm. In the modern literature, there is practically no systematization of the approaches used (see [6–8]), for algorithms with the real-number (centroid-based) chromosome encoding.

The first GA for the discrete p-median problem was proposed by Hosage and Goodchild [9]. Algorithm presented in [10] gave more precise results with a very slow convergence. In [11], the authors proposed a faster algorithm with a special "greedy" heuristic crossover operator which is also precise. All these algorithms solve discrete problems (p-median problem on a network) and use a simple binary chromosome encoding (1 for the network nodes selected as the medians and 0 for those not selected).

The mutation operator is devoted to guarantee the GA population diversity [12]. Usually, for the k-means and similar problems, the mutation randomly changes one or many chromosomes, replacing some centroids [12–14] or assignment of an object. For example, in [13] authors proposed the distance-based mutation which changes an allele value (an allele is a part of a chromosome that encodes the assignment an object to a cluster) depending on the distances of the cluster centroids from the corresponding data point. Each allele corresponds to a data point and its value represents the corresponding cluster number. The mutation operator is defined such that the probability of changing an allele value to a cluster number is higher if the corresponding centroid is closer to the data vector. To apply the mutation operator to the allele $s_W(i)$ corresponding to centroid X_i, let us denote $d_j = \|X_i - A_j\|$ where A_j is a data vector. Then, the allele is replaced with a value chosen randomly from the following distribution: $p_j = Pr\{s_W(i) = j\} = (c_m d_{max} - d_j)/(\sum_{i=1}^{K}(c_m d_{max} - d_i))$ where c_m is a constant usually ≤ 1 and $d_{max} = \max\{d_j\}$. In the case of a partition with one or more than one singleton clusters, the above mutation may result in the formation of empty clusters with a non-zero probability. It may be noted that the smaller the number of clusters, the larger is the SSE measure; so empty clusters must be avoided [13].

If the cluster centroids are searched in a continuous space, some GAs still use the binary encoding [15–17]. In Algorithm 1, the initial solutions are usually subsets of the data vectors set. Thus, in the chromosome code, 1 means that the corresponding data vector must be used as the initial centroid, and 0 for those not selected. In this case, some LSA (Algorithm 1 or similar) is used at each iteration of the GA. In the GAs for the k-means and analogous problems, which use the traditional binary chromosome encoding, many mutation techniques can be used. For example, in [18], the authors use binarization and represent the chromosome with binary strings composed of binary-encoded features (coordinates) of the centroids. The mutation operator arbitrarily alters one or more components (binary substrings) of a selected chromosome. In [19,20], authors call their algorithms "Evolutionary k-Means." However, they actually solve an alternative problem related to the k-Means problem aimed to increase clustering stability. This algorithm operates with the binary consensus matrices and uses two types of the mutation operators: cluster split (dissociative) and cluster merge (agglomerative) mutation. In [21], the chromosomes are strings of integers representing the cluster number for each of clustered objects, and the authors solve the k-means problem with simultaneous determining the number of clusters based on the silhouette [22] and David-Bouldin criteria [23] (similar approach is used in a much simpler algorithm X-Means [24]) which are used as the fitness functions. Thus, in [21], authors solve a problem with the mathematical statement other than (1) and use cluster recalculating in accordance with (1) as the mutation operator. Similar encoding is used in [13] where authors propose a mutation operator, which changes the assignment of individual data objects to clusters.

In [14], the authors encode the solutions (chromosomes) in their GA as sets of centroids represented by their coordinates (vectors of real numbers) in a d-dimensional space. The same principle of centroid-based chromosome representation is used in the GAs of the Greedy Heuristic Method [25]. In [14], the mutation procedure is as follows. Randomly generate a number from 0 to 1. If the number is less than mutation probability μ, the chromosome will mutate. The number $b \in (0,1]$ is randomly generated with the uniform distribution. If the position of a centroid is v, the mutation is as follows:

$$v \leftarrow \begin{cases} v \pm 2 \times b \times v, & v \neq 0, \\ v = v \pm 2 \times b, & v = 0. \end{cases} \qquad (2)$$

Signs "+" and "−" have the same probability here [18].

In (2), coordinates of a centroid are shifted randomly. A similar shifting technique with an "amplification factor" was used in [26,27]. However, the local minima distribution among the search space is not uniform [12]: new local minima of (1) can be found with higher probability in some neighborhood of a given local minimum than in a neighborhood of a randomly chosen point (here, we do not mean an ϵ-neighborhood). Thus, mixing the centroids from two local minima must usually outperform the random shift of centroid coordinates. The idea of combining local minima is the basic idea of the GAs with the greedy agglomerative heuristic crossover procedures [25] and other algorithms [28] which use no mutation operator. Such algorithms are able to demonstrate more accurate results in comparison with many other algorithms for many practical problems. However, one of the most important problems of the GAs is the convergence of the entire population into some narrow area (population degeneration) around some local minimum.

The Variable Neighborhood Search algorithms with the greedy agglomerative heuristic procedures proposed in [29,30] demonstrate better results than similar GAs. New randomly generated solutions (local minima) are used to form a randomized neighborhood around the best-achieved solution. This randomized approach provides some neighborhood variety.

The idea of this research is to use GAs with greedy agglomerative and other crossover operators in combination with the new mutation procedures which apply the same algorithm as the crossover procedure to the mutated solution and a randomly generated solution, and thus provide the population diversity.

2   New Crossover-Like Mutation Operator in a One-Point Crossover Genetic Algorithm

The GA in [14] uses the roulette wheel selection without any elitism (i.e., equal probabilities of selecting each of the individuals) and a simple one-point crossover procedure for the chromosomes. This algorithm uses the mutation procedure based on (2) with mutation probability 0.01. In our experiments below, we replaced this mutation procedure with the following algorithm:

Algorithm 2. k-Crossover-like mutation procedure with a one-point crossover

Step 1: Generating a random initial solution $S = \{X_1, ..., X_k\}$;
Step 2: Application of Algorithm 1 to S for obtaining local optimum S;
Step 3: Applying the simple one-point crossover procedure to the mutated individual S' from the population and S for obtaining the new solution S'';
Step 4: Application of Algorithm 1 to S'' for obtaining local optimum S'';
Step 5: If $F(S'') < F(S')$ then $S' \leftarrow S''$.

This new procedure is used with probability equal to 1 after each crossover operator. In our experiments, the population size $N_{POP} = 20$. The results of running the original algorithm described in [14] and its version with Algorithm 2 as the mutation operator are shown in Table 1 and Fig. 1. Our experiments show that the new mutation procedure is faster and more effective.

Table 1. Computational results for Mopsi-Joensuu data set [31] (6014 two-dimensional data vectors), 300 clusters, time limitation 180 s.

GA generations	Result with the original mutation (2)	Result with the mutation (Algorithm 2)	Ordinary k-means in a multi-start mode
10	1697.29	1667.95	1859.06
20	1682.37	1664.78	
50	1679.58	1664.78	
150	1664.81	1664.78	
200	1664.78	1664.78	

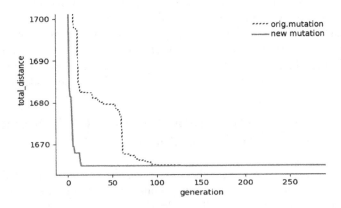

Fig. 1. Two mutation strategies in a one-point crossover GA

3 Known Clustering Algorithms of the Greedy Heuristic Method

The greedy agglomerative heuristic procedure for location problems can be described as an algorithm with two steps. The first step is combining two known ("parent") solutions (individuals) into one invalid intermediate solution with an excessive number of centroids. At the second step, the algorithm eliminates centroids in each iteration so that the removal of the centroid gives us the least significant increase in the value of the objective function (1) [11,16]:

Algorithm 3. Basic Greedy Agglomerative Heuristic Procedure

Require: needed number of clusters k, initial solution $S = \{X_1, ..., X_K\}$, $|S| = K$, $k < K$.
  Step 1: Improve S with Algorithm 1 or other LSA.
  while $K > k$
    for all $i' \in \{\overline{1, K}\}$
      Step 2: Assign $S' \leftarrow S \setminus \{X'_i\}$. Calculate $F'_{i'} \leftarrow F(S')$ where $F(.)$ is the objective function value, (1) for the k-means problem.
    end for
    Step 3: Select a subset S_{elim} of n_{elim} centers, $S_{elim} \subset S$, $|S_{elim}| = n_{elim}$, with the minimal values of corresponding variables $F'_{i'}$. Here, $n_{elim} = \max\{1, 0.2(|S| - k)\}$.
    Step 4: Obtain new solution $S \leftarrow S \setminus S_{elim}$; $K \leftarrow K - 1$, and run an LSA.
  end while

Algorithms 4–5 are known heuristic procedures [11,29,32], which modify some given solution based on the second known solution (see Algorithm 3).

Algorithm 4. Greedy Procedure #1

Require: Two solutions (sets of centroids) $S' = \{X'_1, ..., X'_k\}$ and $S'' = \{X''_1, ..., X''_k\}$.
  for all $i' \in \{\overline{1, k}\}$
    Step 1: Merge S' and one item of the set S'': $S \leftarrow S' \cup \{X''_{i'}\}$.
    Step 2: Run Algorithm 3 with the initial solution S and save the obtained result .
  end for
  Return the best of the solutions obtained on Step 2.

A simpler algorithm below combines the full "parent" solutions.

Algorithm 5. Greedy Procedure #2

  Combine sets $S \leftarrow S' \cup S''$, and run Algorithm 3 with the initial solution S.

These algorithms can be used in various global search strategies as their parts. Sets of solutions derived ("children") from the solution S' formed by combining its items with the items of some solution S'' and running Algorithm 1 are used as the neighborhoods in which a solution is searched. Thus, the second solution S'' is a parameter of the neighborhood selected randomly (randomized) [32].

4 GAs with Greedy Agglomerative Heuristic Procedures for the p-Median and k-Means Problems

The basic genetic algorithm for the k-means problem [7,26] can be described as follows:

Algorithm 6. GA with real-number alphabet for the k-means problem [18,25]

Require: Initial population size N_{POP}.

Step 1: Select N_{POP} initial solutions $S_1, ..., S_{N_{POP}}$ where $|S_i| = k$, and $\{S_1, ..., S_{N_{POP}}\}$ is a randomly chosen subset of the data vectors set. Improve each initial solution with Algorithm 1 and save corresponding obtained values of the objective function (1) as variables $f_k \leftarrow F(S_k)$, $k = \overline{1, N_{POP}}$.

loop

　Step 2: **If** the STOP condition is satisfied **then** STOP; return solution S_{i^*}, $i^* \in \{\overline{1, N_{POP}}\}$ with minimal value of f_{i^*}.

　Step 3: Randomly choose the two indexes $k_1, k_2 \in \{\overline{1, N_{POP}}\}$, $k_1 \neq k_2$.

　Step 4: run the crossover operator : $S_C \leftarrow Crossover(S_{k_1}, S_{k_2})$.

　Step 5: run the mutation operator : $S_C \leftarrow Mutation(S_C)$.

　Step 6: Run a selection procedure to change the population set.

end loop

We used such a tournament selection on Step 6:

Algorithm 7. Tournament selection

Randomly choose two indexes $k_4, k_5 \in \{\overline{1, N_{POP}}\}$, $k_4 \neq k_5$; if $f_{k_4} > f_{k_5}$ **then** $S_{k_4} \leftarrow S_C$, $f_{k_4} \leftarrow F(S_C)$ **else** $S_{k_5} \leftarrow S_C$, $f_{k_5} \leftarrow F(S_C)$.

Other selection methods do not significantly improve the result [11,16,21]. GAs with greedy agglomerative crossover can be described as follows [16,21]:

Algorithm 8. GA with greedy heuristic for the p-median problem and k-means problem (modifications GA-FULL, GA-ONE, and GA-MIX)

Step 1. Assign $N_{iter} \leftarrow 0$; select a set of the initial solutions $\{S_1, ..., S_{N_{POP}}\} \subset \{A_i | i = \overline{1, N}\}$, $|S_i| = k$. Improve each initial solution with Algorithm 1 and save the obtained values of the objective function (1) as variables $f_k \leftarrow F(S_k)$, $k = \overline{1, N_{POP}}$. We used initial populations with $N_{POP} = 5$.

loop

　Step 2: **If** STOP condition is satisfied **then** STOP; return solution S_{i^*}, $i^* \in \{\overline{1, N_{POP}}\}$ with minimal value of f_{i^*} **else** adjust the population size : $N_{iter} \leftarrow N_{iter} + 1$; $N_{POP} \leftarrow \max\{N_{POP}, \lceil \sqrt{1 + N_{iter}} \rceil\}$; if N_{POP} has changed, **then** initialize the new individual $S_{N_{POP}}$ as described in Step 1.

　Step 3: Randomly choose two indexes $k_1, k_2 \in \{\overline{1, N_{POP}}\}$.

　Step 4: Run Algorithm 4 (for GA-ONE modification) or Algorithm 5 (for GA-FULL modification) with "parent" solutions S_{k_1} and S_{k_2}. For the GA-MIX modification, Algorithms 4 or 5 are chosen randomly with equal probabilities. Obtain new solution S_C.

　Step 5: $S_C \leftarrow Mutation(S_C)$. By default, no mutation procedure is used.

　Step 6: Run Algorithm 7.

end loop

This algorithm uses a dynamically growing population [16,17]. In our new version of Step 5, the Crossover-like mutation operator is as follows.

Algorithm 9. Crossover-like mutation operator (Step 5 of Algorithm 8, its modifications GA-FULL-MUT, GA-ONE-MUT, and GA-MIX-MUT)

Run the ALA algorithm (Algorithm 1) for a randomly chosen initial solution to get solution S'.

Run Algorithm 4 (for GA-ONE modification) or Algorithm 5 (for GA-FULL modification) with "parent" solutions S_c and S''. Obtain new solution S'_C.

If $F(S'_C) < F(S_C)$, then $S_C \leftarrow S'_C$.

Our computational experiments (the next Section) show that new GAs with Algorithm 9 as the mutation operator are able to outperform both the original GAs with greedy agglomerative crossover operator (Algorithm 8) and the Variable Neighborhood Search with randomized neighborhoods (k-GH-VNS1 k-GH-VNS2, k-GH-VNS2) in some practically important problems.

5 Computational Experiments

We used data sets from the UCI (Machine Learning Repository) and the Clustering Basic Benchmark repositories [31,33] and the results of the non-destructive tests of prefabricated production batches of electronic radio components conducted in a specialized test center of JSC "TTC - NPO PM" used for the spacecraft equipment manufacturing [34]. The problem here is to divide a given mixed lot of radio components into clusters of similar devices manufactured from the same raw materials as a single homogeneous production batch. The test system consisted of Intel Core 2 Duo E8400CPU, 16 GB RAM. For all data sets, 30 attempts were made to run each of the algorithms. The j-means and k-means algorithms were launched in a multi-start mode [29,32].

Our new modifications (GA-xxx-MUT, Algorithm 8) of three GAs (Tables 2, 3) were compared to other known algorithms, such as corresponding known GAs without mutation (GA-FULL, GA-ONE, GA-MIX, see [25,30]). Variable Neighborhood Search with randomized neighborhoods (k-GH-VNS1, k-GH-VNS2, k-GH-VNS2, see [29,32]), their combinations with the j-means algorithm (j-means-GH-VNSx, see [29]), known GAs with the greedy agglomerative crossover procedures (GA-FULL, GA-ONE, GA-MIX modifications, see Algorithm 8), and other known algorithms (j-means and k-means, see [29,32]) in a multi-start mode.

The best-achieved values of the objective function (1) (its minimum value, mean value, and standard deviation) are underlined; the best values of new algorithms are given in a bold font, the best values of the known algorithms are given in italic. The results of the best of new algorithms (a sample of 30 results) were compared with the best of known tested algorithms (also 30 results) to prove the statistical significance of the advantage or disadvantage of new algorithms. We used the Mann-Whitney U-test and the t-test (significance level 0.01 for both tests).

Table 2. Computational experiment results for various data sets

Algorithm	Objective function (1) value			
	Min	Max	Average	Std.Dev
Europe data set (169309 data vectors of dimensionality 2) 30 clusters, 4 h				
j-means	7.51477E + 12	7.60536E + 12	7.56092E + 12	29.764E + 9
k-means	7.54811E + 12	7.57894E + 12	7.56331E + 12	13.560E + 9
k-GH-VNS1	*7.49180E + 12*	7.49201E + 12	*7.49185E + 12*	*0.073E + 9*
k-GH-VNS2	7.49488E + 12	7.52282E + 12	7.50082E + 12	9.989E + 9
k-GH-VNS3	7.49180E + 12	7.51326E + 12	7.49976E + 12	9.459E + 9
j-means-GH-VNS1	7.49180E + 12	7.49211E + 12	7.49185E + 12	0.112E + 9
j-means-GH-VNS2	7.49187E + 12	7.51455E + 12	7.4962E + 12	8.213E + 9
GA-FULL-MUT*	7.49293E + 12	7.49528E + 12	7.49417E + 12	0.934E + 9
GA-MIX-MUT*	7.49177E + 12	7.49211E + 12	7.49186E + 12	0.117E + 9
GA-ONE-MUT*↑⇑	**7.49177E + 12**	7.49188E + 12	**7.49182E + 12**	**0.042E + 9**
Testing results of the integrated circuits 5514BC1T2-9A5 (91 data vectors of dimensionality 173), grouping into 10 homogeneous batches (clusters), 2 min				
j-means	7 060.45	7 085.67	7 073.55	8.5951
k-means	7 046.33	7 070.83	7 060.11	8.8727
k-GH-VNS1	7 001.12	7 009.53	7 004.48	4.3453
k-GH-VNS2	7 001.12	7 010.59	7 002.26	2.9880
k-GH-VNS3	7 001.12	7 009.53	7 003.01	3.1694
j-means-GH-VNS1	*7 001.12*	7 001.12	*7 001.12*	*0.0000*
j-means-GH-VNS2	7 001.12	7 011.94	7 003.88	4.4990
GA-FULL-MUT*	7 001.12	7 001.27	7 001.24	0.0559
GA-MIX-MUT*↕⇕	**7 001.12**	7 001.12	**7 001.12**	**0.0000**
GA-ONE-MUT*↕⇕	7 001.12	7 001.12	7 001.12	0.0000
Ionosphere data set (351 data vectors of dimensionality 35, 10 clusters,1 min, Mahalanobis distance metric [35]				
k-means	9 253.2467	9 304.2923	9 275.3296	13.5569
k-GH-VNS1	9 083.6662	9 153.0192	9 121.0728	20.6875
k-GH-VNS2	9 085.8065	9 144.3779	9 112.2959	14.6803
k-GH-VNS3	9 090.5465	9 128.4111	9 109.8492	10.2740
GA-FULL	9 117.5695	9 175.1517	9 142.8457	15.3522
GA-FULL-MUT*	9 098.8748	9 157.0265	9 136.6556	16.1343
GA-MIX	9 095.1540	9 141.6417	9 114.1280	13.0638
GA-MIX-MUT*	9 102.8695	9 138.2243	9 113.4571	**8.6361**
GA-ONE	*9 078.6460*	9 115.9342	*9 099.7687*	*10.1104*
GA-ONE-MUT*↕⇕	**9 073.1919**	9 120.1842	*9 101.6286*	*12.8542*

Note: "*": new algorithm; "↑", "⇑": the advantage of the best of new algorithms over known algorithms is statistically significant ("↑" for t-test and "⇑" for Mann–Whitney U test), "↓", "⇓": the disadvantage of the best of new algorithms over known algorithms is statistically significant; "↕", "⇕": the advantage or disadvantage is statistically insignificant.

Table 3. Computational experiment results for various data sets

Algorithm	Objective function (1) value			
	Min	Max	Average	Std.Dev

Results of testing the integrated circuits 5514BC1T2-9A5 (91 data vectors of dimensionality 173), grouping into 10 homogeneous batches (clusters), 2 min, Mahalanobis distance [35]

Algorithm	Min	Max	Average	Std.Dev
k-means	7 289.7935	7 289.8545	7 289.8296	0.0153
k-GH-VNS1	7 289.7366	7 289.8024	7 289.7648	0.0147
k-GH-VNS2	7 289.7472	7 289.8021	7 289.7792	0.0153
GA-FULL	7 289.7474	7 289.8134	7 289.7742	0.0175
GA-FULL-MUT*	**7 289.7062**	7 289.7754	7 289.7508	0.0181
GA-MIX	7 289.7319	7 289.7771	7 289.7501	*0.0133*
GA-MIX-MUT*	7 289.7227	7 289.7772	7 289.7496	**0.0149**
GA-ONE	*7 289.7228*	7 289.7796	*7 289.7494*	0.0159
GA-ONE-MUT*⇕	7 289.7147	7 289.7752	**7 289.7466**	0.0165

Results of testing the integrated circuits 5514BC1T2-9A5 (1234 data vectors of dimensionality 157), grouping into 10 homogeneous batches (clusters), 2 min

Algorithm	Min	Max	Average	Std.Dev
j-means	43 841.97	43 843.51	43 842.59	0.4487
k-means	43 842.10	43 844.66	43 843.38	0.8346
k-GH-VNS1	43 841.97	43 844.18	43 842.34	0.9000
k-GH-VNS2	43 841.97	43 844.18	43 843.46	1.0817
k-GH-VNS3	43 841.97	43 842.10	43 841.99	0.0424
j-means-GH-VNS1	43 841.97	43 841.97	*43 841.97*	*0.0000*
j-means-GH-VNS2	43 841.97	43 844.18	43 842.19	0.6971
GA-FULL-MUT*	43 841.97	45 009.09	44 620.29	569.14
GA-MIX-MUT*	43 841.97	45 009.09	44 542.31	591.74
GA-ONE-MUT*↓⇓	**43 841.97**	45 009.09	**44 363.83**	**583.63**

6 Conclusion

We proposed a new approach to the design of Genetic Algorithms for the k-means problem with real-number (centroid-based) chromosome encoding, where the same procedure is used as both crossover and the mutation operators. Our experiments show that the GAs with one-point and greedy agglomerative crossover operators built in accordance with this idea outperform the algorithms without any mutation procedure and algorithms with the uniform random mutation by the obtained objective function value (SSE). In further research, our new approach can be applied to other problems such as p-median with various distance measures, k-medoids, mix probability distribution separation, etc. The efficiency of the GAs with greedy agglomerative crossover operators and Variable

Neighborhood Search algorithms with the randomized neighborhoods formed by greedy agglomerative procedures give us a reasonable hope for the successful application of our new idea for such problems.

References

1. Farahani, R., Hekmatfar, M.: Facility Location: Concepts, Models, Algorithms and Case Studies. Contributions to Management Science. Springer, Heidelberg (2009). https://doi.org/10.1007/978-3-7908-2151-2
2. Shirkhorshidi, A.S., Aghabozorgi, S., Wah, T.Y.: A comparison study on similarity and dissimilarity measures in clustering continuous data. PLoS ONE **10**(12) (2015). https://doi.org/10.1371/journal.pone.0144059
3. Lloyd, S.P.: Least squares quantization in PCM. IEEE Trans. Inf. Theory **28**(2), 129–137 (1982). https://doi.org/10.1109/TIT.1982.1056489
4. MacQueen, J. B.: Some methods for classification and analysis of multivariate observations. In: Proceedings of 5th Berkeley Symposium on Mathematical Statistics and Probability, vol. 1, pp. 281–297 (1967)
5. Rand, W.M.: Objective criteria for the evaluation of clustering methods. J. Am. Stat. Assoc. **66**(336), 846–850 (1971). https://doi.org/10.1080/01621459.1971.10482356
6. Hruschka, E., Campello, R., Freitas, A., de Carvalho, A.: A survey of evolutionary algorithms for clustering. IEEE Trans. Syst. Man Cybern. Part C Appl. Rev. **39**, 133–155 (2009). https://doi.org/10.1109/TSMCC.2008.2007252
7. Freitas, A.A.: A review of evolutionary algorithms for data mining. In: Maimon, O., Rokach, L. (eds.) Data Mining and Knowledge Discovery Handbook, pp. 371–400. Springer, Boston (2009). https://doi.org/10.1007/978-0-387-09823-419
8. Zeebaree, D.Q., Haron, H., Abdulazeez, A.M. and Zeebaree, S.R.M.: Combination of K-means clustering with genetic algorithm: a review. Int. J. Appl. Eng. Res. **12**(24), 14238–14245 (2017). https://www.ripublication.com/ijaer17/ijaerv12n24_35.pdf
9. Hosage, C.M., Goodchild, M.F.: Discrete space location-allocation solutions from genetic algorithms. Ann. Oper. Res. J. **6**, 35–46 (1986). https://doi.org/10.1007/bf02027381
10. Bozkaya, B., Zhang, J., Erkut, E.: A genetic algorithm for the p-median problem. In: Drezner, Z., Hamacher, H. (eds.) Facility Location: Applications and Theory. Springer, Heidelberg (2002)
11. Alp, O., Erkut, E., Drezner, Z.: An efficient genetic algorithm for the p-median problem. Ann. Oper. Res. **122**, 21–42 (2003). https://doi.org/10.1023/A:1026130003508
12. Eremeev, A.V.: Genetic algorithm with tournament selection as a local search method. Discrete Anal. Oper. Res. **19**(2), 41–53 (2012)
13. Krishna, K., Murty, M.M.: Genetic K-Means algorithm. IEEE Trans. Syst. Man Cybern. Part B (Cybernetics) **29**(3), 433–439 (1999). https://doi.org/10.1109/3477.764879
14. Maulik, U., Bandyopadhyay, S.: Genetic algorithm-based clustering technique. Pattern Recogn. J. **33**(9), 1455–1465 (2000). https://doi.org/10.1016/S0031-3203(99)00137-5
15. Neema, M.N., Maniruzzaman, K.M., Ohgai, A.: New genetic algorithms based approaches to continuous p-median problem. Netw. Spat. Econ. **11**, 83–99 (2011). https://doi.org/10.1007/s11067-008-9084-5

16. Kazakovtsev, L.A., Antamoshkin, A.N.: Genetic algorithm with fast greedy heuristic for clustering and location problems. Informatica **38**(3), 229–240 (2014). http://www.informatica.si/index.php/informatica/article/view/704/574

17. Kwedlo, W., Iwanowicz, P.: Using genetic algorithm for selection of initial cluster centers for the k-means method. In: ICAISC 2010: Artificial Intelligence and Soft Computing, pp. 165–172 (2010). https://doi.org/10.1007/978-3-642-13232-2_20

18. Kim, K., Ahn, H.: A recommender system using GA K-means clustering in an online shopping market. Expert Syst. Appl. **34**(2), 1200–1209 (2008)

19. He, Z., Yu, C.: Clustering stability-based evolutionary k-means. Soft Comput. **23**(1), 305–321 (2018). https://doi.org/10.1007/s00500-018-3280-0

20. Naldi, M.C., Campello, R.J.G.B., Hruschka, E.R., Carvalho, A.C.P.L.F.: Efficiency issues of evolutionary k-means. Appl. Soft Comput. **11**(2), 1938–1952 (2011). https://doi.org/10.1016/j.asoc.2010.06.010

21. Graña, M., López-Guede, J.M., Etxaniz, O., Herrero, Á., Quintián, H., Corchado, E. (eds.): SOCO/CISIS/ICEUTE -2016. AISC, vol. 527. Springer, Cham (2017). https://doi.org/10.1007/978-3-319-47364-2

22. Rousseeuw, P.J.: Silhouettes: a graphical aid to the interpretation and validation of cluster analysis. J. Comput. Appl. Math. **20**, 53–65 (1987). https://doi.org/10.1016/0377-0427(87)90125-7

23. Davies, D.L., Bouldin, D.W.: A cluster separation measure. IEEE Trans. Pattern Anal. Mach. Intell. **1**(2), 224–227 (1979). https://doi.org/10.1109/TPAMI.1979.4766909

24. Pelleg, D., Moore A.: X-means: extending K-means with efficient estimation of the number of clusters. In: Proceedings of the 17th International conference on Machine Learning, pp. 727–734 (2000)

25. Kazakovtsev, L.A., Antamoshkin, A.N.: Greedy heuristic method for location problems. Vestnik SibGAU **16**(2), 317–325 (2015). https://cyberleninka.ru/article/n/greedy-heuristic-method-for-location-problems

26. Kwedlo, W.: A clustering method combining differential evolution with the k-means algorithm. Pattern Recogn. Lett. **32**(12), 1613–1621 (2011). https://doi.org/10.1016/j.patrec.2011.05.010

27. Chang, D.-X., Zhang, X.-D., Zheng, C.-W.: A genetic algorithm with gene rearrangement for k-means clustering. Pattern Recogn. **42**(7), 1210–1222 (2009). https://doi.org/10.1016/j.patcog.2008.11.006

28. Brimberg, J., Drezner, Z., Mladenovic, N., Salhi, S.: A new local search for continuous location problems. Eur. J. Oper. Res. **232**(2), 256–265 (2014)

29. V I Orlov, V.I., Kazakovtsev, L.A., Rozhnov, I.P., Popov, N.A., Fedosov, V.V.: Variable neighborhood search algorithm for k-means clustering. IOP Conf. Ser. Mater. Sci. Eng. 450, 022035 (2018). https://doi.org/10.1088/1757-899X/450/2/022035

30. Kazakovtsev, L., Stashkov, D., Gudyma, M., Kazakovtsev, V.: Algorithms with greedy heuristic procedures for mixture probability distribution separation. Yugoslav J. Oper. Res. **29**(1), 51–67 (2019). https://doi.org/10.2298/YJOR1711

31. Clustering basic benchmark. http://cs.joensuu.fi/sipu/datasets

32. Orlov, V.I., Rozhnov, I.P., Kazakovtsev, L.A., Lapunova, E.V.: An approach to the development of clustering algorithms with a combined use of the variable neighborhood search and greedy heuristic method. IOP Conf. Ser. **1399** (2019). https://doi.org/10.1088/1742-6596/1399/3/033049

33. UCI Machine Learning Repository. http://archive.ics.uci.edu/ml

34. Rozhnov, I., Orlov, V. Kazakovtsev, L.: Ensembles of clustering algorithms for problem of detection of homogeneous production batches of semiconductor devices. In: CEUR Workshop Proceedings OPTA-SCL 2018. Proceedings of the School-Seminar on Optimization Problems and their Applications. CEUR-WS 2098, pp. 338–348 (2018). http://ceur-ws.org/Vol-2098/paper29.pdf
35. McLachlan, G.J.: Mahalanobis distance. Resonance 4(20), 1–26 (1999). https://doi.org/10.1007/BF02834632

Using Merging Variables-Based Local Search to Solve Special Variants of MaxSAT Problem

Ilya V. Otpuschennikov[ID] and Alexander A. Semenov[(✉)][ID]

ISDCT SB RAS, Irkutsk, Russia
otilya@yandex.ru, biclop.rambler@yandex.ru

Abstract. In this paper we study the inversion of discrete functions associated with some hard combinatorial problems. Inversion of such a function is considered in the form of a special variant of the well-known MaxSAT problem. To solve the latter we apply the previously developed local search method based on the Merging Variables Principle (MVP). The main novelty is that we combine MVP with evolutionary strategies to leave local extrema generated by Merging Variables Hill Climbing algorithm. The results of computational experiments show the effectiveness of the proposed technique in application to inversion of several cryptographic hash functions and to one problem of combinatorial optimization, which is a variant of the Facility Location Problem.

Keywords: Pseudo-boolean optimization · SAT · MaxSAT · Local search · Evolutionary algorithms · Cryptographic hash functions · Branch Location Problem

1 Introduction

In the present paper we study one particular class of the well-known MaxSAT problem [7]. The main feature of this class is that the combinatorial dimension of the considered problem can be significantly less than the number of variables that occur in a Boolean formula corresponding to this problem. Using this fact we describe a special strategy for solving MaxSAT from the mentioned class, which is built on the so-called Merging Variables Principle (MVP) proposed in [40]. The main goal of the present paper is to demonstrate the applicability of MVP to MaxSAT when the ratio between the combinatorial dimension of the considered problem and the number of variables in the corresponding Boolean formula is small.

In the next section we provide general information necessary for understanding the main results of the paper. Section 3 contains the main theoretical contribution of this paper. Namely, we introduce here a special class of MaxSAT problem, which is associated with the problems of finding preimages of quickly computable discrete functions. Essentially, we provide the theoretical basis that

Y. Kochetov et al. (Eds.): MOTOR 2020, CCIS 1275, pp. 363–378, 2020.
https://doi.org/10.1007/978-3-030-58657-7_29

allows us to apply MVP to the specified class of MaxSAT. Section 4 contains a short description of Merging Variables Principle as well as a description of several strategies used to exit from strong local extrema points. In Sect. 5 we present the main experimental contribution of the paper. We consider two families of MaxSAT benchmarks which can be attributed to the class introduced in Sect. 3. We demonstrate that our algorithms (based on MVP) outperform some of the best solvers that are usually applied to such problems. In Sect. 6 we briefly summarize the main results and talk about our future work.

2 Preliminaries

The variables that take values from the set $\{0, 1\}$ are called Boolean. A Boolean formula (or a propositional formula) is an expression constructed according to special pre-defined rules over the alphabet that includes Boolean variables and special symbols called Boolean connectives (according to [29]). For an arbitrary $k \in \mathbb{N}$ let us denote by $\{0, 1\}^k$ the set of all binary words (Boolean vectors) of length k. We refer to $\{0, 1\}^k$ as to a Boolean hypercube or a Boolean cube.

Let $X = \{x_1, \ldots, x_k\}$ be a set of Boolean variables. Then we can consider an arbitrary Boolean vector $\alpha \in \{0, 1\}^k$ as an assignment of variables from X (assuming that there is a fixed mapping between the coordinates of α and the variables from X). Now let F be a Boolean formula over the set of Boolean variables $X = \{x_1, \ldots, x_k\}$ and $\alpha \in \{0, 1\}^k$ be an arbitrary assignment of variables from X. We can define in a standard manner (according to [10]) a substitution of α into F. Boolean Satisfiability Problem (SAT) is formulated as the following question: if there exists such an $\alpha \in \{0, 1\}^k$, the substitution of which to F results in true. Below we will refer to it as $F(\alpha) = 1$. If such an α exists, then F is called satisfiable, otherwise – unsatisfiable. It is commonly known that SAT for any Boolean formula F can be reduced in polynomial time to SAT for formula in Conjunctive Normal Form (CNF) using Tseitin transformations [41].

SAT is a classical NP-complete problem [11]. It is often the case in practice that one not only needs to determine the satisfiability of CNF C, but also to find its satisfying assignment if C is satisfiable. In this formulation SAT is an NP-hard problem [17]. Despite being so hard in the worst-case scenario, SAT can be solved quite effectively for a wide class of application.

MaxSAT is the optimization variant of SAT. Traditionally, this problem is formulated as follows. Given a CNF C over a set of Boolean variables $X = \{x_1, \ldots, x_k\}$ to find as assignment $\alpha \in \{0, 1\}^k$ of variables from X such that the number of clauses satisfied by α in C is maximal. It is clear that α is a satisfying assignment of C if and only if it satisfies all clauses in C. Thus, MaxSAT is NP-hard. During the last 10 years there have been published a huge number of papers that study different aspects of MaxSAT. One of the main strategies for solving MaxSAT consists in iteratively invoking a SAT solver as an oracle [2,12,16,20–22,32,35], etc. Usually, a SAT solver based on a CDCL algorithm [28] is used for this purpose.

In the present paper we describe the algorithms for solving MaxSAT that are based on a completely different foundation. In particular, we consider MaxSAT as a maximization problem for a pseudo-Boolean functions of a special kind and solve it using a variant of local search methodology. Note that the basic idea of the corresponding approach is far from being novel: in this context we would like to cite an extensive review [19] and the related citations. The novelty of the approach proposed in the present paper is formed by the following components:

1. We work with a class of CNFs for which it is possible to significantly reduce the search space over which the local search operates when solving MaxSAT.
2. We combine the well-known Hill Climbing algorithm, as a basic local search scheme, with Merging Variables technique described in [40].

Hereinafter, consider the maximization problem for a pseudo-Boolean function, i.e. a function of the following kind [8]

$$f : \{0,1\}^n \to \mathbb{R}. \tag{1}$$

There is a large number of problems, in which little to nothing is known about the analytical properties of a function f. Below we assume that the only thing we know for sure about f is that for an arbitrary $\alpha \in \{0,1\}^n$ we can effectively compute the value $f(\alpha)$. In other words, we suppose that for each $\alpha \in \{0,1\}^n$ the value $f(\alpha)$ is produced by an oracle O_f. In these conditions, the search for a maximum of function (1) transforms into a problem of traversal of $\{0,1\}^n$ in accordance with some rational scenario. Usually, one of the most used scenario of such kind is the local search. To use local search it is necessary to specify a neighborhood function [9]

$$\aleph : \{0,1\}^n \to 2^{\{0,1\}^n}, \tag{2}$$

which defines some neighborhood structure on $\{0,1\}^n$. For an arbitrary $\alpha \in \{0,1\}^n$ the value of (2) for this α is denoted by $\aleph(\alpha)$ and is called the neighborhood of α.

The simplest example of the strategy for maximization of (1) that is based on the local search scenario can be given by the algorithm, which is often referred to as Hill Climbing (HC) [37]. Below we present the pseudocode of HC.

Input: an arbitrary point $\alpha \in \{0,1\}^n$, a value $f(\alpha)$;
1. α is a current point;
2. traverse the points from $\aleph(\alpha)\backslash\{\alpha\}$, computing for each point α' from this set a value $f(\alpha')$ (using the oracle O_f). If there is such a point α' that $f(\alpha') > f(\alpha)$, then go to step 3, otherwise go to step 4;
3. $\alpha \leftarrow \alpha'$, $f(\alpha) \leftarrow f(\alpha')$, go to step 1;
4. $\alpha^* \leftarrow \alpha'$; $(\alpha^*, f(\alpha^*))$ is a local maximum of (1) on $\{0,1\}^n$;
Output: $(\alpha^*, f(\alpha^*))$.

The neighborhoods in $\{0,1\}^n$ (or in any other search space) can be defined using different methods. One of the simplest is to use the Hamming neighborhoods in $\{0,1\}^n$: for an arbitrary $\alpha \in \{0,1\}^n$ its Hamming neighborhood of radius R is the following set $\aleph_R(\alpha) = \{\alpha' | d_H(\alpha, \alpha') \leq R\}$, where by $d_H(\alpha, \alpha')$ we mean the Hamming distance between vectors α and α' [27].

Now let us consider an arbitrary CNF C over a set of Boolean variables $X = \{x_1, \ldots, x_k\}$. Recall, that C is a formula of the following kind: $C = D_1 \wedge \ldots \wedge D_m$, where D_j, $j \in \{1, \ldots, m\}$ are the Boolean formulas called clauses. For example, the CNF $(x_1 \vee \neg x_3) \wedge (\neg x_1 \vee x_2 \vee \neg x_3) \wedge (\neg x_1 \vee \neg x_2 \vee x_3)$ is defined over the set $X = \{x_1, x_2, x_3\}$ and contains 3 clauses. With an arbitrary CNF C over $X = \{x_1, \ldots, x_k\}$, which consists of m clauses let us associate the following function of the kind (1):

$$f_C : \{0,1\}^k \to \{0, 1, \ldots, m\} \tag{3}$$

For an arbitrary $\alpha \in \{0,1\}^k$ the value $f_C(\alpha)$ is the number of clauses satisfied by the assignment α. Thus, an arbitrary C is satisfiable if and only if $max_{\{0,1\}^k} f_C = m$. The MaxSAT problem consists in the following: for an arbitrary CNF C over the set of Boolean variables X, $|X| = k$ to maximize the function of the kind (3) over a Boolean hypercube $\{0,1\}^k$.

3   On One Special Case of MaxSAT Problem

In this section we introduce such a subclass of CNFs for which it is possible to effectively move from the maximization problem for (3) over $\{0,1\}^k$ to the maximization problem for a special function

$$g_C : \{0,1\}^q \to \{0, 1, \ldots, m\} \tag{4}$$

over $\{0,1\}^q$, where $q << k$. To better describe the proposed approach, we will require some additional comments.

As we noted above, both SAT and MaxSAT are NP-hard problems. It means that a wide spectrum of combinatorial problems can be reduced to them. When constructing such a reduction it is often necessary to introduce a large number of new variables that play an auxiliary role and are not directly connected to an original combinatorial problem. Below we describe in more detail one class of CNFs related to mentioned situation.

Consider an arbitrary algorithm that transforms binary words of length q into binary words of length u. Such an algorithm specifies (or computes) a discrete function of the kind

$$h : \{0,1\}^q \to \{0,1\}^u. \tag{5}$$

Given an arbitrary $\gamma \in Range\ h \subseteq \{0,1\}^u$ the problem of inversion of h is to find such $\beta \in \{0,1\}^q$, that $h(\beta) = \gamma$. Many cryptanalysis problems can be viewed as inversion problems of functions that are specified by corresponding cryptographic algorithms.

The corollary of the well known Cook-Levin theorem [11,18] is the fact that for an algorithm A_h defining h of the kind (5) it is possible to effectively construct CNF C_h, that contains all the information about how A_h works on arbitrary inputs from $\{0,1\}^q$. The technique used to construct CNFs of the kind C_h

contains two stages: on the first stage a Boolean circuit S_h over an arbitrary complete basis is constructed (usually a $\{\wedge, \neg\}$-basis is used). Then for circuit S_h a CNF C_h is constructed using the so-called Tseitin transformations [41]. In the process of these transformations, the inputs, outputs and inner nodes of circuit S_h are associated with Boolean variables. In particular, with the input of S_h the Boolean variables forming the set $X^{in} = \{x_1, \ldots, x_q\}$ are associated. The remaining variables are auxiliary. The set of auxiliary variables contains the set $Y = \{y_1, \ldots, y_u\}$ formed by the variables corresponding to the outputs of circuit S_h. For an arbitrary $\gamma \in \{0,1\}^u$ let us denote by $C_h(\gamma)$ the CNF, formed from C_h as a result of substituting γ into C_h (as an assignment of variables from Y).

Below we will use the notion of Strong Backdoor Set first introduced in [43] in the context of Constraint Satisfaction Problem (CSP). Let us provide the definition of this notion in relation to SAT which is a special case of CSP.

Let C be an arbitrary CNF over the set of Boolean variables X and let B be an arbitrary subset of X. Suppose that β is an arbitrary assignment of variables from B and let us denote this fact as $\beta \in \{0,1\}^{|B|}$. Denote by $C[\beta/B]$ the CNF obtained from C as a result of substituting a Boolean vector β into C (as an assignment of variables from B) and performing all possible elementary transformations.

Definition 1 (see [43]). *Let C be an arbitrary CNF formula over a set of variables X, and let A be a polynomial-time algorithm. A non-empty set B, $B \subseteq X$, is a Strong Backdoor Set for C w.r.t. algorithm A if for each $\beta \in \{0,1\}^{|B|}$ algorithm A results an answer for SAT of the CNF $C[\beta/B]$.*

Note that the trivial SAT solving algorithm for an arbitrary CNF C over X consists in traversing all possible vectors from $\{0,1\}^{|X|}$, substituting them into C and computing the value of a Boolean function defined by C. Let us call such an algorithm a brute force algorithm. It is easy to see that its complexity is $2^{|X|} \cdot p(|C|)$, where $p(\cdot)$ is a polynomial and $|C|$ is the length of CNF C description. Due to the Definition 1, if B is a Strong Backdoor Set B: $|B| << |X|$, then there exists an algorithm for solving SAT for C with the complexity of $2^{|B|} \cdot q(|C|)$, where $q(\cdot)$ is some polynomial. In this algorithm for each assignment $\beta \in \{0,1\}^{|B|}$ of variables from B the polynomial algorithm A is invoked in application to CNF $C[\beta/B]$. The complexity of the resulting algorithm may be significantly lower than the complexity of the brute force algorithm.

The important fact related to arbitrary CNFs of the kind $C_h(\gamma)$ consists in that the set X^{in} is the Strong Backdoor Set w.r.t a simple algorithm called the Unit Propagation rule (UP) [14, 28].

Recall that an arbitrary literal over X is a formula of the kind $l_\lambda(x)$, $x \in X$, $\lambda \in \{0, 1\}$, where:

$$l_\lambda(x) = \begin{cases} \neg x, \ \lambda = 0 \\ x, \ \lambda = 1 \end{cases}$$

The Unit Propagation rule in application to an arbitrary CNF C over X and a set of literals $L = \{l_{\lambda_1}(x_1'), \ldots, l_{\lambda_s}(x_s')\}$, $\{x_1', \ldots, x_s'\} \subseteq X$ works as follows.

Consider a CNF

$$l_{\lambda_1}(x'_1) \wedge \ldots \wedge l_{\lambda_s}(x'_s) \wedge C,$$

for an arbitrary $j \in \{1, \ldots, s\}$ first all clauses in C that contain $l_{\lambda_j}(x'_j)$ are removed, and from each clause containing the literal $\neg l_{\lambda_j}(x'_j)$ this literal is removed. Note that as a result of UP application it is possible to derive new unit clauses, i.e. the clauses that contain a single literal. Then one can apply the Unit Propagation rule to these newly derived literals, etc. In [14] there was proposed the algorithm implementing UP that has a linear complexity in the size of description of C. Below we employ the following important fact.

Lemma 1. *Let C_h be a CNF over a set of Boolean variables X, constructed according to the method outlined above based on the algorithm specifying function h of the kind (5). Let $\beta = (\beta_1, \ldots, \beta_q)$ be an arbitrary assignment of variables from $X^{in} = \{x_1, \ldots, x_q\}$. Then the result of application of only the Unit Propagation rule to CNF*

$$l_{\beta_1}(x_1) \wedge \ldots \wedge l_{\beta_q}(x_q) \wedge C_h \qquad (6)$$

is the derivation of the values of all variables from $X \setminus X^{in}$ including an assignment of variables from $Y = \{y_1, \ldots, y_u\}$ such that $y_1 = \gamma_1, \ldots, y_u = \gamma_u$ and $h(\beta) = \gamma$, $\gamma = (\gamma_1, \ldots, \gamma_u)$.

Definition 2. *Assume that all conditions of Lemma 1 are satisfied and α is an assignment of variables from X, which was derived as the result of application of UP to CNF (6). Then we will say that α is induced by β.*

The statements that are close to the Lemma 1 in spirit have been independently proven in several papers [5,38]. The direct corrolary of Lemma 1 is the fact that for an arbitrary CNF of the kind $C_h(\gamma)$ constructed for an algorithm defining a function $h : \{0,1\}^q \to \{0,1\}^u$, the set X^{in} is a Strong Backdoor Set w.r.t. Unit Propagation rule. Any set of this kind is referred to as Strong Unit Propagation Backdoor Set (SUPBS).

Once again, let C_h be a CNF over a set of Boolean variables X constructed for the algorithm defining a function of the kind (5). Fix an arbitrary $\gamma \in \{0,1\}^u$ and consider CNF $C = C_h(\gamma)$. Assume that the number of clauses in C is m. Let us introduce the following function

$$g_C : \{0,1\}^q \to \{0, 1, \ldots, m\} \qquad (7)$$

defined as follows. For an arbitrary $\beta \in \{0,1\}^q$, $\beta = (\beta_1, \ldots, \beta_q)$, consider it as an assignment for variables from $X^{in} = \{x_1, \ldots, x_q\}$ and construct a CNF (6). Then apply to this CNF the Unit Propagation rule. Let α be an assignment of all variables from X, that is induced by β. Then we define $g_C(\beta)$ as the number of clauses in C that are satisfied by vector α. The following fact holds.

Theorem 1. *Consider an arbitrary function h of the kind (5) and CNF C_h constructed using the algorithm specifying h. Let us fix an arbitrary $\gamma \in \text{Range } h \subseteq$*

$\{0,1\}^u$ and consider CNF $C = C_h(\gamma)$. Assume that $X : |X| = k$ is the set of variables occurring in C. Consider two problems: in the first one we need to maximize the function f_C of the kind (3) over $\{0,1\}^k$. In the second the goal is to maximize function g_C of the kind (7) over $\{0,1\}^q$. Let $\beta^* = argmax_{\{0,1\}^q} g_C$ and α^* be an assignment of variables from X induced by β^*. Then $\alpha^* = argmax_{\{0,1\}^k} f_C$.

Proof (Sketch proof). The key aspect of the proof consists in the fact that γ is chosen from the range of the function h. In this case it follows from the Cook-Levin theorem and the properties of Tseitin transformations that CNF $C = C_h(\gamma)$ is satisfiable. The detailed description of this fact can be found in [39]. Thus, in the considered case there exists an assignment that satisfies C and, therefore, $max_{\{0,1\}^k} f_C = m$. Again, using the technique described in [39] it is possible to show that each assignment α that satisfies C corresponds to exactly one Boolean vector $\beta \in \{0,1\}^q$ that induces α in the sense of Definition 2. This fact means that the Theorem 1 holds.

As a conclusion of this section, we would like to note that in the context of the proposed approach we can view the problem of inversion of an arbitrary function (5) as a special variant of the MaxSAT problem. Essentially, we maximize the number of satisfied clauses in a CNF $C_h(\gamma)$, but thanks to the above it is possible to work only with the vectors from the Boolean hypercube $\{0,1\}^q$, considering them as assignments of variables from SUPBS X^{in}. Note that in the cases that are interesting from the practical point of view, X^{in} may contain tens or hundreds of variables while set X over which CNF $C_h(\gamma)$ is defined may consist of tens of thousand variables.

4   Merging Variables Principle

Merging Variables Principle (MVP) was proposed in the paper [40], though closely related approaches appeared in much earlier works [1,3]. MVP is aimed at pseudo-Boolean optimization. It is based on a very simple idea: as it was shown in [40], one can effectively construct a one-to-one correspondence between $\{0,1\}^n$ and a special metric space, which is denoted as D^μ. Let us connect with $\{0,1\}^n$ the set of Boolean variables $X = \{x_1, \ldots, x_n\}$, and let the set of variables $Y = \{y_1, \ldots, y_r\}$, $1 \le r < n$ be connected with D^μ. Additionally, we assume that some surjection mapping $\mu : X \to Y$ (merging mapping) is specified. Then

$$D^\mu = D_1 \times \ldots \times D_r$$

where D_j, $j \in \{1, \ldots, r\}$ are domains for variables from Y. Let us establish a one-to-one correspondence

$$\tau_\mu : D^\mu \to \{0,1\}^n$$

between D^μ and $\{0,1\}^n$. Finally, let us move from maximization of pseudo-Boolean function $f : \{0,1\}^n \to \mathbb{R}$ to maximization of μ-conjugated function $F_{f,\mu} : D^\mu \to \mathbb{R}$ (see [40]). We can define the Hamming metric over space D^μ and specify a neighborhood structure formed by neighborhoods of radius 1. Actually for an arbitrary μ the corresponding neighborhood structure in D^μ generates special neighborhood structure in $\{0,1\}^n$ and the latter is formed by the neighborhoods which are, generally speaking, not Hamming neighborhoods. It is easy to construct examples when f reaches a local maximum in $\alpha \in \{0,1\}^n$, but $\tau_\mu^{-1}(\alpha)$ is not local maximum of $F_{f,\mu}$ in D^μ. However, the following holds: $max_{\{0,1\}^n} f = max_{D^\mu} F_{f,\mu}$. In [40] we described the Hill Climbing algorithm augmented with Merging Variables strategy, to which we further refer as Merging Variables Hill Climbing (MVHC). Also, in [40] it was noted that we can view MVP as a special variant of the well known Variable Neighborhood Search strategy [31]. Using MVP we can effectively generate many different neighborhood structures in $\{0,1\}^n$.

An arbitrary merging mapping μ can be constructed as an example of classic distribution of particles to boxes. In all computational experiments we use the so-called uniform merging mapping ([40], definition 7). In this case for a fixed r we pursue the goal to obtain r boxes which are filled by particles almost uniformly. We achieve this by filling the boxes one-by-one: the first box on the first step, the second box on the second step, and so on. As a result we have the situation when the number of variables in each box is either $\lfloor n/r \rfloor$ or $\lfloor n/r \rfloor + 1$. Each box represents the set of preimages (w.r.t μ) of a corresponding variable from Y. Let us denote the maximum number of preimages for variables from Y as l. As it was noted in [40], MVHC with uniform merging mapping can be naturally implemented in form of multi-thread application: distinct threads process distinct neighborhood fragments which correspond to domains of specific merged variables. In this case all the threads will perform almost equal amount of work because all the domains have almost equal power.

Definition 3. *Assume that MVHC found such an $\alpha^* \in \{0,1\}^n$ that each $\beta_i = \tau_{\mu_i}^{-1}(\alpha^*), i \in \{1,\dots,t\}$ is a local maximum of F_{f,μ_i} in D^{μ_i}. Let us refer to such an α^* as strong local maximum of f w.r.t. mappings μ_i, $i \in \{1,\dots,t\}$.*

In the experiments presented in the next section we use several evolutionary strategies to escape strong local maxima to which MVHC falls in. Let us give a short description of these algorithms. The most simple example of such strategy is the $(1+1)$-Evolutionary Algorithm $((1+1)$-EA$)$ [33]. This is a simple algorithm based on the concept of a random mutation. In the case of $(1+1)$-EA a single random mutation of an arbitrary $\alpha \in \{0,1\}^n$ consists in a series of n independent Bernoulli trials with probability $p = \frac{1}{n}$. If $i, i \in \{1,\dots,n\}$ is the number of successful trial, then the bit number i in α is flipped (changed from 0 to 1 or from 1 to 0). $(1+1)$-EA is extremely ineffective in theory [15], but often shows quite good performance in practice. As it was noted in [42], this can be attributed to the fact that on average $(1+1)$-EA acts in a fashion similar to Hill Climbing (because the expected value of the number of flipped bit after a single random mutation is 1). At the same time, unlike Hill Climbing, $(1+1)$-EA can move from α to any other point from $\{0,1\}^n$ with a non-zero probability.

There are several techniques that make it possible to reduce the worst-case estimation of $(1+1)$-EA effectiveness (in the context of the complexity measure introduced in [15]) by changing the mutation rate in its base schema. One of such algorithms is the $(1+1) - FEA_\beta$ described in [13]. This algorithm is based on the idea of employing the heavy-tailed mutation operator: such an operator flips an arbitrary bit in α with probability $\frac{\lambda}{n}$ instead of $\frac{1}{n}$ in case of the standard $(1 + 1)$-EA. Here, λ is a value of random variable that has the so-called Power-law distribution $D_{\frac{n}{2}}^\beta$ with the parameter β [13]. The worst case estimation of this algorithm is $O(n^\beta \cdot 2^n)$ for any pseudo-Boolean function instead of n^n in case of the standard $(1 + 1)$-EA. In our computational experiments we used the $(1 + 1) - FEA_\beta$ variant with $\beta = 3$ to exit local extrema since in this case (according to [13]) the expected value of the number of flipped bits in one random mutation is $b \rightarrow 1.3685\ldots$ when $n \rightarrow \infty$ (for $\beta \leq 2$ this value grows to infinity with the increase of n).

We also performed the generation of a new starting point based on several strong local maxima, using a special variant of a genetic algorithm described in [36]. In this algorithm a population is the set of different strong local maxima constructed by MVHC during its work. Denote a current population by $P_{current}$ and a new population by P_{next}. Hereinafter, $|P_{current}| = |P_{next}| = N$ for some fixed N. Assume that $P_{current} = \{I_1, \ldots, I_N\}$. Let us associate with $P_{current}$ the distribution $D_{current} = \{p_1, \ldots, p_N\}$, where $p_i = \frac{f(I_i)}{\sum_{i=1}^N f(I_i)}$, $i \in \{1, \ldots, N\}$, where f is our objective function. A new population P_{next} is formed as a result of several actions. First, choose from $P_{current}$ E individuals with maximal weight (the value of f) and add them to P_{next} (it corresponds to the so-called elitism concept [26]). At the second step, choose from $P_{current}$ H individuals in accordance with the distribution $D_{current}$ and apply to each of them the standard $(1 + 1)$-random mutation by flipping each bit with probability $\frac{1}{n}$. The resulting individuals are added to P_{next}. Finally, choose individuals from $P_{current}$ according to the fixed distribution and perform standard two-point crossover to them [26]. The G individuals formed that way are also added to P_{next}. It is necessary that $E + H + G = N$.

The decision to generate a new starting point is made when we have s strong local maxima w.r.t t uniform merging mappings $\mu_1, \ldots, \mu_t$. The parameters s and t are tuned in the course of testing. Assume that using one of the methods outlined above we generated a new point α'. Then, in accordance to the ideas from [40] the point α' is checked to see if it is located at the Hamming distance $\geq l + 1$ from the current set of strong local maximums. If this condition is held, then α' is taken as a new starting point, otherwise the process of generation α' is repeated.

5   Computational Experiments

In our computational experiments we considered two classes of tests. The first class is related to problems of finding preimages of several known cryptographic hash functions. In particular, we studied the MD-5, SHA-1 and SHA-256 hash functions. According to [30] a cryptographic hash function is a function of the kind $\chi : \{0,1\}^* \rightarrow \{0,1\}^c$, where $\{0,1\}^*$ is a Kleene star over the set $\{0,1\}$

and c is a constant. Preimage attacks for χ implies finding a preimage of some hash value from $\{0,1\}^c$. Usually, one searches for a preimage of a fixed length which as a rule of thumb coincides with the size of a single block of an input message. For the MD5, SHA-1 and SHA-256 hash functions this size is equal to 512 bits, and c is 128 for MD5, 160 for SHA-1 and 256 for SHA-256. For all considered functions there are no known algorithms that could find a preimage (e.g. a 512-bit input block) for a given hash image in realistic time. We can consider the problem of finding a 512-bit input the hash value of that contains a particular number of first zero bits. In this case the hardness of the resulting problems depends on the number of the zero bits. This very argumentation is used to justify the resistance of the majority of today's cryptocurrency protocols [34]. The problems of finding 512-bit inputs, the hash images of which contain K first zero bits, formed the first class of considered tests. Therefore, the name of the test is name of `function_K_launch_number` (for example: MD5_20_3).

In Table 1 column named msl-MV corresponds to regular MVHC with parameter l (the maximum number of preimages for variables from Y); msl-FEA corresponds to MVHC, complemented by strategies for exiting from strong local maxima, based on $(1+1)$-Fast Evolutionary Algorithm; msl-GA corresponds to the MVHC variant, which uses the Genetic Algorithm described above to exit strong local maxima.

We compared the efficiency of algorithms based on MVP with two multi-threaded SAT solvers PLINGELING [6] and PAINLESS [24] (`Pling` and `Painl` in Table 1 respectively). For a more detailed presentation of the results, we built cactus plots in the style which is commonly used in the SAT Race and SAT Competition competitions. To build cactus plots we used the MKPLOT tool[1].

It should be noted that some tests have not been solved in a time limit, which was equal to 5000 s as standard value (to each such a fact in Table 1 corresponds an empty cell). In the construction of cactuses, we take this into account by applying the PAR scoring system 2 used in the SAT Race and SAT Competition[2]: i.e. for an instance, which has not been resolved in the time limit, a decision time was chosen as a double time limit. Also the results of this series of experiments we demonstrate in Fig. 1. To plot the lines we take all runtimes of the corresponding algorithm on the considered instances that are below the time limit, order them in accordance with the increasing value (individually for each line) and plot the corresponding graph. It is commonly referred to as cactus plot.

As a final remark, we would like to note that the best results on the considered class of tests in accordance with the described form of presentation showed MVHC, complemented by strategies for exiting from strong local maxima, based on $(1+1)$-Fast Evolutionary Algorithm $(1+1)-FEA_\beta$ with $\beta = 3$. Each of the tested algorithms was launched in parallel mode on one node of the computing cluster of "Academician V.M. Matrosov"[3], where each node is equipped with

---

[1] https://github.com/alexeyignatiev/mkplot.

[2] http://sat-race-2019.ciirc.cvut.cz/.

[3] http://hpc.icc.ru.

Table 1. Results of solving inversion problems of cryptographic hash functions

Test	ms8-MV	ms8-GA	ms8-FEA	ms10-MV	ms10-GA	ms10-FEA	Pling	Painl
MD5_18_1	349	219	1380	803	524	164	-	905
MD5_18_2	215	57	384	269	49	87	2108	294
MD5_18_3	122	86	52	238	92	69	765	3036
MD5_20_1	40	1407	2093	1226	1521	175	3479	–
MD5_20_2	757	46	585	233	777	512	–	–
MD5_20_3	603	121	1782	450	670	1772	–	–
MD5_22_1	555	1143	611	940	–	–	–	–
MD5_22_2	1278	4998	–	651	4743	2662	–	–
MD5_22_3	2563	478	–	2079	3888	389	3425	–
SHA-1_18_1	117	200	267	95	497	585	190	1407
SHA-1_18_2	123	765	730	703	136	199	1061	3080
SHA-1_18_3	327	128	78	450	503	279	855	3077
SHA-1_20_1	273	2949	282	1930	538	779	537	288
SHA-1_20_2	69	1032	2488	511	2139	267	1007	1124
SHA-1_20_3	76	2004	4390	826	70	1595	2602	483
SHA-1_22_1	1204	–	1304	–	1654	925	2950	–
SHA-1_22_2	–	–	–	–	1150	4119	–	–
SHA-1_22_3	–	2973	2720	–	–	3629	–	1954
SHA-2_18_1	89	2396	330	26	26	26	44	3253
SHA-2_18_2	94	696	112	26	26	26	204	1994
SHA-2_18_3	220	201	233	26	26	26	763	2750
SHA-2_20_1	729	518	890	–	217	481	3637	–
SHA-2_20_2	125	1903	3101	105	720	4386	3211	–
SHA-2_20_3	751	124	1239	838	2913	1267	2147	414
SHA-2_22_1	–	497	3881	335	–	2039	–	–
SHA-2_22_2	–	–	–	4168	642	2400	–	–
SHA-2_22_3	4042	–	–	4487	–	2662	2290	–

two 18-core Intel Xeon E5-2695 CPUs. The results of experiments show that the best multithreaded solvers based on the CDCL algorithm (PLINGELING and PAINLESS), on the considered class of tests, significantly lost against all variants of Merging Variables Hill Climbing.

The second class of tests for which computational experiments were carried out was formed by MaxSAT variants of the Branch Location Problem (BLP), which can be considered as a special case of the Facility Location Problem [23]. The original formulation of the BLP uses weighted undirected complete bipartite graph $G(U, V, E, w)$, where U is a set of customers, V is a set of facilities, $E = U \times V$ is a set of edges. Function $w : E \to \mathbb{R}_{\geq 0}$ defines edge weights for this graph. The number of facilities, that should be deleted is denoted by $k, k \leq |V|$. Also the parameter Δ, $\Delta \geq 0$ is introduced. This parameter is a threshold by which

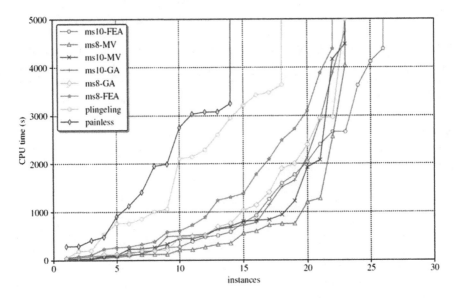

Fig. 1. Comparison of different algorithms efficiency on the inversion hash functions problems

the minimum edge weight of each customer can be increased when we delete some facilities vertices. The BLP P^Δ is formulated in the following manner [44].

Definition 4. *(Definition 1 from [44]). Given a tuple $P^\Delta = (G(U, V, E, w), k, \Delta)$ to find a subset $V', V' \subseteq V, |V'| = k$ such that if V' is deleted from V, the number of customers (from U), whose minimum edge weight is increased by more than Δ, is minimal.*

In [44] this problem was reduced to MaxSAT formulated by using so-called hard and soft clauses (see [25]). In such formulations, the set of clauses of the considered CNF is divided into two subsets: clauses, marked as 'hard' and 'soft'. Hard clauses must be satisfied and it is required to maximize the number of satisfied soft clauses under these conditions (or, which is the same thing, to minimize the number of unsatisfied soft clauses).

In our computational experiments we used the MaxSAT tests constructed in [44]. The peculiarity of this class of tests is that for each specific CNF formed by hard clauses a corresponding SUPBS $\tilde{X}$ is known. Thus, we can run Merging Variables Hill Climbing on this SUPBS. In addition, we took into account that the substitution of values of variables from $\tilde{X}$ can lead to a conflict on the set of hard clauses: in each such case, we determined the value of the objective function equal to the number of satisfied hard clauses.

For each assignment of variables from $\tilde{X}$ that satisfies all hard clauses, the value of the objective function was determined as the number of unsatisfied soft clauses and this value was minimized: this is a common form of representing results of testing MaxSAT solvers.

Algorithms for solving MaxSAT problems with hard and soft clause using the Merging Variables Principle are incomplete, unlike algorithms based on CDCL. These two classes of algorithms w.r.t. this variant of MaxSAT work according to completely different schemes, thus it is incorrect to compare them with each other directly. That is why we compared our algorithms for solving these cases of MaxSAT with incomplete SAT solver LOANDRA [4], which won the competition MaxSAT Evaluation 2019[4] among incomplete algorithms. In this series of experiments we used a personal computer of the following configuration: Intel Core i7-6700 CPU 3.40 GHz, 16 Gb RAM. It should be noted that several complete MaxSAT solvers which we used in our experiments weren't able to cope with the tests from the considered class for $k > 10$.

Table 2. Results of solving Branch Location Problem as MaxSAT

Test	ms8-GA	Loandra	Test	ms8-GA	Loandra	Test	ms8-GA	Loandra
k10_d100	8538	9826	k16_d300	14538	21830	k22_d500	22373	36099
k10_d200	6779	7197	k16_d400	12012	15653	k24_d100	45613	82174
k10_d300	5287	5287	k16_d500	10144	14325	k24_d200	39759	64742
k10_d400	4189	4189	k18_d100	25963	46157	k24_d300	34061	54062
k10_d500	3368	3368	k18_d200	21937	36753	k24_d400	30573	49746
k12_d100	11961	22247	k18_d300	18799	28302	k24_d500	27280	43962
k12_d200	9593	10474	k18_d400	15995	27131	k26_d100	53902	89965
k12_d300	7702	7845	k18_d500	13946	16530	k26_d200	47102	73874
k12_d400	6124	7301	k20_d100	31722	53121	k26_d300	40802	64648
k12_d500	5074	5074	k20_d200	27376	44737	k26_d400	36620	64986
k14_d100	15866	28993	k20_d300	23357	36773	k26_d500	32801	57051
k14_d200	12908	22168	k20_d400	20254	25355	k28_d100	62303	110191
k14_d300	10647	11344	k20_d500	18021	31686	k28_d200	55116	84828
k14_d400	8685	11286	k22_d100	38083	64766	k28_d300	48291	74623
k14_d500	7311	7556	k22_d200	33293	51940	k28_d400	43205	68754
k16_d100	20642	40328	k22_d300	28400	42631	k28_d500	38815	68217
k16_d200	17024	26589	k22_d400	25194	40265			

In Table 2 we present the results of experiments for the second class of tests (related to BLB). In all these problems $|V| = 58$. The test name kN_dM corresponds to the problem for $k = $ N and $\Delta = $ M. For these experiments (based on preliminary testing), we chosen the MVP variant with $l = 8$ and the procedure for exiting local maxima based on the Genetic Algorithm. Both MVHC and LOANDRA were launched with a time limit of 1000 s. Each Best Known Value, presented in Table 2, is the smallest value of the objective function, which is equal to the number of unsatisfied soft clauses, found in time limit (the smaller this value, the better the found point). From Table 2 it can be seen that on

---

[4] https://maxsat-evaluations.github.io/2019/.

the considered test class the used MVHC variant always finds a point at which
the value of the objective function is no worse than at the point found by the
LOANDRA solver. For most tests, the value of the objective function found by
MVHC is better (sometimes quite significant).

6   Conclusion and Future Work

In this paper we presented the results of application of the Merging Variables
Principle (MVP) metaheuristic strategy to one special class of MaxSAT problem
that consists in the maximization of number of satisfied clauses in an arbitrary
Conjunctive Normal Form (CNF). We showed that for a considered subclass of
MaxSAT the dimension of the search space is significantly less than the number
of variables in CNF. We used a maximization strategy based on MVP on such
a reduced space. To exit strong local maxima (this concept was introduced in
[40]) we applied two evolutionary strategies, namely, $(1 + 1) - FEA_\beta$ described
in [13] and one variant of Genetic Algorithm presented in [36]. As test sets we
considered the preimage finding problems of cryptographic hash functions (so-
called preimage attacks) and MaxSAT variants of the Branch Location Problem,
which is a special case of the Facility Location Problem. On the considered
classes of tests algorithms that are based on MVP showed higher efficiency in
comparison with the algorithms that won the SAT competitions, SAT Races and
MaxSAT Evaluation competitions in recent years. In the nearest future we plan
to investigate the possibility of applying MVP-based computational schemes to
the search problems of some combinatorial designs.

Acknowledgements. We express our deep gratitude to Dr. Oleg Zaikin for his valuable advice and help. The research was funded by Russian Science Foundation (project
No. 16-11-10046).

References

1. Ahuja, R.K., Ergun, O., Orlin, J.B., Punnen, A.P.: A survey of very large-scale
   neighborhood search techniques. Discrete Appl. Math. **123**(1–3), 75–102 (2002)
2. Ansotegui, C., Bonet, M.L., Levy, J.: Sat-based MaxSAT algorithms. Artif. Intell.
   196, 77–105 (2013)
3. Avella, P., D'Auria, B., Salerno, S., Vasil'ev, I.: A computational study of local
   search algorithms for Italian high-school timetabling. J. Heuristics **13**(6), 543–556
   (2007)
4. Berg, J., Demirović, E., Stuckey, P.J.: Core-boosted linear search for incomplete MaxSAT. In: Rousseau, L.-M., Stergiou, K. (eds.) CPAIOR 2019. LNCS,
   vol. 11494, pp. 39–56. Springer, Cham (2019). https://doi.org/10.1007/978-3-030-
   19212-9_3
5. Bessiere, C., Katsirelos, G., Narodytska, N., Walsh, T.: Circuit complexity and
   decompositions of global constraints. In: IJCAI 2009, Proceedings of the 21st International Joint Conference on Artificial Intelligence, Pasadena, California, USA,
   11–17 July 2009, pp. 412–418 (2009)

6. Biere, A.: CaDiCaL, lingeling, plingeling, treengeling, YalSAT entering the SAT competition 2017. In: Balyo, T., Heule, M.J.H., Järvisalo, M. (eds.) SAT Competition 2017, vol. B-2017-1, pp. 14–15 (2017)
7. Biere, A., Heule, M., van Maaren, H., Walsh, T. (eds.) Handbook of Satisfiability, vol. 185. IOS Press (2009)
8. Boros, E., Hammer, P.L.: Pseudo-boolean optimization. Discrete Appl. Math. **123**(1–3), 155–225 (2002)
9. Burke, E., Kendall, G.: Search Methodologies, 2nd edn. Springer, New York (2014). https://doi.org/10.1007/978-1-4614-6940-7
10. Chang, C.L., Lee, R.C.T.: Symbolic Logic and Mechanical Theorem Proving, 1st edn. Academic Press Inc, Orlando (1997)
11. Cook, S.A.: The complexity of theorem-proving procedures. In: Proceedings of the 3rd Annual ACM Symposium on Theory of Computing, 3–5 May 1971, Shaker Heights, Ohio, USA, pp. 151–158 (1971)
12. Davies, J., Bacchus, F.: Solving MAXSAT by solving a sequence of simpler SAT instances. In: Lee, J. (ed.) CP 2011. LNCS, vol. 6876, pp. 225–239. Springer, Heidelberg (2011). https://doi.org/10.1007/978-3-642-23786-7_19
13. Doerr, B., Le, H.P., Makhmara, R., Nguyen, T.D.: Fast genetic algorithms. In: Proceedings of the Genetic and Evolutionary Computation Conference. GECCO 2017, pp. 777–784. Association for Computing Machinery (2017)
14. Dowling, W.F., Gallier, J.H.: Linear-time algorithms for testing the satisfiability of propositional horn formulae. J. Log. Program. **1**(3), 267–284 (1984)
15. Droste, S., Jansen, T., Wegener, I.: On the analysis of the $(1 + 1)$ evolutionary algorithm. Theor. Comput. Sci. **276**(1–2), 51–81 (2002)
16. Bacchus, F.: CSPs: adding structure to SAT. In: Biere, A., Gomes, C.P. (eds.) SAT 2006. LNCS, vol. 4121, pp. 10–10. Springer, Heidelberg (2006). https://doi.org/10.1007/11814948_2
17. Garey, M.R., Johnson, D.S.: Computers and Intractability: A Guide to the Theory of NP-Completeness. W. H. Freeman & Co., New York (1979)
18. Goldreich, O.: Computational Complexity: A Conceptual Perspective, 1st edn. Cambridge University Press, New York (2008)
19. Gu, J., Purdom, P.W., Franco, J., Wah, B.W.: Algorithms for the satisfiability (sat) problem: a survey. In: DIMACS Series in Discrete Mathematics and Theoretical Computer Science, pp. 19–152. American Mathematical Society (1996)
20. Ignatiev, A., Morgado, A., Manquinho, V., Lynce, I., Marques-Silva, J.: Progression in maximum satisfiability. In: Proceedings of the Twenty-First European Conference on Artificial Intelligence. ECAI 2014, pp. 453–458. IOS Press (2014)
21. Ignatiev, A., Morgado, A., Marques-Silva, J.: RC2: an efficient MaxSAT solver. J. Satisf. Boolean Model. Comput. **11**(1), 53–64 (2019)
22. Ignatiev, A., Morgado, A., Marques-Silva, J.: PySAT: a python toolkit for prototyping with SAT oracles. In: Beyersdorff, O., Wintersteiger, C.M. (eds.) SAT 2018. LNCS, vol. 10929, pp. 428–437. Springer, Cham (2018). https://doi.org/10.1007/978-3-319-94144-8_26
23. Laporte, G., Nickel, S., Saldanha da Gama, F. (eds.): Location Science. Springer, Switzerland (2015). https://doi.org/10.1007/978-3-319-13111-5
24. Le Frioux, L., Baarir, S., Sopena, J., Kordon, F.: PaInleSS: a framework for parallel SAT solving. In: Gaspers, S., Walsh, T. (eds.) SAT 2017. LNCS, vol. 10491, pp. 233–250. Springer, Cham (2017). https://doi.org/10.1007/978-3-319-66263-3_15
25. Li, C.M., Manyà, F.: MaxSAT, hard and soft constraints. In: Biere, A., Heule, M., van Maaren, H., Walsh, T. (eds.) Handbook of Satisfiability, Frontiers in Artificial Intelligence and Applications, vol. 185, pp. 613–631. IOS Press (2009)

26. Luke, S.: Essentials of Metaheuristics, 2nd edn. George Mason University (2015)
27. MacWilliams, F., Sloane, N.: The Theory of Error-Correcting Codes. North Holland (1983)
28. Marques-Silva, J.P., Lynce, I., Malik, S.: Conflict-driven clause learning SAT solvers. In: Biere, A., Heule, M., van Maaren, H., Walsh, T. (eds.) Handbook of Satisfiability, Frontiers in Artificial Intelligence and Applications, vol. 185, pp. 131–153. IOS Press (2009)
29. Mendelson, E.: Introduction to Mathematical Logic, 4th edn. Chapman and Hall (1997)
30. Menezes, A.J., Vanstone, S.A., Oorschot, P.C.V.: Handbook of Applied Cryptography, 1st edn. CRC Press Inc, Boca Raton (1996)
31. Mladenović, N., Hansen, P.: Variable neighborhood search. Comput. Oper. Res. **24**(11), 1097–1100 (1997)
32. Morgado, A., Heras, F., Liffiton, M.H., Planes, J., Marques-Silva, J.: Iterative and core-guided MaxSAT solving: a survey and assessment. Constraints Int. J. **18**(4), 478–534 (2013)
33. Mühlenbein, H.: How genetic algorithms really work: mutation and hillclimbing. In: Parallel Problem Solving from Nature 2, PPSN-II, 28–30 September 1992, Brussels, Belgium, pp. 15–26 (1992)
34. Nakamoto, S.: Bitcoin: a peer-to-peer electronic cash system (2019)
35. Narodytska, N., Bacchus, F.: Maximum satisfiability using core-guided MaxSAT resolution. In: Proceedings of the Twenty-Eighth AAAI Conference on Artificial Intelligence. AAAI 2014, pp. 2717–2723. AAAI Press (2014)
36. Pavlenko, A., Semenov, A., Ulyantsev, V.: Evolutionary computation techniques for constructing SAT-based attacks in algebraic cryptanalysis. In: Kaufmann, P., Castillo, P.A. (eds.) EvoApplications 2019. LNCS, vol. 11454, pp. 237–253. Springer, Cham (2019). https://doi.org/10.1007/978-3-030-16692-2_16
37. Russell, S.J., Norvig, P.: Artificial Intelligence - A Modern Approach, Third International Edition. Pearson Education (2010)
38. Semenov, A.A.: Decomposition representations of logical equations in problems of inversion of discrete functions. J. Comput. Syst. Sci. Int. **48**, 718–731 (2009)
39. Semenov, A., Otpuschennikov, I., Gribanova, I., Zaikin, O., Kochemazov, S.: Translation of algorithmic descriptions of discrete functions to SAT with applications to cryptanalysis problems. Logical Methods Comput. Sci. **16**(1), 1–42 (2020)
40. Semenov, A.A.: Merging variables: one technique of search in pseudo-boolean optimization. In: Mathematical Optimization Theory and Operations Research. Communications in Computer and Information Science, vol. 1090, pp. 86–102 (2019)
41. Tseitin, G.S.: On the complexity of derivation in propositional calculus. In: Studies in Constructive Mathematics and Mathematical Logic, part II, Seminars in Mathematics, pp. 115–125 (1970)
42. Wegener, I.: Theoretical aspects of evolutionary algorithms. In: Orejas, F., Spirakis, P.G., van Leeuwen, J. (eds.) ICALP 2001. LNCS, vol. 2076, pp. 64–78. Springer, Heidelberg (2001). https://doi.org/10.1007/3-540-48224-5_6
43. Williams, R., Gomes, C.P., Selman, B.: Backdoors to typical case complexity. In: The 18th International Joint Conference on Artificial Intelligence (IJCAI 2003), pp. 1173–1178 (2003)
44. Zaikin, O., Ignatiev, A., Marques-Silva, J.: Branch location problem with maximum satisfiability. In: ECAI 2020 (2020, in press)

Fast Heuristic for Vehicle Routing Problem on Trees

Irina Utkina, Olga Bukanova, and Mikhail V. Batsyn[(✉)] [ID]

Laboratory of Algorithms and Technologies for Network Analysis,
National Research University Higher School of Economics,
136 Rodionova Street, Niznhy Novgorod, Russia
mbatsyn@hse.ru

Abstract. In this paper we propose an efficient heuristic for the Vehicle Routing Problem on Trees (TVRP). An initial solution is constructed with a greedy algorithm based on the Depth-First Search (DFS) approach. To optimize initial solutions obtained by our DFS heuristic, Ruin-and-Recreate (RR) method is then applied. For diversification purposes a randomization mechanism is added to the construction of initial solutions and DFS+RR algorithm is executed multiple times until the best found solution stops changing. The results of our approach are compared with the solutions obtained by the exact model of Chandran & Raghavan (2008). The computational experiments show that the suggested heuristic is fast and finds solutions which differ from optimal ones less than by 1% in average.

Keywords: Vehicle Routing Problem on Trees · Depth-First Search · Ruin-and-Recreate · Large neighbourhood search

1 Introduction

In this paper we consider the Vehicle Routing Problem on Trees (TVRP), which arises in transportation logistics when roads represent a graph with a tree structure. For example, it can be rail and water routes, long roads between cities and towns, and other cases which have a high cost of new road construction.

Vehicle Routing Problems are very popular in literature due to the vast variety of applications in real practice [9,10]. Many algorithms are developed for such problems including exact approaches and heuristics [6]. The Vehicle Routing Problem on Trees (TVRP) was proposed by Labbe, Laporte and Mercure in 1991 [4]. They developed a branch-and-bound algorithm for the TVRP. An initial solution was obtained by a linear-time heuristic and lower bounds were based on solving of the Bin Packing Problem (BPP). The authors also proved the NP-hardness of the problem by reduction from the BPP as well as the APX-hardness by showing that the suggested heuristic is a 2-approximation algorithm. Mbaraga et al. extended this branch-and-bound algorithm in 1999 to the TVRP with an additional constraint on a route duration [5]. These authors

© Springer Nature Switzerland AG 2020
Y. Kochetov et al. (Eds.): MOTOR 2020, CCIS 1275, pp. 379–386, 2020.
https://doi.org/10.1007/978-3-030-58657-7_30

also suggested a column generation approach which showed a better performance on tightly constrained instances, difficult for the branch-and-bound method. In 2008 Chandran & Raghavan proposed an efficient integer programming model for the TVRP which showed high performance [2].

In 1995 Rennie developed a sequential greedy heuristic which was more flexible and had better results than Labbe's linear heuristic [7]. Basnet et al. suggested two heuristics for the TVRP in 1999 [1]. They considered the TVRP for transportation of milk in New Zealand, where due to mountainous terrain construction of a new road is costly. The first heuristic H1 was based on Clarke and Wright algorithm [3]. It built separate routes, one for each customer, and then sequentially combined them minimizing the sum of route lengths on every step. The second heuristic H2 started with one big route and then divided it into small ones to satisfy capacity constraints. H1 was slower than other heuristics, but provided better results.

In this paper we develop a fast heuristic algorithm combined of a greedy randomized heuristic based on the Depth-First Search (DFS) method and a large neighbourhood search heuristic based on the Ruin-and-Recreate (RR) approach. We provide computational results for DFS, DFS+RR, and Randomized DFS+RR (RDFS+RR) combinations comparing the obtained solutions with the exact ones found by Chandran & Raghavan model [2]. DFS and DFS+RR are deterministic algorithms which computational time on all instances is never greater than 1 s. RDFS+RR algorithm performs many iterations before the best found solution stops changing and its computational times reach 45 s on the largest instances. At the same time RDFS+RR provides high-quality solutions which differ from optimal ones less than by 1% in average.

2 Mathematical Model

The following model is based on the model proposed by Chandran & Raghavan [2]. We assume that the number of vehicles is not limited. So a vehicle index k can have any value from 1 to K, where K is the maximum possible number of vehicles. We consider the symmetric case in which distance d_{ij} is equal to d_{ji}. So in this model we take into account only down-arcs going from a parent to a child, because for each down-arc (i, j) in a route there should be a returning up-arc (j, i) with the same distance d_{ij}. Thus up-arcs only double the value of the objective function and could be skipped.

Parameters:
d_{ij} – the travel distance of arc (i, j)
q_i – the demand of customer i
$p(i)$ – the parent of vertex i in the road tree
V – the set of customer vertices $\{1, ..., n\}$ (the depot vertex has index 0)
Q – the capacity of each vehicle
A – the set of down-arcs (from parents to children) including the depot arcs
K – the maximal possible number of vehicles

Decision Variables:
$x_{ijk} = 1$, if arc (i, j) belongs to the route of vehicle k, otherwise 0
$y_{ik} = 1$, if vehicle k serves customer i, otherwise 0

Objective function:

$$\sum_{k=1}^{K} \sum_{(i,j) \in A} d_{ij} x_{ijk} \to \min \tag{1}$$

Constraints:

$$x_{p(i)ik} \geq x_{ijk} \quad \forall k = 1, ..., K, \ \forall (i, j) \in A, \ i \neq 0 \tag{2}$$

$$x_{p(i)ik} \geq y_{ik} \quad \forall k = 1, ..., K, \ \forall i \in V \tag{3}$$

$$\sum_{i \in V} q_i y_{ik} \leq Q \quad \forall k = 1, ..., K \tag{4}$$

$$\sum_{k=1}^{K} y_{ik} = 1 \quad \forall i \in V \tag{5}$$

$$y_{ik}, x_{ijk} \in \{0, 1\} \quad \forall k = 1, ..., K, \ \forall (i, j) \in A \tag{6}$$

The objective is to minimize the total travelled distance of all vehicles. Constraints (2) guarantee that if a vehicle has arc (i, j) in its route, then this route must also contain the arc from the parent of i to i. The next constraint (3) states that if a vehicle serves customer i, then it should come to i from his parent $p(i)$. Constraint (4) prevents an overload of a vehicle. The last constraint (5) requires every customer to be served by one vehicle. The maximal possible number of vehicles K is taken equal to the upper bound $K = \lceil 2 \sum_{i \in V} q_i / Q \rceil$ [2].

3   Initial Heuristic Based on DFS

The main idea for our initial heuristic is to apply the DFS approach. In the DFS we can traverse vertices in different order: first children, then parent; first parent, then children; first one child, then parent, then another child, then parent and so on. Our experiments show that for the TVRP it is more efficient to use the third variant in our heuristic. We traverse vertices in the described order and add a vertex to the first route in which the remaining capacity of the vehicle allows it. If no route can take the current vertex, then we start a new route with this vertex.

To explain our heuristic in detail we provide a small TVRP example in Fig. 1. We assume for simplicity that all edges have distance 1 and all vehicles can carry 20 units. Bold numbers show demands and numbers in circles – the index of the vertex in our DFS order. According to our algorithm we put vertices 1, 2, 3, 4 to the first route with the total load equal to $3 + 8 + 5 + 4 = 20$. Then vertices 5,

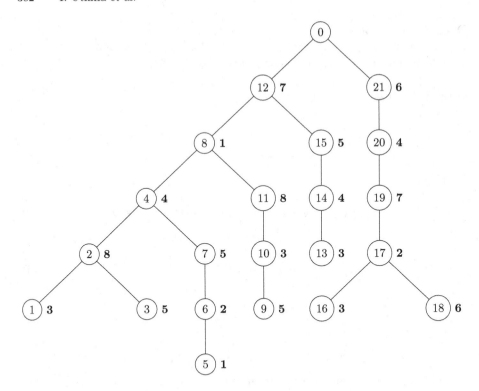

Fig. 1. TVRP example

6, 7, 8, 9, 10 will go to route 2 giving the total load of $1 + 2 + 5 + 1 + 5 + 3 = 17$. Vertices 11 and 12 cannot be added to route 2 because of the capacity limit 20. So we add them to a new route 3, and vertex 13 with demand 3 completes route 2: $\{5, 6, 7, 8, 9, 10, 13\}$. Route 3 is then completed with vertex 14: $\{11, 12, 14\}$ and the total load of $8 + 7 + 4 = 19$. Customers 15, 16, 17, 18 with the total load of 16 go to a new route 4. Customer 19 has to start a new route 5, and customer 20 completes route 4: $\{15, 16, 17, 18, 20\}$. Finally the last route 5 becomes $\{19, 21\}$. The constructed routes have distances, computed only by down-arcs, equal to 6, 12, 5, 8, and 3 correspondingly. The total travelled distance by down-arcs is 34.

For computational experiments we randomly generated 5 trees for each number of vertices: 10, 20, 30, 50, 100, and set the time limit for CPLEX solver to 300 s. All exact solutions have been obtained by CPLEX only for trees with 10, 20, and 30 vertices. And our greedy algorithm was able to find exact solutions only for all trees with 10 and 20 vertices (see Table 1).

Table 1. Computational results for exact model and DFS heuristic

Number of nodes	CPLEX average time (sec)	Number of exact solutions	DFS heuristic average time (sec)	Number of exact solutions
10	0.1	5	0.01	5
20	0.9	5	0.01	5
30	2.0	5	0.04	3
50	184.7	4	0.16	2
100	300.0	0	2.15	0

4   Ruin-and-Recreate Algorithm

Ruin-and-Recreate is a well-known approach from the family of large neighbourhood search methods [8]. In our algorithm we suggest the following Ruin-and-Recreate procedures for improvement of initial solutions:

- **Ruin**: At every crossroad we divide a route into parts, based on the number of children the node has. The main part will still have all its customers and other parts will be cut from it into separate routes.
- **Recreate**: For every two routes we calculate the difference their merge will make in the objective function. At each step we look at every possible merge and apply the best one based on the objective function decrease.

Due to the deterministic ruin procedure dividing all routes at crossroads and the recreate procedure that chooses the best combination of two routes on each step, the Ruin-and-Recreate algorithm is executed only once.

After the ruin algorithm applied to the initial solution from Fig. 1, our recreate algorithm will create the following 6 routes:

- Route 1 $\{1, 2, 3, 4\}$: we merge parts $\{1, 2, 4\}$ and $\{3\}$.
- Route 2 $\{5, 6, 7\}$: this part $\{5, 6, 7\}$ remains the same after the ruin stage and is not merged with any other part during the recreate stage.
- Route 3 $\{8, 9, 10, 11\}$: parts $\{9, 10\}$, $\{11\}$, and $\{8\}$ are merged into this route.
- Route 4 $\{12, 13, 14, 15\}$: this route is combined from parts $\{13\}$, $\{14\}$, $\{15\}$, and $\{12\}$.
- Route 5 $\{16, 17, 18, 19\}$: we join parts $\{16, 17\}$, $\{18\}$, and $\{19\}$.
- Route 6 $\{20, 21\}$: parts $\{20\}$ and $\{21\}$ form this route.

This solution has a 15% smaller total travelled distance – 29 instead of 34.

Due to the greediness in construction of an initial solution, there are cases in which Ruin-and-Recreate approach could not construct an optimal solution starting from a bad initial one. To provide diversification during solution search and obtain better results we have added a randomization mechanism to our initial DFS heuristic. When going to each next child after the parent vertex we now skip this road for the current route with probability $1/2$. This randomization technique allows generating many different initial solutions and leads to a final solution of higher quality.

5   Computational Results

Chandran & Raghavan tested their exact model on randomly generated test instances with the following parameters:

- Number of nodes: 20, 40, 60, 80, and 100.
- For each number of nodes customer demands are generated from different intervals with uniform distribution: [1, 10], [1, 30], [1, 50], [1, 100], [10, 10], [10, 90], [20, 20], [20, 80], [30, 30], [30, 70].
- For each number of nodes and each demand interval 10 different input trees are generated having from 1 to 5 children in each node with probability 1/5.
- The travel distance of each edge is generated uniformly from [1].
- The capacity of vehicles is taken equal to 100.

We compared the solutions obtained by the exact Chandran & Raghavan model on such instances with the results of our three algorithms: DFS heuristic (DFS), DFS heuristic followed by Ruin-and-Recreate algorithm (DFS+RR), and multi-start Randomized DFS heuristic plus Ruin-and-Recreate algorithm (RDFS+RR). RDFS+RR for each input instance is run 10 times and average results are calculated. CPLEX time limit is set to 300 s, so not all of the results obtained by CPLEX are exact. Most of the solutions for large trees are the best results that CPLEX computed within this time limit. That is why our algorithms sometimes have better solutions, and the reported deviation from CPLEX solution is negative. The source code of our algorithm and instances generator can be found at https://nnov.hse.ru/en/latna/benchmarks.

As it is shown in Table 2 our solutions differ from optimal ones not more than by 5%. CPLEX has found optimal solutions for all instances only with 20 and 40 nodes. On these instances the average error of RDFS+RR algorithm is 0.765%. Average computational times are provided in Table 3. DFS needs only several milliseconds, DFS+RR times reach one second, and RDFS+RR spends not more than a minute on each instance.

Funding. The paper is prepared within the framework of the Basic Research Program at the National Research University Higher School of Economics (NRU HSE).

Table 2. Deviation in percent from CPLEX solutions

Number of nodes	Demand intervals	DFS	DFS+RR	RDFS+RR
20	1–10	0.0	0.0	0.0
20	1–30	2.3	2.3	1.1
20	1–50	1.8	1.8	1.2
20	1–100	1.5	1.5	0.3
20	10–10	0.0	0.0	0.0
20	10–90	2.7	2.4	0.7
20	20–20	0.0	0.0	0.0
20	20–80	2.1	2.1	0.3
20	30–30	0.0	0.0	0.0
20	30–70	2.3	1.9	0.6
40	1–10	0.0	0.0	0.0
40	1–30	1.3	1.0	0.7
40	1–50	5.4	4.6	2.8
40	1–100	1.7	1.5	1.2
40	10–10	0.0	0.0	0.0
40	10–90	2.7	2.7	1.4
40	20–20	1.8	1.8	0.4
40	20–80	5.1	5.1	2.9
40	30–30	0.0	0.0	0.0
40	30–70	3.8	3.7	2.0
60	1–10	0.0	0.0	0.0
60	1–30	2.3	2.3	1.6
60	1–50	5.1	4.8	4.3
60	1–100	3.4	3.4	1.7
60	10–10	0.0	0.0	0.0
60	10–90	2.9	2.9	1.5
60	20–20	0.0	0.0	-8.8^*
60	20–80	3.6	3.6	2.8
60	30–30	0.0	0.0	0.0
60	30–70	3.9	3.9	2.8
80	1–10	0.3	0.3	0.0
80	1–30	2.1	1.8	1.2
80	1–50	1.3	1.3	1.1
80	1–100	1.7	1.7	0.6
80	10–10	0.0	0.0	0.0
80	10–90	2.6	2.6	1.2
80	20–20	0.0	0.0	0.0
80	20–80	2.3	2.3	1.0
80	30–30	-0.4^*	-0.4^*	-0.4^*
80	30–70	2.3	2.3	1.1
100	1–10	0.5	0.5	0.4
100	1–30	1.2	1.2	1.1
100	1–50	-0.4^*	-0.4^*	-0.7^*
100	1–100	0.2	0.2	-0.6^*
100	10–10	0.6	0.6	0.2
100	10–90	-1.8^*	-1.8^*	-2.5^*
100	20–20	-0.7^*	-0.7^*	-0.8^*
100	20–80	-2.2^*	-2.2^*	-3.8^*
100	30–30	-3.1^*	-3.1^*	-3.3^*
100	30–70	-0.9^*	-0.9^*	-1.4^*

* On most instances starting from 60 nodes CPLEX did not find optimal solutions

Table 3. Average computational time in seconds

Number of nodes	CPLEX average time (sec)	DFS (sec)	DFS+RR (sec)	RDFS+RR (sec)
20	0.9	0.002	0.01	0.3
40	134.9	0.005	0.08	2.5
60	202.8	0.008	0.18	7.8
80	240.5	0.013	0.57	25.1
100	300.0	0.018	1.05	44.2

References

1. Basnet, C., Foulds, L.R., Wilson, J.M.: Heuristics for vehicle routing on tree-like networks. J. Oper. Res. Soc. **50**, 627–635 (1999)
2. Chandran, B., Raghavan, S.: Modeling and solving the capacitated vehicle routing problem on trees. In: Golden, B., Raghavan, S., Wasil, E. (eds.) The Vehicle Routing Problem: Latest Advances and New Challenges. Operations Research/Computer Science Interfaces, vol. 43, pp. 239–261. Springer, Boston (2008). https://doi.org/10.1007/978-0-387-77778-8_11
3. Clarke, G., Wright, J.: Scheduling of vehicles from a central depot to a number of delivery points. Oper. Res. **12**, 568–581 (1964)
4. Labbe, M., Laporte, G., Mercure, H.: Capacitated vehicle routing on trees. Oper. Res. **39**(4), 616–622 (1991)
5. Mbaraga, P., Langevin, A., Laporte, G.: Two exact algorithms for the vehicle routing problem on trees. Naval Res. Logistics **46**, 75–89 (1999)
6. Rafael, M., Pardalos, P.M., Resende, M.G.: Handbook of Heuristics. Springer, Cham (2018). https://doi.org/10.1007/978-3-319-07153-4
7. Rennie S.: Optimal dispatching and routing of milk tankers for northland dairy board. In: Proceedings of the 30th Annual Conference of the Operational Research Society of New Zealand. ORSNZ, Wellington, New Zealand, pp. 95–102 (1995)
8. Schrimpf, G., Schneider, J., Stamm-Wilbrandt, H., Dueck, G.: Record breaking optimization results using the ruin and recreate principle. J. Comput. Phys. **159**(2), 139–171 (2000)
9. Toth, P., Vigo, D.: The vehicle routing problem. In: SIAM Monographs on Discrete Mathematics and Applications, Philadelphia (2002)
10. Toth, P., Vigo, D.: Vehicle routing: problems, methods, and applications. Second Edition. In: MOS-SIAM Series on Optimization, Philadelphia (2014)

Machine Learning and Data Analysis

Network Distances for Weighted Digraphs

Ilaria Granata[1] , Mario Rosario Guarracino[1,2](✉) , Lucia Maddalena[1] ,
and Ichcha Manipur[1]

[1] Institute for High-Performance Computing and Networking,
National Research Council, Naples, Italy
{ilaria.granata,lucia.maddalena,ichcha.manipur}@icar.cnr.it
[2] University of Cassino and Southern Lazio, Cassino, Italy
mario.guarracino@unicas.it

Abstract. The interpretation of the biological mechanisms through the systems biology approach involves the representation of the molecular components in an integrated system, namely a network, where the interactions among them are much more informative than the single components. The definition of the dissimilarity between complex biological networks is fundamental to understand differences between conditions, states, and treatments. It is, therefore, challenging to identify the most suitable distance measures for this kind of analysis. In this work, we aim at testing several measures to define the distance among sample- and condition-specific metabolic networks. The networks are represented as directed, weighted graphs, due to the nature of the metabolic reactions. We used four different case studies and exploited Support Vector Machine classification to define the performance of each measure.

Keywords: Metabolic networks · Network simplification · Network distances

1 Introduction

When analysing multivariate data, network models can capture complex relations existing among variables. In cellular biological modeling of molecules and their interactions, networks play a central role. Indeed, the multitude of complex interactions taking place within a single cell, and among cells, can only be captured with network-based modeling.

The abundance and accessibility of real experimental data, together with metadata describing the biological experiments and conditions, as well as the availability of analytical models and prior knowledge in form of ontologies, have attracted the interest of data scientists in developing novel methodologies might overcome the limits in existing data analysis techniques. The interpretation of complex biological networks through the comparison of different states and the understanding of similar/dissimilar characteristics as well as modules and patterns, is an ongoing area of research, not only in biology [12,20]. The increasing

© Springer Nature Switzerland AG 2020
Y. Kochetov et al. (Eds.): MOTOR 2020, CCIS 1275, pp. 389–408, 2020.
https://doi.org/10.1007/978-3-030-58657-7_31

size of biological data, due to the continuous development of omics science technologies, makes the network approach indispensable. Unfortunately, this huge amount of omics data does not correspond to the availability of datasets composed of biological networks. Indeed, these datasets are rarely available compared to social, economics, finance networks, which are widely used by researchers to test new algorithms and approaches [26,27]. The imbalance is mainly due to the fact that the general problem of comparing two networks is supposed to be NP-complete. In biological applications, the networks edges and nodes have a biological meaning, and therefore any pair of networks representing the same phenomenon in two different conditions are already aligned, and the similarity of networks can be computed in polynomial time. Indeed, different representations of the networks and measures might highlight different aspects. For this reason, it becomes crucial to understand which are the available options and how they can be used to understand a specific problem at hand. Thus, all subsequent data analysis tasks will be based on the network representation and choices of similarity measures.

Many measures for analyzing differences or similarities between graphs have been adopted in the literature, as surveyed in [11,12,14,31,34], and most of them have been evaluated on different types of biological networks, including protein-protein interaction [2,34], biochemical [34], transcriptional regulation [34], signal transduction [34], co-methylation [37], and metabolic networks [31,34,39].

We focus our attention on metabolic networks, as the metabolic mechanisms and their alteration are subject of great attention in the context of systems biology and precision medicine. Indeed, the metabolism takes part to all the physiological and pathological processes and suffers a big impairment in cancer [19,30], although the cause-effect relationship has not been completely clarified. We build these networks by integrating transcriptomic data of cancer patients with publicly available metabolic models. The similarity measures are therefore performed on directed, weighted, and structurally similar networks.

In [18], we presented a survey on network distances in the context of directed, weighted, and structurally similar biological networks. We summarized the underlying mechanisms and features exploited by distances adopted in the literature, which help to bring out the differences or similarities between networks. For comparing these distances, we deliberately exploited an extremely easy dataset, consisting of four tumor metabolic networks, belonging to two very different breast tumor sub-classes, in order to carefully analyze their distance matrices.

The main contribution of the present research is an extensive comparison of distances suited for evaluating similarities/dissimilarities among weighted digraphs sharing the same set of nodes. A subjective evaluation based on the visual inspection of their distance matrices is coupled with an objective evaluation based on the performance of classification results that they allow to obtain.

The rest of this article is organized as follows. A brief description of the distances adopted for assessing dissimilarity of weighted digraphs is provided in Sect. 2, subdividing them according to the use of the distance between nodes, to clustering properties of the nodes, and to distance of distributions extracted from the networks. The results of those measures on metabolic networks for different kinds of cancer are reported and thoroughly compared in Sect. 3. Section 4 draws conclusions and highlights future research directions.

2   Network Distances

Many measurements for graphs are based on parameters describing them in terms of the distances between their nodes [11,14,22,25,28,34], in terms of their clustering coefficients [6,15,32,34,38], or in terms of distances of their probability distributions [2,9,14,16,29,39]. Here, after introducing some basic notations and definitions, we summarize some of the network distances tailored for weighted digraphs extensively described in [18] and adopted for our experiments (see Sect. 3). The distance computations were implemented using the R packages in [5,8,23,41].

2.1   Basic Notations and Definitions

A *weighted directed graph* (digraph) $\mathcal{G}$ can be defined as a triplet (V, E, W), where V is a set of vertices representing the graph nodes, $E \subseteq V \times V$ is the set of edges representing the connections between the nodes, consisting of ordered pairs of elements of V, and W is a set of real numbers, called *weights*, such that for each $e \in E$ there exists a $w(e) \in W$. The graph will be denoted as $\mathcal{G}(V, E, W)$.

A *walk* from node i to node j of a graph $\mathcal{G}(V, E, W)$ is a sequence $p = <v_0, v_1, \ldots, v_k>$ of nodes such that $i = v_0$, $j = v_k$, $v_i \in V$, $(v_{i-1}, v_i) \in E$, $i = 1, \ldots, k$ [7]. A *path* is a walk where all nodes and edges along it are distinct. A path p from i to j will be denoted as $i \overset{p}{\rightsquigarrow} j$. The *weight of a path* $p = <v_0, v_1, \ldots, v_k>$ is given by the sum of the weights of its edges [7]

$$weight(p) = \sum_{i=1}^{k} w(e_{v_{i-1}, v_i}). \tag{1}$$

Given a weighted digraph $\mathcal{G}(V, E, W)$, the *distance* of its nodes i and $j \in V$ can be defined as the weight of the shortest path p from i to j; if no such path exists, then it is set to ∞

$$\delta(i, j) = \begin{cases} \min\{weight(p) : i \overset{p}{\rightsquigarrow} j\} & \text{if } p \text{ exists} \\ \infty & \text{otherwise} \end{cases} \tag{2}$$

2.2   Network Distances Based on Distances Between Nodes

Let $\mathcal{G}^p$ and $\mathcal{G}^q$ be two weighted digraphs on the same set of nodes V, with weights W^p and W^q, respectively. Several network distances of $\mathcal{G}^p$ and $\mathcal{G}^q$ can be defined based on the distances $\delta(i, j)$ between their nodes i and j defined in Eq. (2), some of which are summarized in Table 1. These network distances provide a local measure of dissimilarity between networks, considering edges as independent entities, while disregarding the overall structure [11,25].

Table 1. Network distances of graphs $\mathcal{G}^p$ and $\mathcal{G}^q$ based on distances between nodes.

Name	Definition	Reference
Average path length difference	$d_{AvgPL}(\mathcal{G}^p, \mathcal{G}^q) = \dfrac{\left\lvert \sum\limits_{i,j \in \mathcal{G}^p} \delta(i,j) - \sum\limits_{i,j \in \mathcal{G}^q} \delta(i,j) \right\rvert}{\lvert V \rvert (\lvert V \rvert - 1)}$	[14,34]
Global efficiency difference	$d_{GE}(\mathcal{G}^p, \mathcal{G}^q) = \dfrac{\left\lvert \sum\limits_{i,j \in \mathcal{G}^p} \dfrac{1}{\delta(i,j)} - \sum\limits_{i,j \in \mathcal{G}^q} \dfrac{1}{\delta(i,j)} \right\rvert}{\lvert V \rvert (\lvert V \rvert - 1)}$	[14,28]
Difference of harmonic means of geodesic distances	$d_{hGE}(\mathcal{G}^p, \mathcal{G}^q) = \dfrac{\lvert V \rvert (\lvert V \rvert - 1)}{\left\lvert \sum\limits_{i,j \in \mathcal{G}^p} \dfrac{1}{\delta(i,j)} - \sum\limits_{i,j \in \mathcal{G}^q} \dfrac{1}{\delta(i,j)} \right\rvert}$	[14]
Hamming	$d_{Ham}(\mathcal{G}^p, \mathcal{G}^q) = \dfrac{\sum\limits_{i,j} \lvert w_{i,j}^p - w_{i,j}^q \rvert}{\lvert V \rvert (\lvert V \rvert - 1)}$	[11]
Jaccard	$d_J(\mathcal{G}^p, \mathcal{G}^q) = 1 - \dfrac{\sum\limits_{i,j} \min(w_{i,j}^p, w_{i,j}^q)}{\sum\limits_{i,j} \max(w_{i,j}^p, w_{i,j}^q)}$	[11]
Normalized edge difference	$d_{nEDD}(\mathcal{G}^p, \mathcal{G}^q) = \dfrac{\sqrt{\sum\limits_{i,j} (w_{i,j}^p - w_{i,j}^q)^2}}{\sum\limits_{i,j} \max(w_{i,j}^p, w_{i,j}^q)}$	[18,42]

Table 2. Network distance of graphs $\mathcal{G}^p$ and $\mathcal{G}^q$ based on the clustering coefficients C_i defined in Eq. (3).

Name	Definition	Reference
Clustering coefficient difference	$d_{CC}(\mathcal{G}^p, \mathcal{G}^q) = \left\lvert \sum\limits_i C_i(\mathcal{G}^p) - \sum\limits_i C_i(\mathcal{G}^q) \right\rvert$	[6]

2.3   Network Distance Based on Clustering Coefficients

Clustering coefficients show the tendency of a graph to form tightly connected neighborhoods [15], i.e., to be divided into *clusters*, intended as subsets of vertices that contain many edges connecting these vertices to each other [34]. The considered network distance summarized in Table 2 is based on the local clustering coefficient C_i for node i of a weighted digraph $\mathcal{G}$ introduced in [6]

$$C_i(\mathcal{G}) = \alpha_1 C_i^{in}(\mathcal{G}) + \alpha_2 C_i^{out}(\mathcal{G}) + \alpha_3 C_i^{cyc}(\mathcal{G}) + \alpha_4 C_i^{mid}(\mathcal{G}). \tag{3}$$

It is the weighted average of four different components that separately consider different link patterns of triangles (closed triplets of nodes, centred on one node) that node i can be part of. $C_i^{in}(\mathcal{G})$ deals with triangles where there are two edges incoming into node i; $C_i^{out}(\mathcal{G})$ with triangles where there are two edges coming out of node i; $C_i^{cyc}(\mathcal{G})$ with triangles where all the edges have the same direction; $C_i^{mid}(\mathcal{G})$ deals with remaining triangles. Coefficients $\alpha_1, \ldots, \alpha_4$ are defined in terms of in- and out- strengths and degrees of node i (see [6] for the complete definition).

2.4   Network Distances Based on Probability Distributions

Other distances between networks can be obtained by describing them through probability distributions and adopting different distances of probability distributions for evaluating their dissimilarity [2,9,14,16,29,39].

Network Probability Distributions. Several probability distributions have been considered for describing local and global topological properties of each node of a graph.

Node Distance Distribution (NDD). The NDD $\mathcal{N}_i^r$ of node i in graph $\mathcal{G}^r$ has as its generic element $\mathcal{N}_i^r(h)$ the fraction of nodes in $\mathcal{G}^r$ having distance h from node i [2,16]

$$\mathcal{N}_i^r(h) = \frac{|\{j \in V : \delta(i,j) \in [h, h+1)\}|}{|V| - 1}, \quad h = 0, 1, \ldots, \lceil diam \rceil,$$

where $diam$ indicates the diameter (i.e., the longest shortest path) of $\mathcal{G}^r$. The set of all NDDs $\{\mathcal{N}_1^r, \ldots, \mathcal{N}_{|V|}^r\}$ contains information about the global topology of the graph $\mathcal{G}^r$.

Transition Matrices (TMs). The TM $T^r(s)$ of order s for graph $\mathcal{G}^r$ has as its generic element $T_{i,j}^r(s)$ the probability for node i of reaching node j by a random walker in s steps [2,16]. The TMs $T^r(1)$ and $T^r(2)$ contain local information about the connectivity of the graph $\mathcal{G}^r$.

Clustering Coefficient Distribution (CCD). Based on local clustering coefficients of Eq. (3), the generic element $\mathcal{P}_i^{CC}$ of the CCD $\mathcal{P}^{CC}$ is computed as the clustering coefficient probability for node i of a graph $\mathcal{G}^r$, defined as

$$\mathcal{P}_i^{CC} = \frac{C_i(\mathcal{G}^r)}{\sum\limits_{j=1}^{|V|} C_j(\mathcal{G}^r)}, \quad i = 1, \ldots, |V|. \tag{4}$$

Distance Measures Between Probability Distributions. Several distance measures are adopted in the literature for comparing two probability distributions [3]. Given two discrete distributions $P = \{P_1, \ldots, P_d\}$ and $Q = \{Q_1, \ldots, Q_d\}$, we consider the distribution distances summarized in Table 3.

Table 3. Distances between probability distributions $P = \{P_1, \ldots, P_d\}$ and $Q = \{Q_1, \ldots, Q_d\}$.

Name	Definition	Reference
Euclidean	$d_{Euc}(P,Q) = \sqrt{\sum_{i=1}^{d}(P_i - Q_i)^2}$	
Jaccard	$d_{Jac}(P,Q) = \dfrac{\sum\limits_{i=1}^{d}(P_i - Q_i)^2}{\sum\limits_{i=1}^{d}P_i^2 + \sum\limits_{i=1}^{d}Q_i^2 - \sum\limits_{i=1}^{d}P_iQ_i}$	[24]
Hellinger	$d_{Hel}(P,Q) = \sqrt{1 - \sum\limits_{i=1}^{d}\sqrt{P_iQ_i}}$	[10]
Jensen-Shannon	$d_{JS}(P,Q) = \sqrt{\mathcal{J}(P,Q)}$	[13]

$\mathcal{J}(P,Q) = $ *Jensen-Shannon divergence of P and Q*

Distribution-Based Network Distances. Given two weighted digraphs $\mathcal{G}^p$ and $\mathcal{G}^q$, for each of the described network probability distributions $\mathcal{P}_i^{1,r} = \mathcal{N}_i^r$, $\mathcal{P}_i^{2,r} = \mathcal{T}_i^r(1)$, and $\mathcal{P}_i^{3,r} = \mathcal{T}_i^r(2)$ for node i in graph $\mathcal{G}^r$, $r = p, q$, and for each of the distribution distances $d_l, l \in \{Euc, Jac, Hel, JS\}$ described in Table 3, we consider the network distance

$$\mathcal{M}_l^k(\mathcal{G}^p, \mathcal{G}^q) = \frac{1}{|V|}\sum_{i=1}^{|V|} d_l(\mathcal{P}_i^{k,p}, \mathcal{P}_i^{k,q}), \tag{5}$$

obtained by averaging over all the $|V|$ nodes the distances of the probability distributions of their nodes. Moreover, we also consider two further network distances, given as combinations of the above ones [16]

$$\mathcal{D}_l^k(\mathcal{G}^p, \mathcal{G}^q) = \frac{1}{k}\sum_{i=1}^{k} \mathcal{M}_l^k(\mathcal{G}^p, \mathcal{G}^q), \quad k = 2, 3. \tag{6}$$

Finally, using the CCD defined in Eq. (4) and any of the distribution distances $d_l, l \in \{Euc, Jac, Hel, JS\}$, we evaluate the distance of two graphs $\mathcal{G}^p$ and $\mathcal{G}^q$ in terms of the distributions $\mathcal{P}^{CC}$. This network distance will be denoted as

$$\mathcal{M}_l^{CC}(\mathcal{G}^p, \mathcal{G}^q) = d_l(\mathcal{P}^{CC,p}, \mathcal{P}^{CC,q}), \tag{7}$$

for $l \in \{Euc, Jac, Hel, JS\}$, where $\mathcal{P}^{CC,p}$ and $\mathcal{P}^{CC,q}$ indicate the probability distribution $\mathcal{P}^{CC}$ for graphs $\mathcal{G}^p$ and $\mathcal{G}^q$, respectively.

3   Experimental Results

3.1   Data

In our experiments, we consider sets of weighted digraphs representing metabolic networks, constructed and simplified as described in [17]. Given samples of gene expression data from patients affected by different types of cancer and available correspondent metabolic models, the resulting weighted digraphs consist of nodes representing the involved metabolites; the directed edges connect reagent and product metabolites and their weights are obtained by combinations of expression values of the enzymes catalyzing the reactions in which the couple of metabolites is involved.

RNA sequencing data of breast (Project TCGA-BRCA), lung (Projects TCGA-LUSC and TCGA-LUAD), kidney (Projects TCGA-KIRC and TCGA-KIRP), and brain cancers (Projects TCGA-GBM and TCGA-LGG) collected into the Genomic Data Commons data portal (https://portal.gdc.cancer.gov) were downloaded in the form of FPKM (fragments per kilobase per million reads mapped) normalized read counts. The specific metabolic models for each tumor type were downloaded from the Metabolic Atlas (https://metabolicatlas.org): INIT cancer model for breast cancer data, tissue specific models for kidney, lung, and brain cancer data.

Table 4. Number of samples (#) per class for all the datasets.

	Class 1	#	Class 2	#	Total
Kidney	ccRCC	159	PRCC	90	249
Lung	Adenocarcinoma	159	Squamous carcinoma	150	309
Breast	Basal-like	175	Luminal A	542	717
Brain	GBM	161	LGG	511	672

As summarized in Table 4, for each dataset we considered only two classes. For the Kidney dataset, we only considered samples from clear cell Renal Cell Carcinoma (ccRCC or KIRC) and Papillary Renal Cell Carcinoma (PRCC or KIRP), excluding the solid tissue normal control samples. Indeed, control samples have features strongly different from those of cancerous samples, thus simplifying too much the classification task for this class. Analogous reasoning lead us to exclude the solid tissue normal control samples from the Lung dataset, taking into account only the Adenocarcinoma and Squamous carcinoma samples. In the case of the Breast dataset, we considered only the two most different intrinsic molecular subtypes based on PAM50 classification [33] (Basal-like and Luminal A), so as to reduce in our analysis the influence of uncertainty in the ground truth data. Indeed, Bartlet et al. [1] investigated the classification of breast cancer into intrinsic molecular subtypes, showing that the classifications obtained using different tests were discordant in 40.7% of the studied cases. For the Brain dataset, we considered the two available classes of GlioBlastoma Multiforme (GBM) and Low-Grade Glioma (LGG).

For each dataset, we considered two network simplifications based on eigencentrality, as we have shown [17] strongly reduced execution times at the price of slightly reduced classification accuracy. Simplification 1 is obtained by retaining 9.2% of the nodes chosen among the eigen-top nodes, together with their neighbours; this percentage of eigen-top nodes leads to an only mild decrease in classification accuracy for the reduced problem [17]. Simplification 2 is obtained by retaining only the nodes having eigen-centrality higher than 0.1, together with their neighbours; it leads to oversimplified networks, at the price of further reduction in classification accuracy. Figure 1 shows an example network and the nodes selected to give rise to the two subsequent simplifications. The number of nodes in the original networks and in the simplified networks for all the datasets are reported in Table 5.

3.2   Evaluation

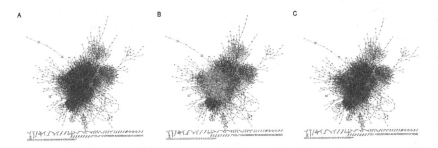

Fig. 1. Representation of the Breast network through Cytoscape v3.6.1 application and its clusterMaker plugin: A) whole network (green nodes), B) Simplification 1 (yellow nodes), C) Simplification 2 (yellow nodes). (Color figure online)

Table 5. Number of nodes in the whole and simplified networks for all the datasets.

	Whole network	Simplification 1	Simplification 2
Kidney	4022	1034	441
Lung	3959	1017	312
Breast	3380	733	58
Brain	3911	989	305

To analyze the role of the network distances described in Sect. 2 and their ability to highlight differences between networks, for any of them we compute the distance matrix, i.e., the symmetric square matrix containing the distances, taken pairwise, between the networks of a given dataset. Specifically, given a dataset consisting of n weighted digraphs $\mathcal{G}^1, \ldots, \mathcal{G}^n$, the (i, j)-th element of the distance matrix for a network distance d is given by $d(\mathcal{G}^i, \mathcal{G}^j), i, j = 1, \ldots, n$. The obtained distance matrices can be represented through heatmaps, such as those in Fig. 2 reporting the distances of the Simplification 1 of the Kidney dataset. Samples are ordered by class, i.e., top left and bottom right matrix blocks include distances of samples belonging to the same class (light blue values indicate low distances, while dark blue values indicate high distances). Therefore, visually inspecting the heatmaps, it is expected that a good network distance produces heatmaps with two light blue diagonal blocks, indicating a low inter-class distance and a good intra-class closeness, respectively. For example, from Fig. 2 we can observe that network distances d_J, $\mathcal{M}_l^2$ and $\mathcal{M}_l^3$ for any l appear to be quite accurate, while the suitability of the remaining network distances needs to be ascertained.

To provide an objective means for comparing the network measures, we consider a *graph classification* problem, that consists in building a model to predict the class label of a whole graph (as opposed to the *label propagation* task, that consists in predicting the class labels of the nodes in a graph) [40] and in its general setting can be formalized as follows. Let $(\mathcal{G}^i, y_i)$, $i = 1, \ldots, n$, be the set of n weighted digraphs over a common set of nodes in $\mathcal{G}$, each associated with a class label $y_i \in K = \{1, \ldots, k\}$. Let $f : \mathcal{G} \longrightarrow K$, be a function such that $f(\mathcal{G}^i) \sim y_i$, for $f \in \mathcal{H}$, the hypothesis space, and let $d(f(\mathcal{G}), y)$ be a metric for evaluating the difference between the predicted value $f(\mathcal{G})$ and the actual value y. The aim of graph classification is to find the function $f \in \mathcal{H}$ that minimizes the empirical risk:

$$I[f] = \frac{1}{n} \sum_{i=1}^{n} d(f(\mathcal{G}^i), y_i). \tag{8}$$

In the experiments, each network in the dataset is represented by the vector containing the distances from all other elements (i.e., the corresponding row of the distance matrix) and the classification is obtained using the sequential minimal optimization (SMO) [35] implementation of Support Vector Machine present in the Weka software [21]. The statistical validation is obtained using a ten-fold cross validation to ensure that the results are not biased to a specific training set. Average accuracy results computed using the Weka Experiment Environment are obtained over ten iterations of ten-fold cross validation:

$$Acc = \frac{1}{10} \sum_{i=1}^{10} Acc_i = \frac{1}{10} \sum_{i=1}^{10} \frac{\#\text{of correct predictions at iteration } i}{\text{total}\#\text{of predictions}}. \tag{9}$$

In Tables 6(a) and (b) we report average accuracy results obtained by all the network distances described in the previous section on the Simplification 1 of the Kidney dataset. Here, we can observe that high accuracy values correspond to visually correct heatmaps in Fig. 2.

Due to low accuracy values obtained for the Kidney dataset, in the following we will disregard d_{AvgPL}, d_{GE}, d_{hGE}, d_{nEDD}, and d_{CC} network distances, concentrating on the remaining distances. In Fig. 3(a), we plot performance results obtained using the selected subset of network distances over Simplification 1 of all datasets. Overall, we observe that lower average accuracy is generally obtained by the network distances based on node distances d_{Ham} and d_J, as well as by the distribution-based network distances $\mathcal{M}_l^{CC}$ and $\mathcal{M}_l^1$ for any distribution distance l. Higher accuracy is obtained by network distances $\mathcal{M}^3$ and

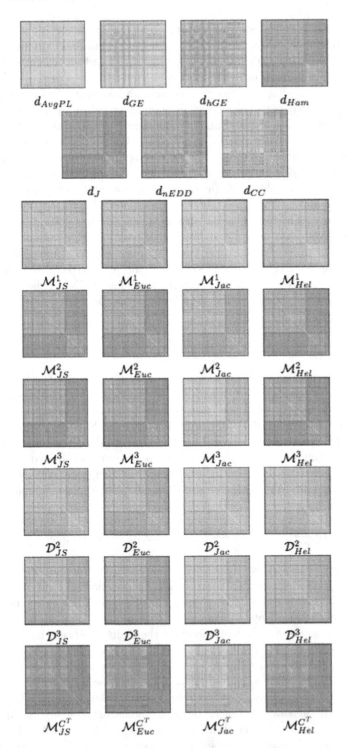

Fig. 2. Heatmaps of the distance matrices obtained by all the network distances for Simplification 1 of the Kidney dataset. (Color figure online)

Table 6. Average accuracy (%) on Simplification 1 of Kidney dataset: (a) network distances based on node distances and on clustering coefficients; (b) distribution-based network distances, varying the probability distribution (along the rows) and the distance distribution (along the columns).

(a)

d_{AvgPL}	d_{GE}	d_{hGE}	d_{Ham}	d_J	d_{nEDD}	d_{CC}
60.59	62.61	63.45	91.49	91.53	81.69	70.55

(b)

	JSD	Euc	Hel	Jac
$\mathcal{M}_l^1$	91.73	90.56	91.49	89.16
$\mathcal{M}_l^2$	92.73	93.53	92.69	92.97
$\mathcal{M}_l^3$	94.94	94.74	94.98	94.94
$\mathcal{D}_l^2$	92.93	93.74	93.09	92.25
$\mathcal{D}_l^3$	93.85	93.90	94.05	93.78
$\mathcal{M}_l^{CC}$	92.33	89.20	92.09	89.96

$\mathcal{D}^3$ for Kidney dataset (around 94%), $\mathcal{M}^2$, $\mathcal{D}^2$, and $\mathcal{D}^3$ for Lung dataset (around 95%), $\mathcal{M}^2$, $\mathcal{M}^3$, $\mathcal{D}^2$, and $\mathcal{D}^3$ for Brain dataset (around 96%), as well as by most of the network distances for Breast dataset (around 99%). Analogous results reported in Fig. 3(b) for Simplification 2 show that, despite the overall slight performance decrease due to the extreme simplification of network data, high accuracy is obtained by network distances $\mathcal{M}^3$, $\mathcal{D}^2$, and $\mathcal{D}^3$ for Kidney dataset (around 93%), $\mathcal{M}^2$, $\mathcal{M}^3$, and $\mathcal{D}^3$ for Lung dataset (around 93%), $\mathcal{M}^2$, $\mathcal{M}^3$, $\mathcal{D}^2$, and $\mathcal{D}^3$ for Breast dataset (around 96%), and $\mathcal{M}^2$, $\mathcal{D}^2$, and $\mathcal{D}^3$ for Brain dataset (around 96%). Detailed numerical results for both simplifications are reported in the Appendix.

To learn more about the role of the distribution distance adopted for computing distribution-based network distances, in Table 7(a) we rearrange average accuracy results obtained over Simplification 1 of all datasets. For a fixed network distance (i.e., fixing a table row), we observe only slight accuracy variations using different distribution distances; the ones leading to highest average accuracy are almost always JS, Euclidean, or Hellinger for all datasets. Table 7(b) reports analogous results based on Simplification 2 of the four datasets. Here, similar conclusions can be drawn, even though the Jaccard distribution distance is more often (five times instead of just one) the one leading to highest accuracy for fixed network distances, while the Hellinger distance is only three times (instead of seven) the one that leads to highest accuracy.

In order to analyze the performance of the distribution-based distances in terms of total time and memory allocation, we focused on the Kidney dataset. We have limited the experiments to the $\mathcal{M}^1$, $\mathcal{M}^2$, $\mathcal{M}^3$ and $\mathcal{M}^{CC}$ distribution distances based on JS, as $\mathcal{D}^2$ and $\mathcal{D}^3$ are derived from $\mathcal{M}^1$, $\mathcal{M}^2$ and $\mathcal{M}^3$. All the

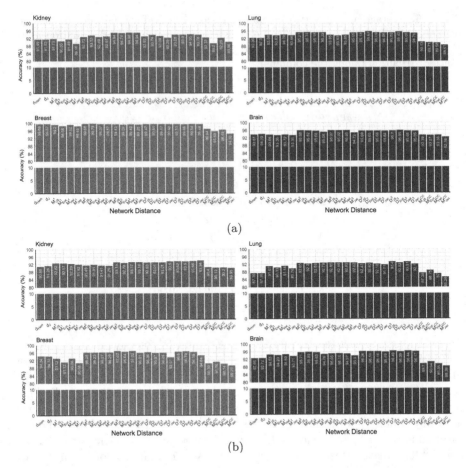

Fig. 3. Average accuracy (%) on (a) Simplification 1 and (b) Simplification 2 of all datasets, varying the distance distribution.

experiments were performed on an Intel i7 quad core 4 GHz CPU, 32 GB RAM, and a 64 bit platform. The implementations of the distance measures were run in R (version 3.6.3) from the RStudio (version 1.2.5033) terminal [36] and the Rprof profiler and profvis R package [4] were used for time and memory profiling at a sampling interval of 10 ms.

We report the total run-time and maximum memory allocation at a sampling instance in Tables 8(a) and (b), respectively. In case of the $\mathcal{M}^1$ and $\mathcal{M}^{CC}$ measures, we observe that there is a decrease in runtime of more than ten times when the number of nodes is reduced from 1034 (Simplification 1) to 441 (Simplification 2). We obtain a three times reduction in run-time with the $\mathcal{M}^2$ and $\mathcal{M}^3$ distances.

The distribution-based distances involve two components: computation of the network probability distributions followed by the measurement of pairwise distances between them. In Fig. 4, we show the run-time and memory allocated separately for these two stages. Here, we observe that for the $\mathcal{M}^{CC}$ distance, computation time is higher during distribution calculation than during distance matrix computation, whereas it is lower in the case of the $\mathcal{M}^2$ and $\mathcal{M}^3$ distances.

The distance matrices for the $\mathcal{M}^1$, $\mathcal{M}^2$ and $\mathcal{M}^3$ distances are built by running four parallel processes. This results in a reduction of total run-time, but increases the memory footprint.

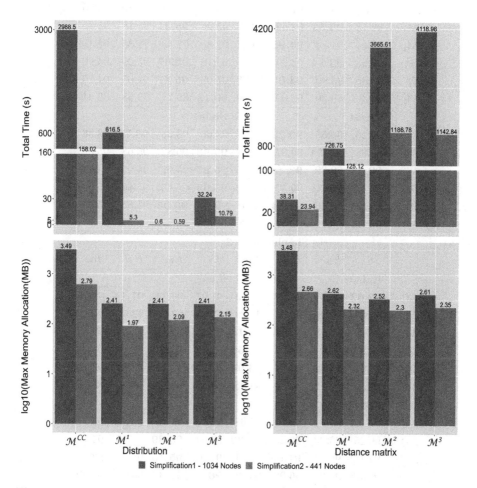

Fig. 4. Total time and maximum memory allocation for computing the distribution (Left panel) and distance components (Right panel) of the distribution-based distance calculation with the JS distance for the Kidney dataset.

Overall, from the above evaluation of the considered network distances, we can conclude that generally

Table 7. Average accuracy (%) on the two simplifications of all datasets, varying the probability distribution (along the rows) and the distance distribution (along the columns). Highest Acc among the four different distribution distances for each network distance are boldface.

(a) Simplification 1

	JS	Euc	Hel	Jac	JS	Euc	Hel	Jac
	Kidney				Lung			
$\mathcal{M}_l^1$	**91.73**	90.56	91.49	89.16	93.98	93.63	**94.08**	93.66
$\mathcal{M}_l^2$	92.73	**93.53**	92.69	92.97	95.21	**95.31**	95.18	95.22
$\mathcal{M}_l^3$	94.94	94.74	**94.98**	94.94	93.85	93.43	93.85	**95.05**
$\mathcal{D}_l^2$	92.93	**93.74**	93.09	92.25	95.18	**95.67**	95.09	94.96
$\mathcal{D}_l^3$	93.85	93.90	**94.05**	93.78	95.02	**95.47**	95.05	94.76
$\mathcal{M}_l^{CC}$	**92.33**	89.20	92.09	89.96	**90.13**	89.71	89.58	88.41
	Breast				Brain			
$\mathcal{M}_l^1$	99.20	98.58	**99.22**	98.45	93.66	93.36	**93.72**	93.36
$\mathcal{M}_l^2$	99.68	**99.79**	99.57	99.57	**95.98**	95.82	95.68	95.08
$\mathcal{M}_l^3$	**99.43**	99.29	**99.43**	99.25	**95.97**	95.92	95.95	94.72
$\mathcal{D}_l^2$	99.47	**99.51**	99.47	99.48	**95.69**	95.65	95.66	95.64
$\mathcal{D}_l^3$	99.53	99.46	**99.54**	99.48	95.70	**95.95**	95.65	95.55
$\mathcal{M}_l^{CC}$	**96.94**	95.69	96.48	94.36	**93.39**	93.27	93.38	92.38

(b) Simplification 2

	JS	Euc	Hel	Jac	JS	Euc	Hel	Jac
	Kidney				Lung			
$\mathcal{M}_l^1$	**92.62**	92.57	92.29	91.93	91.07	90.33	**91.14**	89.78
$\mathcal{M}_l^2$	91.45	**91.81**	91.41	91.29	92.65	92.30	92.59	**92.78**
$\mathcal{M}_l^3$	93.13	92.78	93.17	**93.18**	92.85	**92.88**	92.85	92.82
$\mathcal{D}_l^2$	92.93	93.01	93.01	**93.50**	92.43	**92.63**	92.18	91.71
$\mathcal{D}_l^3$	93.62	93.50	93.66	**93.94**	**93.40**	92.82	93.31	92.10
$\mathcal{M}_l^{CC}$	**90.40**	90.12	90.20	89.95	87.38	**88.75**	87.00	85.25
	Breast				Brain			
$\mathcal{M}_l^1$	**93.12**	91.12	93.08	90.68	94.53	94.23	**94.69**	93.71
$\mathcal{M}_l^2$	96.25	96.23	96.27	**96.29**	95.71	**95.89**	95.65	94.76
$\mathcal{M}_l^3$	**97.02**	96.64	**97.02**	96.01	95.05	**95.15**	94.97	93.90
$\mathcal{D}_l^2$	**96.36**	95.97	96.08	93.61	**96.26**	96.16	96.16	95.89
$\mathcal{D}_l^3$	96.60	**96.79**	96.39	94.64	**96.49**	96.09	96.44	96.17
$\mathcal{M}_l^{CC}$	90.39	**91.16**	90.14	89.23	89.70	**90.64**	89.55	88.39

Table 8. Total time and maximum memory allocation for the distribution-based distances for the Kidney dataset.

(a) Total time (seconds)

	Simplification 1	Simplification 2
$\mathcal{M}^1_{JSD}$	1343.25	130.42
$\mathcal{M}^2_{JSD}$	3666.21	1187.36
$\mathcal{M}^3_{JSD}$	4151.21	1153.62
$\mathcal{M}^{CC}_{JSD}$	3026.81	181.96

(b) Maximum Memory Allocation (MB)

	Simplification 1	Simplification 2
$\mathcal{M}^1_{JSD}$	676.17	299.55
$\mathcal{M}^2_{JSD}$	586.75	320.55
$\mathcal{M}^3_{JSD}$	663.54	367.69
$\mathcal{M}^{CC}_{JSD}$	6131.83	1078.65

- network distances based on probability distributions are to be preferred, as they succeed in taking into account the contribution of each single node. A clear example is given by the $\mathcal{M}^{CC}$ distance based on distributions of clustering coefficients defined in Eq. (7), that achieves much higher accuracy than the clustering coefficient-based d_{CC} distance defined in Table 2, as shown by the results of Tables 6(a) and (b);
- among distribution-based network distances, distances $\mathcal{M}^2$, $\mathcal{M}^3$, $\mathcal{D}^2$, and $\mathcal{D}^3$ lead to higher classification accuracy, for both simplified and over-simplified networks, as shown by the plots of Fig. 3;
- only slight accuracy variations can be observed varying the adopted distance between probability distributions (JS, Euclidean, Hellinger, or Jaccard), as shown by the results of Table 7;
- the simplification of the network structure leads to extreme reduction in terms of both execution times and memory allocation, at the price of slight reduction in classification performance.

4   Conclusions

We have presented an extensive comparison of distances suited for evaluating similarities/dissimilarities among weighted digraphs sharing the same set of nodes. A subjective evaluation based on the visual inspection of their distance matrices is coupled with an objective evaluation based on the performance of classification results that they allow to obtain. The analysis, carried out on four datasets of tumor metabolic networks, has shown that some network distances based on probability distributions describing them generally lead to more accurate results, showing only slight variations based on the adopted distribution distance. Due to the limited availability of datasets of weighted directed networks, as well as for the sake of reproducibility of our results, we are planning to provide the network scientific community with datasets of this specific type, extracted from publicly available data for various diseases.

5   Availability and Implementation

We have provided R packages to construct the metabolic networks used in this study in https://github.com/cds-group/MetabolitesGraphs and to compute the distribution based graph distances in https://github.com/cds-group/GraphDistances.

Acknowledgments. The work was carried out also within the activities of the authors as members of the INdAM Research group GNCS.

Appendix

In Tables 9 and 10, we report detailed numerical performance results obtained using the considered network distances over Simplifications 1 and 2 of all datasets, plotted in Figs. 3(a) and (b), respectively. To provide deeper insight into the performance with respect to each class c, besides Accuracy (Acc) as defined in Eq. (9), we further consider Sensitivity (Se) and Specificity (Sp), defined as

$$Se = \frac{TP_c}{TP_c + FN_c}, \quad Sp = \frac{TN_c}{TN_c + FP_c}.$$

Here, TP_c and FN_c indicate the number of samples belonging to class c that are correctly classified in class c and those that are misclassified, respectively; TN_c and FP_c indicate the number of samples that do not belong to class c that are correctly classified as not belonging to it and those that are misclassified as belonging to it, respectively. Having considered binary classification problems, Se for Class 1 coincides with Sp for Class 2; likewise, Sp for Class 1 coincides with Se for Class 2.

Table 9. Average accuracy (Acc), Sensitivity (Se) and Specificity (Sp) for Class 1 of all datasets (Simplification 1). Highest Acc for each dataset in bold.

	Kidney			Lung			Breast			Brain		
Network distance	Acc	Se	Sp	Acc	Se	Sp	Acc	Se	Sp	Acc	Se	Sp
d_{Ham}	91.49	93.97	87.11	92.36	92.57	92.13	99.68	98.80	99.96	93.81	87.73	95.73
d_J	91.53	94.03	87.11	92.30	91.57	93.07	99.67	98.74	99.96	94.09	88.11	95.99
$\mathcal{M}^1_{JS}$	91.73	94.46	86.89	93.98	94.35	93.60	99.20	98.00	99.59	93.66	88.97	95.15
$\mathcal{M}^1_{Euc}$	90.56	95.02	82.67	93.63	94.04	93.20	98.58	97.43	98.95	93.36	90.38	94.30

(*continued*)

Table 9. (*continued*)

Network distance	Kidney			Lung			Breast			Brain		
	Acc	Se	Sp	Acc	Se	Sp	Acc	Se	Sp	Acc	Se	Sp
$\mathcal{M}^1_{Hel}$	91.49	94.39	86.33	94.08	94.35	93.80	99.22	97.94	99.63	93.72	88.72	95.30
$\mathcal{M}^1_{Jac}$	89.16	94.14	80.33	93.66	94.60	92.67	98.45	96.80	98.99	93.36	90.13	94.38
$\mathcal{M}^2_{JS}$	92.73	93.64	91.11	95.21	96.55	93.80	99.68	99.37	99.78	**95.98**	94.52	96.44
$\mathcal{M}^2_{Euc}$	93.53	94.20	92.33	95.31	96.42	94.13	**99.79**	99.48	99.89	95.82	93.79	96.46
$\mathcal{M}^2_{Hel}$	92.69	93.58	91.11	95.18	96.61	93.67	99.57	99.37	99.63	95.68	94.15	96.16
$\mathcal{M}^2_{Jac}$	92.97	93.14	92.67	95.22	97.05	93.27	99.57	99.31	99.65	95.08	92.30	95.95
$\mathcal{M}^3_{JS}$	94.94	95.03	94.78	93.85	94.40	93.27	99.43	98.74	99.65	95.97	89.76	97.93
$\mathcal{M}^3_{Euc}$	94.74	94.77	94.67	93.43	93.90	92.93	99.29	98.57	99.52	95.92	89.95	97.81
$\mathcal{M}^3_{Hel}$	**94.98**	95.03	94.89	93.85	94.40	93.27	99.43	98.74	99.65	95.95	89.76	97.91
$\mathcal{M}^3_{Jac}$	94.94	95.66	93.67	95.05	96.22	93.80	99.25	98.07	99.63	94.72	87.28	97.07
$\mathcal{D}^2_{JS}$	92.93	94.59	90.00	95.18	96.55	93.73	99.47	98.86	99.67	95.69	91.70	96.95
$\mathcal{D}^2_{Euc}$	93.74	94.91	91.67	**95.67**	97.87	93.33	99.51	98.80	99.74	95.65	92.31	96.71
$\mathcal{D}^2_{Hel}$	93.09	95.02	89.67	95.09	96.30	93.80	99.47	98.80	99.69	95.66	91.88	96.85
$\mathcal{D}^2_{Jac}$	92.25	94.14	88.89	94.96	96.05	93.80	99.48	98.75	99.72	95.64	92.55	96.61
$\mathcal{D}^3_{JS}$	93.85	94.65	92.44	95.02	96.80	93.13	99.53	98.86	99.74	95.70	90.70	97.28
$\mathcal{D}^3_{Euc}$	93.90	94.41	93.00	95.47	97.61	93.20	99.46	98.86	99.65	95.95	91.82	97.26
$\mathcal{D}^3_{Hel}$	94.05	95.09	92.22	95.05	96.55	93.47	99.54	98.92	99.74	95.65	90.95	97.14
$\mathcal{D}^3_{Jac}$	93.78	95.41	90.89	94.76	96.54	92.87	99.48	98.57	99.78	95.55	91.76	96.75
$\mathcal{M}^{CC}_{JS}$	92.33	93.08	91.00	90.13	93.32	86.73	96.94	93.13	98.17	93.39	86.04	95.71
$\mathcal{M}^{CC}_{Euc}$	89.20	91.26	85.56	89.71	93.07	86.13	95.69	89.26	97.77	93.27	85.85	95.62
$\mathcal{M}^{CC}_{Hel}$	92.09	93.02	90.44	89.58	92.69	86.27	96.48	91.94	97.95	93.38	85.79	95.77
$\mathcal{M}^{CC}_{Jac}$	89.96	94.47	82.00	88.41	90.67	86.00	94.36	85.81	97.12	92.38	82.26	95.58

Table 10. Average accuracy (Acc), Sensitivity (Se) and Specificity (Sp) for Class 1 of all datasets (Simplification 2). Highest Acc for each dataset in bold.

Network distance	Kidney			Lung			Breast			Brain		
	Acc	Se	Sp	Acc	Se	Sp	Acc	Se	Sp	Acc	Se	Sp
d_{Ham}	90.88	93.96	85.44	87.35	87.98	86.67	94.34	87.43	96.57	92.66	86.43	94.64
d_J	91.26	95.29	84.11	87.35	86.91	87.80	94.32	86.39	96.88	92.72	86.36	94.74
$\mathcal{M}^1_{JSD}$	92.62	95.66	87.22	91.07	91.21	90.93	93.12	85.26	95.66	94.53	90.32	95.85
$\mathcal{M}^1_{Euc}$	92.57	96.03	86.44	90.33	91.33	89.27	91.12	77.70	95.46	94.23	90.80	95.30
$\mathcal{M}^1_{Hel}$	92.29	95.41	86.78	91.14	90.95	91.33	93.08	84.87	95.74	94.69	89.76	96.24
$\mathcal{M}^1_{Jac}$	91.93	94.52	87.33	89.78	90.14	89.40	90.68	74.36	95.96	93.71	88.96	95.21
$\mathcal{M}^2_{JSD}$	91.45	92.95	88.78	92.65	91.11	94.27	96.25	91.48	97.78	95.71	93.23	96.50
$\mathcal{M}^2_{Euc}$	91.81	93.57	88.67	92.30	91.25	93.40	96.23	90.48	98.10	95.89	93.43	96.67
$\mathcal{M}^2_{Hel}$	91.41	93.20	88.22	92.59	90.99	94.27	96.27	91.48	97.82	95.65	93.17	96.44

(*continued*)

Table 10. (*continued*)

Network distance	Kidney			Lung			Breast			Brain		
	Acc	Se	Sp	Acc	Se	Sp	Acc	Se	Sp	Acc	Se	Sp
$\mathcal{M}^2_{Jac}$	91.29	92.82	88.56	92.78	92.26	93.33	96.29	91.49	97.84	94.76	92.99	95.32
$\mathcal{M}^3_{JSD}$	93.13	93.46	92.56	92.85	93.08	92.60	**97.02**	93.65	98.10	95.05	87.90	97.30
$\mathcal{M}^3_{Euc}$	92.78	93.72	91.11	92.88	92.96	92.80	96.64	93.13	97.76	95.15	89.39	96.97
$\mathcal{M}^3_{Hel}$	93.17	93.59	92.44	92.85	93.08	92.60	**97.02**	93.65	98.10	94.97	87.83	97.22
$\mathcal{M}^3_{Jac}$	93.18	94.09	91.56	92.82	92.71	92.93	96.01	91.42	97.49	93.90	87.52	95.91
$\mathcal{D}^2_{JSD}$	92.93	95.72	88.00	92.43	92.72	92.13	96.36	92.81	97.51	96.26	95.59	96.48
$\mathcal{D}^2_{Euc}$	93.01	95.60	88.44	92.63	94.42	90.73	95.97	90.38	97.79	96.16	95.39	96.40
$\mathcal{D}^2_{Hel}$	93.01	95.72	88.22	92.18	92.09	92.27	96.08	92.30	97.30	96.16	95.84	96.26
$\mathcal{D}^2_{Jac}$	93.50	95.85	89.33	91.71	92.57	90.80	93.61	82.53	97.20	95.89	95.46	96.03
$\mathcal{D}^3_{JSD}$	93.62	95.60	90.11	**93.40**	93.46	93.33	96.60	94.34	97.32	**96.49**	95.10	96.93
$\mathcal{D}^3_{Euc}$	93.50	95.53	89.89	92.82	92.71	92.93	96.79	93.23	97.95	96.09	94.46	96.60
$\mathcal{D}^3_{Hel}$	93.66	95.65	90.11	93.31	93.33	93.27	96.39	94.00	97.16	96.44	95.84	96.63
$\mathcal{D}^3_{Jac}$	**93.94**	95.97	90.33	92.10	93.71	90.40	94.64	86.19	97.38	96.17	94.71	96.63
$\mathcal{M}^{CC}_{JSD}$	90.40	89.18	92.56	87.38	89.44	85.20	90.39	76.16	94.98	89.70	81.73	92.23
$\mathcal{M}^{CC}_{Euc}$	90.12	89.06	92.00	88.75	92.34	84.93	91.16	81.00	94.43	90.64	80.17	93.95
$\mathcal{M}^{CC}_{Hel}$	90.20	89.06	92.22	87.00	89.13	84.73	90.14	75.42	94.89	89.55	80.49	92.43
$\mathcal{M}^{CC}_{Jac}$	89.95	89.37	91.00	85.25	87.37	83.00	89.23	72.95	94.48	88.39	68.27	94.75

References

1. Bartlett, J., et al.: Comparing breast cancer multiparameter tests in the OPTIMA prelim trial: no test is more equal than the others. JNCI J. Natl. Cancer Inst. **108**(9), djw050 (2016)

2. Carpi, L., et al.: Assessing diversity in multiplex networks. Sci. Rep. **9**(1), 1–12 (2019)

3. Cha, S.H.: Comprehensive survey on distance/similarity measures between probability density functions. Int. J. Math. Models Methods Appl. Sci. **1**(4), 300–307 (2007)

4. Chang, W., Luraschi, J., Mastny, T.: profvis: Interactive Visualizations for Profiling R Code (2019). https://CRAN.R-project.org/package=profvis. r package version 0.3.6

5. Clemente, G.P., Grassi, R.: DirectedClustering: Directed Weighted Clustering Coefficient (2018). https://CRAN.R-project.org/package=DirectedClustering. r package version 0.1.1

6. Clemente, G., Grassi, R.: Directed clustering in weighted networks: a new perspective. Chaos Solitons Fractals **107**, 26–38 (2018)

7. Cormen, T.H., Stein, C., Rivest, R.L., Leiserson, C.E.: Introduction to Algorithms, 2nd edn. McGraw-Hill Higher Education, New York (2001)

8. Csardi, G., Nepusz, T.: The igraph software package for complex network research. InterJournal Complex Syst. **1695** (2006). http://igraph.org

9. Dehmer, M., Mowshowitz, A.: A history of graph entropy measures. Inf. Sci. **181**(1), 57–78 (2011)

10. Deza, E., Deza, M.M. (eds.): Dictionary of Distances. Elsevier, Amsterdam (2006)
11. Donnat, C., Holmes, S.: Tracking network dynamics: a survey using graph distances. Ann. Appl. Stat. **12**(2), 971–1012 (2018)
12. Emmert-Streib, F., Dehmer, M., Shi, Y.: Fifty years of graph matching, network alignment and network comparison. Inf. Sci. **346**(C), 180–197 (2016)
13. Endres, D.M., Schindelin, J.E.: A new metric for probability distributions. IEEE Trans. Inf. Theory **49**(7), 1858–1860 (2003)
14. Costa, L.d.F., Rodrigues, F.A., Travieso, G., Boas, P.R.V.: Characterization of complex networks: a survey of measurements. Adv. Phys. **56**, 167–242 (2007)
15. Fagiolo, G.: Clustering in complex directed networks. Phys. Rev. E **76**, 026107 (2007)
16. Granata, I., Guarracino, M., Kalyagin, V., Maddalena, L., Manipur, I., Pardalos, P.: Supervised classification of metabolic networks. In: IEEE International Conference on Bioinformatics and Biomedicine, BIBM 2018, Madrid, Spain, 3–6 December 2018, pp. 2688–2693 (2018)
17. Granata, I., Guarracino, M.R., Kalyagin, V.A., Maddalena, L., Manipur, I., Pardalos, P.M.: Model simplification for supervised classification of metabolic networks. Ann. Math. Artif. Intell. **88**(1), 91–104 (2019). https://doi.org/10.1007/s10472-019-09640-y
18. Granata, I., Guarracino, M.R., Maddalena, L., Manipur, I., Pardalos, P.M.: On network similarities and their applications. In: Mondaini, R.P. (ed.) BIOMAT 2019, pp. 23–41. Springer, Cham (2020). https://doi.org/10.1007/978-3-030-46306-9_3
19. Granata, I., Troiano, E., Sangiovanni, M., Guarracino, M.: Integration of transcriptomic data in a genome-scale metabolic model to investigate the link between obesity and breast cancer. BMC Bioinformatics **20**(4), 162 (2019)
20. Guzzi, P., Milenković, T.: Survey of local and global biological network alignment: the need to reconcile the two sides of the same coin. Brief. Bioinform. **19**(3), 472–481 (2017)
21. Hall, M., Frank, E., Holmes, G., Pfahringer, B., Reutemann, P., Witten, I.H.: The WEKA data mining software: an update. SIGKDD Explor. **11**(1), 10–18 (2009)
22. Hammond, D.K., Gur, Y., Johnson, C.R.: Graph diffusion distance: a difference measure for weighted graphs based on the graph Laplacian exponential kernel. In: 2013 IEEE Global Conference on Signal and Information Processing, pp. 419–422, December 2013
23. Dorst, H.G.: Philentropy: information theory and distance quantification with R. J. Open Source Softw. **3**(26), 765 (2018). http://joss.theoj.org/papers/10.21105/joss.00765
24. Jaccard, P.: Étude comparative de la distribution florale dans une portion des alpes et des jura. Bulletin del la Société Vaudoise des Sciences Naturelles **37**, 547–579 (1901)
25. Jurman, G., Visintainer, R., Filosi, M., Riccadonna, S., Furlanello, C.: The HIM glocal metric and kernel for network comparison and classification. In: 2015 IEEE International Conference on Data Science and Advanced Analytics (DSAA), pp. 1–10 (2015)
26. Kalyagin, V.A., Pardalos, P.M., Rassias, T.M. (eds.): Network Models in Economics and Finance. SOIA, vol. 100. Springer, Cham (2014). https://doi.org/10.1007/978-3-319-09683-4
27. Konstantinos, G., et al.: Network Design and Optimization for Smart Cities, vol. 8. World Scientific, Singapore (2017)
28. Latora, V., Marchiori, M.: Efficient behavior of small-world networks. Phys. Rev. Lett. **87**, 198701 (2001)

29. Liu, Q., Dong, Z., Wang, E.: Cut based method for comparing complex networks. Sci. Rep. **8**(1), 1–11 (2018)
30. Maiorano, F., Ambrosino, L., Guarracino, M.R.: The MetaboX library: building metabolic networks from KEGG database. In: Ortuño, F., Rojas, I. (eds.) IWBBIO 2015. LNCS, vol. 9043, pp. 565–576. Springer, Cham (2015). https://doi.org/10.1007/978-3-319-16483-0_55
31. Mueller, L.A.J., Dehmer, M., Emmert-Streib, F.: Comparing biological networks: a survey on graph classifying techniques. In: Prokop, A., Csukás, B. (eds.) Systems Biology, pp. 43–63. Springer, Dordrecht (2013). https://doi.org/10.1007/978-94-007-6803-1_2
32. Opsahl, T., Panzarasa, P.: Clustering in weighted networks. Soc. Netw. **31**(2), 155–163 (2009)
33. Parker, J.S., et al.: Supervised risk predictor of breast cancer based on intrinsic subtypes. J. Clin. Oncol. **27**(8), 1160 (2009)
34. Pavlopoulos, G.A., et al.: Using graph theory to analyze biological networks. Bio-Data Min. **4**(1), 10 (2011)
35. Platt, J.: Fast training of support vector machines using sequential minimal optimization. In: Schoelkopf, B., Burges, C., Smola, A. (eds.) Advances in Kernel Methods - Support Vector Learning. MIT Press, Cambridge (1998)
36. RStudio Team: RStudio: Integrated Development Environment for R. RStudio Inc, Boston (2019). http://www.rstudio.com/
37. Ruan, D., Young, A., Montana, G.: Differential analysis of biological networks. BMC Bioinformatics **16**, 1–13 (2015)
38. Saramäki, J., Kivelä, M., Onnela, J.P., Kaski, K., Kertész, J.: Generalizations of the clustering coefficient to weighted complex networks. Phys. Rev. E **75**, 027105 (2007)
39. Schieber, T., Carpi, L., Díaz-Guilera, A., Pardalos, P., Masoller, C., Ravetti, M.: Quantification of network structural dissimilarities. Nat. Commun. **8**(1), 1–10 (2017)
40. Tsuda, K., Saigo, H.: Graph classification. In: Aggarwal, C., Wang, H. (eds.) Managing and Mining Graph Data. Advances in Database Systems, vol. 40, pp. 337–363. Springer, Boston (2010). https://doi.org/10.1007/978-1-4419-6045-0_11
41. You, K.: NetworkDistance: Distance Measures for Networks (2019). https://CRAN.R-project.org/package=NetworkDistance. r package version 0.3.2
42. Zvaifler, N.J., Burger, J.A., Marinova-Mutafchieva, L., Taylor, P., Maini, R.N.: Mesenchymal cells, stromal derived factor-1 and rheumatoid arthritis [abstract]. Arthritis Rheum. **42**, s250 (1999)

Decomposition/Aggregation K-means for Big Data

Alexander Krassovitskiy[1]([✉])[iD], Nenad Mladenovic[1,2][iD],
and Rustam Mussabayev[1]([✉])[iD]

[1] Institute of Information and Computational Technologies, Pushkin Str. 125,
Almaty, Kazakhstan
akrassovitskiy@gmail.com, rmusab@gmail.com
[2] Khalifa University, Abu Dhabi 41009, United Arab Emirates
nenad.mladenovic@ku.ac.ae

Abstract. Well-known and widely applied k-means clustering heuristic
is used for solving Minimum Sum-of-Square Clustering problem. In solving large size problems, there are two major drawbacks of this technique:
(i) since it has to process the large input dataset, it has heavy computational costs and (ii) it has a tendency to converge to one of the local
minima of poor quality. In order to reduce the computational complexity, we propose a clustering technique that works on subsets of the entire
dataset in a stream like fashion. Using different heuristics the algorithm
transforms the Big Data into Small Data, clusters it and uses obtained
centroids to initialize the original Big Data. It is especially sensitive for
Big Data as the better initialization gives the faster convergence. This
approach allows effective parallelization. The proposed technique evaluates dynamically parameters of clusters from sequential data portions
(windows) by aggregating corresponding criteria estimates. With fixed
clustering time our approach makes progress through a number of partial
solutions and aggregates them in a better one. This is done in comparing
to a single solution which can be obtained by regular k-means-type clustering on the whole dataset in the same time limits. Promising results
are reported on instances from the literature and synthetically generated
data with several millions of entities.

Keywords: k-means · Parallel · Clustering · Big Data · MSSC ·
Dataset · Decomposition · Aggregation

1 Introduction

Recently, clustering methods have attracted much attention as effective tools in
theoretical and applied problems of machine learning that allows to detect patterns in raw/poorly structured data. Another motivation is the need to process

---

This research is conducted within the framework of the grant num. BR05236839 "Development of information technologies and systems for stimulation of personality's sustainable development as one of the bases of development of digital Kazakhstan".

Y. Kochetov et al. (Eds.): MOTOR 2020, CCIS 1275, pp. 409–420, 2020.
https://doi.org/10.1007/978-3-030-58657-7_32

large data sets to obtain a natural grouping of data. Therefore, one of the main aspects for clustering methods is their scalability [9,15].

There are a number of studies aimed at improving clustering quality using methods with high cost and time complexity [15–17,19]. This type of methods usually has a significant drawback: it is practically intractable to cluster medium and large data sets (approximately 10^5–10^7 objects or more). These methods cannot work with huge databases, because the computational complexity in time/space (memory) growths (polynomially) very rapidly. Therefore, it has a sense to look for algorithms with reasonable trade-offs between effective scalability and the quality of clustering [7,8,10,11].

One of the known methods of data clustering is the k-means algorithm, which is widely used due to its simplicity and good characteristics [18,20]. A number of algorithms and technologies have improved this method by clustering input data objects in portions. There are algorithms that use the data stream clustering or decomposition approach, for example, mini-batch k-means [3,8,13]. It has a weighted version of k-means algorithm with many applications [10].

Meta-heuristics can be of great help when the exact solution is difficult or expensive in terms of used computation time and space. Some heuristics for k-means that accelerate calculations have been developed and implemented: part of the heuristics is devoted to accelerating the convergence of the method, another discards redundant or insignificant intermediate calculations. The following meta-heuristics have shown their effectiveness in clustering big data:

- deletion at each iteration of data patterns that are unlikely to change their membership in a particular cluster, as in [11,12];
- using the triangle inequality in [14];
- combinations of various techniques [1,4].

For many machine learning algorithms, processing of big data is problematic and severely limits functionality usage. Our approach is directed to make an advantage out of this drawback, i.e., the more data is given, the better estimates can be obtained. The k-means is one of the fastest algorithms, so we use it as the underlying basis in our approach. In this paper, we use the k-means++ modification to build an algorithmic meta-heuristic that uses some subsets from the entire dataset at each step. We note that ++ version of the k-means has a special initialization of centroids [6].

Formally, given a set of objects $X = \{x_1, ..., x_N\}$ in Euclidean space to be clustered and a set of corresponding weights $\{w_1, ..., w_N\}$, for $w_l \in \mathbf{R}_+$, $l \in 1, ..., N$. Then $\{C_1, ..., C_k\}$ is a partition of X in k clusters if it satisfies (i) $C_i \neq \emptyset$, (ii) $C_i \cap C_j = \emptyset, i \neq j$, $i, j = 1, 2, ..., k$, and (iii) $\bigcup C_i = X$. Then, minimum sum-of-squared clustering problem is defined as following:

$$MSSD = \min_{C_1, ..., C_k} \sum_{l=1}^{N} w_l \min_{j=1,...,k} ||x_l - c_j||^2,$$

where centroids $c_j = \sum_{l \in arg(C_j)} w_l x_l / \sum_{l \in arg(C_j)} w_l$. Correspondingly, SSD criteria gives an estimate on a particular clustering partition:

$$SSD(C_1, ..., C_k) = \sum_{l=1}^{N} w_l \min_{j=1,...,k} ||x_l - c_j||^2.$$

In case the weights of objects are not specified, then $w_l = 1$ for $l \in 1, ..., N$.

Decomposing the dataset can be technically realized by a stream like methods. Streaming to process a window may be considered as searching dependencies between the obtained essential information and the one gathered previously by the computational model. The principal goal of the study is to investigate methods of dataset decomposition in a stream-like fashion for computing k-means centroid initialization of clustering that produces convincing results regarding MSSD criteria (*Minimum* Sum-of-Squared Distance) [5]. Shortly, methods for finding close to optimal k-means initialization, while having fast computational speed.

The idea of merging clusters obtained by partial clusterings is known in the literature: there are formal approximations that guarantee certain estimates on performance, and quality [9]. Known clustering algorithm STREAM with k-median l1-metric in [13] weights each center by the number of points assigned to it. The stream clustering usually assumes processing the input data in sequential order. Unlike this we use decomposition that may use essential the parallelization of the clustering of the input dataset portions. In our algorithm we add additional heuristic SSD estimates to the weighting. This algorithm can be used in cases the dataset is replenished dynamically, on fly, and possibly in real time. The clustering of additional portions clarifies the clustering structure.

The goal of this work is to create a decomposition method for the k-means algorithm on large-scale datasets to initialize centroids in order to obtain qualitative results with respect to the $MSSD$ criteria. In other words, we use the method of finding the initialization of k-means so that it is close to optimal while having a high calculation speed. Different types of meta-heuristics are used in the task of clustering k-means by processing the obtained data in a secondary (high-level) clustering procedure. Another goal of this work is to study the influence of meta-parameters to the algorithm behaviour with glance to the time and the SSD (Sum-of-Squared Distance) criterion minimization. Another purpose of our research is to define the bounds of such algorithm efficiency and its behaviour regarding meta-parameters.

2   Algorithm

The idea of the algorithm is to make a partition of the dataset into smaller portions, then find the corresponding clustering structure and replace them by single centroid points. On this stage we obtain compact representation of initial dataset that preserves its most essential structural information. Then aggregate and clusterize these centroids in different possible ways getting new heuristic for generalized centroids. Shortly, transform the Big Data into Small Data, cluster them and use obtained centroids to initialize the original Big Data.

More formally, by this approach, first, we decompose the entire dataset entries shuffled randomly on subsets of fixed size (taking either all, or representative portion of elements). Next step is to do k-means clustering of some of these subsets (batches/windows). We use the term 'window' (along the term 'batch' from the literature) to stress that the data subsets are taken in sizes proportional to the entire dataset.

(Meta-) parameters of the algorithm:

- k is the number of required clusters.
- N is the number of objects in the entire data set.
- d is the window size (number of objects in one window). The sizes of the windows are chosen in proportion to the entire dataset. E.g., taking 5 wins decomposition means taking the size of the windows equal to $[N/5]$.
- n is the number of windows used for independent initialization of k-means during Phase 1, see next section.
- $m(\geq n)$ is the total number of windows used for the clustering. The union of m windows may or may not cover the entire dataset.

By using SSD estimates on their corresponding clusterings we make heuristics for better initialization of the algorithm on the entire dataset.

We considered the following two modes for the windows (wins) generation: 1. Segmentation of the entire data set on windows, then a random permutation of objects in the data set is created. The data set is segmented into successive windows of size d. We refer to this as uniform window decomposition mode. 2. For each window, d random objects are selected from the entire set. By repeating this, the required number of windows is generated (objects may be picked repeatedly in different wins). We refer to this mode as random window generation mode.

In order to simplify description we distinguish centroids according to algorithmic steps at which they appear:

- centroids that results from k-means++ on separate windows and used for subsequent initializations we call *local* centroids;
- set of *generalized* centroids is obtained by gathering (uniting) resulted local centroids from clusterings on windows;
- *basis* centroids are obtained by k-means++ clusterings of the set of generalized centroids (considered as a small dataset of higher level representation of windows);
- *final* centroids are obtained by computing k-means on the entire datasest, initialized by basis centroids.

2.1   Phase 1: Aggregation of Centroids

An independent application of k-means++ algorithm on a fixed number n of windows in order to obtain local centroids with following aggregation to generalized set of centroids. The scheme for the algorithms is shown in Fig. 1. This centroids are considered as higher level representation of clustered windows. Each object of generalized set of centroids is assigned to the weight corresponding to the normalized SSD value for the window in which it is calculated as centroid. The weight of i-th object is calculated as follows:

$$w_i = 1 - (SSD_i - SSD_{min})/(SSD_{max} - SSD_{min}), \qquad (1)$$

where SSD_i is the SSD value for such window from which the i-th centroid is taken as an object. Then, using k-means, the new dataset of generalized centroids is divided into k-clusters, taking into account the weights w_i of the objects. In the case of degeneration, k-means is reinitialized. The resulting (basis) centroids are used for:

1. initialization of k-means on the Input Dataset in order to obtain final centroids;
2. evaluation of the SSD on the Input Dataset;
3. initialization Phase 2 of the algorithm described in the following sections.

Alternatively, during processing subsequent windows $n + 1$, $n + 2$, ..., m we have considered the following options in Phase 2: parallel option, straightforward option, and sequential option.

2.2   Phase 2: Parallel Option

The centroids obtained in the previous Phase 1 are used to initialize k-means on each subsequent window $n + 1$, $n + 2$, ..., m. The resulting (local) centroids and SSD estimates are stored if there is no centroid degeneration. The stopping condition is the specified limit either on the computation time or on the number of windows being processed. Similar to Phase 1 we do the clustering on the generalized set of centroids. Subsequent use of its results is similar to clauses 1.1 and 1.2 of Phase 1. Both Phase 1. and parallel option of the Phase 2 are unified in Fig. 1.

2.3   Phase 2: Straightforward Option

An alternative heuristic of splitting the entire dataset and an alternative way of choosing centroids for the clustering initialization is used (see Fig. 2).

The idea of this heuristics is to evaluate and to use the best centroids regarding SSD criteria for initialization of k-means on the subsequent window. While each window is clustered the best obtained centroids are accumulated to process them for final clustering, like in Phase 1.

Algorithm Sketch:

1. Make dataset decomposition on subsets $win_0, win_1, ..., win_l, ...$ of equal size.
2. Obtain list of initial centroids $cent_0$ either by k-means++ on the first window win_0 or by Phase 1. Assign $AC \leftarrow [cent_0]$, $c \leftarrow cent_0$, $BestSSD \leftarrow SSD_0$, $l = 1$.

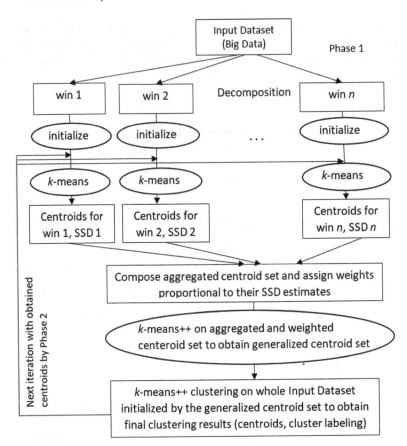

Fig. 1. Scheme for the decomposition/aggregation clustering method. In Phase 1 k-means++ initialization is performed independently on windows $1, ..., n$. Resulted final centroids are used for initialization during Phase 2 on windows $n + 1, ..., m$.

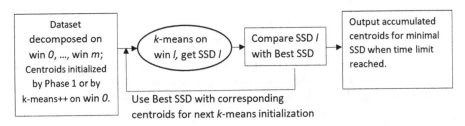

Fig. 2. Direct subsequent use of the best obtained SSD

3. (Start iteration) Use centroids c to initialize k-means with the next window win_l. Calculate $cent_l$ and SSD_l.

4. If degeneracy in the clustering is presented (i.e., the number of obtained non-trivial clusters less then k, $|cent_l| < k$) then withdraw win_l and continue from step 3 for the next $l \leftarrow l + 1$.
5. If its clustering SSD is within the previously obtained or best SSD then $AC \leftarrow AC \cup cent_l$.
6. If its clustering SSD gives better score then mark it as the best and use for the following initializations, i.e., $BestSSD \leftarrow SSD_l$, $c \leftarrow cent_l$.
7. Repeat from step 3 until all windows have been processed or, time bounding condition is satisfied.
8. The accumulated centroids AC are considered as elements for additional clustering, while their SSD values are used to calculate corresponding weights like in Phase 1.
9. The obtained centroids AC are used for the clustering (like in the previous part) and its final SSD value has been compared with SSD of k-means++ on the entire dataset.

2.4   Phase 2: Sequential Option

The following is the sequential version of the algorithm. It is schematically represented in Fig. 3.

Algorithm Sketch:

1. $l = 1$, $init \leftarrow$ centroids from Phase 1, m is the fixed parameter
2. k-means clustering on the window $m + l$ with initialization $init$.
3. If there is no degeneration during clustering then memorize the resulting (local) centroids and the corresponding SSD values.
4. In order to obtain new centroids, we carry out clustering with weights on the united set of centroids (similarly to Phase 1).
5. If the time limit has not been exhausted then $init \leftarrow$ centroids obtained in step 4, $l = l + 1$ and go to step 2, otherwise step 6.
6. Subsequent usage of obtained centroids is similar to clauses 1 and 2 of Phase 1.

3   Computational Experiments

In this section we show the testing results of our algorithm from Sect. 2.4 on three datasets. We only present the results of computation by the sequential version described in Phase 2, with the initial centroids precomputed (sequentially) according to Phase 1. We do not include parallel version of Phase 2 as it distinguishes in the way windows are clustered and it requires additionally efforts in order to compare computational times (taking into account parallelism). The straightforward case can be seen as a particular case of the window aggregation.

Table 1 summarizes clustering estimates of used datasets. Results of computations on various meta-parameter sets from Table 2 are compared regarding SSD/time estimates to the ones obtained by k-means++ and summarized in Table 3. Each line of Table 3 corresponds to unique meta-parameter's set and

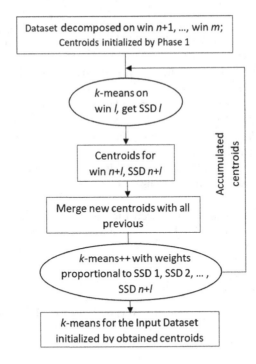

Fig. 3. Sequential aggregation (accumulation) of the heuristically optimal SSD and centroids

includes two SSD estimates, average time per clustering an proportion between computation time of proposed decomposition algorithm and k-means++. The first SSD estimate is obtained as following: we cluster corresponding entire dataset by k-means++ with corresponding parameters and consider obtained SSD criteria as a baseline value. Then, we do independent clusterings by our algorithms on ranges of experiments to calculate basis centroids and compare whether obtained SSD (on the basis centroids) values improves baseline values. The rates are presented for the cases our algorithm finds better solution. We present it in order to show what approximation our algorithm gives if the entire dataset have not been involved. We note that in order to obtain basis centroids we only need to process separate windows. The second SSD estimate is calculated in the same way with addition of one more step. Specifically, k-means is processed on the entire dataset while initialized by the basis centroids. Comparing these two columns of SSD estimates in Table 3 on various parameters and datasets allows us to consider obtained basis centroids as reasonable approximation to MSSD on the entire dataset.

Datasets Description:
We used three datasets DS1, DS2, DS3 of real numbers.

Table 1. SSDs and computation times for datasets DS1, DS2, DS3. The k-means++ is performed with default parameters from the programming library sklearn [1], i.e., 10 separate initializations are executed, the result of the best is presented.

Dataset	Num Clusters	SSD k-means++	Time (sec) k-means++	SSD Ours
DS1	5	3.0625×10^8	389.69	3.0622×10^8
DS1	10	2.7270×10^8	617.70	2.7209×10^8
DS1	20	2.2960×10^8	836.44	2.2740×10^8
DS1	30	1.9482×10^8	677.52	1.9449×10^8
DS2	5	1.5947×10^8	118.23	1.5947×10^8
DS2	10	1.1111×10^8	198.92	1.1106×10^8
DS2	20	7.5041×10^7	627.57	7.4883×10^7
DS2	30	7.3088×10^7	3312.62	7.3080×10^7
DS3	5	1.5734×10^6	312.77	1.5734×10^6
DS3	10	1.2657×10^6	776.62	1.2657×10^6
DS3	20	1.0151×10^6	1279.21	1.0151×10^6
DS3	30	9.1251×10^5	2785.85	9.1173×10^5

Table 2. Meta-parameters of experiments from Table 3. Window sizes are taken in the ranges $(N/20, N/25, ..., N/100)$, $(N/100, N/110, ..., N/250)$ and $(N/10, N/20, ..., N/150)$. 'Allow repeats in windows' refers to the mode how windows are generated. Random window generation mode allows repeats in data objects (TRUE), while uniform window decomposition does not allow it (FALSE).

Param id	Window sizes ($\times N$)	Allow repeats in windows	Number of clusters	Time limit (sec)
1	1/20, ..., 1/100; step 1/5	FALSE	5	30
2	1/20, ..., 1/100; step 1/5	FALSE	10	30
3	1/20, ..., 1/100; step 1/5	FALSE	20	30
4	1/20, ..., 1/100; step 1/5	FALSE	30	30
5	1/100, ..., 1/250; step 1/10	FALSE	5	45
6	1/100, ..., 1/250; step 1/10	FALSE	10	45
7	1/100, ..., 1/250; step 1/10	FALSE	20	45
8	1/100, ..., 1/250; step 1/10	FALSE	30	45
9	1/10, ..., 1/150; step 1/10	TRUE	5	60
10	1/10, ..., 1/150; step 1/10	TRUE	10	60
11	1/10, ..., 1/150; step 1/10	TRUE	20	60
12	1/10, ..., 1/150; step 1/10	TRUE	30	60

Table 3. Experiments on different parameter sets from Table 2. SSD criteria and computation times are presented. Two SSD estimates are considered: 1. the rates our algorithm improves* computed SSD in regards to k-means++ (baseline), where centroids are resulted from aggregation step, i.e., the criteria is estimated on basis centroids; 2. the rate our algorithm improves** the baseline with final centroids, i.e., after additional step with k-means initialized by basis centroids. The improvement rates are given regarding k-means++ on windows from Table 1. Average times per clustering procedure are given.

Dataset	Param id	Improves SSD rate %		Avg. time in seconds	$\frac{time(ours)}{time(kmeans++)}$
		*direct	**global step		
DS1	1	11.8	52.9	42.0	0.108
DS1	2	58.8	100.0	54.8	0.089
DS1	3	41.2	88.2	56.1	0.067
DS1	4	0.0	2 3.5	54.9	0.081
DS2	1	0.0	58.8	21.2	0.179
DS2	2	11.8	64.7	29.0	0.146
DS2	3	76.5	100.0	45.2	0.072
DS2	4	0.0	5.9	458.2	0.138
DS3	1	0.0	17.6	28.7	0.092
DS3	2	0.0	29.4	56.9	0.073
DS3	3	0.0	23.5	69.8	0.055
DS3	4	5.9	41.2	196.3	0.070
DS1	5	0.0	68.8	39.7	0.102
DS1	6	0.0	87.5	56.0	0.091
DS1	7	0.0	93.8	61.0	0.073
DS1	8	0.0	68.8	49.2	0.073
DS2	5	0.0	43.8	21.2	0.179
DS2	6	0.0	68.8	26.1	0.131
DS2	7	87.5	93.8	39.9	0.064
DS2	8	0.0	25.0	445.2	0.134
DS3	5	0.0	6.2	27.9	0.089
DS3	6	0.0	37.5	60.8	0.078
DS3	7	0.0	25.0	53.8	0.042
DS3	8	0.0	18.8	200.4	0.072
DS1	9	6.7	66.7	53.7	0.138
DS1	10	26.7	93.3	76.8	0.124
DS1	11	20.0	100.0	82.0	0.098
DS1	12	0.0	26.7	72.1	0.106
DS2	9	0.0	46.7	26.9	0.227
DS2	10	0.0	66.7	35.7	0.179
DS2	11	80.0	86.7	55.8	0.089
DS2	12	0.0	6.7	519.2	0.157
DS3	9	0.0	6.7	39.4	0.126
DS3	10	0.0	26.7	72.1	0.093
DS3	11	0.0	26.7	85.6	0.067
DS3	12	6.7	60.0	229.6	0.082

- DS1 contains 4×10^6 objects and number of attributes (features) is 25. The structure of the data: 50 synthetic blobs having Gaussian distribution, each having the same number of elements and the same standard deviation value. There is no overlaps in separate blobs.
- DS2 contains 4×10^6 objects and number of attributes (features) is 20. The structure of the data: 20 synthetic blobs having Gaussian distribution, each blob has variable number of objects (from 10^4 to 40×10^4) and variable standard deviation values (distributed randomly in the range from 0.5 to 1.5).
- DS3 is SUSY dataset from open UCI database [2]. The number of attributes is 18 and the number of objects is 5×10^6. In our study we do not take into account the true labelling provided by the database, i.e., the given predictions for two known classes. The purpose of using such dataset is to search for internal structure in the data. This dataset is preprocessed by normalization prior the clustering.

4 Conclusions

In this approach we show that it is possible to achieve better results in the meaning of SSD criteria by applying iteratively the clustering procedure on subsets of the dataset. Obtained centroids are processed again by (meta-) clustering, resulting to the final solution.

Some observations:

- One promising result is that centroids calculated by shown method on large datasets provide reasonable good quality SSD values even without clustering on the whole dataset. Step 6 in Part 2.2 and step 9 in Part 2.3 in many cases may be omitted giving essential advantage in computational speed.
- It is observed there is no sense in splitting the dataset for a huge number of windows as the number of degenerated clusters growths as well.
- Slight improvement is detected on normalized data and small number of clusters.
- Our experiments mostly support the idea that quality and precision of clustering results are highly dependant on the dataset-size and its internal data structure, while it does not strongly depend on the clustering window/batch size, as far as the majority of windows represents the clustering structure of the entire dataset.

References

1. Comparing different clustering algorithms on toy datasets. http://scikit-learn.org/stable/auto_examples/cluster/plot_cluster_comparison.html
2. Online clustering data sets uci. https://archive.ics.uci.edu/ml/datasets.html
3. Comparison of the k-means and minibatch-kmeans clustering algorithms, January 2020. https://scikit-learn.org/stable/auto_examples/cluster/plot_mini_batch_kmeans.html

4. Abbas, O.A.: Comparisons between data clustering algorithms. Int. Arab J. Inf. Technol. **5**(3), 320–325 (2008). http://iajit.org/index.php?option=com_content& task=blogcategory&id=58&Itemid=281

5. Alguwaizani, A., Hansen, P., Mladenovic, N., Ngai, E.: Variable neighborhood search for harmonic means clustering. Appl. Math. Model. **35**, 2688–2694 (2011). https://doi.org/10.1016/j.apm.2010.11.032

6. Arthur, D., Vassilvitskii, S.: K-means++: the advantages of careful seeding. In: Proceedings of the 18th Annual ACM-SIAM Symposium on Discrete Algorithms (2007)

7. Bagirov, A.M.: Modified global k-means algorithm for minimum sum-of-squares clustering problems. Pattern Recogn. **41**(10), 3192–3199 (2008). https://doi. org/10.1016/j.patcog.2008.04.004. http://www.sciencedirect.com/science/article/ pii/S0031320308001362

8. Bahmani, B., Moseley, B., Vattani, A., Kumar, R., Vassilvitskii, S.: Scalable k-means++. Proc. VLDB Endow. **5** (2012). https://doi.org/10.14778/2180912. 2180915

9. HajKacem, M.A.B., N'Cir, C.-E.B., Essoussi, N.: Overview of scalable partitional methods for big data clustering. In: Nasraoui, O., Ben N'Cir, C.-E. (eds.) Clustering Methods for Big Data Analytics. USL, pp. 1–23. Springer, Cham (2019). https:// doi.org/10.1007/978-3-319-97864-2_1

10. Capo, M., Pérez, A., Lozano, J.: An efficient approximation to the k-means clustering for massive data. Knowl.-Based Syst. **117** (2016). https://doi.org/10.1016/ j.knosys.2016.06.031

11. Chiang, M.C., Tsai, C.W., Yang, C.S.: A time-efficient pattern reduction algorithm for k-means clustering. Inf. Sci. **181**, 716–731 (2011). https://doi.org/10.1016/j.ins. 2010.10.008

12. Cui, X., Zhu, P., Yang, X., Li, K., Ji, C.: Optimized big data K-means clustering using MapReduce. J. Supercomput. **70**(3), 1249–1259 (2014). https://doi.org/10. 1007/s11227-014-1225-7

13. Guha, S., Mishra, N., Motwani, R.: Clustering data streams. In: Annual Symposium on Foundations of Computer Science - Proceedings, pp. 169–186, October 2000. https://doi.org/10.1007/978-0-387-30164-8_127

14. Heneghan, C.: A method for initialising the k-means clustering algorithm using kd-trees. Pattern Recogn. Lett. **28**, 965–973 (2007). https://doi.org/10.1016/j.patrec. 2007.01.001

15. Karlsson, C. (ed.): Handbook of Research on Cluster Theory. Edward Elgar Publishing, Cheltenham (2010)

16. Krassovitskiy, A., Mussabayev, R.: Energy-based centroid identification and cluster propagation with noise detection. In: Nguyen, N.T., Pimenidis, E., Khan, Z., Trawiński, B. (eds.) ICCCI 2018. LNCS (LNAI), vol. 11055, pp. 523–533. Springer, Cham (2018). https://doi.org/10.1007/978-3-319-98443-8_48

17. Mladenovic, N., Hansen, P., Brimberg, J.: Sequential clustering with radius and split criteria. Cent. Eur. J. Oper. Res. **21**, 95–115 (2013). https://doi.org/10.1007/ s10100-012-0258-3

18. Sculley, D.: Web-scale k-means clustering. In: Proceedings of the 19th International Conference on World Wide Web, WWW 2010, pp. 1177–1178. Association for Computing Machinery, January 2010. https://doi.org/10.1145/1772690.1772862

19. Shah, S.A., Koltun, V.: Robust continuous clustering. Proc. Natl. Acad. Sci. **114**(37), 9814–9819 (2017)

20. Wu, X., et al.: Top 10 algorithms in data mining. Knowl. Inf. Syst. **14**, 1–37 (2007). https://doi.org/10.1007/s10115-007-0114-2

On the Optimization Models for Automatic Grouping of Industrial Products by Homogeneous Production Batches

Guzel Sh. Shkaberina[1] , Viktor I. Orlov[1,2], Elena M. Tovbis[1] ,
and Lev A. Kazakovtsev[1,3](✉)

[1] Reshetnev Siberian State University of Science and Technology,
prosp. Krasnoyarskiy Rabochiy 31, Krasnoyarsk 660031, Russia
`levk@bk.ru`
[2] Testing and Technical Center – NPO PM,
20, Molodezhnaya Street, Zheleznogorsk 662970, Russia
[3] Siberian Federal University,
prosp. Svobodny 79, Krasnoyarsk 660041, Russia

Abstract. We propose an optimization model of automatic grouping (clustering) based on the k-means model with the Mahalanobis distance measure. This model uses training (parameterization) procedure for the Mahalanobis distance measure by calculating the averaged estimation of the covariance matrix for a training sample. In this work, we investigate the application of the k-means algorithm for the problem of automatic grouping of devices, each of which is described by a large number of measured parameters, with various distance measures: Euclidean, Manhattan, Mahalanobis. If we have a sample with the composition known in advance, we use it as a training (parameterizing) sample from which we can calculate the averaged estimation of the covariance matrix of homogeneous production batches using the Mahalanobis distance. We propose a new clustering model based on the k-means algorithm with the Mahalanobis distance with the averaged (weighted average) estimation of the covariance matrix. We used various optimization models based on the k-means model in our computational experiments for the automatic grouping (clustering) of electronic radio components based on data from their non-destructive testing results. As a result, our new model of automatic grouping allows us to reach the highest accuracy by the Rand index.

Keywords: K-means · Electronic radio components · Clustering

1 Introduction

The increasing complexity of modern technology leads to an increase in the requirements for the quality, of industrial products reliability and durability.

© Springer Nature Switzerland AG 2020
Y. Kochetov et al. (Eds.): MOTOR 2020, CCIS 1275, pp. 421–436, 2020.
https://doi.org/10.1007/978-3-030-58657-7_33

Determination of product quality is carried out by production tests. The quality of products within a single production batch is determined by the stability of the product parameters. Moreover, an increase in the stability of product parameters in manufactured batches can be achieved by increasing the stability of the technological process.

In order to exclude the possibility of potentially unreliable electronic and radio components (ERC) intended to be installed in the onboard equipment of a spacecraft with a long period of active existence, the entire electronic component base passes through specialized technical test centers [1,2]. These centers carry out operations of the total input control of the ERC, total additional screening tests, total diagnostic non-destructive testing and the selective destructive physical analysis (DPA). To expand the results of the DPA to the entire batch of products obtained, we must be sure that the products are manufactured from a single batch of raw materials. Therefore, the identification of the original homogeneous ERC production batches from the shipped lots of the ERC is one of the most important steps during testing [1].

The k-means model in this problem is well established [1,3–10]. Its application allows us to achieve a sufficiently high accuracy of splitting the shipped lots into homogeneous production batches. The problem is solved as a k-means problem [11]. The aim is to find k points (centers or centroids) $X_1, \ldots, X_k$ in a d-dimensional space, such that the sum of the squared distances from known points (data vectors) $A_1, \ldots, A_N$ to the nearest of the required points reaches its minimum (1):

$$arg\,minF(X_1, \ldots, X_k) = \sum_{i=1}^{N} min_{j \in \{\overline{1,k}\}} \| X_j - Aj \|^2. \tag{1}$$

Factor analysis methods do not significantly reduce the dimension of the space without loss of accuracy in solving problems [12]. However, in some cases, the accuracy of partitioning into homogeneous batches (the proportion of objects correctly assigned to "their" cluster representing a homogeneous batch of products) can be significantly improved, especially for samples containing more than 2 or 3 homogeneous batches. In addition, the methods of factor analysis, although they do not significantly reduce the dimension of the search space, show the presence of linear statistical dependencies (correlations) between the parameters of the ERC in a homogeneous batch.

A slight increase in accuracy is achieved by using an ensemble of models [3]. We also applied some other clustering models, such as the Expectation-Maximization (EM) model and Self-organized Cohonen Maps (COM) [12].

Distance measure used in practical tasks of automatic objects grouping in real space depends on the features of space. Changing distance measures can improve the accuracy of automatic ERC grouping.

The idea of this work is to use the Mahalanobis distance measure in the k-means problem and study the accuracy of clustering results. We proposed a new algorithm, based on k-means model using the Mahalanobis distance measure with an averaged estimation of the covariance matrix.

2   Mahalanobis Distance

In k-means, k-median [13–15] and k-medoid [16–18] models, various distance measures may be applied [19,20]. The use of correlation dependencies can be involved by moving from a search in space with a Euclidean or rectangular distance to a search in space with a Mahalanobis distance [21–24]. The square of the Mahalanobis distance D_M defined as follows (2):

$$D_M(X) = \sum_{i=1}^{n} (X - \mu)^T C^{-1} (X - \mu),\qquad(2)$$

where X is vector of values of measured parameters, μ is vector of coordinate values of the cluster center point (or cluster center), C is the covariance matrix.

Experiments on automatic ERC grouping with the k-medoid and k-median models using the Mahalanobis distance show a slight increase in the clustering accuracy in simple cases (with 2–4 clusters) [25].

3   Data and Preprocessing

In this study, we used data of test results performed in the testing center for the batches of integrated circuits (microchips) [26]. The source data is a set of some ERC parameters measured during the mandatory tests. The sample (mixed lot) was originally composed of data on products belonging to different homogeneous batches (in accordance with the manufacturer's markup). The total amount of ERC is 3987 devices. Batch 1 contains 71 device, 116 devices for Batch 2, 1867 for Batch 3, 1250 for Batch 4, 146 for batch 5, 113 for Batch 6, 424 for Batch 7. The items (devices) in each batch are described by 205 input measured parameters.

Computationally, the k-means problem, in which the sum of squared distances acts as the minimized objective function, is more convenient than the k-median model using the sum of distances, because when using the sum of the squared distances, the center point of the cluster (the centroid) coincides with the average coordinate value of all objects in the cluster. When passing to the sum of squared Mahalanobis distances, this property is preserved.

Nevertheless, the use of the Mahalanobis distance in the problem of automatic ERC grouping in many cases leads to accuracy decrease in comparison with the results achieved with the Euclidean distance due to the loss of the advantage of the special data normalization approach (Table 1, hit percentage computed as the sum of hits of algorithm (True Positives) in every batch divided by number of products in the mixed lot).

The assumption that the statistical dependences of the parameter values appear in different batches of ERC in a similar way has experimental grounds. As can be seen from Fig. 1, the span and variance of the parameters of different batches vary significantly. Even if the difference in the magnitude of the span and variance of any parameters is insignificant among separate batches, they differ significantly from the span and variance of the entire mixed lot (Fig. 2).

Table 1. Comparison of the clustering results with different measures of distance, number of exact hits (proportion of hits)

Batches	Squared Euclidean distance	Squared Mahalanobis distance	Rectangular (Manhattan) distance	Cosine distance	Correlation distance
Four-batch mixed lot (n = 446)					
Batch 1 (n = 71)	70 (0.99)	47 (0.66)	71 (1.00)	70 (0.99)	70 (0.99)
Batch 2 (n = 116)	78 (0.67)	83 (0.72)	64 (0.55)	78 (0.67)	84 (0.72)
Batch 5 (n = 146)	96 (0.66)	88 (0.60)	105 (0.72)	96 (0.66)	104 (0.71)
Batch 6 (n = 113)	44 (0.39)	91 (0.81)	50 (0.44)	44 (0.39)	38 (0.37)
Average	**0.65**	**0.69**	**0.65**	**0.65**	**0.66**
Sum of distances	473.174	26146.350	401.4	0.0012	0.0011
Full mixed lot (n = 3987)					
Batch 1 (n = 71)	67 (0.94)	70 (0.99)	68 (0.96)	67 (0.94)	71 (1.00)
Batch 2 (n = 116)	4 (0.03)	4 (0.03)	4 (0.03)	4 (0.03)	78 (0.67)
Batch 3 (n = 1867)	578 (0.31)	223 (0.12)	558 (0.30)	578 (0.31)	0 (0.00)
Batch 4 (n = 1250)	403 (0.32)	127 (0.11)	446 (0.36)	406 (0.33)	227 (0.18)
Batch 5 (n = 146)	66 (0.45)	81 (0.55)	63 (0.43)	64 (0.44)	78 (0.53)
Batch 6 (n = 113)	88 (0.78)	113 (1.00)	82 (0.73)	88 (0.78)	32 (0.28)
Batch 7 (n = 424)	311 (0.73)	404 (0.95)	303 (0.72)	311 (0.73)	314 (0.74)
Average	**0.38**	**0.26**	**0.38**	**0.38**	**0.20**
Sum of distances	5008.127	248808.6	1755.8	0.007	0.004

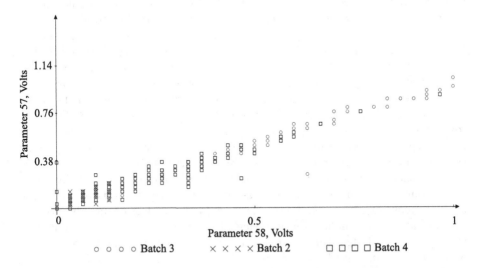

Fig. 1. Statistical dependence of the ERC parameters 57, 58

Thus, it is erroneous to take the variance and covariance coefficients in each of the homogeneous batches equal to the variance and covariance coefficients for the whole sample. Experiments with the automatic grouping model based

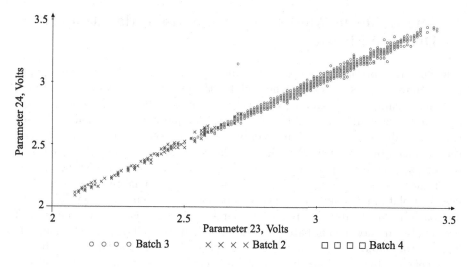

Fig. 2. Statistical dependence of the ERC parameters 23, 24

on a mixture of Gaussian distributions by maximizing the likelihood function by the EM algorithm [27] show a relatively high model adequacy only when using diagonal covariance matrices (i.e. uncorrelated distributions), moreover, equal for all distributions. Apparent correlations between the parameters are not taken into account.

Mahalanobis distance is scale invariant [28]. Due to this property, data normalization does not matter if this distance is applied. At the same time, binding of the boundaries of the parameters to the boundaries, determined by their physical nature, sets a scale proportional to the permissible fluctuations of these parameters under operating conditions, without reference to the span and variance of these values in a particular production batch. The solution to the problem of preserving the scale could be to use the Mahalanobis distance with the correlation matrix R instead of the covariance matrix C (3):

$$D_M(X) = \sum_{i=1}^{n} (X - \mu)^T R^{-1} (X - \mu). \tag{3}$$

Each element of the matrix R is calculated as follows (4):

$$r_{XY} = \frac{\sum_{i=1}^{N}(X_i - \overline{X})(Y_i - \overline{Y})}{(N - 1)S_X S_Y}, \tag{4}$$

where S_X and S_Y are standard deviations of parameters X and Y, $\overline{X}$ and $\overline{Y}$ are their average values.

As shown by experiments, the results of which are given below, this approach does not show advantages compared to other methods.

4   The K-Means Model with Supervised Mahalanobis Distance Measure

The clustering problem is a classical example of the unsupervised learning approach. However, in some cases, when solving the problem of automatic grouping, we have a sample of a known composition. This sample can serve as a training (parameterizing) sample. In this case, a unique covariance matrix C (see (2)) is calculated on this training sample and then used on other data. We call the Mahalanobis distance (2) with the covariance matrix C pre-calculated on a training sample the supervised (or parameterized) Mahalanobis distance.

If there is no training sample, well-known cluster analysis models can be used to isolate presumably homogeneous batches with some accuracy. With this approach, a presumably heterogeneous batch can be divided into the number of presumably homogeneous batches, determined by the silhouette criterion [29–31]. At the same time, a mixed lot can be divided into a larger number of homogeneous batches than it actually is: smaller clusters are more likely to contain data of the same class, i.e. the probability of false assignment of objects of different classes to one cluster reduces. The proportion of objects of the same class, falsely assigned to different classes, is not so important for assessing the statistical characteristics of homogeneous groups of objects.

In the next experiment, there were training sample contains 6 batches: Batch 1 (71 device), Batch 2 (116 devices), Batch 4 (1250 devices), Batch 5 (146 devices), Batch 6 (113 devices), Batch 7 (424 devices). Using covariance matrix C, datasets contain 2 batches in all combinations were clustered with the use of various distance measure. The result was compared with the traditional k-means clustering method with the squared Mahalanobis distance (unsupervised squared Mahalanobis distance, Tables 2, 3, 4 proportion of hits computed as the sum of hits of algorithm in every batch divided by number of products in the batch), and with Euclidean and rectangular distances. For each model, we performed 5 experiments. Average clustering results are shown in Tables 2, 3, 4.

Table 2. Comparison of the clustering results with different measures of distance, number of exact hits (proportion of hits) (Part 1)

Batches	Supervised squared Mahalanobis distance	Unsupervised squared Mahalanobis distance	Squared Euclidean distance	Rectangular (Manhattan) distance
Batch 4 (n = 1250)	850 (0.68)	685 (0.55)	741 (0.59)	895 (0.72)
Batch 7 (n = 424)	390 (0.92)	256 (0.60)	228 (0.54)	423 (1.00)
Average	**0.74**	**0.56**	**0.58**	**0.79**
Avg. total squared distance	94467	100898	7119	12272
Batch 7 (n = 424)	253 (0.60)	Singular	416 (0.98)	415 (0.98)
Batch 1 (n = 71)	71 (1.00)	Matrix	71 (1.00)	71 (1.00)
Average	**0.65**	-	**0.98**	**0.98**
Avg. total squared distance	17551	-	1233	2795

The experiment showed that the results of solving the k-means problem with a supervised Mahalanobis distance measure are higher in comparison with the results of a model with unsupervised Mahalanobis distance, however, it is still lower than in case of Euclidean and rectangular distances.

5   The K-Means Model with Supervised Mahalanobis Distance Measure Based on Averaged Estimation of the Covariance Matrix

Since the original covariance matrices are of the same dimension, we are able to calculate the average estimation of the covariance matrix among all homogeneous batches of products in the training (parameterizing) sample:

$$C = \frac{1}{n} \sum_{j=1}^{k} C_j n_j, \tag{5}$$

where n_j is number of objects (components) in jth production batch, n is total sample size, C_j are covariance matrices calculated on separate production batches, each of which can be calculated by (6):

$$C_j = E[(X - EX)(E - EY)^T]. \tag{6}$$

We propose the k-means algorithm using the Mahalanobis distance measure with averaged estimation of the covariance matrix. Convergence of the k-means algorithm using a Mahalanobis distance reviewed in [32]. Optimal k value was found by silhouette criterion [30]:

Algorithm 1

Step 1. Divide randomly initial sample into k clusters.

Step 2. Calculate for each cluster a centroid μ_i. A centroid is defined as the arithmetic mean of all points in a cluster (7):

$$\mu_i = \frac{1}{m} \sum_{j=1}^{m} X_{ji} \tag{7}$$

where m is number of points, X_j is vector of measured parameter values ($j = 1..m$), $i = 1..n$ (n is a number of parameters).

Step 3. Calculate the averaged estimation of the covariance matrix (5). If the averaged estimation of the covariance matrix is singular, then proceed to Step 4, else proceed to step 5.

Step 4. Increase the number of clusters by $(k + 1)$ and repeat steps 1 and 2. Form new clusters with squared Euclidean distance measure (8):

$$D(X_j, \mu_i) = \sum_{i=1}^{n} (X_{ji} - \mu_i)^2 \tag{8}$$

where n is a number of parameters.

Return to step 3 with new training sample.

Step 5. Assign each point to the nearest centroid using the squared Mahalanobis distance with averaged estimation of the covariance matrix to form new clusters.

Step 6. Repeat algorithm from step 2 until clusters do not change.

Table 3. Comparison of the clustering results with different measures of distance, number of exact hits (proportion of hits) (Part 2)

Batches	Supervised squared Mahalanobis distance	Unsupervised squared Mahalanobis distance	Squared Euclidean distance	Rectangular (Manhattan) distance
Batch 7 (n = 424)	223 (0.53)	Singular	244 (0.58)	282 (0.67)
Batch 6 (n = 113)	113 (1.00)	Matrix	92 (0.81)	84 (0.75)
Average	0.63	-	0.63	0.68
Avg. total squared distance	18190	-	1396	3300
Batch 7 (n = 424)	216 (0.51)	Singular	217 (0.51)	274 (0.65)
Batch 2 (n = 116)	116 (1.00)	Matrix	95 (0.82)	97 (0.84)
Average	0.62	-	0.58	0.69
Avg. total squared distance	18190	-	1123	3090
Batch 7 (n = 424)	424 (1.00)	218 (0.51)	380 (0.90)	385 (0.91)
Batch 5 (n = 146)	136 (0.93)	85 (0.58)	146 (1.00)	146 (1.00)
Average	0.98	0.53	0.92	0.93
Avg. total squared distance	34385	34282	1250	3202
Batch 1 (n = 71)	71 (1.00)	47 (0.66)	71 (1.00)	71 (1.00)
Batch 4 (n = 1250)	471 (0.38)	653 (0.52)	772 (0.62)	642 (0.51)
Average	0.41	0.53	0.64	0.54
Avg. total squared distance	82458	79599	7237	11120
Batch 4 (n = 1250)	410 (0.33)	648 (0.52)	735 (0.59)	570 (0.46)
Batch 6 (n = 113)	102 (0.90)	59 (0.52)	67 (0.59)	85 (0.75)
Average	0.38	0.52	0.59	0.48
Avg. total squared distance	82649	82054	5452	10014
Batch 4 (n = 1250)	412 (0.33)	622 (0.50)	769 (0.62)	485 (0.39)
Batch 2 (n = 116)	98 (0.85)	69 (0.59)	76 (0.66)	96 (0.82)
Average	0.37	0.51	0.62	0.43
Avg. total squared distance	82693	82318	5410	9996
Batch 4 (n = 1250)	953 (0.76)	772 (0.62)	772 (0.62)	873 (0.70)
Batch 5 (n = 146)	91 (0.62)	91 (0.62)	146 (1.00)	146 (1.00)
Average	0.75	0.62	0.66	0.73
Avg. total squared distance	99605	83963	6689	11619
Batch 1 (n = 71)	71 (1.00)	Singular	71 (1.00)	71 (1.00)
Batch 6 (n = 113)	111 (0.98)	Matrix	113 (1.00)	113 (1.00)
Average	0.99	-	1.00	1.00
Avg. total squared distance	6500	-	354	797

(*continued*)

Table 3. (continued)

Batches	Supervised squared Mahalanobis distance	Unsupervised squared Mahalanobis distance	Squared Euclidean distance	Rectangular (Manhattan) distance
Batch 1 (n = 71)	71 (1.00)	Singular	71 (1.00)	71 (1.00)
Batch 2 (n = 116)	116 (1.00)	Matrix	112 (0.97)	114 (0.98)
Average	**1.00**	-	**0.98**	**0.99**
Avg. total squared distance	6481	-	325	747
Batch 1 (n = 71)	71 (1.00)	39 (0.56)	70 (0.99)	71 (1.00)
Batch 5 (n = 146)	84 (0.58)	80 (0.55)	99 (0.68)	108 (0.74)
Average	**0.71**	**0.55**	**0.78**	**0.83**
Avg. total squared distance	22199	13004	223	841
Batch 2 (n = 116)	91 (0.78)	Singular	89 (0.77)	70 (0.60)
Batch 6 (n = 113)	87 (0.77)	Matrix	37 (0.33)	48 (0.42)
Average	**0.78**	-	**0.55**	**0.52**
Avg. total squared distance	7319	-	282	903

Table 4. Comparison of the clustering results with different measures of distance, number of exact hits (proportion of hits) (Part 3)

Batches	Supervised squared Mahalanobis distance	Unsupervised squared Mahalanobis distance	Squared Euclidean distance	Rectangular (Manhattan) distance
Batch 5 (n = 146)	96 (0.66)	81 (0.55)	146 (1.00)	146 (1.00)
Batch 6 (n = 113)	113 (1.00)	66 (0.59)	105 (0.93)	109 (0.75)
Average	**0.81**	**0.57**	**0.97**	**0.99**
Avg. total squared distance	23172	6564	512	1246
Batch 2 (n = 116)	116 (1.00)	67 (0.57)	108 (0.93)	109 (0.94)
Batch 5 (n = 146)	78 (0.54)	80 (0.55)	146 (1.00)	146 (1.00)
Average	**0.74**	**0.56**	**0.97**	**0.97**
Avg. total squared distance	23070	15710	458	1175

6   Computational Experiments

A series of experiments was carried out on the data set described above. This mixed lot is convenient due to its composition is known in advance, which allows us to evaluate the accuracy of the applied clustering models. Moreover, this data set is difficult for grouping by well-known models: some homogeneous batches in its composition are practically indistinguishable from each other, and the accuracy of known clustering models on this sample is low [12,33].

As a measure of the clustering accuracy, we use the Rand Index (RI) [34], which determines the proportion of objects for which the reference and resulting cluster splitting are similar.

To train the model with the averaged Mahalanobis distance measure from the components of the mixed lot, new combinations of batches were compiled containing devices belonging to different homogeneous batches. New combinations consists of 2–7 homogeneous batches. Training sample include the entire data from each batch.

Experiments conducted with 5 different clustering models:

Model DM1: K-means with the Mahalanobis distance measure, the estimation of the covariance matrix calculates for the entire training sample. The objective function defines as the sum of the squared distances.

Model DC: K-means with a distance measure similar to the Mahalanobis distance, but using a correlation matrix instead of a covariance matrix (3). The objective function defines as the sum of the squared distances.

Model DM2: K-means algorithm with Mahalanobis distance measure based on averaged estimation of the covariance matrix (4). The objective function defines as the sum of the squared distances.

Model DR: K-means with Manhattan distance measure. The objective function defines as the sum of the distances.

Model DE: K-means with Euclidean distance measure. The objective function defines as the sum of the squared distances.

This paper presents the results of three groups of experiments. In each of the groups of experiments, for each working sample, the k-means algorithm was run 30 times with each of the five studied clustering models. In these groups of experiments the highest RI value was shown by K-means algorithm with Mahalanobis distance measure based on averaged estimation of the covariance matrix.

First Group. The training set corresponds to the working sample for which clustering was carried out. Five series of experiments were carried out. In each series of experiments, the sample is composed of a combination of products belonging to 2–7 homogeneous batches. Table 5 presents the maximum, minimum, mean

Table 5. An experiment of the 1st group

	Rand index					Objective function				
	DM1	DC	DM2	DR	DE	DM1	DC	DM2	DR	DE
Max	0.755	0.66	0.822	0.739	0.745	255921	3843	2645	18902	6008
Min	0.560	0.64	0.732	0.702	0.704	250558	3706	2600	17785	5010
Mean	0.627	0.65	0.771	0.716	0.721	253041	372289	261582	18225	5298
σ	0.051	0.00	0.024	0.010	0.009	1178	261.01	989.3	433.12	290.276
V						0.466	0.701	0.378	2.377	5.479
R						5363	1369	4517	1117	998

value and standard deviation for the Rand index and objective function for the 7-batches sample. For objective function also calculated the coefficient of variation (V) and span factor (R, where $R = Max - Min$).

Second Group. Training and work samples do not match. In practice, the test center can use retrospective data from the supply and testing of products of the same type as a training sample. In this series of experiments, no more than seven homogeneous batches are presented in the training set. The working sample is represented by a new combination of products belonging to different homogeneous batches. In Table 6 represented results for 5-batches working set and 7-batches training set.

Table 6. An experiment of the 2nd group

	Rand index					Objective function				
	DM1	DC	DM2	DR	DE	DM1	DC	DM2	DR	DE
Max	0.7490	0.645	0.8524	0.7337	0.73567	254822	38704	263405	20509	9194.61
Min	0.4312	0.631	0.7470	0.6955	0.68932	249355	37856	257534	19408	6554.1
Mean	0.5660	0.636	0.8117	0.7079	0.71919	251694	37982	259689	19674	7119.85
σ	0.0519	0.003	0.0324	0.0153	0.01002	1462.8	203.55	1502.09	289.63	571.119
V						0.581	0.536	0.578	1.472	8.022
R						5467	848	5871	1102	2641

Third Group. The training and working samples also do not match, but the results of the automatic product grouping were used as the training sample (k-means in multistart mode with Euclidean distance measure). In each series of experiments, the training set consists of 10 batches, which in turn are the result of applying the k-means algorithm to the training set containing the entire sample. The working sample is represented by a new combination of products belonging to different homogeneous batches. In Table 7 showed results for 7-batches working set.

Table 7. An experiment of the 3rd group

	Rand index					Objective function				
	DM1	DC	DM2	DR	DE	DM1	DC	DM2	DR	DE
Max	0.7672	0.6579	0.7489	0.73969	0.73456	255886	379167	281265	18897	6495
Min	0.5618	0.6453	0.6958	0.70286	0.70466	250839	36997	274506	17785	5009
Mean	0.6317	0.6499	0.7246	0.71359	0.71935	252877	37178	277892	18240	5250
σ	0.0468	0.0032	0.0160	0.0081	0.0063	1164.5	152.84	2358.92	452.73	367.5
V						0.461	0.411	0.849	2.482	6.981
R						5047	920	6759	1112	1485

In most cases, the coefficient of variation of the objective function values is highest for the DE model, where the Euclidean distance measure used. The span factor of the objective function, in the opposite, has most high values for the DM2 model, where the Mahalanobis distance measure with the average estimation of the covariance matrix used. Therefore, obtaining consistently good values of the objective function requires multiple attempts to run the k-means algorithm, or using other algorithms based on the k-means model, such as j-means [35] or greedy heuristic algorithms [36] or others.

According to Rand index, DM2 model shows the best accuracy among the presented models (Fig. 3(a)–3(c)) in almost all series of experiments. And in all cases, the DM2 model surpasses the traditional DE model, where Euclidean distance measure used (Fig. 3(b), 3(c)).

Experiments showed that there is no correlation between the values of the objective function and the Rand index in series of experiments with model DM1 in any combinations of training and working samples (Fig. 4(a)). In other models with an increase the volume of training and working samples (n_t and n_w, respectively), the clustering accuracy becomes constant (Fig. 4(b)). For DM2 model there is an inverse correlation between the achieved value of the objective function and the clustering accuracy RI on a small sample (Fig. 5(a)).

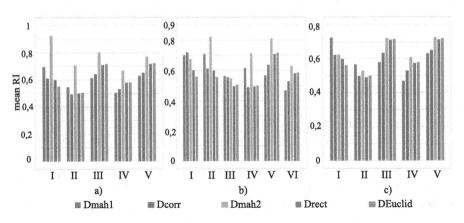

Fig. 3. The mean value of the Rand index for a) 1st group; b) 2nd group; c) 3rd group

In addition, the fact deserves attention that when applying the Euclidean distance measure, the best (smaller) values of the objective function do not correspond to the best (large) accuracy values. (Fig. 5(b)). This fact shows that the model with the Euclidean distance measure is not quite adequate: the most compact clusters do not exactly correspond to homogeneous batches.

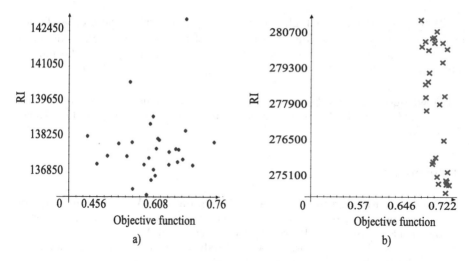

Fig. 4. Dependence of the Rand index on the value of the objective function for a) DM1 model ($n_t = 3987$, $n_w = 2054$); b) DM2 model ($n_t = 3987$, $n_w = 3987$)

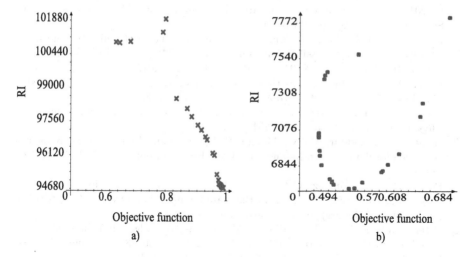

Fig. 5. Dependence of the Rand index on the value of the objective function for a) DM2 model ($n_t = 187$, $n_w = 187$); b) DE model ($n_t = 187$, $n_w = 187$)

7 Conclusion

The proposed clustering model and algorithm which uses the k-means model with Mahalanobis distance and an averaged (weighted average) estimation of the covariance matrix was compared with the k-means model with the Euclidean and rectangular distances in solving the problem of automatic grouping of industrial products by homogeneous production batches.

Taking into account the higher average Rand Index value, the proposed optimization model and algorithm applied for the electronic radio components clustering by homogeneous production batches has an advantage over the models with traditionally used Euclidean and rectangular (Manhattan) metrics.

Acknowledgement. Results were obtained within the framework of the State Task FEFE-2020-0013 of the Ministry of Science and Higher Education of the Russian Federation.

References

1. Orlov, V.I., Kazakovtsev, L.A., Masich, I.S., Stashkov, D.V.: Algorithmic support of decision-making on selection of microelectronics products for space industry. Siberian State Aerospace University, Krasnoyarsk (2017)
2. Kazakovtsev, L.A., Antamoshkin, A.N.: Greedy heuristic method for location problems. Vestnik SibGAU **16**(2), 317–325 (2015)
3. Rozhnov, I., Orlov, V., Kazakovtsev, L.: Ensembles of clustering algorithms for problem of detection of homogeneous production batches of semiconductor devices. In: 2018 School-Seminar on Optimization Problems and their Applications, OPTA-SCL 2018, vol. 2098, pp. 338–348 (2018)
4. Kazakovtsev, L.A., Antamoshkin, A.N., Masich, I.S.: Fast deterministic algorithm for EEE components classification. IOP Conf. Ser. Mater. Sci. Eng. **94**. https://doi.org/10.1088/1757-899X/04/1012015. Article ID 012015
5. Li, Y., Wu, H.: A clustering method based on K-means algorithm. Phys. Procedia **25**, 1104–1109 (2012). https://doi.org/10.1016/j.phpro.2012.03.206
6. Ansari, S.A., et al.: Using K-means clustering to cluster provinces in Indonesia. J. Phys. Conf. Ser. **1028**, 521–526 (2018). 012006
7. Hossain, Md., Akhtar, Md.N., Ahmad, R.B., Rahman, M.: A dynamic K-means clustering for data mining. Indones. J. Electr. Eng. Comput. Sci. **13**(521), 521–526 (2019)
8. Perez-Ortega, J., Almanza-Ortega, N.N., Romero, D.: Balancing effort and benefit of K-means clustering algorithms in Big Data realms. PLoS ONE **13**(9), e0201874 (2018). https://doi.org/10.1371/journal.pone.0201874
9. Patel, V.R., Mehta, R.G.: Modified k-Means clustering algorithm. In: Das, V.V., Thankachan, N. (eds.) CIIT 2011. CCIS, vol. 250, pp. 307–312. Springer, Heidelberg (2011). https://doi.org/10.1007/978-3-642-25734-6_46
10. Na, S., Xumin, L., Yong, G.: Research on k-means clustering algorithm: an improved k-means clustering algorithm. In: 2010 Third International Symposium on Intelligent Information Technology and Security Informatics, Jinggangshan, pp. 63–67 (2010)
11. MacQueen, J.: Some methods for classification and analysis of multivariate observations. In: Proceedings of the Fifth Berkeley Symposium on Mathematical Statistics and Probability, vol. 1, pp. 281–297 (1967)
12. Shkaberina, G.S., Orlov, V.I., Tovbis, E.M., Kazakovtsev, L.A.: Identification of the optimal set of informative features for the problem of separating of mixed production batch of semiconductor devices for the space industry. In: Bykadorov, I., Strusevich, V., Tchemisova, T. (eds.) MOTOR 2019. CCIS, vol. 1090, pp. 408–421. Springer, Cham (2019). https://doi.org/10.1007/978-3-030-33394-2_32

13. Jain, A.K., Dubes, R.C.: Algorithms for Clustering Data. Prentice-Hall, Englewood Cliffs (1981)
14. Bradley, P.S., Mangasarian, O.L., Street, W.N.: Clustering via concave minimization. In: Advances in Neural Information Processing Systems, vol. 9, pp. 368–374 (1997)
15. Har-Peled, S., Mazumdar, S.: Coresets for k-Means and k-Median clustering and their applications. In: Proceedings of the 36th Annual ACM Symposium on Theory of Computing, pp. 291–300 (2003)
16. Maranzana, F.E.: On the location of supply points to minimize transportation costs. IBM Syst. J. **2**(2), 129–135 (1963). https://doi.org/10.1147/sj.22.0129
17. Kaufman, L., Rousseeuw, P.J.: Clustering by means of Medoids. In: Dodge, Y. (ed.) Statistical Data Analysis Based on the L1-Norm and Related Methods, pp. 405–416. North-Holland, Amsterdam (1987)
18. Park, H.-S., Jun, C.-H.: A simple and fast algorithm for K-medoids clustering. Expert Syst. Appl. **36**(2), 3336–3341 (2009). https://doi.org/10.1016/j.eswa.2008.01.039
19. Davies, D.L., Bouldin, D.W.: A cluster Separation measure. IEEE Trans. Pattern Anal. Mach. Intell. **PAMI–1**(2), 224–227 (1979)
20. Deza, M.M., Deza, E.: Metrics on normed structures. In: Encyclopedia of Distances, pp. 89–99. Springer, Heidelberg (2013) https://doi.org/10.1007/978-3-642-30958-8_5
21. De Maesschalck, R., Jouan-Rimbaud, D., Massart, D.L.: The Mahalanobis distance. Chem. Intell. Lab. Syst. **50**(1), 1–18 (2000). https://doi.org/10.1016/S0169-7439(99)00047-7
22. McLachlan, G.J.: Mahalanobis distance. Resonance **4**(20), 1–26 (1999). https://doi.org/10.1007/BF02834632
23. Xing, E.P., Jordan, M.I., Russell, S.J., Ng, A.Y.: Distance metric learning with application to clustering with side-information. In: Advances in Neural Information Processing Systems, vol. 15, pp. 521–528 (2003)
24. Arathiand, M., Govardhan, A.: Performance of Mahalanobis distance in time series classification using shapelets. Int. J. Mach. Learn. Comput. **4**(4), 339–345 (2014)
25. Orlov, V.I., Shkaberina, G.S., Rozhnov, I.P., Stupina, A.A., Kazakovtsev, L.A.: Application of clustering algorithms with special distance measures for the problem of automatic grouping of radio products. Sistemy upravleniia I informacionnye tekhnologii **3**(77), 42–46 (2019)
26. Orlov, V.I., Fedosov, V.V.: ERC clustering dataset (2016). http://levk.info/data1526.zip
27. Kazakovtsev, L.A., Orlov, V.I., Stashkov, D.V., Antamoshkin, A.N., Masich, I.S.: Improved model for detection of homogeneous production batches of electronic components. IOP Conf. Ser. Mater. Sci. Eng. **255** (2017). https://doi.org/10.1088/1757-899x/255/1/012004
28. Shumskaia, A.O.: Evaluation of the effectiveness of Euclidean distance metrics and Mahalanobis distance metrics in identifying the origin of text. Doklady Tomskogo gosudarstvennogo universiteta system upravleniia I radioelektroniki **3**(29), 141–145 (2013)
29. Kaufman, L., Rousseeuw, P.: Finding Groups in Data: An Introduction to Cluster Analysis. Wiley, New York (1990)
30. Rousseeuw, P.: Silhouettes: a graphical aid to the interpretation and validation of cluster analysis. J. Comput. Appl. Math. **20**, 53–65 (1987)

31. Golovanov, S.M., Orlov, V.I., Kazakovtsev, L.A.: Recursive clustering algorithm based on silhouette criterion maximization for sorting semiconductor devices by homogeneous batches. IOP Conf. Ser. Mater. Sci. Eng. **537** (2019). 022035
32. Lapidot, I.: Convergence problems of Mahalanobis distance-based k-means clustering. In: IEEE International Conference on the Science of Electrical Engineering in Israel (ICSEE) (2018). https://doi.org/10.1109/icsee.2018.8646138
33. Shkaberina, G.Sh., Orlov, V.I., Tovbis, E.M., Sugak, E.V., Kazakovtsev, L.A.: Estimation of the impact of semiconductor device parameters on the accuracy of separating a mixed production batch. IOP Conf. Ser. Mater. Sci. Eng. **537** (2019). https://doi.org/10.1088/1757-899X/537/3/032088. 032088
34. Rand, W.M.: Objective criteria for the evaluation of clustering methods. J. Am. Stat. Assoc. **66**(336), 846–850 (1971). https://doi.org/10.1080/01621459.1971.10482356
35. Hansen, P., Mladenovic, N.: J-means: a new local search heuristic for minimum sum of squares clustering. Pattern Recogn. **34**(2), 405–413 (2001). https://doi.org/10.1016/S0031-3203(99)00216-2
36. Kazakovtsev, L.A., Antamoshkin, A.N.: Genetic algorithm with fast greedy heuristic for clustering and location problems. Informatica **38**(3), 229–240 (2014)

Author Index

Printed in the United States
By Bookmasters